LECTURES ON
DEVELOPMENTAL PHYSIOLOGY

LECTURES ON
DEVELOPMENTAL PHYSIOLOGY

ALFRED KÜHN

Translated by
ROGER MILKMAN

Second, expanded edition
With 620 illustrations

SPRINGER-VERLAG
NEW YORK · HEIDELBERG · BERLIN
1971

ISBN 0-387-05304-2 Springer-Verlag New York · Heidelberg · Berlin
ISBN 3-540-05304-2 Springer-Verlag Berlin · Heidelberg · New York

From the Preface to the First Edition

"I find published lectures insipid."
ALEXANDER VON HUMBOLDT *to* LOUIS AGASSIZ, *January 15, 1840*

Many things speak against the publication of lectures. The effect of the spoken word is very different from that of the written word. Repeated recapitulations, which help the listener understand, are unnecessary, and the overtones of speech disappear in print. A lecture, particularly when not meant for beginners and not offered in preparation for a test, permits the privilege of subjective choice, and when published it is rightly open to criticism. So it has not been easy for me to determine to publish these lectures, which I have given in various forms in Göttingen and then again in Tübingen. In the end, the deciding factors were the urging of my students, the encouragement of colleagues, many of whom I was delighted to see among my listeners, and the wish of my publisher, with whom I have enjoyed a friendly relationship for a long time. The final form of the text is basically a copy, and thus real lectures are the heart of it. If the lectures had not been taken down repeatedly, this book would not have appeared; I could not have undertaken a textbook-sized compilation of this material. With all the changes, the organization, the line of thought, and often the composition of the lectures themselves were retained. They may recall to my listeners of old the hours we spent together.

I have always used these lectures with pleasure in order to confront myself constantly with the problem of developmental physiology. It is my intention to place this problem in all its breadth and depth before young specialists in the modern areas of biological investigation and by my efforts win a few converts, whose insight appears to me to be important and informative.

The selection from the gigantic theater of developmental physiology must remain incomplete, even though I have taken my examples from the three kingdoms of animals, plants, and protists. The bibliography is naturally incomplete also. Nevertheless, the interested reader will find marked with an asterisk review articles, newer works in which the older literature is cited, and work stemming from those cited, and thus be able to go further without much difficulty. In the text, references are numbered. If reference is made to a particular page, it is given in italics.

I have found it necessary to add many pictures to the book, even for things that could be drawn on the blackboard in a lecture. The problems of developmental physiology spring from visible forms; even when we seek the basis of development at the physico-chemical level, end products are nevertheless structures. Most of our experiments remain, moreover, in the realm of distinctions between visible structural components of developing organisms, and they are therefore in evidence.

Illustrations from the work of other authors are, insofar as they were not able to be copied directly, almost all line sketches in a uniform style. Only a few photomicrographs and electron micrographs have been reproduced. Most of the sketches have been carried out with clarity by Mr. ERICH FREIBERG, who thus relieved me of a tremendous burden.

A lecture is always an exhortation to the listeners with the object of winning them over to one's own point of view, and to critical colleagues to whom one wishes to give intellectual satisfaction while setting forth his viewpoint. So I dedicate this book to the two friends, KARL HENKE and GEORG MELCHERS, to whom my addresses, whether they were present or not, were directed especially often.

A. KÜHN

Tübingen, Max-Planck-Institut für Biologie
October, 1954

vi

Preface to the Second Edition

The "*Vorlesungen*" have been kindly received, and now I am taking the liberty of offering them in a second edition. In the ten years that have passed since the publication of the first edition, a great number of problems that at the time seemed very important to me have been solved or brought near solution, and whole new areas of research have opened up. So almost every lecture has taken a new turn. A few themes have been taken up anew. Regeneration phenomena have been covered in two lectures; and I have also tried to come to terms with the problem of "developmental physiology and evolution."

The character of the lectures remains the same. They came into being, in their entirety, in the presence of the listeners, and this form has permitted me the free choice of themes and examples. I have once again made use of this freedom. I could not try for completeness. The generous suggestions to assemble the material into a "Lehrbuch," I would not and could not follow.

A treasured colleague, himself a renowned teacher, once said in a review that he could never impose this much on his students in one lecture hour. So I must admit that I always lectured for two hours without a break, recognizing that this was really an imposition on the listeners; but it gave me the satisfaction of going over a subject without losing continuity.

Reviews in journals, personal letters, and conversations have all been encouraging, and I am grateful for them. I am especially obliged to my colleagues all over the world who have sent me their reprints and so helped me to keep up with the entire literature. I want especially to thank Mr. FREIBERG for preparing the new figures and Mr. and Mrs. EGELHAAF and Mr. and Mrs. JUNG for all kinds of help.

A. KÜHN

Tübingen, Max-Planck-Institut für Biologie
June, 1965

Translator's Preface

This volume is an unaltered translation of the Second Edition of Professor KÜHN's *Vorlesungen über Entwicklungsphysiologie*, which was completed in 1965. Because any individual efforts to update the contents might give the false impression that all recent developments have been taken into account, no changes have been made.

The translator's virtue is fidelity, and I have done my best at it. I hope the reader will find a great deal of ALFRED KÜHN in the pages to follow, and nothing of ROGER MILKMAN. But on *this* page I will speak personally. I translated this work as a student of development for other students of development. It is one of the two great courses in embryology I have taken. The other was at the Marine Biological Laboratory at Woods Hole.

I have often thought that developmental biologists feel especially warmly toward their subject and the systems they study. This warmth, which is easily communicated, made translating the "*Vorlesungen*" a pleasure. Finally, I should like to thank Professor ERNST CASPARI, who suggested that I do the translation, and whose extensive criticism and coaching, gentle but incisive, helped me a great deal.

ROGER MILKMAN

Iowa City
December 7, 1970

Contents

LECTURES ON
DEVELOPMENTAL PHYSIOLOGY

Lecture 1

The problem of developmental physiology is typically biological—not necessarily the most important problem, but certainly the most typical.

Life is a property only of individuals, organisms. These consist of diverse parts in which a great diversity of events takes place. The organization of its parts into a structure and of its processes into an integrated operation is responsible for its own maintenance. We can consider individuals under set conditions as stationary: while the exchange of substances and energy (metabolism) proceeds, the arrangement of the parts and their substances and energy remain constant. Synthetic processes (anabolism) and breakdown processes (catabolism) remain in balance. But in each individual this stationary state obtains only for a certain period of time. In many plants and animals, we observe periodic changes such as the fall of leaves and the growth of new leaves in trees, and the reproductive periods in long-lived animals. As cyclical processes, these lead back during the course of time to an identical condition. But every organism also pursues a one-way road of life: it arises in a certain incipient state and experiences a series of changes until the final state is reached. The tempo and course of this journey are not independent of external influences but the choices, the forks in the road, are restricted by the nature of the organism itself. The parts and processes in the organism, in addition to their spatial organization, have an orderly temporal relation to the one-way road.

One-way roads are found in the inorganic world as well: cosmogony presents the transformation of the universe as a unidirectional process. Each heavenly body pursues a one-way path, and radioactive substances change in only one direction. But the one-way road of the organism is different: In the life of each individual substances of complex atomic structure and high energy content are not only replaced but created in abundance, and the course of life leads to an increase in number of individuals; increase in substance is always bound to the formation of particular structures characteristic of the organism. By this diagnostic property, limitless production of organized substances within limited lifetimes of reproducing individuals, the living continuum can be distinguished from all the events in the nonliving world.

The task of developmental physiology is the investigation of the laws of the course of individual life and of the continuation of life from generation to generation: the study of the laws of the growth, the morphogenesis, and the reproduction of organisms.

There is infinite variety among life histories, but the individuals, which in a living continuum are connected to the chain of reproduction, repeat similar courses of development when conditions remain the same. Reproduction thus transmits a specific basis for development to the new individual. This is related to a "specific structure" (GEORG KLEBS, 1903), which remains constant during the unbroken continuity of generations. Concerning "specific structure" we can make two general statements: it must replicate identically, and it must result in a particular repetitive course of development.

The specific structure is contained within the physical fabric of the cell. It represents biological substance in its elementary form. Events in development stem from events in cells. The cell fabric does not constitute the specific structure, but contains it. The cell as a whole is a system of substances and forces, maintained by the specific structure and changing under its control. Cytology has revealed a variety of fundamental structures in cells which have the property of self-replication. Thus, specific structures can now be described morphologically and to some extent defined chemically. Genetics has, by means of crossing and mutation experiments, established the existence of individual, separable genes which determine specific metabolic and morphogenetic processes; and the union of these two fields, cytogenetics, has been able to identify certain genes with particular self-replicating cell structures. The "specific structure" emerges as the genome or idiotype, the entire hereditary constitution. The problem which developmental physiology must solve is this: How does the continuous, specific structure work to create periodically in each individual life a new specific diversity. How does a complex organism, with a striking variety of parts, regularly emerge from a particular species of zygote, and how are more of these zygotes, with their remarkably varied potencies, produced from parts of these organisms?

The specific structure of the cells of a particular species, the "Artzellen" (species-specific cells, OSKAR HERTWIG), does not automatically determine the course of development of an organism, but rather establishes a range of developmental possibilities. The onset of particular developmental processes is always a response to particular developmental influences. The genetic material establishes the norm of reaction of the species-specific cells, fitting their response to a certain environment in which development normally takes place. For developmental processes, as for all living processes, there are certain preliminary requirements of energy and matter. In addition, however, certain specific influences (formative or morphogenetic stimuli) are needed to evoke particular developmental reactions. The developmental stimuli bring about certain functional states in the cells leading to the synthesis of particular molecules and the genesis of particular forms. Throughout the whole array of developmental reactions, the hereditary material, as far as we know, is not altered; but, within the organism, by means of the substances and structures formed, a more or less lasting order of internal conditions is established. These trigger new reactions, lead to certain reactive competences, or restrict developmental potentialities by closing off some pathways. Even as the cell grows, internal conditions change the relations to the outer environment, altering the functional conditions of the whole cell.

The reactive character of the developmental processes, and their determination by specific structure and by exogenous and endogenous stimuli was first placed in clear relief by GEORG KLEBS[339, 7], one of the first students of developmental physiology in plants, who said: "We have one class of constants, the specific competences, and two classes of variables, the internal and external conditions."

The various sorts of differentiation which an organism or one of its parts can undergo in response to developmental influences are called *modifications*. Modifiability in response to external influences means that a given germ cell can give rise to any one of a number of quite disparate individuals according to the particular developmental influences which act during sensitive periods in development. For example, there are the various castes of an ant—or termite—colony and the large, highly differentiated *Bonellia* female, so different from the tiny, degenerate, essentially parasitic male.

The differentiation of cells in the embryonic parts and tissues of multicellular plants and animals constitutes modification of the species-specific cell by locally disparate developmental influences within the multicellular fabric. Certain developmental stages can—given the fundamental necessities for development—run their courses without modification, as we see in the development of an extremely mosaic embryo. But in this case the initial stage of this period of development has already been established as a specific modification of the species-specific cell with the spatial organization of a variety of internal conditions.

The establishment of a functional condition that leads to a particular developmental process is called *determination*. This can be definitive and stable, so that only one reactive pathway is left open to a cell or to an embryonic part; indeed, the automatic unfolding of a particular kind of differentiation may be prescribed. Determination may also be labile: In this case the tendency for a particular developmental career is established without eliminating the possibility of alternatives. The spatial and temporal distribution of determination to the parts of a developing organism can be of various sorts. The determinative state of a certain part can be established if it differentiates autonomously when isolated in an "indifferent" environment, which is to say that it contains all the necessary instructions for its development up to a given point. The array of reactive competences, or potencies, of the part may be catalogued by exposing it to a variety of developmental influences. Two extremes are possible here: On the one hand, the differentiation of the part may require the constant influence of a developmental stimulus. Thus, under normal circumstances the part is influenced by its surroundings to undergo dependent differentiation. After isolation or disorganization of parts *regulation* can take place—developmental processes leading back toward normal organization. On the other hand, the autonomous differentiation of isolated parts may be exactly the same as what takes place in the intact organism. The organism develops through the *self-differentiation* of its parts. Here, development is a *mosaic* process which permits no regulation. In pure form these extremes rarely, if ever, persist throughout the entire course of development. In general, determination passes during the course of development from a period of interdependence of parts to a period of self-differentiation of the parts. However, the change may proceed in the opposite direction as well.

While the nature of the determining and determined conditions within the cells is still inaccessible to us, we can study them operationally in terms of regular responses to specific influences.

The component processes of development, the replication of the hereditary material, growth, and morphogenesis must in the last analysis be physicochemical events. And the aim of developmental physiology is to define all the component processes of development in physicochemical terms. Only a tiny part of this task has been accomplished. Everywhere there are encouraging beginnings, but the job is an endless one. Nevertheless, the nature of the development of an organism lies not with these component processes but in their organization. Normal development (normogenesis) depends continuously on an ordered interplay of particular and diverse morphogenetic events; it is always "teamwork" (Kombinative Einheitsleistung—F. E. Lehmann), whose characteristics lie in the area between mosaic spatiotemporal coincidences of individual events and a reactive collaboration of component processes.

The science of physiology of the mature organism has been very successful in reducing the individual functions of the organs and organ systems to the identification of the physicochemical bases of these functions; as, for instance, the understanding of digestion in terms of hydrolysis and, more recently, of nerve conduction in terms of local electrical currents and ion permeability changes. But the chemical reactions in the digestive tract have a biological significance only in the particular context of the whole mode and mechanism of feeding and ultimately in relation to the totality of animal behavior. The transmission, gradation, sequencing, and distribution of particular signals of particular strengths by means of nerve impulses and hormones gain biological meaning only in the entire fabric of the organism. The total organism as a self-maintaining unit is taken for granted by organismic physiologists. Its formation during development, the self-construction of the organism, is the subject of developmental physiology; and since this meaningful integration of events is the hallmark of an organism, the task of developmental physiology is the most characteristic task of biology.

All of the individual processes of morphogenesis are *means* toward the harmonic and species-specific course of development, the process of construction; and their precondition is the norm of reaction which is determined by the genetic material. The individual genes, which we can isolate from this continuous structure, also are characterized as being the means through which, in collaboration with other genes, the biologically significant reactivities of the species-specific cell types are created. The genome at the outset of development already has the character of a meaningful structure, which can be understood only in the context of evolutionary history.

The transformation of the structural and the functional plans of species in the course of evolution comprises the reconstruction of the organs accompanying complex functional changes. It requires a change in the norm of reaction affecting the individual developmental processes, which

3

in turn change and must be integrated anew. Between the transformation of the genome and the new, definitive organization falls the whole mechanism of development. Seen from this vantage point, the problem of evolution becomes a developmental problem.

When we speak of a "task," of "biological meaning," of a "process," of "means," "construction," and "reconstruction," we only wish to indicate the self-maintaining nature of the organization and structures of the organism. But this sort of consideration of events in the organism does not provide us with a scientific understanding. The "purposiveness" itself requires an explanation. Kant, whose *Critique of Teleological Judgment* established the ineluctability of the consideration of organisms as purposive, also marked out the limits of applicability of this concept. It is a "regulative," "heuristic" principle of inquiry: "But if we invoked a cause working toward each goal in order to explain the forms of the objects of experience, because we believe we find purposiveness in them, our explanation would be entirely tautological, for we would deceive reason with words" (§ 78).

Developmental physiology, like every science, can only be a causal inquiry; for this reason it has been called "causal morphology." Its task is to analyze the complex net of developmental pathways into causal chains and then to investigate their interrelations in the total mechanism. Where this causal inquiry reaches a limit, there is either a still unsolved problem or indeed the unanswerable fundamental mystery of life and experience.

The fundamental questions of developmental physiology are thus clear in the light of what we have said.

1. How do the continuous structures of the cells behave during development? What are the conditions for their identical replication? What is their chemical nature?

2. What developmental possibilities does the norm of reaction of a species-specific cell comprise? Under what circumstances are they expressed?

3. What are the physicochemical components of a developmental process?

4. How do the temporal sequence and spatial organization of those conditions come about which lead to a typical, normal developmental career?

5. What makes the developmental reactions possible? In other words, how does the specific structure, the genetic material, determine the norm of reaction, and how are individual genes involved in particular developmental events?

6. How is the genetic material changed, and how are the altered individual events that result rewoven into a new typical course of development?

We are concerned with the answers to these questions. We search for observations and experimental results which can lead to these answers. As yet, not a single one of these fundamental questions has received a satisfactory answer, even for a single species; nevertheless, there is an abundance of relevant information for all sorts of living things. We shall not strive for completeness in these lectures. Rather, we shall take instructive examples where we find them, always keeping these questions in mind. Sometimes our choices may be arbitrary. Basically it is the experiment which tells us most. But in many cases the right experiments have not been done, and sometimes we do not know how to design productive ones; in these cases we can often make use of "experiments of nature." "Chance" changes in the normal course of development are as informative as intentional experiments, when we know what has been changed and when we can establish the consequences of this change. Changes in the genetic material (mutations) whose "random" occurrence we can make more frequent, although we cannot yet direct the induction of specific ones, offer us an array of useful aids to choose from. Then again, the great experiment of nature in the creation of the diversity of living forms and life cycles lets us discover the recurrent fundamental processes and shows us possibilities for altering them and using them in various combinations which far exceed our experimental possibilities.

4

We cannot offer a valid *theory* covering the phenomena of development; hypotheses serve us only when, as "working hypotheses," they lead us to certain experiments in which we can decide whether a certain assumption is correct or not. Our discussions will often lead us not to a conclusion, but to a question, partly because we will have reached the limits of what is already known, but also partly because our own knowledge of the broad areas that have already been investigated is too restricted or because we have not been able to see clearly enough to grasp the relationships contained in seemingly disparate experimental results.

The fundamental questions have a general application. The answers may be valid only for particular species, due to the diversity of morphogenesis and developmental stages. Nevertheless, in all cellular organisms there are certain basic processes, namely those of cellular organization and cellular activities, which form the basis of the most diverse developmental processes. And certain morphogenetic principles, ways of forming organisms, recur in the establishment of the basic plan and the elaboration of the most diverse structural types, such as the use of cell division as a means of differentiation; the appearance of polarity in one or more axes, the genesis of qualitative differences from quantitative gradations in external or internal conditions due to alternative modifiability; induction; self-organizing fields, inhibitory fields, and others. These morphogenetic principles can only be characterized in biological terms at present. Their conceptual definition implies no causal explanation or physicochemical hypothesis. But naturally the question arises as to whether, and to what extent, similar or analogous means are used in a particular morphogenetic pattern for the differentiation of an originally uniform entity, a cytoplasmic mass or an aggregate of identical morphogenetic cells (in a meristem or a blastema). How extensive is this similarity, and are their concordant similarities in the same general theater? This always leads to another question: What determines the particular nature of the end result? In the laying out of higher organization, various developmental principles always work together in time and space. This in itself is a principle, the principle of synergism, which applies as well to human technology and to society.

Now we turn to the developmental physiology of the cell. The basic processes in which the continuous structures of cells participate during cell generations and generations of individuals are cell division, fertilization, and meiosis, which is complementary to fertilization. We can conclude only from the behavior of the cell structures in these events and from the related phenomena of heredity which structures are continuous; that is, which structures are part of the "specific structure," the genetic material. It is obvious morphologically today that the chromosomes of the cell nucleus are common to all animal, plant, and protistan cells, and in higher plant cells there are always plastids. The presence of additional, continuous elements in the cytoplasm has been demonstrated by genetic experiments and by developmental observation; but we have not yet been able to establish anything more specific about these "biosomes" (F. E. LEHMANN) in the cytoplasm. Centrioles, mitochondria, elements of the Golgi apparatus, and more or less exactly defined granules have frequently been followed over cell generations; but there is no proof that they replicate only autonomously, that is, that their structure is determined by the structure of the material already present. As to the physiological role that these cytoplasmic structures play in the formation of the cell, much can be said. Here we shall consider only the properties of the chromosomes, since these are universal and certainly self-replicating. As late as 1932 a book on theoretical biology could still state "the persistence of the chromosomes" is not a material continuity but rather the persistence of the integrated conditions of a dynamic order. "The chromosomes are not, they happen."[43, 223] Today it is indisputable that the chromosomes do not arise *de novo* at the onset of each mitosis, but rather remain intact throughout the life of the nucleus. Their basic structure duplicates identically between nuclear divisions, and they form the major complement of the genetic material of the cell. At fertilization, two sets of chromosomes, or genomes, unite. And as a correlate to this doubling of the chromosomal content, there is a reduction at meiosis from the diploid condition of the cell to the haploid. The essence of fertilization is a simple fusion of two haploid nuclei. The maneuvers which bring the nuclei of different sexes together are extraordinarily diverse in the domain of living things. The behavior of the chromosomes in the mitotic cycles of cell division, and in meiosis, which involves chromosomal pairing and two maturation divisions, is much the same for protistans, plants, and animals. Diversity is found only in the accessory apparatus which effects the distribution of the chromosomes to the daughter nuclei.

In the mitotic cycle, which includes interphase (the "resting nucleus") and the stages of prophase, prometaphase, metaphase, anaphase, and telophase, the chromosomes undergo a change in form which is closely bound to chemical alterations. A permanent structure which persists unaltered through these changes in each chromosome has been shown cytologically to be a thread with a certain linear architecture, called the chromonema. With respect to the chromonema, two contrasting stages can be distinguished in the mitotic cycle: a stage of maximal contraction during metaphase, and a stage of maximum extension during interphase or at the beginning of prophase. The particular character of each individual chromosome of a genome is seen most clearly in metaphase.

Each chromosome at the equatorial plate is a cylindrical body of a particular size and shape. Almost always there is a characteristic place where it becomes attached to the spindle fibers, the kinetochore (centromere). At this place, the chromosome is constricted and usually bent. The two "arms" into which the kinetochore divides the chromosome can be of similar or very different lengths. The site of spindle fiber attachment is almost never at the end of the chromosome. Even in apparently "rod-shaped" chromosomes there is almost always a tiny arm at the other side of the kinetochore. The sequence of morphological changes in the chromosomes during the mitotic cycle is now in large part clear, although certain questions have not been answered conclusively: the contraction and stretching of the chromonema are now seen to depend upon the pitch of their helical coiling ("spiralization"). The metaphase form of the chromosomes represents a particular degree of coiling of the long threadlike chromonema.

The coiled structure of chromosomes has been seen in animals, plants, and protists, both in living cells and in fixed preparations, and has been confirmed by countless excellent photomicrographs. Figure 1 illustrates diagrammatically the sequence of changes in coiling. In early prophase the chromosomes already are seen to be split longitudinally; the two daughter strands are chromatids which are loosely coiled together (Fig. 1a). The coiling now becomes tighter (Fig. 1b). The diameter of each coil gradually increases, and the turns move closer together (Fig. 1c) until the final metaphase condition is reached (Fig. 1d). In anaphase (Fig. 1e) each coiled chromatid separates from its sister, and uncoiling begins in the telophase nucleus (Fig. 1f). The closed cylindrical form of the metaphase chromosomes results from the tightness of the coiling and from the matrix which covers the coiled chromonema. This matrix, or calymma, can be distinguished from the chromonema itself by special positive or negative staining at prophase (Fig. 2).

The characteristic dimensions of a metaphase chromosome, leaving out the surrounding matrix, are the diameter and pitch of the helix. These values are constant for any particular chromosomal type in a genome as long as the cell is kept under constant conditions. In a metaphase chromosome of *Tradescantia reflexa*, for example, twenty-five coils were counted with a diameter from 1/8 to 1/2 micron and a distance between successive turns of 1/2 to 3/4 micron.

The direction of coiling (left or right) is not characteristic for a particular chromosomal type or for an arm. The two homologous chromosomes of a pair frequently show differences during the first meiotic metaphase; and the direction of coiling can change along the chromosome (Fig. 3).

6

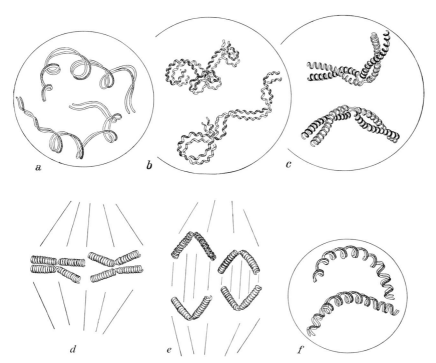

Fig. 1. Diagram of chromosomal coiling in the course of nuclear division. a–c prophase; d metaphase; e anaphase; f telophase-interphase. (After STRAUB, 1938)

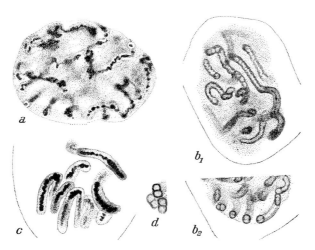

Fig.2. Mitotic chromosomes in epithelial cells of the cornea of young *Salamandra maculosa* larvae; fixation after SANFELICE, staining with 1/600 M methylene blue at various pHs. a early prophase pH 3.6; b_1 late prophase, two focal levels of the same nucleus; b_2 shows optical cross section of the chromosomes, pH 2.9; c metaphase, pH 3.6; d metaphase; optical cross section through the chromosomes, pH 2.9. (Sketched after microphotographs by ZEIGER, 1934)

The mechanism of coiling is a good first developmental question.

One is first inclined to think of a mechanical effect of the matrix; if this contracts longitudinally an elastic fiber can be forced into a coil. And experiments do point to a role for the matrix in the maintenance of the coils: Hydration with tap water and exposure to various chemicals (e.g., sugar solutions containing calcium nitrate or ammonium chloride, or ammonia vapor) results in the loosening of the coils and the extension of the chromosomes into simple threads. This phenomenon has been ascribed to the dissolving of the matrix. But we cannot exclude the possibility that the change in form of the chromosomes, the coiling and uncoiling, results from changes in the

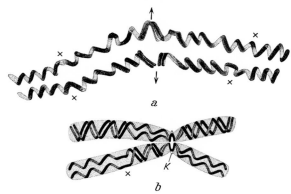

Fig. 3. Chromosomal coiling in first meiotic metaphase of *Trillium*. a *T. erectum* 4,400× ; b *T. kamtschaticum*. Diagram; "×" marks changes in direction of coiling. *K* kinetochore. (a after HUSKINS and SMITH, 1941; b after MATSURA, 1941)

Fig. 4. Diagram of a chromosome with major and minor coils. *K* kinetochore; *M* matrix

chromonemata themselves. In fact, such changes are generally assumed, although we are not sure what these changes are, and attempts to describe them are entirely conjectural.

The question also remains as to whether changes in the length of the chromosomes can be understood completely in terms of the coiling and stretching of the chromonemata. The shortening of the prophase chromosomes into the metaphase chromosomes during somatic division is three- to six-fold. These values can be obtained in model experiments with simple coiling. But the leptotene chromosomes in meiosis (Figs. 39a, 40a) are always much longer than the somatic prophase chromosomes. They are always completely uncoiled, and they shorten to 1/10 their former length or less at meiotic metaphase. And in meiosis in many forms a dual coiling pattern involving large coils and small coils (or primary and secondary coils) has been demonstrated or at least suggested (Fig. 4). It is seen in the pollen mother cells of plants during the experimental loosening of their chromosomal coils. It is also seen very clearly in the chromosomes of the Foraminiferan *Patellina corrugata* at the end of the first meiotic division (Fig. 5e) and in the two or three giant chromosomes of various gigantic Hypermastigophora which are clearly visible during the entire nuclear cycle (Fig. 6). This twofold coiling can account for the tremendous difference in length between meiotic chromosomes in leptotene and in metaphase, as model experiments demonstrate.

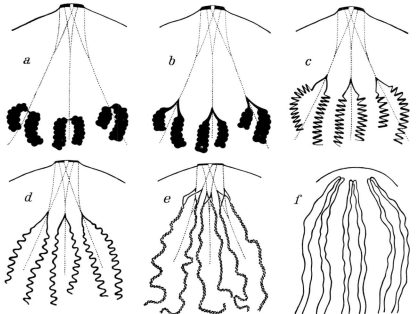

Fig. 5. End of the first meiotic division in *Patellina corrugata* (Foraminifera), diagrammatic. From *d* on, the full length of the chromosomes is not shown. (After LE CALVEZ, 1938)

The next question concerns the doubling of the chromonemata which must precede the longitudinal division of the chromosomes. In early prophase the two chromonemata are always already present; these then separate as daughter chromatids (Fig. 1). Frequently they still lie close together, but the appropriate fixation and staining always reveals a complete longitudinal separation. When does the duplication take place? In many plants and animals two coiled chromatids can be seen in each anaphase or telophase chromosome; each of the two chromatids thus enters the next with two "half chromatids" which are separated only at metaphase nuclear division. Double chromonemata are thus present during interphase also; at this time or during the microscopically indistinguishable transition to early prophase each chromonema replicates (Fig. 7c). A few cytologists have even described four strands in anaphase so that the chromonemal replication is accomplished two divisions in advance.

A new question now arises: What causes the separation of the sister chromonemata?

The distinct half-chromatids lie close together in the coiling pattern so that they form common coils (Fig. 7d–f). Their separability must be assured either by the nature of the coiling or by complete uncoiling before the next prophase. Two strands can be coiled together in two different ways: 1. If both ends can move freely with respect to one another while coiling takes place, then the two strands wind around one another like the strands in a cable and cannot be separated transversely (Fig. 8a); or 2. if the ends are held fast so that they

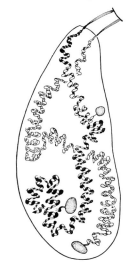

Fig. 6. Chromosomes in an anaphase nucleus of *Holomastigotoides tusitala* (2,000×). (After CLEVELAND, 1949)

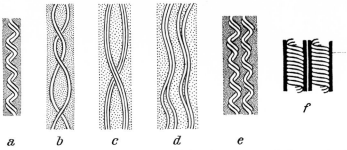

Fig. 7. Diagram of coiling, uncoiling, and duplication of the chromonema during a mitotic cycle. a telophase chromosome; b–d interphase to early prophase; e late prophase; f metaphase chromatids

do not change their relative positions during coiling, then the two strands are freely separable (Fig. 8b). One is always to the right, the other left, as is often seen, for example, in the filaments of a light bulb. In the first case it becomes obvious that the half-chromatids of the metaphase and anaphase chromosomes, as assumed in the diagrams of Figures 7a, e, and f, must uncoil during interphase or early prophase to such an extent that the chromonemata which are to separate in the coming mitosis can coil independently of each other (Fig. 7c–e). The microscopic

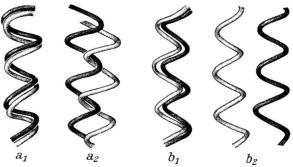

Fig. 8. The two ways in which two coiled strands can coil together: plectonemic and paranemic coiling, showing the separability of paranemic coils. (After Kuwada, modified from Straub, 1938)

evidence leads to no single clear conclusion. In *Patellina*, the two sister chromatids which have been coiled together are in the anaphase chromosomes and separate in the second meiotic division after complete uncoiling (Fig. 5e, f).

At any rate, each chromatid acquires its own matrix in the course of prophase, and this matrix encloses the double coils of its two half-chromatids, which may even be further subdivided (Fig. 7e, f; Fig. $2b_2$, d). Shortening by means of tight coiling is a suitable mechanism for disentangling the long chromosomes in the relatively restricted cell space and for distributing them without complications. This mechanical significance of the spiral structure is suggestive. Under conditions where a shortening of the chromatid threads is not necessitated by spatial considerations, the organism occasionally does not bother with coiling during metaphase and anaphase, as we see in the nuclear divisions at the cytoplasmic surface of the large sporonts of a Coccidian (Fig. 9). The uncoiling at prophase (Fig. 9b, c) which in this case continues into anaphase (Fig. 9d), must apparently permit the smooth separation of the extraordinarily long sister chromatids. Only at the end of anaphase do the chromosomes re-form tight coils, permitting packing into the daughter nuclei (Fig. 9e) within which they loosen once again, having evidently

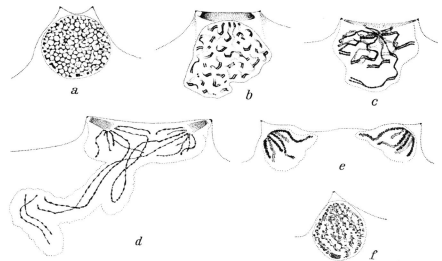

Fig. 9. Nuclear division in *Aggregata eberthi* sporonts. a interphase; b prophase; c, d anaphase; e, f telophase. (After BELAR, 1926)

replicated once more (Fig. 9f). In the Foraminifera complete coiling is used only to arrange the numerous chromosomes on the spindles, and uncoiling begins in anaphase (Fig. 10).

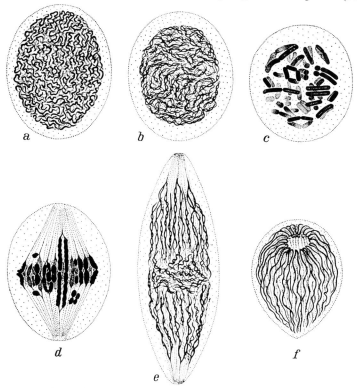

Fig. 10. Some stages in the first meiotic division of *Patellina corrugata*. (After GRELL, 1959)

Lecture 2

Functional differences between the individual chromosomes of a genome and differentiation along the length of a chromonema — other than the kinetochore already described — are strikingly seen in the nucleolus-organizing regions. Specific nucleolar chromosomes which occur in plants, metazoans, and protozoans (Figs. 11–14) are generally recognized during metaphase and anaphase by a constriction or by an actual gap bridged by a thin, weakly staining (or nonstaining) thread,

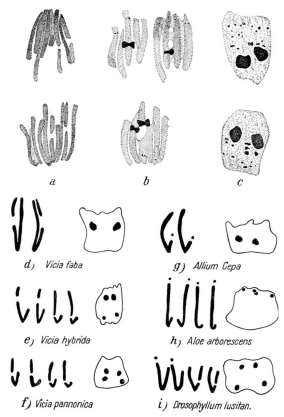

Fig. 11. Nucleolar chromosomes in flowering plants. a–c *Vicia faba*; a anaphase; b early telophase, onset of nucleolus formation; c resting nucleus with nucleoli and numerous little heterochromatic regions (chromosomal centers), some of which have already appeared in b; d–i nucleolar chromosomes in anaphase and position of the nucleoli in resting nuclei of various plants. (After HEITZ, 1931)

the nucleolar fiber. These places are the sites of the formation or condensation of the nucleolar substance. The position and the number of nucleoli in the resting nuclei correspond to the position and the number of these chromosomal constrictions, which are constant for a given species (Fig. 11); thus they are identical among sister nuclei (Fig. 11a–c). The changes in form of the nucleolar chromosomes, the appearance of their nucleoli in telophase, and the disappearance of these nucleoli in prophase have been followed in several cases throughout the entire nuclear cycle (Fig. 12). The nucleolar fiber may be far from the end of the chromosome (Fig. 11a, b, d–f); frequently it does lie near the end, and the rest of the arm hangs on as a little "satellite"

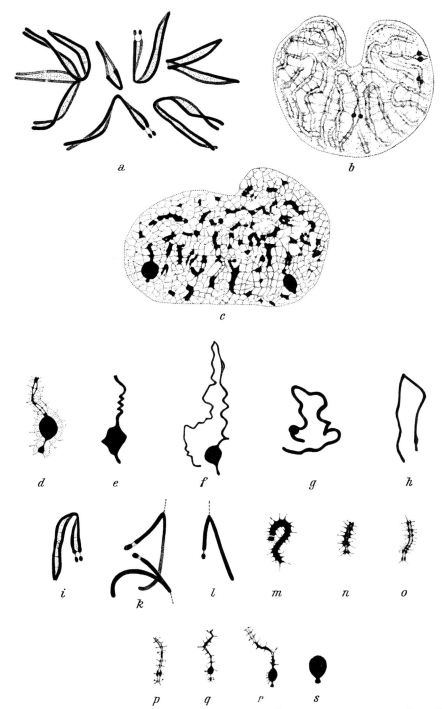

Fig. 12. Structural changes in the nucleolar chromosomes in *Ambystoma tigrinum*. a metaphase; b telophase; c interphase; d–s an individual satellite chromosome; d–h prophase; i metaphase; k, l anaphase; m–r telophase; s nucleolus with satellite in resting nucleus. (After DEARING, 1934)

13

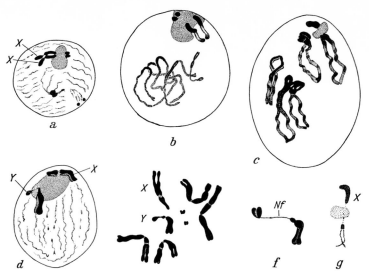

Fig. 13. Nucleolar chromosomes (X and Y) in somatic cells (ganglion cells) of *Drosophila* larvae, about 2400×. a–f *D. melanogaster*; a–c female; d–f male; a, d interphase; b, c prophase; e equatorial plate; f Y-chromosome during prophase with attenuated nucleolar fibre; g X-chromosome of *D. virilis* in prophase with nucleolus. (a–f after KAUFMANN, 1934, g after HEITZ, 1934)

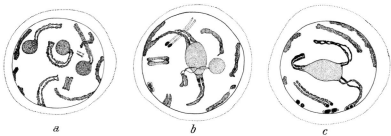

Fig. 14. Oocytes of *Mytilicola intestinalis* (Copepoda). Three pairs of nucleolar chromosomes; nucleoli separate, two fused, or all three fused. (After AHRENS, 1939)

(Figs. 11 g–i, 12, 13, 16d, f, 23e). These satellite chromosomes are generally the most easily recognized in any karyotype.

The nucleoli may remain attached to the chromosome continuously, extending the satellite fiber greatly (Figs. 13f, g, 16f, g, 23e), or the nucleolar mass may surround its site of formation like a thick sleeve; frequently the nucleolus is attached only laterally to the satellite region, or it separates entirely from the chromosome and lies free in the nuclear sap. Rarely a nucleolus forms at the end of a chromosome and remains attached there (Fig. 14a). The nucleolar chromosomes may be autosomes or sex chromosomes. In various *Drosophila* species the nucleolar organizing region is located on the Y chromosome and the homologous portion of the X (Fig. 13). There may be one or several nucleolar chromosomes in the genome (Figs. 11 e, f, h, i, 14, 23e). The nucleoli of a pair of nucleolar chromosomes may remain separated (Figs. 11, 12), or they may fuse sooner or later (Fig. 14).

The significance of the satellite chromosomes in nucleolus formation is shown by experiments on maize: the normal satellite chromosome (number *VI*) has a knob at the place where the satel-

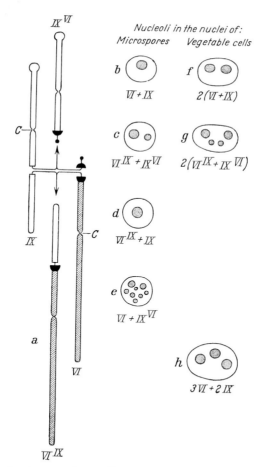

Fig. 15. Diagram of a translocation between the satellite Chromosome *VI* and chromosome *IX* after a break in the region near the nucleolar fiber of chromosome *VI*; formation of the nucleoli in various chromosomal combinations in *Zea mays*. (After the experiments of McCLINTOCK, 1934)

lite is connected to the fiber (Fig. 16a). A translocation after chromosomal breakage moved one half of the knob together with the satellite to the end of a piece of chromosome *IX* (Fig. 15, IX^{VI}), and the other half of the knob into the middle of an arm of a chromosome composed of portions of chromosomes *VI* and *IX* (Fig. 15, VI^{IX}). When these translocation chromosomes VI^{IX} and IX^{VI} were combined in homozygous condition in a microspore mother cell, sometimes both pairs of chromosomes were attached to one large nucleolus (Fig. 16b), and sometimes each pair formed its own nucleolus. In this latter case, the nucleolus formed by IX^{VI} (the chromosome with the satellite fiber) was larger than that formed by VI^{IX} (Fig. 16c). The nucleoli in various chromosomal combinations of microspores with VI^{IX} and IX^{VI} (Figs. 15c, 16d, e) and somatic cells (Fig. 15g) present a similar picture. If VI^{IX} is present in addition to *IX*, one large nucleolus arises (Figs. 15d, 16g). If a somatic cell has three chromosomes *VI* instead of two, three nucleoli are formed (Fig. 15h). In the combination with the translocation chromosome IX^{VI} several additional small nucleoli appear surprisingly in other places (Fig. 15e) in addition to the nucleoli on the satellite fibers (Fig. 16h). This suggests that the segment of chromosome IX missing in this

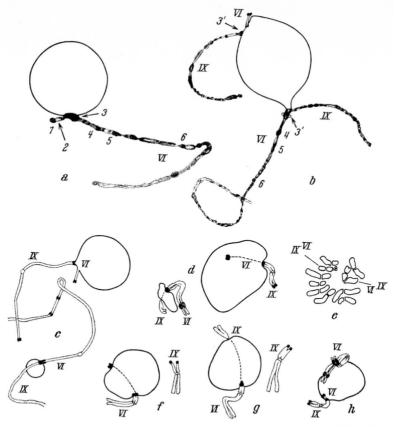

Fig. 16. Translocation after break in the nucleolus-forming region of chromosome *VI* in *Zea mays*. a normal chromosome *VI* (satellite chromosome), pachytene of the microspore mother cell with nucleolus, *1–6* easily recognizable regions of the chromosome, (*1*) satellite, (*2*) attachment fiber, (*3*) large strongly staining body (heterochromatic) at one end of the nucleolar fiber; b same stage, homozygous for the translocation (see Fig. 15a); c preparation similar to b, separate nucleoli formed by the translocation chromosomes; d–h microspores; d, e microspores from translocation homozygotes; f, g combinations of heterozygotes; d, f–h prophase, chromosomes split longitudinally; d one nucleus each formed by *VI*^{IX}and *IX*^{VI}; f combination of *VI* and *IX* (= normal); g combination *IX*^{VI} and *VI*. (After McCLINTOCK, 1934, a and b drawn from microphotographs, remainder of chromosomes omitted)

combination with VI (cf. Fig. 15a) is also involved in the regulation of nucleolus formation. If the entire nucleolus-forming end of chromosome VI is removed by X-irradiation, numerous small nucleoli appear in the microspores which have received the defective chromosome.

These experiments show that the satellite regions of certain chromosomes – not only the nucleolar fibers – are normal centers of nucleolus formation, but that they do not contain all the factors necessary for nucleolus formation in the intact nuclear system.

In addition to the nucleolus-forming sites, other regions of the chromosomes show peculiar behavior in the mitotic cycle: most chromosomes become more delicate in telophase as they uncoil, staining more lightly. Appropriate fixation and staining reveals the so-called nuclear reticulum of the resting nucleus, where occasionally tangled threads or rows of granules can be recognized. But certain whole chromosomes or pieces of chromosomes do not undergo this loosening characteristic of the *euchromatic* parts of the chromosome, the *euchromatin*, but remain

16

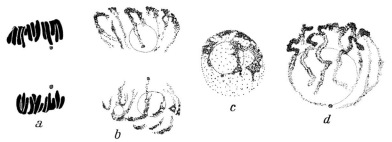

Fig. 17. Heterochromatin in *Collinsia bicolor* (Angiosperm); mitosis in the root tip. a anaphase; b telophase; c interphase; d prophase. (After HEITZ from GEITLER, 1939)

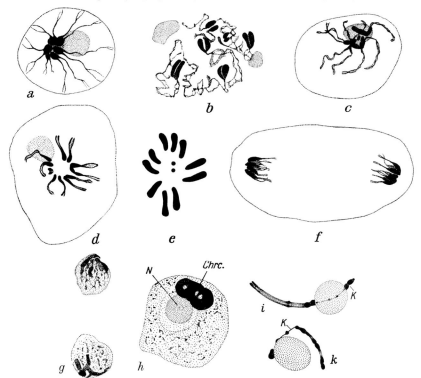

Fig. 18. Chromosomes in the nuclear division of larval neuroblasts of *Drosophila*. a–h *D. virilis*; a–d prophase, emergence of the heterochromatic regions of the chromosomes from the chromocenter (*Chrc.*), shortening of the euchromatic regions; e equatorial plate; f early telophase; g late telophase; h interphase nucleus; i, k *D. melanogaster* X and Y chromosomes in late prophase; i X; k Y; the heterochromatic regions are black. *K* kinetochore, *N* nucleolus; about 2,400× (a–f after MAKINO, 1940; g, h after HEITZ, 1934; i, k after COOPER, 1959)

denser and strongly staining. These chromosomal regions are called *heterochromatic*, and their substance is called *heterochromatin*. Its unusual behavior makes particular chromosomes or regions stand out in telophase (Figs. 17b, 18f, g) as well as in prophase. It can be followed until the euchromatic regions coil tightly and become as compact and strongly staining as the heterochromatic region (Figs. 17d, 18a–e). In the resting nucleus the heterochromatic regions of the chromosomes may remain distinct, as compact or fairly loose individual chromocenters, or they may fuse into a common chromocenter (Fig. 18h), from which they emerge at prophase (Fig. 18a).

The sex chromosomes are frequently heterochromatic entirely or in large part (Fig. 13a, d). Shorter or longer heterochromosomal regions can lie at the end of chromosomes (Fig. 18a–g) or at various places within the chromosomes (Fig. 11). Frequently heterochromatic regions are situated near the kinetochore (Figs. 23c, 41a) in both animal and plant cells. The compactness and strong stainability, or heterochromasia, appear in many cases to stem from the fact that the heterochromatic regions do not uncoil (or do not uncoil as much as the euchromatic regions). Their matrix also appears to be different. This does not mean, however, that the heterochromatin remains unchanged and without function in the interphase nuclei.

Nucleolar fibers are always connected on one or both sides to heterochromatic regions of the chromosomes (Figs. 13, 15, 16, 18i, k). In resting nuclei one often sees chromocenters adjoining

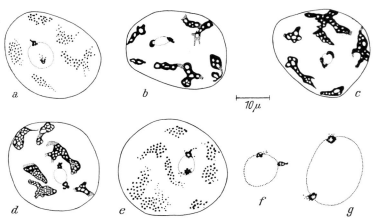

Fig. 19. Nuclei from *Sauromatum guttatum* (Araceae) cells. a–e alterations in chromocenters in the epidermis of the appendix of the flower stalk during the rise in respiration at flowering; a three days before blooming; b at the time of the greatest (endogenous) rise in temperature; c–e reversal of the structural changes over 24 hours; f, g nucleoli with the satellite chromocenters from raphis cells. (After GRAFL, 1940)

the nucleoli, and they grow as the nucleoli grow (Fig. 9f, g). In addition, chromocenters which are not closely connected with the nucleolus can show striking structural changes according to the functional conditions of the cell; for example, in the epidermal cells during the flowering of a member of the family Araceae (Fig. 19a–e), which is accompanied by an increase in respiration. In the cells of *Drosera* tentacles the chromocenters enlarge during digestion. The X-chromomeres of locusts and hemipterans are alternately "positively" and "negatively" heterochromatic: according to the developmental stage, irrespective of sex, they may be more highly condensed and stainable than the autosomes or less so. In many insects heterochromasia does not occur at all in the somatic cells but only in certain stages of gametogenesis. In *Perla marginata* it is seen in spermatogenesis, but not in oogenesis. All these observations show that the ability to assume heterochromatic characteristics stems from the structure of the chromonema throughout the chromosome or in particular regions, but that the expression of this characteristic depends upon the functional state of the nucleus.

The conditions for the specific changes in chromosomal form, as well as for their activities in cell metabolism and in development, must lie in the chemical nature of the chromosomes.

The essential components of chromosomes are nucleic acids and proteins. The structure of the nucleic acids in so far as is necessary for the interpretation of cytochemical observations is shown in Figure 20. The nucleic acids are polymers and chains of mononucleotides, each of which in turn is composed of a phosphate, a pentose sugar, and a purine or pyrimidine base (Fig. 20a).

18

The pentose of the nucleotide in a nucleic acid is either deoxyribose (Fig. 20b) or ribose (Fig. 20c). The purine bases are adenine and guanine. The pyrimidine bases are cytosine and thymine in the case of deoxyribonucleic acid (DNA), and cytosine and uracil in ribonucleic acid (RNA). DNA, which has long been extracted from cell nuclei, is bound in the chromosomes with protein to form a nucleoprotein.

DNA can fortunately be distinguished from RNA by specific staining reactions. The Feulgen reaction stains DNA bright red in microscopic preparations.

The Feulgen reaction is a true chemical reaction: After acid hydrolysis of the purine bases, the aldehyde groups of the deoxyribose combine with Schiff's reagent (a fuchsin sulfonate) to form a magenta compound. A double staining method with methyl green and pyronine (BRACHET)

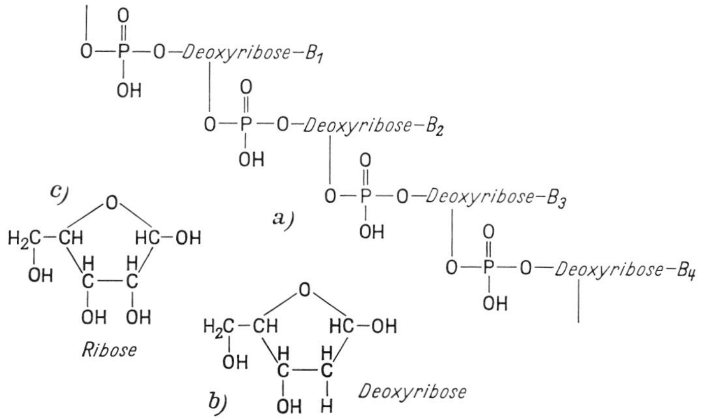

Fig. 20. a diagram of the structure of a nucleic acid; b, c the pentoses; B_1–B_4, nitrogen bases

reveals both DNA and RNA in microscopic preparations: DNA selectively binds the methyl green and RNA binds the pyronine and becomes red. In the nuclei the chromatin structures are green and the nucleoli are red. When both DNA and RNA are found close together in a particular cell component, they can be separated by specific hydrolytic enzymes. Thus, it can be shown that metaphase chromosomes which show intense Feulgen staining also contain RNA. They stain with both methyl green and pyronine, but after treatment with ribonuclease, which hydrolyzes RNA specifically, they stain pure green. The overall nucleic acid content can be determined by the measurement of ultraviolet absorption, which reaches a peak at about 260 millimicrons (Fig. 21 a) due to the resonance frequencies of the purines and pyrimidines. The absorption curves of meta-phase chromosomes obtained by means of the sophisticated microphotometric technique of CASPERSSON show a similar peak (Fig. 21 a). The greater absorption at the lower wavelengths by the chromosomes, as compared to pure nucleic acids, can be attributed to their protein content. In the interphase nucleus (Fig. 27) the nucleolus contains much nucleic acid and protein. Since the latter has an absorption peak above 280 millimicrons, it is probably a histone. In addition, microchemical reactions indicate the presence in nucleoli of acidic, tryptophan-containing proteins. The nucleoli do not stain with the Feulgen reaction, but are often stained by pyronine; this suggests that the nucleolar nucleic acid is RNA. The very flat absorption curve of the nuclear sap is indicative of the presence of a variety of proteins.

Chromosomes can be liberated from blood and tissue cells after the destruction of the cell body and nucleus by relatively gentle solvents. They can then be isolated by centrifugation and so made available for chemical analysis[452]. According to such analyses, lymphocyte chromosomes are composed of 90–92 per cent relatively soluble nucleoproteins which in turn consist of 45 per cent DNA and a remainder of relatively low-molecular-weight protein of the histone type. After these DNA-histones are dissolved, some coiled fibers remain which scarcely stain with the Feulgen reaction, but which stain clearly with pyronine. Chemical analysis shows that more than 80 per cent of this remainder consists of a single high-molecular-weight protein, containing 1.36 per cent tryptophan, together with 12–14 per cent RNA. This large residual protein, which is not a histone, evidently maintains the longitudinal structure of the chromosome.

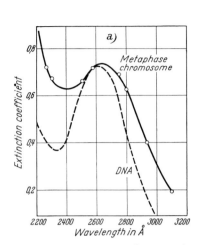
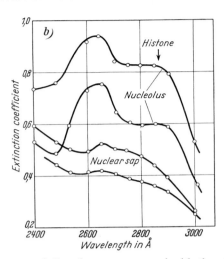

Fig. 21. UV absorption curves. a of a metaphase chromosome of *Gomphocerus*, compared with the absorption spectrum of DNA; b of nucleoli and nuclear sap in *Arenicola* egg cells. (After CASPERSSON, 1939–40)

During metaphase and anaphase in the chromosomes of a variety of species an additional component can be seen. In addition to the protein- and RNA-containing matrix, there is a lipid-containing "pellicle" which stains black with osmic acid (Fig. 22a) and which leaves an empty space around the chromosomes after fixation in lipid solvents (Fig. 22b). This lipid coat can be demonstrated as early as prophase; its origin and its fate in telophase are unknown.

Most of what we know about the architecture of chromosomes and their chemical differences comes from the giant chromosomes of the salivary glands of Dipterans. These chromosomes exhibit a regular banding pattern *in vivo*. The highly refractile discs, or "bands," stain strongly with basic dyes (Fig. 23a), while the less refractile regions do not, so that the chromosomes present the same pattern in phase microscopy and in light microscopy.

a   b

Fig. 22. Equatorial plate of the first spermatocyte division of the bug *Palomena viridissima*. a after osmic fixation; b after fixation with Fleming's solution and staining with Heidenhain's iron hematoxylin. (Sketched after microphotographs by HIRSCHLER, 1942)

This linear organization of the giant chromosomes does not consist of the repetition of identical regions. The bands, or chromomeres, vary in their thickness, in their distance from one another, and in their structural characteristics (Figs. 23, 25). Stretched chromosomes and UV photography show that what appears to be a single thick band is often a group of fine bands lying close to one another. In each chromosome of the genome the same pattern of chromomeres occurs in all cells of the salivary glands of a species, so that chromomeres in the chromosomes of *Drosophila* and other Dipterans can be mapped and

20

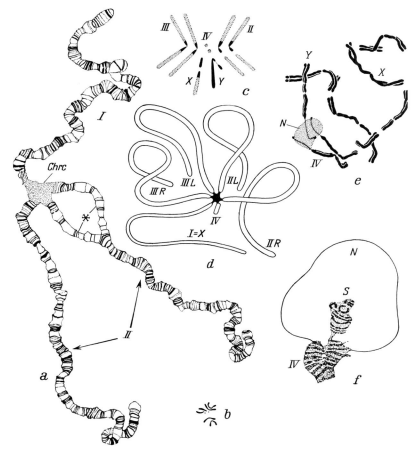

Fig. 23. *Drosophila* chromosomes. a–d *melanogaster*; a first and second chromosomes in a salivary gland cell of a female, absence of pairing at * due to local nonhomologies; b mitotic chromosomes in an oogonium, same magnification as a; c diagram of the distribution of euchromatic (light) and heterochromatic (dark) regions in the chromosomes of a male; d diagram of the arrangement of the chromosomes around the chromocenter in a salivary gland nucleus; e, f *ananassae*; e prophase in a larval neurocyte, about 2,900×; f end of the fourth chromosome in the salivary gland, about 1,900×. *Chrc.* Chromocenter; *N* nucleolus; *S* satellite; *L, R,* left and right arms of the respective chromosomes. (a, b after PAINTER, 1937; c after HEITZ, 1934; d after PÄTAU, 1935; e, f after KAUFMANN, 1937)

designated according to region and position (see Fig. 613, for example). The X chromosome of *Drosophila melanogaster* contains at least a thousand bands which can be seen in the light microscope. Electron microscope studies show that the apparently empty spaces between the bands may contain additional bands at submicroscopic levels.

Appropriate preparations employing special techniques of fixation and staining show clearly that the giant chromosomes are actually bundles of identical strands, which evidently correspond to individual chromonemata. The number of chromonemata in a giant chromosome cannot be determined precisely, but it is very large. In various dipteran tissues, smaller chromosomes of a similar nature are formed. In the ovarian nurse cells of the Muscidae (house flies and their relatives) similar giant chromosomes later break up into individual rods undergoing at the same time

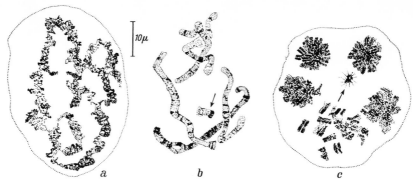

Fig. 24. Emergence of typical giant chromosomes and their fragmentation into individual chromosomes in nurse cell nuclei of *Calliphora erythrocephala*. a coiled stage; b polytene chromosomes; c metaphase chromosomes; arrow points to X-chromosome. (After BIER, 1957)

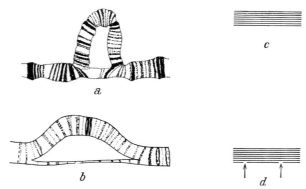

Fig. 25. Salivary chromosomes of *Drosophila melanogaster*. a somatic pairing of a normal X-chromosome and one with a deletion; b deletion in some of the chromonemata resulting from X-irradiation during chromonemal replication (5–12 hours after egg deposition); c–d diagram of the origin of the deletion in the 8 chromonema stage. (a after PAINTER, 1933; b–d after SLIZYNSKI, 1947)

longitudinal contraction, so that they now correspond to metaphase chromosomes (Fig. 24). These are already very numerous in the relatively small nuclei; the salivary gland chromosomes must contain many more chromonemata, certainly hundreds (cf. p. 8f). These bundles of chromonemata arise through repeated longitudinal divisions of the chromonemata of both homologous chromosomes, which in many dipteran tissues lie together in tight somatic pairing. The chromosomal bands of the polytene chromosomes are composed of aggregates of homologous chromomeres in register.

The pairing of homologous chromosomes is disturbed when the chromomeres of one partner have no homologue in the other, due either to a physical separation of certain regions (Fig. 23a at *) or to a deficiency (Fig. 25a). Chromosomal deficiencies can arise spontaneously and can be produced by X-irradiation. If embryos are irradiated, the gradual formation of the bundle of chromonemata can be followed through successive divisions. In salivary chromosomes a deficiency might appear in one part of the cable (Fig. 25b). The earlier the cell has been irradiated, the larger a part of the chromosome (up to half) exhibits the deficiency. If a piece is knocked out of one of a very few chromonemata (Fig. 25c, d), the deficiency exhibited by all its descendants becomes quite obvious.

22

In the giant chromosomes heterochromatic regions are always clearly seen. Frequently large heterochromatic regions of various chromosomes fuse more or less completely (Fig. 26a). In *Drosophila melanogaster* they all form one large chromocenter (Fig. 23a, d). Heterochromatic pieces are often intercalated among the chromomeres. They can be distinguished from the euchromatic bands by their granular or vacuolar structure.

Certain heterochromatic regions are sites of nucleolus formation. In the salivary glands of *Drosophila melanogaster* a large nucleolus is attached to the common chromocenter. At the nucleolar organizing site there is an accumulation of "achromatic" material either adjoining (Fig. 26b), or within (Fig. 26c) the chromosome. The bands are forced apart and the nucleolar

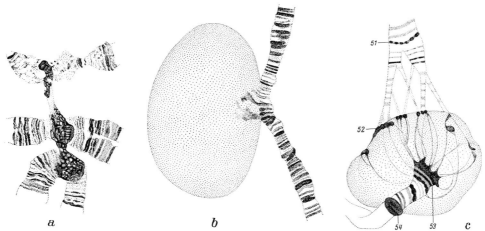

Fig. 26. Chromosome regions from salivary glands of Chironomid larvae. a *Chironomus melanotus*, incomplete fusion of the heterochromatic regions of three chromosomes; b *Chironomus pallidivittatus*, nucleolar organizing region on one of two homologous chromosomes, with attached nucleolus; c *Acricotopus lucidus*, nucleolus within the chromosome. (a after KEYL, 1961; b after BEERMANN, 1960; c after MECHELKE, 1953)

mass is pervaded by the strands connecting the chromomeres, recalling the nucleolar fibers in satellite chromosomes (Fig. 23e). In many species of Chironomids nuclear organizing regions are present in several chromosomes. Crosses between *Chironomus* species show that the nucleoli are vital components of the nucleus. In *Chironomus tentans* and *Chironomus pallidivittatus*, which produce fertile offspring, the nucleolar organizing regions are not in homologous chromosomes, and chromosomal combinations emerge in the F_2 which lack both nucleoli. The embryos derived from these anucleolate zygotes die in the germ band stage[29].

The diversity of nucleoli that can be distinguished according to structural and staining properties in various differentiated cells leads to the conclusion that nucleoli have a complex relationship with metabolic events in the nucleus and in the cell as a whole.

In euchromatic regions of giant chromosomes, local swellings arise, often accompanied by the release of materials. These are called *puffs* (Figs. 613, 614); they vary strikingly with the state of the cell, and we shall consider them in detail later on. Extremely large and active puffs are called *Balbiani rings*.

Cytological studies on giant chromosomes reveal species-specific patterns of structure and function. Cytogenetic investigations, in which the products of crosses are studied cytologically, show that in particular chromosomal regions and apparently in individual chromatic bands there are factors which govern specific processes in metabolism and in development, and we are led once more to the question of the chemical nature of these microscopically visible structural components.

The distribution of chemically defined substances in the architecture of the giant chromosomes can be established by a variety of methods. Millon staining shows that protein containing tyrosine or a similar component is present in the chromomeres and between them. The UV absorption curves for euchromatic bands (Fig. 27b) are similar to those of metaphase chromosomes (Fig. 21a). Heterochromatic bands, the chromocenter, and puffs exhibit a high nucleic acid and histone content (Fig. 27). The interband regions contain a protein of another type, apparently a globulin.

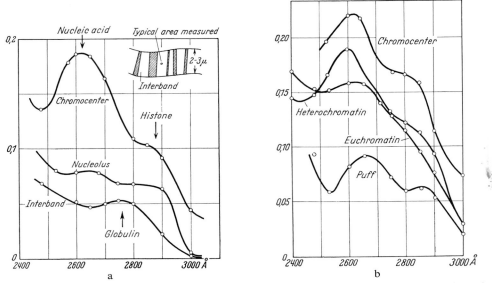

Fig. 27. UV absorption spectra of various nuclear regions of *Drosophila melanogaster* salivary glands. a chromocenter, interband region, and nucleolus, insert shows dimensions; b chromocenter, heterochromatic region near the chromocenter, puff, and euchromatic band. (After CASPERSSON 1939–40, 1941)

A specific arginine reaction results in the intense staining of the chromomeres and somewhat weaker staining (about 70 per cent as strong) between them. These differences correspond approximately to the relative amounts of arginine in histones and in globulins, which contain about 60 per cent as much. After treatment with deoxyribonuclease, a protein strand remains, which runs the length of the chromosome. Feulgen and methyl green-pyronine staining show that both the euchromatic and heterochromatic regions contain DNA. The nucleolar structures are rich in RNA (as shown by pyronine staining) and histones (Fig. 27a).

As to the orientation of the polypeptide chains in the entire chromonema and the nucleotide chains in the chromomeres, the polarizing microscope should be very useful. Unfortunately, its use has not led to unambiguous conclusions. The chromomeres are negatively birefringent along the longitudinal axis of the chromosome. This suggests that the nucleic acid molecules are oriented parallel to the chromosomal axis. Since the DNA molecules reach a length of several thousand Å', they can run through an entire chromomere. Since the protein molecules are positively birefringent with respect to their longitudinal axes, they should reduce the chromomeres' anisotropy and induce positive birefringence in the interband regions, provided they are extended and oriented parallel to the long axis of the chromosome. The interband regions are, however, almost, or entirely, isotropic. The giant chromosomes can be stretched to several times their ordinary length without tearing by the use of fine needles, When this is done, the interband regions alone, or almost alone, are stretched and in this process they become positively birefringent. When released, the chromosomes contract elastically, although admittedly not to their original length, if the stretching has exceeded a certain point. One might suspect that the stretching of the chromosomes corresponds to an unfolding of the polypeptide chains in the interchromomere regions, and that in the chromomeres, on the other hand, the long nucleic acid molecules are tightly bound to polypeptide chains forming a rigid structure which cannot be stretched easily.

The salivary gland chromosomes of the Diptera are nuclear structures of extremely specialized tissue cells; and polytene chromosomes of this type have never been found in other groups of animals in spite of extensive searching. We must therefore ask ourselves to what extent these unusual chromosomal structures can serve as a paradigm for the universal architecture of the chromonema. The longitudinal organization of the chromonemata into Feulgen-positive chromomeres and connecting fibers is also found in animals, plants, and protistans in the extensively uncoiled leptotene chromosomes of the early stages of meiosis (Fig. 28). Since in this case the chromosomes are much more numerous and tangled up inside the nucleus (not being individually demonstrable by squashing the nucleus, as is the case for the giant salivary chromosomes, Fig. 23a), the number of chromomeres in the meiotic chromosomes cannot be determined precisely, although one can say with certainty that the total number for the entire genome is in the thousands.

A question of importance for the formation of the replicative structure as well as for its changes in form, arises from the tremendous difference in length between ordinary metaphase chromosomes and the giant chromosomes of the salivary glands. In *Drosophila* the ratio between the

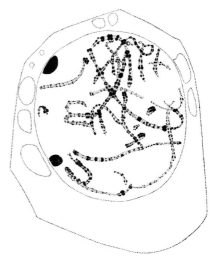

Fig. 28. Section through a pollen mother cell of *Lilium martagon*, pachytene stage (1,400×). (After SCHAFFSTEIN, 1935)

two is about 1:100 (Fig. 23b). In *Drosophila ananassae* the tiny satellite of the metaphase chromosome IV (Fig. 23e) unfolds into a chain of chromomeres containing more than ten chromatic bands (Fig. 23f). The giant chromosomes can be stretched to several times their ordinary length without tearing, so that they now can be measured in millimeters. Although a tenfold change in length between the metaphase chromosomes and the chromosomes of meiotic prophase can still be explained on the basis of the uncoiling of a doubly coiled thread of microscopic dimension, the elongation seen in the giant chromosomes is far too great to explain in this way. It must be explained either by a true intercalary growth (the interposition of material all along the chromosome) or by a lengthening at the molecular level, which still possesses some elastic character. This latter possibility cannot be discounted, since we know that fibrous proteins are capable of a considerable degree of folding and unfolding. We must also consider the possibility that in the tangle of fibers seen in interphase and early prophase nuclei in the light microscope and in recent electron microscope pictures the individual chromosomes are longer than leptotene chromosomes.

A major question intimately related to the chemical composition of the chromonemata concerns the role of nucleic acids in the developmental cycle.

Bacterial genetics and virology have shown that the continuous, self-replicating, specific structure consists of a nucleic acid. Viruses induce the synthesis of specific proteins in plant cells and in the cytoplasm of bacteria and these proteins combine with the nucleic acids to form the mature infective virus particles. But the nucleic acids alone suffice for infection. In eukaryotic chromosomes, on the other hand, DNA is always bound to protein, but there are a variety of reasons for believing that the DNA alone carries the genetic information which is transmitted during cell division and at fertilization.

Table 1: *DNA content (mg $\times$ 10^{-9}) of the nuclei of various types of cell in the domestic fowl.* (From BUTENANDT, 1953)

RBC	Liver	Kidney	Spleen	Heart	Pancreas	Sperms
2.34	2.39					1.26
MIRSKY and RIS (1949)						
2.49	2.56	2.20	2.54	2.45	2.61	
DAVIDSON, LESLIE, SMELLIE and THOMSON (1950)						

The quantitative chemical analysis of isolated nuclei from vertebrate tissues has shown that for a given species all diploid cells have a constant, species-specific amount of DNA which is about twice as great as that in the haploid sperms (Table 1). Such analyses can be made on a suspension of nuclei that can be counted in a cell counter, or by cytophotometry of nuclei in tissue sections.

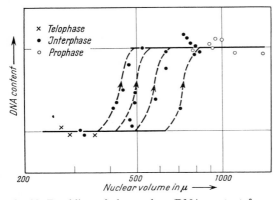

Fig. 29. Doubling of the nuclear DNA content from telophase to the following prophase in growing onion root tips. Microspectrophotometric values from Feulgen preparations: points are average values of four measurements on each nucleus. (After PÄTAU and SRINIVASACHAR, 1959)

If the turnover of various substances is measured with radioactive isotopes, exchange is seen to be rapid in the case of carbohydrates, lipids, proteins, and RNA. Only the DNA of the nucleus does not take part in this rapid turnover. The DNA is synthesized in proportion to the increase in cell number (except in the cases of polyteny, which we have already discussed), and its breakdown is extremely slow in comparison to that of the other cell components, to say the least.

The replication of DNA between two nuclear divisions has been followed in Feulgen preparations of tissues with rapidly dividing cells. During nuclear growth in interphase the DNA content reaches twice the level of anaphase and telophase (Fig. 29). In cells ready to divide there is the same amount of DNA as the total for the metaphase chromosomes.

The DNA is the only nuclear substance known to us which passes unchanged through the gametes to generation after generation. While the DNA passes into the sperm head without loss or change, the protein fraction undergoes profound changes. Histones are replaced by protamines or protamine-like substances, and the nonhistone proteins decrease or disappear completely. Birefringence and X-ray diffraction studies show that in species with long, narrow sperm heads the DNA molecules run essentially parallel to the sperm axis.

That the microscopic structures of chromosomes are not the ultimate structural elements can be seen even by this crude mechanical divisibility: Electron microscope preparations of chromosomes and chromosome fragments, which for example have been isolated from the nuclei of avian blood-cells after plasmylosis and centrifugation, show a splitting of the coiled fibres into coils of several lower orders (Fig. 30), and these can be separated from one another according to the paranemic principle illustrated in Figure 8b.

26

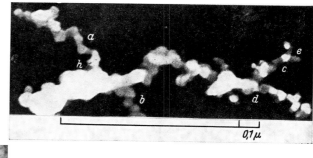

Fig. 30. Electron micrographs of chromosomes from chicken red blood cells isolated with 1 M NaCl. Osmic acid fixation. a, b, c longitudinal components of the chromosome with similar diameters; d, e, h sites of fragmentation. (Hovanitz, Denues, and Sturrock, 1949)

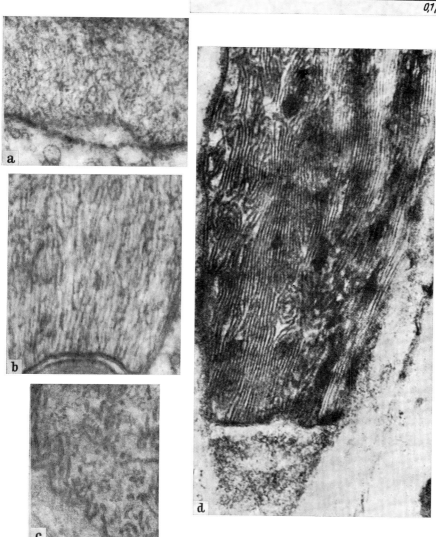

Fig. 31. Chromofibrils in spermatid nuclei. a, b *Octopus vulgaris;* a untangling fibrils; b oriented on the long axis of the nucleus, 19,000×; c *Ranatra* (Hemiptera), oriented 100-Å fibrils of which some can be seen to be aggregates of 40-Å fibrils, 6,400×; d fibrillar arrangement in sperm head of *Helix pomatia*, 40,000×. (a–c Ris, 1959; d Grasse, Carasso, and Favard, 1956 from Serra, 1959)

27

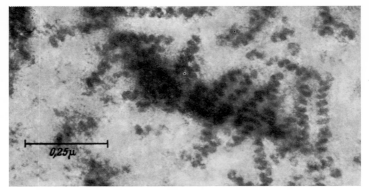

Fig. 32. Section from the Feulgen positive region of the nucleus of *Amoeba proteus*. (PAPPAS and BRANDT, 1960)

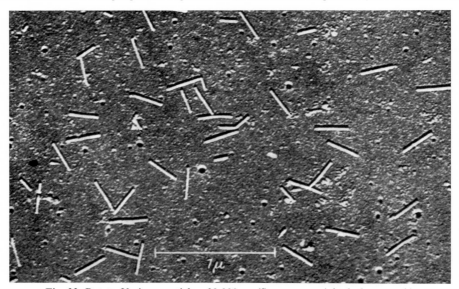

Fig. 33. Potato V-virus particles, 30,000 × . (SCHRAMM, original photograph)

Electron microscope studies on the nuclear structures seen during spermatogenesis lead to more fundamental conclusions. The interphase nuclei of young spermatids are followed with a tangle of fine fibrils. These begin to untangle at the onset of sperm differentiation. They orient in the long axis of the nucleus (Fig. 31 a, b) and join together in groups. As the nucleus elongates in the growing sperm head, the untangling and ordering of the fibrils continues until they are completely parallel, or at least extended in bundles (Fig. 31 d). Evidently the fibrils belonging to one chromonema come together or remain together as the tangle is replaced by a more obvious order. The orienting fibrils are about a hundred Å thick in all sperms examined, and they often turn out to be made up of two smaller fibrils about forty Å thick, which are closely intertwined (Fig. 31 c). Even in functional nuclei containing a good deal of nuclear sap, as in *Amoeba proteus*, helically coiled fibrils of similar dimensions have been found in thin sections in the electron microscope (Fig. 32). At the highest magnifications they appear to be doubly coiled.

All these fibrils are of the same order of size as plant virus particles (Fig. 33), which contain RNA and protein in tight nucleoprotein coils. It would seem safe to conclude that the fibrils in the nuclei (Figs. 31, 32) are the ultimate nucleoprotein units in the chromosomal structure, and that as the units of self-replication they represent the ultimate units of function in the continuity of the chromosomes.

The fibrils exhibit no longitudinal architecture corresponding to the pattern of chromomeres. This pattern, which appears regularly and precisely in the individual chromosomes of particular cell types, must therefore stem from a molecular pattern of nucleoprotein fibrils established in the course of development.

In any event, the chromonema itself has the structure of a cable. When a chromosome divides, the bundle of elementary fibrils, which we can call chromofibrils, must be reorganized for distribution to the two chromatids. What determines the rate of multiplication of the elementary fibrils and their parcelling out into bundles when the chromatids separate? Huskins is right: "The general problem now appears to be not [to] what degree the chromosome is subdivided, but how it comes to behave as a bipartite unit in inheritance." [320, 409]

Lecture 3

The chromosomes are functionally indissociable from the entire cell. Their activities in metabolism and in cell morphogenesis take place in the working nucleus of interphase. In the transition to interphase the nuclear sap and the nuclear membrane are formed. The latter contains protein and lipids; electron microscopy is beginning to tell us about its structure. The membranes of large nuclei from amphibian oocytes and from *Amoeba proteus* have two layers: a porous layer which presumably serves for mechanical support and a continuous layer which may regulate permeability[8]. Now the nucleoli form. Surely these events involve the intense exchange of substance between nucleus and cytoplasm. Differences in the appearance of the nuclei of various tissues and from various functional states of the same cell indicate nucleocytoplasmic interactions. The changing behavior of the heterochromatin has already been mentioned (p. 17). The ratio of protein to DNA in "resting nuclei" in various tissues varies in the few determinations that have been made between 8:1 and 20:1 according to the cell type, while the ratio is about 1.5:1 in isolated chromosomes. The ratio of RNA to DNA also varies in various types of interphase nuclei.

But even in their mitotic transformations, the individual chromosomes are not fully autonomous. The dimensions of the metaphase chromosomes and sometimes other aspects of their structure vary according to certain conditions in the cell and according to the tissue (Fig. 34).

Fig. 34. Metaphase chromosomes in *Kleinia* (*Senecio*) *spinulosa;* a from the shoot meristem; b from the leaf parenchyma; c from the leaf xylem. (After CZEIKA, 1960)

Chromosomal size appears to be influenced by the relative amount of cytoplasm. In the course of cleavage divisions which place the nuclei in ever smaller cells the chromosomes generally become smaller also. Haploid cells in *Triturus* have longer chromosomes in a given cell generation than have diploid cells. The average volume of the chromosomes in the first meiotic metaphase of the tetraploid form in *Antirrhinum* is smaller than in the diploid form. Even the composition of the chromosome sets in the cell has an influence on their size. In a cross between two species of fish with large and with small chromosomes the size difference is generally retained, and, indeed, this provided an early example of evidence for the individuality of the chromosomes; yet in some cases the chromosome size of one species is dominant. In plants we know of similar cases. In crosses between various species of *Crepis* the size of the chromosomes of both species

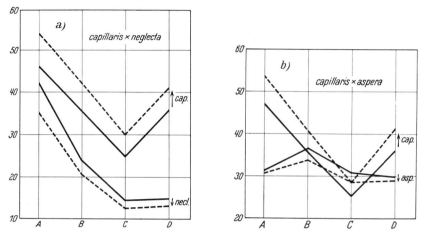

Fig. 35. Lengths of chromosomes *A*, *B*, *C*, *D* in pure species (——) and hybrids (---) of *Crepis*. *C. Capillaris* lacks the B chromosome. (cf. Fig. 38b). (After NAVASHIN, 1934)

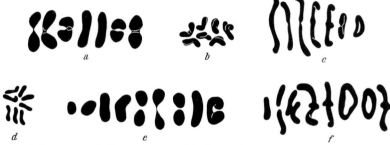

Fig. 36. Dependence of chromosomal dimensions on alleles in stock (*Matthiola*). a, b wild type; c, d mutant strain; e, f F₂ plants, e with short chromosomes, f with long chromosomes; a, c, e, f first meiotic metaphase; b, d polar view of second meiotic metaphase. (After LESLEY and FROST, 1927)

is altered in the hybrid; the changes are not in the direction of compromise, rather the differences are increased. In *C. capillaris × neglecta* and *C. capillaris × aspera*, the *capillaris* chromosomes always become larger and those of the other species smaller. Thus, in the first cross the size difference between all the chromosomes becomes larger (Fig. 35a); in the second cross the increase is seen only for A and D; in B a difference arises in the hybrid, and in C the original difference is eliminated (Fig. 35b). Chromosome size can be influenced by individual genes. The metaphase chromosomes of *Matthiola incana* are in general short and compact (Fig. 36a, b); in one race, Snowflake, they are longer and thinner in all individuals (Fig. 36c, d). This difference in chromosomal shape is inherited as a simple Mendelian trait. In the F_1 the chromosomes are short, and in the F_2 individuals with short (Fig. 36e) and long (Fig. 36f) chromosomes segregate in a ratio of roughly 3:1 (45:17). This character therefore appears to be controlled by a single allelic difference.

Environmental conditions, too, can influence the size of metaphase chromosomes, presumably indirectly. A chromosome may be maximally long at a moderate temperature and shorter at either extreme (Fig. 37). The chromosomal thickness varies with marked differences in length, but thickness is not always constant even when the length is constant.

It seems plausible that such differences in dimensions stem from differences in coiling. Volume differences might be due to differences in the amount, or the degree of hydration, of the matrix which surrounds the chromonemal coil and probably pervades it. From the electron microscopic studies previously discussed (p. 28), we can add another possibility — variations in the number of elementary fibrils.

$<-7°$ $-7°$ to $-6°$ $-6°$ to $8°$ $8°$ to $25°$ $25°$ to $27°$ $>27°$

Fig. 37. Changes in chromosomal dimensions and in the basic staining of the special segment of a *Trillium sessile* chromosome according to temperature. (After Resende, De Lemos-Pereira, and Cabral, 1944)

In many species of plants, special structural differentiations appear in the chromosomes if they are raised at low temperatures. These are called special segments (Fig. 37). They are characteristic for a given chromosome; frequently but not always they are located at the end of an arm[577]. Their weak basic staining shows that they have a lower concentration of nucleic acids[163]. Frequently they are also narrower: they can, however, be as wide as the rest of the chromosome or even wider. The conditions which induce them do not act on the fully developed chromosomes but rather on the interphase nucleus or during prophase. If "cold plants" with special segments (seen in mitosis in the calyx tissues) are placed at 15–17° C., the special segments are no longer clear after three hours; after six hours they have disappeared. Plants transferred in the reverse direction exhibit special segments at the onset of the next mitotic cycle. That the reaction of certain chromosomal regions to cold has real functional significance can be concluded from the fact that the special segments are found in plants which must live at low temperatures for a certain part of the year; in the absence throughout the year of cold, *Trillium* species bloom badly or not at all. If there is no period of cold for several years, the plants die[537]. Special segments are not restricted to heterochromatic regions.

In certain *Crepis* crosses the behavior of the satellite chromosome is influenced by the partner's genome (Fig. 38); in *C. capillaris × parviflora* hybrids the nucleolar fiber of *capillaris* disappears (Fig. 38d): in *C. capillaris × dioscorides* crosses the satellite of *dioscorides* remains contiguous with the main body of the chromosome, and the nucleolar fiber of the *capillaris* D-chromosome is drawn out to an unusual extent (Fig. 38e).

Some mutations can have a profound impact on the structural changes of chromosomes. In maize[20] the recessive gene *st* (sticky chromosomes) disturbs the behavior of the chromosomes in somatic divisions and even more so in the meiotic divisions: the chromosomes stick to one another in anaphase, nonhomologous as well as homologous chromatids, so that the individual chromosomes often can no longer be identified as such. The sticky chromosomal masses appear to be pulled apart mechanically during anaphase. This mutation evidently results in a colloidal change, ostensibly in the matrix. Similar disturbances, chromosomal clumping at the end of prophase or in early metaphase, can be induced by certain poisons (trypaflavin) and can be traced to changes in the matrix[283].

Meiosis emerges from a transformed mitotic cycle which takes place in certain cells and therefore under certain intracellular conditions. Here the chromosomes replicate as is always the case

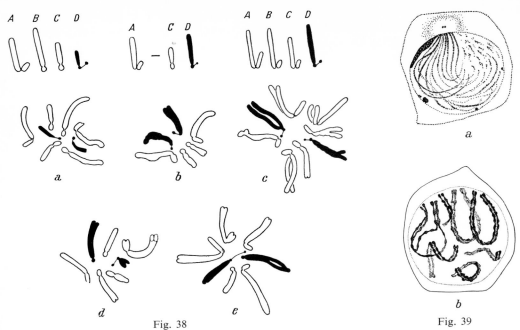

Fig. 38

Fig. 39

Fig. 38. Chromosomes of *Crepis* species. a–c haploid chromosome set and equatorial plates of meristematic cells; d, e hybrids. Satellite chromosomes black; a *C. parviflora*; b *C. capillaris*; c *C. dioscorides*; d *capillaris* × *parviflora* (the satellite of the *capillaris* chromosome has disappeared); e *capillaris* × *dioscorides* (the satellite of the *dioscorides* chromosome has disappeared, that of the *capillaris* chromosome has a long fiber). (After NAVASHIN, 1934)

Fig. 39. a leptotene (bouquet stage) of a spermatocyte from the locust *Stenobothrus lineatus*; b Pachytene of a spermatocyte of the snail *Paludina vivipara*; interference with conjugation in one chromosome pair. (After BELAR, 1928, 1929)

between two nuclear divisions; but now they pair. The lengthwise pairing of the replicated chromonemata results in groups of four homologous chromatids each. Two successive nuclear divisions now follow, through which each tetrad of four chromatids is distributed among four haploid nuclei. This chromosomal reduction takes place in all eukaryotic organisms whose life cycle includes fertilization; but meiosis is incorporated into various stages of the life cycle depending upon the phyletic group. It can follow directly after fertilization, so that only the zygote nucleus is diploid, with all cell generations before and after fertilization possessing only one set of chromosomes (many algae and fungi, some protozoa). In contrast to these haplonts, all diplonts (most protozoans, all metazoans) have haploid gametes. All the other cells are diploid. Here reduction takes place during gametogenesis. Some algae and all higher plants are diplont-haplonts: a diploid generation, the sporophyte, alternates with a haploid generation, the gametophyte, and reduction takes place during the formation of the spore which produces the gametophyte.

The nuclear events in meiosis are always the same, whether they take place in the zygote, during gametogenesis, or in the spore mother cell. The nucleus enlarges, and within it the chromosomes appear as thin threads which usually show no signs of having doubled. In this *leptotene* stage the elongated, uncoiled chromosomes generally orient in a certain way; the thin chromosomes have one or both ends close to the nuclear membrane in a small region near the centriole (Fig. 39a) or, in cells without centrosomes, near a region of polar cytoplasm. This so-called "bouquet" stage signals the onset of chromosomal pairing in many protistans, plants, and animals: the homologous chromosomes approximate their polar ends and gradually come to lie close together throughout their length. In this *zygotene* stage (Fig. 39a, b) the chromomeres are in register, and the pair of chromosomes gradually becomes thicker and often comes to conceal its dual nature in the pachytene stage (Figs. 28, 39b, 40c, 41a, e).

The homologous chromosomes have not always been situated close together; they must find one another, and pairing gradually spreads out from the places where they meet. So it can happen that chromosomal pairs become interlocked, resulting in the prevention of complete pairing (Fig. 39b).

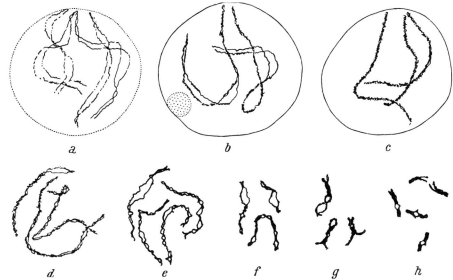

Fig. 40. Chromosome pairing in *Actinophrys sol* (Heliozoa). a bouquet stage (leptotene); b zygotene; c pachytene; d–g diplotene ("strepsitene"); h diakinesis (2,600×). (After Belar, 1922)

The pachytene chromosomes shorten and thicken further: a visible coiling begins during which each member of a pair becomes more distinct from its partner again (Figs. 40, 41). In this *diplotene* stage the chromosomes are wound around each other. One recognizes now that four chromatids (two pairs, of which each pair has resulted from the replication of one chromosome) take part in the pairing (Fig. 42). The tetrads continue to shorten (Figs. 42, 43). The places where the polytene chromatids appear to cross over, the chiasmata, move towards the ends of the chromosomes (terminalize); as they do this their number becomes reduced (Figs. 40, 41, 43). In the stage of maximal shortening (diakinesis), in which they approach the equatorial plate of the first maturation division, it is often only the ends of the chromosomes which touch their homologues (Fig. 41d, k).

Heterochromatic regions of some length influence pairing behavior. Completely heterochromatic chromosomes (sex chromosomes) frequently do not pair tightly, but rather remain in a sort of distant conjugation. Heterochromatic pieces of chromosomes separate at diplotene sooner than euchromatic regions; this appears to be related to the fact that heterochromatic regions do not form chiasmata. The shortening of the heterochromatic regions starts earlier but does not proceed so far as in the euchromatic regions. In *Streptocarpus wendlandii* the chromosomes shorten by the time of diakinesis to 1/8 of their pachytene length. While the heterochromatic central segment undergoes only a sixfold shortening, the euchromatic ends of the chromosome shorten by a factor of about fourteen. So the euchromatic regions are three times as long as the heterochromatic region in diakinesis, as compared to seven and a half times in pachytene[482].

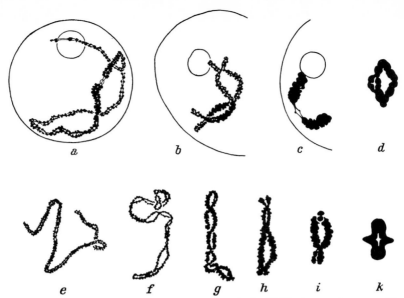

Fig. 41. a–d Pollen mother cell chromosomes from *Kniphofia aloides* (1750×). a pachytene, coiling in the hetero-chromatic regions near the kinetochore; b diplotene; c, d shortening to metaphase chromosomes; e–k meiotic chromosomes of *Anemone baicalensis*; reduction in the number of chiasmata (terminalization) from pachytene to metaphase in the first maturation division. (a–d after DARLINGTON, 1933, e–k after MOFFETT, 1933)

Fig. 42. Diplotene ("Strepsitene") chromosomes from locust spermatocytes. a *Mecostethus grossus*; b, c *Steno-bothrus parallelus*, 2,000×. (After JANSSENS, 1924)

During the maturation divisions the genes situated in the individual chromosomes are distributed to the gametes. While the chromosomes are still paired, crossing over, the exchange of genes between homologous linkage groups takes place. This requires a physical exchange of chromosomal segments. Genetic experiments show that this exchange takes place between chromatids at the four-strand stage rather than between the two chromosomes before replication (Fig. 43): From the four spores resulting from a spore mother cell, four different combinations of alleles can emerge. The exchange can therefore not have taken place before the four-strand stage; if it had, only two combinations would have been possible, and each would have been represented twice. The assumption is thus justified that in pachytene the longitudinally split chromosomes wind around one another, and that at the site of crossing over the chromatids break and rejoin the corresponding breaks in their homologues. The diagram in Figure 43 corresponds to chromosomal exchanges observed in the microscope, showing the passage of given chromatids from one pair to the other pair during diplotene (Fig. 42).

Meiosis is basic to some of the central events in heredity. The first question is: what makes the chromosomes pair? Certainly a requirement for pairing must be the basic structure of the chromonemata: this also makes possible somatic pairing which is seen in certain Dipteran cells (p. 22). A working hypothesis tries to relate the conditions of pairing between homologous chromosomes to physicochemical forces dependent on their known chemical structures. When in the chromonemata the nucleic acids of the chromomeres form salt bridges with basic protein sites (p. 19), dipole moments normal to the long axis will arise, and a certain charge distribution will result characteristic for the chromomere sequence of each chromonema only. Only in the case of a symmetrical arrangement (Fig. 44a) can we expect a uniform attraction of all the parts. Pairing due to these hypothetical forces is possible only between chains of molecules whose homologous atomic configurations can approach one another sufficiently closely. A complete (linear) uncoiling need not take place; paranemic coils can mesh as well as separate (Fig. 8b). After all, one must expect that the uncoiling of the major coils produces conditions well suited for the sort of pairing mechanism can be assumed for the leptotene chromosomes. The dipole forces in the nucleus can, at any rate, exert an attractive force only over relatively short distances, since presumably the free ions in solution restrict the effective range of the electrostatic forces. When, however, the homologous chromosomal ends have met one another, it is possible to extend the junction by dipole forces like a zipper. This conception corresponds in detail to the microscopic picture when the leptotene chromosomes are zipped together from their polarized ends (Figs. 30a, 40a, b).

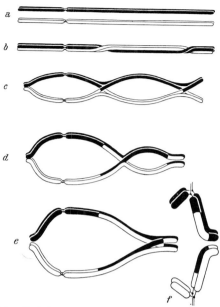

Fig. 43. Diagram of chromosomal exchange in the four-strand stage and of the terminalization of the chiasmata. a leptotene; b pachytene; c, d diplotene; e diakinesis and metaphase; f anaphase. (After KÜHN, 1965)

Although the basic structure of the chromonemata provides a basis for pairing, the state of the cell is critical, since pairing takes place only in spore-forming cells, gamete-forming cells, and zygotes. The nature of this state is unknown; in unicellular organisms it can be induced experimentally, as we shall see.

The further course of meiosis, especially the mutual coiling of the chromosomes and chiasma formation, can be influenced by various conditions. Certain genes are also involved: in an asynaptic strain of maize[91] the homologous chromosomes begin to pair normally but separate again in pachytene without forming chiasmata. Frequently the chromosomes take their places on the spindle individually. The number of unpaired chromosomes in metaphase corresponds to the minimum number of chiasmata in diplotene. This observation supports Darlington's conjecture

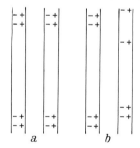

Fig. 44. Like (a) and unlike (b) dipole patterns in two parallel strands. (After FRIEDRICH-FREKSA, 1940)

that the chromosomes are held together up to the time of diakinesis by the chiasmata. In *Datura stramonium* a recessive allele causes the complete absence of pairing in the pollen mother cells and in the macrospore mother cells. The unpaired chromosomes are distributed irregularly in the first division; in the second division a normal segregation of the chromatids takes place. The number of chiasmata within the normal range is also genetically determined; it varies under constant growing conditions among different lines of *Vicia faba*. In *Oenothera* the normal terminal junctions in diakinesis are reduced to a particular degree by certain gene combinations (Fig. 45 b). If the dominant allele (*Co*) for small flowers in *Oenothera suaveolens* is crossed into *Oenothera hookeri*, the tendency for pairing is reduced and the percentage of asynapsis at diakinesis rises (Table 2). A comparison of numerous genome combinations placed in different cytoplasmic contexts by reciprocal crosses shows that the pairing tendency is determined by the interaction of a particular genome with a particular cytoplasm (Table 2).

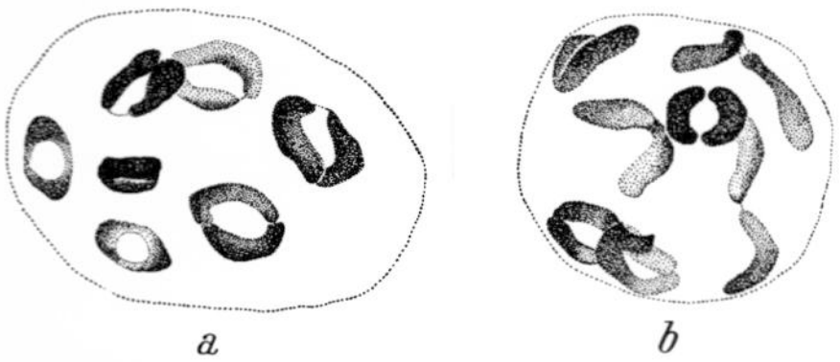

Among the variable internal conditions which affect pairing, the greening ability of the plastids plays a role [481]. This ability depends on the culture method and differs between *O. hookeri* and *O. suaveolens*. The reduction in chlorophyll relative to the yellow pigments results in an increase in asynapsis during diakinesis in the hybrids. Placement of the flower-bearing stems in glucose solution increases the chiasma number, substituting for the anabolic activities of the plastids.

Among the external influences on chiasma number, temperature is the most important. In *Vicia faba* the number of chiasmata drops sharply with duration at high temperature (Fig. 46). After two days at 34° C., pairing may be completely absent (Fig. 46 c). Here again it appears that asynapsis results from a lack of chiasmata. In *Rhoeo discolor* the absence of terminal junctions in diakinesis is raised to 82 per cent by low temperatures: in *Campanularia persicifolia* water content is critical. With increasing osmotic strength in the flower stalk, asynapsis rises. The behavior of the chromosomes in meiosis is thus influenced by metabolic events in the plant.

Fig. 45. Diakinesis in *Oenothera hookeri*. a normal; b absence of some junctions. (After OEHLKERS, 1937)

Table 2: *Dependence of pairing on certain alleles in Oenothera.* (After OEHLKERS, 1937, pooled data)

	% Asynapsis
I. In *hookeri* cytoplasm	
co co	0.64
Co co	3.07
II. In *suaveolens* cytoplasm	
co co	2.85
Co co	8.33
Co Co	7.37
III. In *biennis* cytoplasm	
Co co	8.50

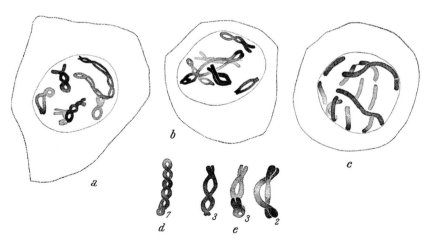

Fig. 46. Diplotene of *Vicia faba*. a normal at 20°; b after one day at 34°; c after two days at 34°, almost complete asynapsis; d, e the largest chromosome (*M*); d at 20° with 7 chiasmata; e after one day at 32° with reduced chiasma number; 1,500×. (After STRAUB, 1937)

Cytogenetic experiments on *Oenothera* and on *Antirrhinum* have shown that the experimentally induced reduction in chiasma number is reflected in a reduction in the recombination frequencies between certain pairs of genes. This agrees with the notion that the chiasmata are the sites of physical exchange of chromosome segments.

Lecture 4

In nuclear division there is a fundamental connection between the division cycle of the chromosomes and the mechanism of distribution of the chromatids to the daughter nuclei. While the chromosome cycle is strikingly uniform throughout the plant, animal, and protistan kingdoms, the distribution mechanism is extraordinarily diverse. There is scarcely any other fundamental process of development which so well exhibits the "heretical nature of all living things" (Hopkins) and which makes such a mockery of any attempt at a unified scheme.

One structure that is almost universal in division is the bipolar spindle. In its equator the metaphase chromosomes become arranged, and within the spindle or at its surface the chromatids move toward their respective poles. But this is about the only generalization one can make with respect to the remarkably varied kinds of division seen in eukaryotic living cells.

In most of the cells of multicellular organisms, in many protozoans, and in many lower plants, there are at the poles of the spindles very small organelles called centrioles. These are self-replicating structures. During interphase they lie near the nucleus, sometimes surrounded by a halo of cytoplasm. They multiply by dividing in two, and in prophase the daughter centrioles move to the poles of the dividing cell. In fertilization in metazoans they are always transmitted to the zygote by the sperm.

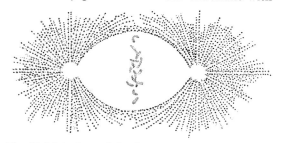

Fig. 47. Metaphase of the first cleavage division in the living egg of *Rhabditis*. (After F. SCHRADER, 1944)

Around the centrioles at the poles the many-rayed astral sphere grows out into the cytoplasm. Within it, as a rule, a homogeneous mass of cytoplasm, the centrosome, collects around the centriole. In many species the centrosome undergoes cyclical growth and transformation [71, 698]. It grows vigorously, then it is surrounded by rays, and finally a new, little centrosome forms around the centriole and grows in turn. The spindles and polar structures are called "the achromatic

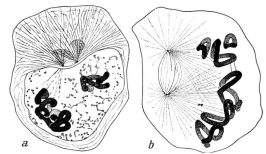

Fig. 48. Prophase of the first meiotic division of *Salamandra maculosa*. (After MEVES, 1896)

figure" or "the achromatic apparatus." In living cells, the spindles, the centrosomes, and the astral rays are visible because of the exclusion of cytoplasmic granules (Fig. 47).

In the higher plants, in some protozoans, and in the maturation spindles of most egg cells there are no centrosomes at the poles, and there is no indication of submicroscopic centrioles, since the spindles do not point directly toward the poles, but rather assume a barrel or sheaf

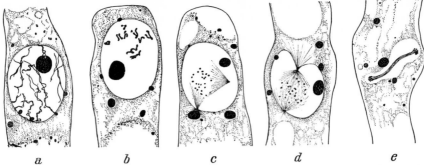

$$a \qquad b \qquad c \qquad d \qquad e$$

Fig. 49. Intranuclear mitosis (first meiotic division) in the ascus of *Morchella deliciosa*. The centriole lies on the nuclear surface. (After WAKAYAMA, 1931)

shape. At the poles of such spindles there are frequently polar caps of dense cytoplasm. The assumption that they are composed of a special kind of cytoplasm is pure speculation.

The spindles can form in a variety of ways, and we shall consider just a few major types. Frequently the spindle arises immediately after centriole division; it extends between the two daughter centrioles and elongates as they move farther and farther apart (Fig. 48). Very often, however, half-spindles are formed emanating from daughter centrioles as they move toward the poles (Fig. 49), or from the polar caps in divisions not involving centrioles. The spindle can arise in the

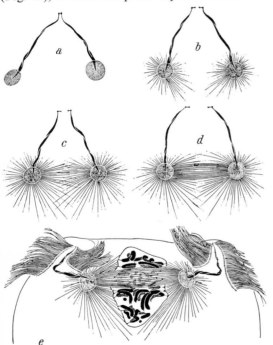

cytoplasm while the nuclear membrane is still intact. Sometimes it forms far from the nucleus and approaches it only afterwards (Fig. 50). On the other hand, it can form entirely within the nucleus (Fig. 49). Again, half-spindles may push through the cytoplasm from the poles to the nucleus, meet in the nuclear space after the nuclear membrane has dissolved, and unite into a single spindle (Fig. 51). Whether in this process they grow through the nuclear space has not been determined. This question, moreover, appears not to be really fundamental: we do know of some spindles that are formed from cytoplasm and others that are formed from nuclear contents.

The spindles and astral rays emerging from the centrioles are functional structures which form, undergo specific changes, and then disappear. The career of this apparatus is in step with the phases of the chromosome cycle, but the two processes can also proceed independently of one another. There are various means to prevent the formation of the division apparatus while permitting the normal chromosomal division to take place; in such cases the chromatids are not

Fig. 50. Division phenomena in the flagellate *Barbulanympha*. a interphase centrosomes; b–d formation of asters and the spindle; e union of the division apparatus with the nucleus, intra- and extranuclear chromosomal spindle fibers 770×. (After CLEVELAND, 1938)

38

distributed to daughter nuclei; rather, a single polyploid nucleus is formed. In cells which have received no nucleus as a result of some disturbance of division, centriole division, astral sphere formation, and sometimes even spindle formation and cytoplasmic division can take place repeatedly and rhythmically (Fig. 89). Normally, therefore, the synchrony of the events is due to the matched temporal characteristics of the individual processes.

We must try to separate individual events from the total picture in order to understand their physicochemical nature, their causality, and the means by which they collaborate to produce a unified process. We know the course of nuclear division mainly from the ordering of stages in fixed, stained cells. But nuclear division can also be followed in certain living animal and plant

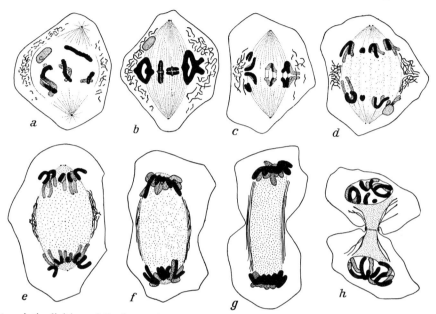

Fig. 51. First meiotic division of the locust *Stenobothrus lineatus*. a prophase; b metaphase; c early; d middle; e late anaphase; f–h telophase. (After BELAR, 1927)

cells, especially since the development of phase contrast microscopy. The polar rays and the spindle appear and the chromosomes first move along the spindle to their equatorial positions (prometaphase movement) and then toward the poles (anaphase movement) while the spindle lengthens.

Mitosis may take anywhere from a few minutes to hours. The duration can be measured, for example, in cleavage stages by the number of nuclei formed during a certain period following fertilization. In this way, one concludes that each mitotic cycle in a *Drosophila* egg takes about ten minutes. Direct observation of embryonic chick mesenchyme cells in tissue culture gave times of seventy to two hundred minutes, from the first signs of prophase to the end of telophase (complete reconstruction of the interphase nucleus). Nevertheless, the beginning and end of the mitotic changes are difficult to establish in living cells.

The relative durations of individual phases can be estimated both from multiplying cells of unicellular organisms in culture and from fixed preparations of tissue cells; the duration of a particular phase is proportional to the frequency with which it is found in the preparation (Fig. 52). Observations of living material and time-lapse photography provide absolute determinations of duration for particular phases. Anaphase is always a relatively fast process. One consequence of this is that in binucleate amoebae it is only in this phase that the two nuclei may be seen to be in different stages (Fig. 94), while they always appear completely in step during the other phases of division. The equatorial plate phase (metaphase) can last a long time. Chick cells in tissue culture, for example, can remain in metaphase for hours; many eggs remain in first meiotic metaphase while awaiting fertilization.

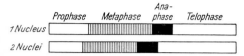

Fig. 52. Frequency (percentage) of the phases of nuclear division in an amoeba (*Naegleria bistadialis*) in preparations of uninucleate amoebae ($n=204$) and binucleate amoeba ($n=77$) from agar cultures with bacteria. (After KÜHN, unpublished)

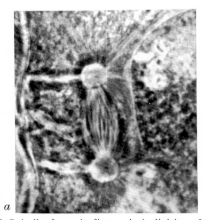

Fig. 53. Spindles from the first meiotic division of *Barbula-nympha*, microphotographs of living cells. a prophase, 1,050 × ; b anaphase 1,600 × . (CLEVELAND, 1953)

The spindle is almost never visible *in vivo*. In fixed and stained preparations it is pervaded by longitudinal central fibers which connect the poles ("continuous fibers"); in addition, fibers run from the chromosomes to the poles ("chromosomal fibers"), and these often stand out because of their stronger staining (Fig. 54a).

There has been much dispute as to whether the spindle fibers are present in the living cell or whether they are artifacts of fixation. Even if they were structures whose bulky appearance stemmed from protein denaturation, a longitudinal orientation of the spindle material would have to underlie their regular precipitation. This longitudinal structure is seen in the polarizing microscope: the spindle is clearly positively birefringent in the long axis, which suggests a parallel orientation of polypeptide chains. Indeed, true and relatively coarse fibers can be seen in the spindles of some living cells: in the giant cells of Hypermastigotes (Trichonymphids) they can be seen in remarkable microphotographs (Fig. 53). There is even proof, however, that optically homogeneous spindles contain differentiated fibers rather than a mere general longitudinal orientation of a morphologically uniform material. When cells are centrifuged during mitosis and then fixed, central fibers and chromosomal fibers are seen in various unusual orientations (Fig. 54 b, c). Individual chromosomes are displaced, together with their distorted fibers, from the equatorial plate or from the early anaphase figure (Fig. 54 c). Preparations of this type could be obtained only if the fibers were already present before fixation. At higher speeds the chromosomes are moved in a centrifugal direction, and the spindle breaks (Fig. 55 b) or tears at the equator (Fig. 55 c); or some of the metaphase or anaphase chromosomes may become separated from the spindle

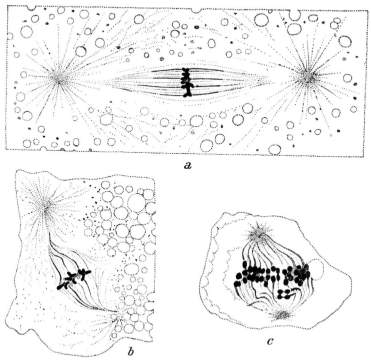

Fig. 54. First cleavage metaphase of *Cyclops americanus*. a normal; b centrifuged; c centrifuged first spermatocyte division of *Homarus*. (After SCHRADER, 1934)

entirely (Fig. 55 d–f). The preparations of the fixed, centrifuged spindle figures show that the fibers have a certain tensile strength and that the chromosomes are attached to particular fibers. The conclusion of BEAMS and KING (Fig. 55) that the bond with the spindle fibers is so firm that the chromosomes are ripped leaving pieces attached to the spindle does not seem sure to me; for the chromosomes of the chick which they used in their experiments are of such diverse sizes (Fig. 56) that the "pieces" remaining on the spindle may be small chromosomes, while the heavier, large chromosomes may have been spun away. In addition to this—as is always the case in equatorial plates containing chromosomes of very disparate sizes—the small chromosomes are found *within* the spindle, while the large ones are attached peripherally where they can be removed more easily.

Important conclusions concerning the formation of the spindle and its functional changes have been obtained through the use of osmotically induced swelling and shrinking on living plant and animal cells. Locust spermatocytes (Fig. 51) have proved to be especially favorable experimental objects. The cytoplasm of the living cell is viscous; granules and mitochondria never show Brownian movement, but are only moved by cytoplasmic streaming. In hypotonic medium the cell enlarges and rounds up (Fig. 57a). The cytoplasm becomes more fluid; mitochondria and particles exhibit Brownian movement. Further dilution of the medium causes the spindle to swell, and in metaphase granules from the cytoplasm penetrate the spindle and become arranged in long chains parallel to the spindle fibers. Moderate swelling scarcely disturbs cell division, but more extreme swelling stops it irreversibly. One can dissolve the rest of the cytoplasm and so isolate the spindle during metaphase or anaphase. This procedure leaves the chromosomes in a fairly normal arrangement fastened firmly to the spindle.

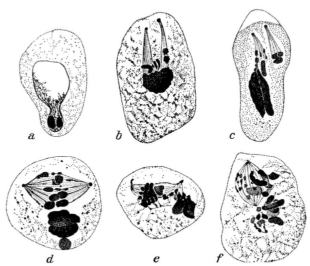

Fig. 55. Mesenchyme cells of 80-hour chick embryos after centrifugation (150,000 × g). Direction of force is downward in all figures. a resting nucleus; b–f metaphase and early anaphase 2,000 ×. (After BEAMS and KING, 1936)

When metaphase spermatocytes are placed in a slightly hypertonic solution, the cell shrinks, and the cytoplasm, which has now lost water of imbibition, lies in a thin coat around the nuclear division figure (Fig. 57c, d). The ability to divide is not disturbed by moderate shrinking. The changes are reversed when the cells are restored to an isotonic medium. At higher osmotic strengths, the water of hydration is removed from the spindle as well. It becomes noticeably narrower but scarcely shortens (Fig. 57c). Visible cracks and crevices appear in the unfixed cell.

Shrunken prophase stages tell us something about the formation of the central spindles. The half-spindles which emanate from the centrioles are independent and change their positions with respect to one another and to the chromosomes in various ways as the cell shrinks (Fig. 58a–c;

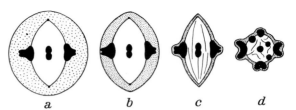

Fig. 56. Chromosomes of *Gallus domesticus*. (After SOKOLOW, TINIAKOW, and TROFIMOW, 1936)

Fig. 57. Effect of gentle dehydration on locust spermatocytes *Chorthippus lineatus* in the equatorial plate stage. d polar view. (After BELAR, 1929)

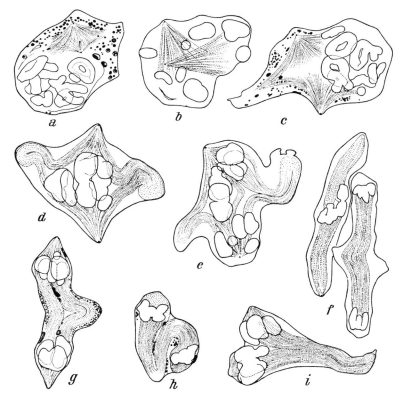

Fig. 58. Effect of strong dehydration on mitotic stages of *Chortippus lineatus*. Treatment begun in various stages: a–c prophase; d metaphase; e early anaphase; f left metaphase, right late anaphase; g–i anaphase; cf. Fig. 51. (After BELAR, 1929)

cf. Fig. 51 a). As soon as a single spindle has been formed, that is, from metaphase on, there occurs an extraordinary elongation of the spindle fibers in a hypertonic medium. The spindle lengthens tremendously compared to normal and becomes correspondingly slender (cf. Fig. 51). It remains straight only rarely (Fig. 58 f); generally it bends this way and that, since the tough coat of cytoplasm which surrounds the spindle like a rubber covering prevents its free stretching. The more extreme the osmotic shrinking, the stronger and more frequent is this folding of the spindle. In metaphase and early anaphase stages the spindle splits generally, and bundles of fibers stick out toward the sides (Fig. 58 d, e). Generally they jackknife in the equatorial zone; perhaps this zone where the half-spindles have united is not quite so rigid. If shrinking is induced in later stages, the spindle as it stretches is generally bent as a whole in the middle of the cell and finally bends double (Figs. 58 i, 59 a). At times it may even bend in two places (Fig. 58 h). The further division has proceeded by the time of the onset of shrinking, the less is the maximal length reached by the stretching spindle fibers in the hypertonic medium. When the normal lengthening is over, hardly any further lengthening can be induced osmotically (Fig. 59 b). The ability to lengthen is thus lost at a particular stage of mitosis. Moreover, since this lengthening takes place only in the presence of oxygen, it is dependent upon metabolic events.

A similar overstretching of the spindle can occur after chance disturbances of the development of the division apparatus. In the first meiotic division during spermatogenesis of *Anisolabis*, bipolarity fails to arise once in a while, and then the chromosomes do not take up positions on the

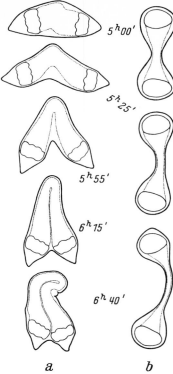

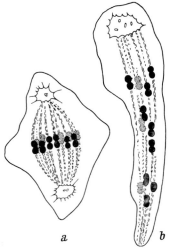

Fig. 59. Effect of strong dehydration on mitotic stages of *Chorthippus lineatus*. Two cells placed in very hypertonic solution at the same time: a beginning of treatment in anaphase (corresponding to Fig. 51e); b beginning of treatment in telophase (corresponding to Fig. 51g). (After Belar, 1929)

Fig. 60. First spermatocyte division in the insect *Anisolabis*. a normal division, a pair of centrioles at each pole; b abnormal division; the centriole pairs have not separated, no equatorial plate has formed, and the homologous chromosomes have not separated, but the spindle has undergone extreme elongation. (After Schrader, 1944)

equator. But the unipolar spindle, over which the chromosome pairs are irregularly distributed, elongates to an extraordinary extent, forcing the cell itself to lengthen (Fig. 60). In certain *Impatiens* individuals, an evidently genetically controlled hyperelongation occurs in the spindle next to the nucleus during an abnormal first meiotic division of the microsporocyte (Fig. 61). The spindle reaches a length comparable to those seen in the osmotic experiments (Fig. 58), and it curls around, being enclosed in a rigid cell wall.

The osmotic experiments with locust spermatocytes suggest that during spindle formation, from the centrioles outward, polypeptide chains become grouped into fibers which continue into the nuclear area. This growth can be thought of as a continuous, terminal addition of protein molecules first to the centrioles and later to the already formed fibers so that an ordering of the structures arises in the region, cytoplasm as well as nucleus, from which the spindle forms. In the equator, the half-spindle fibers meet and join, perhaps by interdigitating, or perhaps by becoming individually continuous. The very forces which order the polypeptide chains into growing fibrils may well unite the molecular chains when they meet. The elongation of the spindle may result from the stretching of previously folded molecular chains. The normal lengthening leads in some dividing cells to extraordinary lengths (Fig. 75), with a substantial increase in volume. One unanswered question is why this osmotic shrinking induces the total elongation of the spindle fibers: Has a substance been withdrawn from the spindle which had previously inhibited the unfolding of the protein molecules, or did the removal of water from the cytoplasm result in the concentration of a substance which promotes elongation?

Staining reactions indicate that RNA has been taken up in the astral spheres and in the spindle, in addition to the polypeptide chains which are responsible for the positive birefringence. In certain cases the incorporation of carbohydrates into the spindle has also been demonstrated. In the sea urchin egg, the birefringence decreases progressively in anaphase, beginning at the equator and continuing toward the poles. In some cells during telophase the dissolution of the spindle proceeds in the same direction. Unfortunately, the formation of the spindle and its transformations have been studied optically and cytochemically in far too few cells until now.

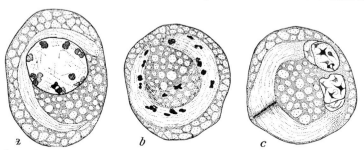

Fig. 61. Abnormal elongation of the spindle in the first meiotic division of *Impatiens pallida*. (After SMITH, 1935)

As the spindle forms and transforms, the chromosomes first move to the equator (prometaphase movement) and then toward the poles (anaphase movement). In the first meiotic division, of course, each daughter chromosome moving toward a pole at anaphase consists of two chromatids connected by a single kinetochore (Fig. 62c).

The kinetochores are of critical importance in chromosomal movements. They lie in a connecting segment between the two chromosomal arms (Fig. 4). In the metaphase chromosomes the kinetochore is lodged in a groove in the matrix (Fig. 62a) or can be seen suspended from connecting fibers which may be of considerable length (Fig. 62b, c). In flowering plants the aggregation of kinetochores has been described. In any event, the kinetochore is a self-replicating structure. After a chromosomal break has occurred, the piece without a kinetochore is lost. In abnormal divisions the two arms of a chromosome can be separated from each other, each attached to one daughter kinetochore (Fig. 63b), so that one arm goes to one daughter cell and the other to the other cell. In the next division, one abnormal chromosome with a terminal kinetochore appears in each of the sister cells. If the two original chromosomal arms were clearly different in length, it is easy to see that they originally belonged together (Fig. 63c).

Observations on living cells make it possible to follow the chromosomal movements directly. In somatic divisions the prophase chromosomes appear in the same general position in which they were last seen in the previous telophase. The curvature of the two-armed chromosomes and their kinetochores are directed toward the pole of the previous division (Fig. 64a). They now reorient so that the kinetochores of the still tightly joined chromatids move toward the equator

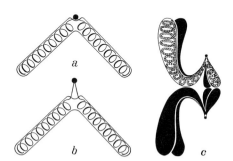

Fig. 62. Diagram of the relationships between the kinetochore and the chromatids. a, b (after SCHRADER, 1949); c (after RESENDE, 1947)

45

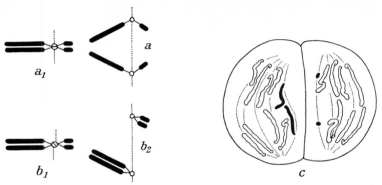

Fig. 63. a, b Diagram of normal and abnormal kinetochore division; c anaphase of the second meiotic division in pollen formation in *Fritillaria kamtschatkensis*, abnormal one-armed chromosomes with terminal kinetochore (as in b) in the previous division. (After DARLINGTON, 1939)

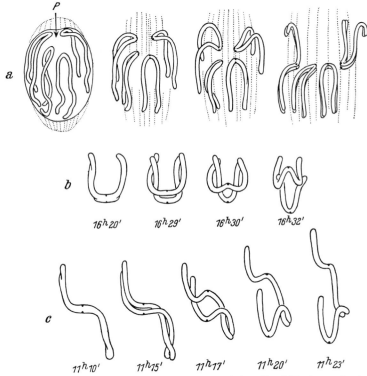

Fig. 64. Chromosomal movements during mitosis in the stamen hair cells of *Tradescantia virginica*. After observations on living cells. a prometaphase movement of two chromosomes; b, c anaphase movement. About 1,400 × .
(After BELAR, 1929)

(Fig. 64a). The chromosome arms are thus, depending upon their position, pushed or pulled along; they do not always lie in the equatorial plane, since in many cells the long arms do not have room there. When there is enough room, they frequently stretch out in the equatorial plane, possibly because of their elasticity and perhaps also in response to cytoplasmic streaming.

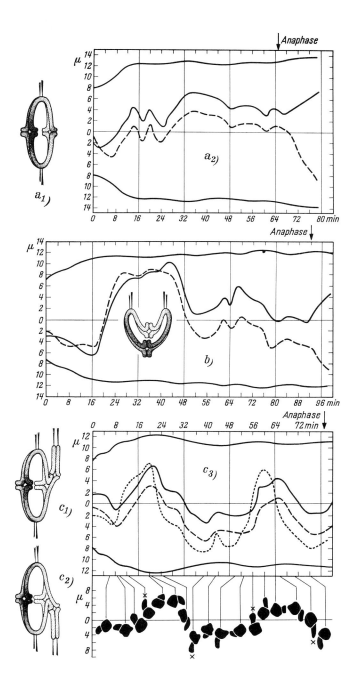

Fig. 65. Prometaphase movements of diakenesis chromosomes of *Tipula oleracea*. Path of the kinetochores with time: a, b normal tetrads, a with bipolar orientation, b with monopolar orientation at first, followed by reorientation of one member of the pair; c trivalent chromosomal aggregate resulting from a translocation between an autosome and the Y chromosome. Reorientation of the Y kinetochore (×) and consequent change of the direction of movement of the chromosomal aggregate to one of the poles. (After BAUER, DIETZ, and RÖBBELEN, 1961)

Much insight into the interplay of forces in chromosomal movements is provided by the analysis of time-lapse phase-contrast movies. Spermatocyte division in animals with few chromosomes, e.g., Tipulids (crane flies), is especially good. The tetrads in diakinesis enter the spindle as bivalent rings in which the members of each pair are joined at their ends (as in Figs. 41 d, k; 43 e). In the spindle the tetrads assume a bipolar orientation in 90 per cent of the cases, so that the sister kinetochores of the two members of a pair face opposite poles (Fig. 65 a). At this time they swing back and forth in the polar axis. The district over which they oscillate out of the equatorial plane is at first fairly large; it becomes smaller until the final position is attained.

In about 10 per cent of the cases, the kinetochores of both members of the pair orient toward the same pole and the whole tetrad is drawn toward this pole (Fig. 65 b). A reorientation of the kinetochore of one member of the pair now takes place, and it begins to move (at about the 48th minute in Fig. 65 b) changing over into the normal bipolar orientation which leads the tetrad into the equatorial plane.

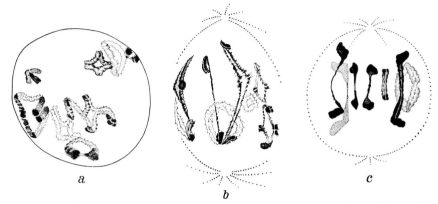

a   *b*   *c*

Fig. 66. First spermatocyte division of the Phasmid *Bostra;* not all of the tetrads are illustrated: a diakinesis; b prometaphase elongation of the chromosomes, a few tetrads are not yet oriented; c metaphase. (After HUGHES-SCHRADER, 1950)

In an extremely ingenious experiment a chromosomal segment rearrangement was used (Fig.65c) which had resulted from an X-ray-induced exchange between an autosome and the Y chromosome which possesses a terminal kinetochore. Now three pairs of kinetochores were present, and two alternate orientations were possible. In each case, the "trivalent" composed of three longitudinally split chromosomes which aggregated at diakinesis approached one of the two poles (Fig. 65c). After a certain time, as a rule, the orientation of the Y kinetochore was reversed, and the aggregate moved toward the opposite pole. The prophase chromosomes are thus brought to the equator as a result of the balance of pulling forces acting on the kinetochores, as seen especially clearly in the stretching of the metaphase chromosomes in many species (Fig. 66). This stretching disappears when the equatorial position is reached. Clearly the pulling force ceases when equilibrium is reached.

With ordinary optics the existence of the chromosomal spindle fibers in living cells can be inferred only from the movements and stretching of the chromosomes; but in polarized light they stand out because of their strong birefringence and, of course, they can be seen in many fixed and stained preparations. They are especially clear in relation to unpaired X chromosomes which move outside the spindle and are attached to a single fiber (Fig. 67). Here the films show the chromosomes fidgeting in the cytoplasm, attached to the chromosomal fibers.

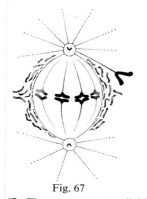

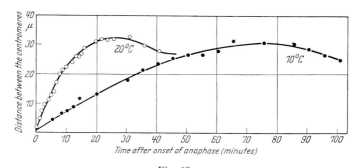

Fig. 67

Fig. 68

Fig. 67. First spermatocyte division of the Mantid *Humbertiella;* the unpaired X chromosome lies outside the spindle on its own chromosomal spindle fiber. (After Schrader, 1951)

Fig. 68. Speed of anaphase movement of the mitotic chromatids in the stamen hair cells of *Tradescantia virginiana* at various temperatures. Position of the kinetochore in μ, measured in photographs made at set times. (After Barber, 1939)

As time studies on living cells and the analyses of films show, anaphase begins suddenly and the movement slows down with the increasing separation of the chromatids, the members of a pair, from one another (Figs. 68, 69). As in the prometaphase movement, the kinetochore is out in front, and the chromosome arms are pulled along behind (Fig. 64).

The stretching of the spindle in anaphase (Fig. 51 e–g) can be inhibited by a certain concentration of chloral hydrate without influencing the anaphase movement of the chromosomes (Fig. 70). The difference between their final positions in these circumstances and their positions at the end of a normal anaphase shows how much further they have been carried apart by the stretching of the spindle (Fig. 70). The cell thus employs two distinct mechanisms to move the daughter nuclei apart.

The separation of the chromatids can also be accomplished without the participation of the division apparatus. This is seen in the occurrence of polyploidy resulting from endomitosis in the development of certain plant and animal tissues (Figs. 100, 101) or produced experimentally. Colchicine prevents spindle formation in plant cells while the division cycles of the chromosomes proceed. Here, a relationship emerges between the absence of spindle formation and the behavior of the kinetochore. While normally the separation of the kinetochores leads to the separation of the chromatids (Fig. 64 b, c), during endomitosis in the presence of col-

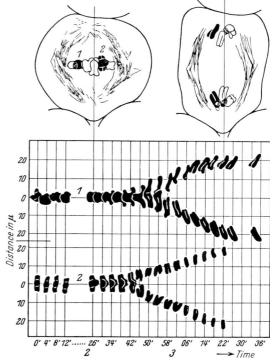

Fig. 69. Anaphase movement of two chromosomes in the first meiotic division of the spermatocytes of the locust *Psophus stridulus,* according to a phase-contrast time-lapse movie. (After Michel, 1943)

49

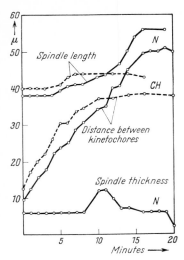

Fig. 70. Chromosomal movement and behavior of the spindle in the first spermatocyte division of the locust *Chorthophaga viridissima* at 30°. Kinetochore position measured each time on a chromosome *in vivo*. N, normal division; CH, course of division in the presence of chloral hydrate. Abscissa: time from the onset of anaphase in minutes. (After RIS, 1949)

chicine the separation, and even the division of the kinetochores, is delayed so that the chromatids remain together for a very long time (Fig. 71 a–d, f). Colchicine has an effect, therefore, on the very division mechanism of the chromosomes themselves. If colchicine is allowed to work for a longer time, several endomitotic cycles can take place (Fig. 71 h), resulting in highly polyploid nuclei (Figs. 100, 101).

A spindle fiber apparatus is necessary for the separation of the divided chromosomes. This need not be a dicentric spindle, however. Various external influences can in the worm *Urechis*, for example, prevent or delay the division of the centrioles. Then one sphere forms around a giant centrosome. The metaphase chromosomes are arranged on its rays in an arch or cap (Fig. 72a), enacting a monocentric mitosis. The chromatids separate as in normal anaphase and move some distance apart from one another along the rays of the sphere (Fig. 72b, c). Here also the kinetochores separate first and pull the arms of the chromosomes along (Fig. 72b, c). Now as the inner chromosomes move closer to the centrosome, the outer chromosomes follow suit. In order to do this they reverse the orientation which previously had their kinetochores pointing away from the centrosome in one nucleus or in several nuclear fragments, which later fuse.

The separation of the chromatids is thus an autonomous process. However, it does not exceed a certain distance, at which the "attraction" of the centrosomes becomes important. The fibers clearly play a role in the migration of the chromosomes, which in the monocentric mitosis runs a course, not between two poles, but radially.

The presence of kinetochores and their function in nuclear division are widespread among animals, plants, and protistans, but this rule is not without exception. In the Hemiptera, strictly localized spindle attachment regions are generally lacking. In many forms fiber bundles are seen to attach to the chromosomes (Figs. 73a, b; 74b). When the chromosomes are long, they lie entirely in the equatorial plane, and the chromatids retain an orientation parallel to this plane as they move toward the poles. Evidently the spindle fibers are pulling the chromatids along their entire length. Experiments on fragments produced by X-rays agree with this interpretation. All fragments, irrespective of size, behave as entire chromosomes through numerous divisions (Fig. 74c, d).

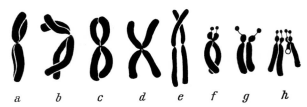

Fig. 71. Chromosomes of a colchicine-induced endomitosis in the root tip of *Allium*. Separation of the daughter chromosomes during delayed kinetochore division. *f–h* SAT-chromosomes; h group of chromosomes descended from a single chromosome in the following mitosis which is still inhibited by colchicine. (After LEVAN, 1938)

50

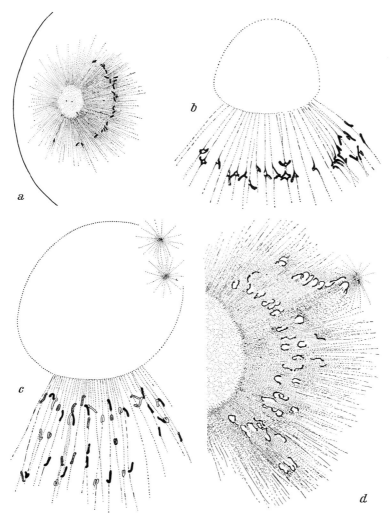

Fig. 72. Monocentric mitosis in the egg of the Echiuroid *Urechis*. a chromosomal metaphase; b separation of the daughter chromosomes; c varying distances separating the daughter chromosomes; d reversal of orientation of the chromosomes farthest from the main pole (left), except for those chromosomes in the sphere of influence of the secondary pole (right above). (After BELAR and HUTH, 1933)

The separation of the daughter chromosomes during mitosis in those Protozoa in which the spindle remains outside the nucleus is still a mystery. In the Trichonymphids the continuous spindle attaches only to the nuclear membrane. The chromosomes, from prophase on, are attached to the nuclear membrane by "connecting fibers" (Figs. 50, 75 b) at whose ends the kinetochores are ostensibly located (Fig. 76). The chromosomes divide longitudinally, and the chromatids move, drawn by the kinetochores, away from one another along the nuclear membrane into the region where chromosomal fibers emanating from the poles attach to the kinetochores (Figs. 50, 75, 76). As the spindle stretches, the nucleus constricts into a dumbbell (Fig. 75c–e).

In the examples of mitosis discussed up to now, common features have been seen in the formation of the continuous spindle fibers: between the centrioles or between the polar caps, oriented

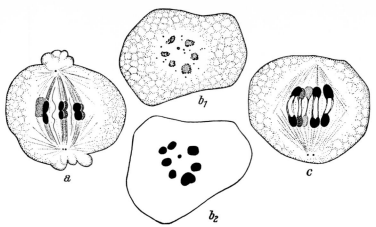

Fig. 73. Spermatocyte division in the Hemiptera. a, b metaphase in *Pachylis gigas;* c anaphase in *Syromastes marginatus*, bundles of chromosomal spindle fibers; b_1 cross section through the half-spindle; b_2 equatorial plate of the same cell; note the connecting bridges between the separating chromosomes in c, apparently consisting of matrix substance. (After Schrader, 1932)

fibers arise. Their formation and stretching is to a great extent independent of the chromosomes (Figs. 48, 50, 58, 60, 75). But the development of the spindle is not completely autonomous: in eggs with haploid, diploid, or polyploid nuclei, the dimensions of the cleavage spindles depend upon the number of chromosomes arriving in the region of the spindle. The origin of the spindle lies nevertheless in a bipolarity determined by the centrioles and independent of the chromosomes. Yet in plant cells evidence for some dependence of spindle formation on the chromosomes is seen at the return of mitotic events after the termination of colchicine treatment. The mass of chromosomes in a highly polyploid nucleus often divides into variously sized groups of chromosomes, for which correspondingly sized spindles form. These distribute the chromosomes to very disparate nuclei which at times even become separated by cell walls.

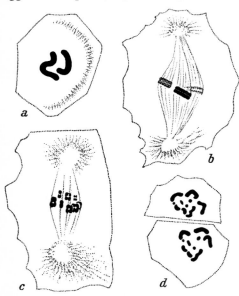

Fig. 74. Chromosomal fragments after X-irradiation of the embryonic cells of the haploid male of *Steatococcus tuberculatus*. a normal equatorial plate ($n=2$); b, c metaphase with slightly (b) and highly (c) fragmented chromosomes; d anaphase with ten fragments in each daughter group. (After Hughes-Schrader and Ris, 1941)

In the Homoptera most unusual spindles are seen in meiosis, in a 1:1 relation with the individual chromosomes. In the Coccid *Llaveia bouvari*, the prophase nucleus pinches into three parts (Fig. 77b, c), each containing a nascent tetrad. Now a sheaf-shaped spindle (Fig. 77d) arises at each tetrad and later also at the X-dyad. The spindle axes are not parallel at first, but they gradually become parallel during the individual anaphases. At this time, the individual spindles of the autosomal tetrads divide longitudinally into individual spindles for each chromatid pair (Fig. 77e, f). When a living cell is caused to burst open, the individual chromosome

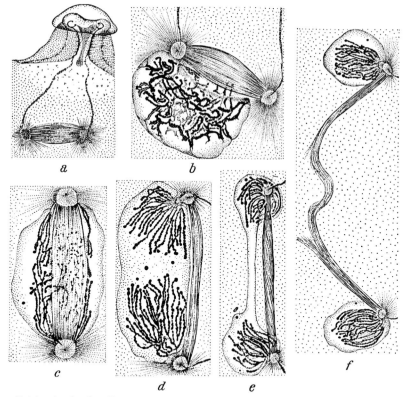

Fig. 75. Nuclear division in the flagellate *Pseudotrichonympha*. a–d 1,100× ; e, f 825× . (After Cleveland, 1935)

spindles move freely apart, carried by streaming. The polar portions of the chromosome spindles shorten increasingly during anaphase and finally disappear entirely. At the same time, central spindle segments appear between the diverging chromatids (Fig. 77e, f). At the beginning of cytokinesis a single spindle is formed by the fusion of the central spindle segments (Fig. 77g), and undergoes extreme elongation as the "stem body" (Fig. 77h).

Most unusual of all is the whole course of "mitosis" during the spermatocyte divisions of the haploid parthenogenetic Hymenopteran male. In the hornet (Fig. 78) in place of the first maturation division only a half-spindle forms which surrounds the nucleus. The cell elongates in the direction of the spindle (Fig. 78a, b). The nucleus does not divide, although the chromosomes undergo the changes associated with prophase. On the side opposite the astral rays, a cytoplasmic lobe is pinched off (Fig. 78c). Then the centriole divides, and the daughter centrioles move apart but do not attach to the nuclear poles. Between them a spindle appears beside the nucleus. The nucleus forms an intranuclear spindle, at whose equator the chromosomes line up (Fig. 78d). Anaphase and telophase are completed within the nuclear membrane. As the cytoplasm divides in two, the now tubular nucleus is also constricted in two (Fig. 78e, g). In the spindle formed beside the

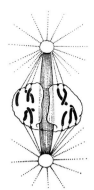

Fig. 76. Diagram of nuclear division in *Barbulanympha* (cf. Fig. 50) after the interpretation of Schrader, 1951

53

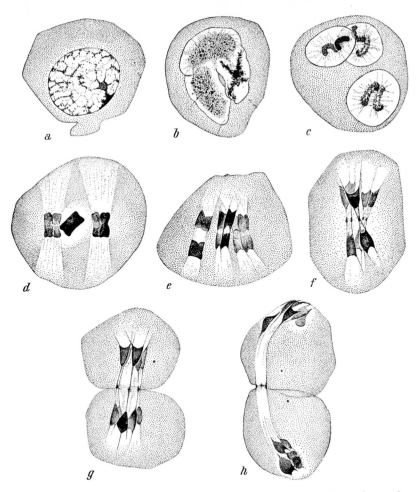

Fig. 77. First meiotic division of the coccid *Llaveia bouvari*. a primary spermatocyte, resting nucleus; b constriction of the nucleus, precocious condensation of the X chromosome; c tetrads in two nuclear vesicles, X chromosomes split lengthwise; d chromosomal spindles formed on both tetrads, X dyad still surrounded by nuclear sap; e–g anaphase, narrowing of the connections between the anaphase chromosomes; g formation of a single spindle as a "stem body," onset of cytoplasmic division; h elongation of the stem body. (After HUGHES-SCHRADER, 1931)

nucleus the mitochondria are lined up and distributed to the spermatids when this spindle elongates and constricts (Fig. 78 e, h, i). Another variant occurs in the honey bee. First an intranuclear bipolar spindle forms, while outside of the nucleus half-spindles grow out from the *centrioles* but do not unite (Fig. 79 b). The one pole now pinches off together with a small cytoplasmic lobe, and the intranuclear spindle is dismantled without the chromosomes having divided (Fig. 79 c). Then a new spindle forms into whose equator the now duplicated chromosomes move after the breakdown of the nuclear membrane. During telophase a second cytoplasmic lobe arises at the same pole as the previous lobe and one daughter nucleus moves into this second lobe (Fig. 79 g). Thus only one spermatid is formed, and it retains all the mitochondria (Fig. 79 h).

Of the two maturation divisions in normal spermatogenesis, one must obviously be left out in the case of these haploid spermatocytes; but as a vestige of this missing division, a rudimentary

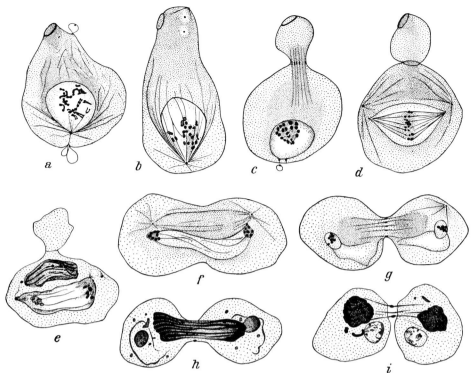

Fig. 78. Spermatogenesis in the hornet. a–c "first maturation division," aster and spindle formation without nuclear division, constriction to form a cytoplasmic lobe; d–i second maturation division, intranuclear spindle, outside the nucleus two polar asters and the spindle devoted to the distribution of the mitochondria; e, h, i mitochondrial staining of stages like f, g. (After MEVES and DUESBERG, 1908)

cycle involving the achromatic apparatus remains. It is as if the cell cannot get rid of it, and must keep the chromosomes away from it in one way or another. And then in the hornet this peculiar occurrence of two spindles: the intranuclear spindle concerned with nuclear division and the special spindle for the distribution of the mitochondria! Individual processes in the chromosome cycle and in the division cycle which are ordinarily tightly coordinated are here separate, altered individually, and combined in what seems to be an arbitrary manner. A reexamination of this remarkable maturation process with modern methods would really be worthwhile!

The most disparate mitoses in protistans, plants, and animals have two things in common: the movement of the chromatids to the pole, and the elongation of the spindle. The anaphase movement, the movement of the chromosomes to the pole, is independent of the spindle elongation, as the observation of diverse normal dividing cells has shown, as well as specific experiments. But the elongation of the central spindle as the "stem body" does influence the final position of the anaphase chromosomes and thus moves the sites of the telophase nuclei further apart.

But the cell can even spare these two processes and employ the achromatic apparatus quite differently. Quite extraordinary monocentric mitoses occur during spermatogenesis in two taxonomically different organisms. In the fly *Sciara* the paternal chromosomes (with the exception of one special chromosome pair) are removed from the total chromosomal complement in the primary spermatocyte without ever having undergone pairing. If the removal of an entire genome on the basis of its ancestry is remarkable, the mode of chromosomal separation is amazing. The

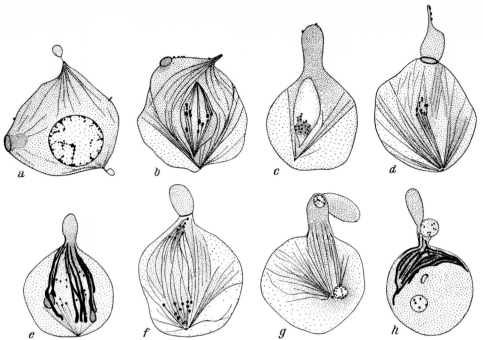

Fig. 79. Honey bee spermatogenesis. a–c first maturation division, asters, intranuclear spindle, breakdown of the spindle without distribution of the chromosomes, constriction to form a cytoplasmic lobe; d–h second maturation division; d nuclear breakdown, metaphase; e,f anaphase; g, h telophase, constriction to form a nucleate cytoplasmic lobe with the mitochondria remaining in the spermatid; e, h mitochondrial staining. (After MEVES, 1907)

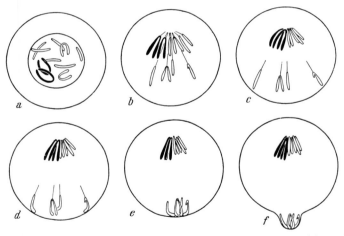

Fig. 80. Monocentric first spermatocyte division of *Sciara coprophilus*. a prophase without chromosomal pairing; b monocentric spindle, two special chromosomes (black) both go to one pole, the rest segregating into two groups. The maternal set goes to the pole, the paternal set moving away from the pole, arms first, fastened to spindle fibers, and finally reaching the cell surface and being pinched off in a lobe of cytoplasm. (After METZ, 1936)

chromosomes which will remain in the cell move toward the only pole that forms, while the others move away from it in various radial directions in the cone-shaped space corresponding to

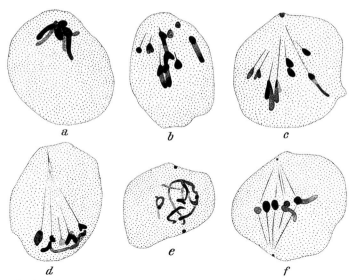

Fig. 81. Spermatogenesis of the haploid parthenogenetic beetle *Micromalthus debilis*. a–d first division, monocentric mitosis, movement of the chromosomes, arms first, toward the cell surface with spindle fibers attached to the centromeres; e, f second division with a more usual course. (After Scott, 1936)

a half-spindle (Fig. 80b, c). Spindle fibers run to these departing chromosomes; the latter are not being pushed by the fibers, however, nor are they sliding along them in the manner of the outer daughter chromosomes in the monocentric mitoses in *Urechis* eggs as they begin their movements (Fig. 72). Rather the chromosomes hang from the spindle fibers by their kinetochores and are moved away from the pole by some unknown force. This can be seen clearly by the shape of the chromosomes: their arms point toward the cell periphery. When they reach the surface opposite the centriole they bend, reflecting the resistance they encounter (Fig. 80 d, e). They then move together and are pinched off in a little lobe.

In the spermatocytes of a beetle with haploid parthenogenetic males, the chromosomes move in a very similar way; here the monocentric mitosis is a vestigial first maturation division. The reduction division is missing, but the centriole is still active. It forms a functional half-spindle, and all the chromosomes are attached to spindle fibers just as the paternal chromosomes in *Sciara*, moving radially away from the centriole until they reach the opposite cell surface (Fig. 81 b–d). The spindle fibers then break down: the chromosomes return to the center of the cell and are all incorporated into one nucleus. The centriole divides and undergoes its second maturation division with a normal spindle (Fig. 81 e, f). As in Hymenopteran spermatogenesis, the cell apparently cannot do without the first centriole cycle and thus makes use of an altered spindle mechanism which does not result in a partition of the chromosomes.

In both these cases of monocentric mitosis this question arises: What moves the chromosomes away from the centriole? Cytoplasmic streaming has not been observed. Does the movement originate in the chromosomes themselves? And if they can move autonomously without being led by the kinetochore, how does the physical process of movement through the viscous cytoplasm come about?

What we have seen in a few natural variants is also shown in experiments: the separation of component processes which are normally seen in lock step. In spermatocytes of the Tipulid *Pales crocata* which are flattened by pressure, the centrosomes with their centrioles which run into the

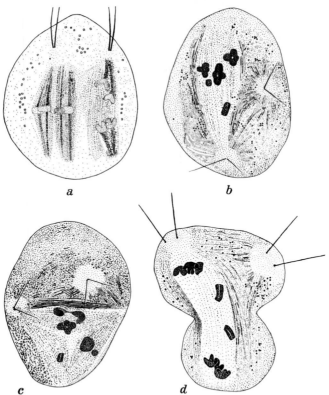

Fig. 82. Primary spermatocytes of the crane fly *Pales crocata* flattened by pressure; reconstruction from movies; a in polarized light; b–d phase contrast. (After original films by DIETZ)

future tail fiber axes of the sperm are isolated from spindle formation (Fig. 82). Now individual spindles arise in conjunction with the individual chromosomal tetrads (as in *Llaveia*, p. 52). At first these are randomly oriented, but then they mostly become parallel and fuse into a single spindle (Fig. 82a, b). But they can also remain apart and support each other in a triangular figure (Fig. 82c). In polarized light the chromosomal fibers (or fiber bundles) which have the strongest birefringence are clearly visible (Fig. 82a). A centrosome can come into contact with a spindle pole, and in this case the spindle fibers converge on it (Fig. 82b, d). Spindle ends without centrosomes are blunt. The centrosomes, which lie in isolation, are surrounded by astral spheres, as can be seen in living cells with phase-contrast, from the orientation of long rod-shaped mito-chondria toward the centrosomes (Fig. 82b–d). Anaphase chromosomes reach the centrosomeless spindle ends just as they normally arrive at the ends with centrosomes. At the onset of cell division in such a flattened cell, the furrow cuts across the spindle equator (Fig. 82d).

We must bear in mind the complex diversity of deviations from the "typical" course of nuclear division in nature and in experiments, in order to comprehend the variety of means employed by cells to achieve the distribution of the daughter chromosomes to two daughter individuals (or to achieve some other related purpose), so that we do not fall into the error of shortsighted over-generalization. A single theory of mitosis embracing all the cases known in nature cannot be constructed. The American cytologist Metz is still right when he says[440, 229]: "What is needed now is not more interpretations, but more evidence, and especially more experimental evidence."

Lecture 5

In general, cytoplasmic division follows closely after nuclear division. There are, to be sure, numerous cases in which the division of the cytoplasm is delayed or entirely absent, but in the typical case cell division is a unit process, and we may inquire how the individual events are related to one another and how the entire process of division is set in motion.

Three major types of cytoplasmic division can be distinguished: (1) constriction, (2) formation of a separation zone in the cytoplasm, and (3) formation of a division plate as a result of differentiation within the spindle. The third form of division is characteristic of plant cells; the first two occur in animal cells where often one or the other can take place even in the same cell type, according to certain developmental conditions; or a combination of the two may occur, in which a more or less deep infolding is accompanied by the formation of a separation zone within the cell.

The process by which a cell pinches in two can be imitated in a model embodying local differences in surface tension: a drop of an oil-chloroform mixture about 1 cm. in diameter is introduced into water on a glass plate. At two opposite poles soda crystals are placed at a distance of several millimeters from the oil droplet. If the soda diffuses to both poles of the oil droplet simultaneously, the surface tension at both poles is reduced and surface streaming spreads toward the equator (Fig. 83a) with a soap membrane at the leading edge. In the axis of the droplet, oil streams toward both poles. When the surface streams coming from each pole meet at the equator, they curve toward the interior, and so a depressed ring forms around the drop, constricting it more and more until two daughter drops form, held together only by a short filament of solid soap

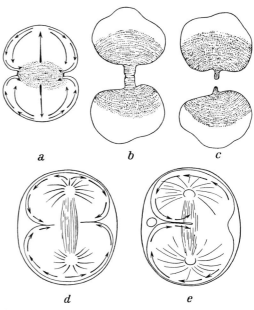

Fig. 83. a–c Constriction of an oil-chloroform drop in water with two centers of streaming; d–e cytoplasmic streaming in the first division of *Rhabditis dolichura;* a constriction on all sides; b on one side. (After SPEK, 1918)

(Fig. 83b); they may even part completely and move away in the directions or reduced surface tension (Fig. 83c). The constriction may proceed slowly or very rapidly, depending upon the viscosity of the oil droplet.

Streaming, as it occurs during the changes in shape and pinching apart of the oil droplet in the model experiment, can also be seen during cell division, where it is reflected in the movements of mitochondria, pigment granules, and other cytoplasmic structures. Streaming can be seen in living cells in the eggs of nematodes, leeches, copepods, and gastropods, as well as in spermatocytes. In fixed preparations the position of the cytoplasmic contents of many cells in a series of stages in cell division permits the reconstruction of a complete picture of streaming. The transparent eggs of various nematodes of the genus *Rhabditis* show streaming remarkably clearly, as well as its relationship to the appearance of the cleavage furrow (Fig. 83d, e). In these living eggs, one sometimes sees stronger streaming on one side; at times the streaming may be restricted entirely to one side, and this is always followed by a unilateral origin of the furrow (Fig. 83e).

In cells, as in oil drops, the constriction proceeds at various rates. For example, the egg of the nematode *Rhabditis pellio*, whose contents have a low viscosity, takes less than two minutes to complete its first cleavage, while the same process requires ten to fifteen minutes from start to finish in the sea urchin *Arbacia*.

We cannot, however, place too much stock in the similarity between the cells and oil droplets; and we certainly cannot equate the liquid interface of an oil droplet with the surface of the cell, so that changes in surface tension cannot play a simple role in cell division. Cytoplasmic streaming takes place beneath a gelatinous cortical layer; and in cell division several factors collaborate which we can infer only from changes in the physical condition of the cytoplasm. An ectoplasmic membrane, or cortex, of varying thickness is not restricted to egg cells, whose cytoplasm is highly differentiated; it can also be seen in other cells, for example, amoebae. Light-polarizing and

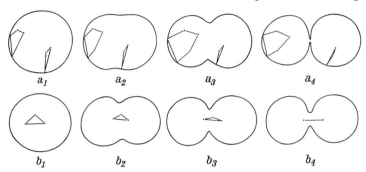

Fig. 84. Changes in the area of surface regions in the sea urchin *Mespilia globulus* during the first cleavage division. a polar; b equatorial region marked. (After DAN and ONO, 1954)

permeability properties show that the membrane is made up of protein and lipids. At the beginning of mitosis the membrane tension increases, reaching its maximum during metaphase. This is expressed in the rounding up of the cells — not only free cells, but also cells in animal tissues (e.g., Figs. 470, 471). In this stage egg cells have the greatest resistance to the flattening effect of centrifugal force, a sign of their tautness. As the spindle develops under the formation of gel structures, the viscosity of the inner cytoplasm (endoplasm) drops; this evidently makes streaming possible. Cell division increases the free surface, and in spherical egg cells and blastomeres this increase amounts to about 26 per cent. The membrane must therefore be stretched passively, or else it must spread actively. An active spreading of the cortex appears to play a major role in the cell division of sea urchin eggs. If several kaolin particles are placed on the surface of an egg whose fertilization membrane has been removed, their changes in position reflect local changes in membrane area. Measurements of these distances during cleavage suggest that a wave of expansion begins at the cell poles during anaphase and spreads toward the equator over the surface of the cell. In the equatorial zone the cortex contracts, and an equatorial furrow forms which gradually constricts the cell in two (Fig. 84). As the membrane expands, a decrease in its birefringence spreads from the poles to the equator. This change is accompanied by an alteration of the response to plasmolysis: if the birefringence is high, the egg surface remains smooth as water is withdrawn; if it is low, the cortex wrinkles. As the membrane begins to stretch at the poles, it becomes thinner and weaker. At this point, eggs placed in dilute sea water burst first at the poles. In many dividing cells polar membrane changes also come to light during anaphase, and the beginning of cleavage as vigorous movements reminiscent of pseudopodium formation produce cytoplasmic bulges or vesicles. This "bubbling" is seen strikingly in time lapse movies of cell division (for example, fibroblasts in tissue culture and spermatocytes) and in the division of amoebae as well. Fixed material often shows the process in an arrested state. On the egg surface the polar regions differ from the site of cleavage in their responses to various chemicals. In the region of the future division furrow, the cortical gel is transformed.

The constriction of the cell equator clearly depends upon active contraction. This is made quite apparent by experiments with "cell models" of fibroblasts in tissue culture. If anaphase stages are cytolysed with glycerol-water solutions, the addition of ATP causes the cells to pinch in two (Fig. 85). The conditions for this constriction are the same as for the contraction of glycerinated muscle fibers: ATP and apparently other nucleoside triphosphates are the only energy sources; Mg^{++} ions must be present, and the same specific inhibitors prevent the muscle fiber contractions and constriction of the fibroblast. So we conclude that the constriction of the cell equator is mediated by a "muscle-like" contraction of cell proteins.

Perhaps the alterations in the cortical layer are triggered by the poles of the spindle, possibly mediated by the astral spheres: during anaphase the astral rays in the sea urchin egg extend gradually to the cell surface and at the same time undergo a structural change. This change is expressed in a reduction in the positive birefringence which the rays previously exhibited in their long axes. One plausible hypothesis is that substances diffuse peripherally from the center of the aster and cause the spreading of the cortex. If these "structural agents" follow the growth of the astral rays, they will first reach the cell poles and ultimately meet at the equatorial zone of the cortex (Fig. 86).

Since the membrane changes coincide with late anaphase, during which stage the structure of the spindle also changes, one may conjecture that the substances which alter the colloidal condition of the mitotic apparatus and of the cortex are released from the chromosomes in anaphase or early telophase.

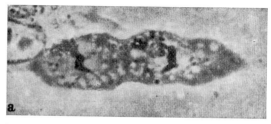
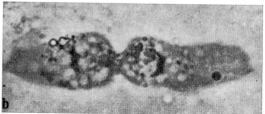
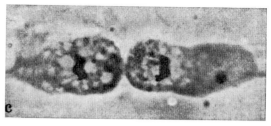

Fig. 85. Equatorial constriction of an anaphase model, tissue cultured fibroblasts extracted in a glycerin-water mixture; five minutes after addition of ATP. (Hoffmann-Berling, 1954)

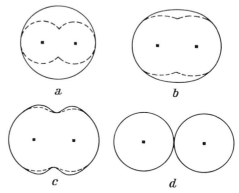

Fig. 86. Diagram of the growth of the astral spheres in the first cleavage division of the sea urchin egg and the spread of a hypothetical substance causing the membrane to extend. (After Mitchison, 1952)

This conjecture is in good agreement with experiments on amoebae growing on an agar surface and feeding on bacteria. When the amoebae live in culture with abundant moisture, nuclear

61

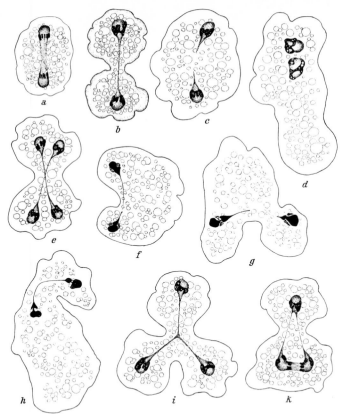

Fig. 87. Amoebae (*Naegleria bistadialis*) in agar culture with bacteria. a, b normal anaphase and early telophase; c–k amoebae flattened under a cover glass; c early telophase; d late telophase without cytoplasmic division; e early telophase in a binucleate amoeba, cytoplasmic constriction in the equator of the two almost parallel stem bodies; f–h cytoplasmic constriction around the peripheral telophase nuclei in amoebae under strong pressure; i, k tripolar nuclear division; i 3-rayed figure, k triangle; a–d, i, k Giemsa stain, f–h Heidenhain stain. 1,000 ×. (After KüHN, 1916)

division is followed closely by cytoplasmic division. The amoebae move around vigorously until the end of the prophase; during metaphase or early anaphase they round up. As the spindle continues to elongate, the cell becomes oval (Fig. 87a). As the constriction of the cytoplasm begins, blunt pseudopodia form at the poles and move vigorously, pulling the two daughter amoebae apart. This will recall the separation of the daughter oil droplets in the model experiment. When the amoebae are raised in the very thin liquid film on top of a relatively dry agar preparation or when they are flattened under the pressure of a cover slip cytoplasmic division frequently does not take place. The connection between the telophase nuclei snaps: the daughter nuclei are carried away by cytoplasmic streaming, and they assume positions at random (Fig. 87c, d). In these binucleate amoebae, no subsequent division into uninucleate products takes place when the nuclei have reached the interphase state. The amoebae can continue to grow and even initiate another nuclear division (Fig. 94). Ordinarily amoebae with four nuclei result. Rarely, if the pressure is released that has kept them flat and the amoebae round up during anaphase, two spindles form in parallel and the cytoplasm constricts in the equator between the two pairs of telophase nuclei (Fig. 87e). Cytoplasmic constriction can occur even in

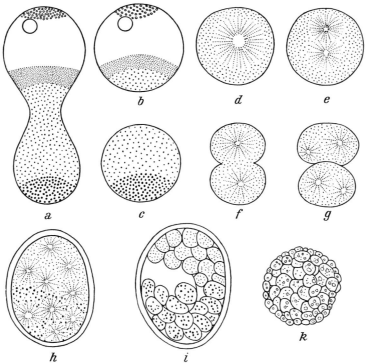

Fig. 88. Cleavage of anucleate egg fragments of the sea urchin *Arbacia punctulata*. a stratification of the unfertilized egg; from the centripetal to the centrifugal pole: oil, clear cytoplasm with the nucleus, mitochondria, pigment; b, c light and heavy halves; d appearance of an aster in an anucleate yolk containing quarter of an egg (resulting from the further division by centrifugation of the heavy half, c, into yolk and pigment portions); e–g replication of the asters and cytoplasmic division; h the heavy half (c) after twelve hours with numerous asters; i like h but with cytoplasmic division; k anucleate "blastula" from the yolk quarter (d–g). After three days about 500 cells. (After HARVEY, 1936)

extremely flattened amoebae if the movements of the cytoplasm carry the telophase spindle to the free margin of this disc-shaped cell. If it lies with both poles close to the edge, a unilateral constriction between the telophase nuclei can take place (Fig. 87f), which once in a great while will lead to a division of the amoebae into two pieces of approximately equal size. More frequently a smaller amount of cytoplasm is pinched off around one daughter nucleus which has come near the edge (Fig. 87g, h). The formation of such midget amoebae can be seen in the living state. Under these experimental conditions, tripolar nuclear divisions are also seen from time to time. If the spindles grow out from a central point, telophase often sees the amoebae constrict into a cloverleaf (Fig. 87i). When the spindles form a triangle, they are frequently not synchronized; in such cases the cytoplasm can constrict around a nucleus which has already reached telophase (Fig. 87k). In normal amoebae the extension of the spindle pushes the telophase nuclei toward the cell poles; in the flattened amoebae the telophase spindle is substantially shorter than the longest diameter of the cell. All these observations suggest that telophase nuclei at a certain stage exercise an effect on the free surface of the cytoplasm, leading to constriction. Whether the agents emanate from the nuclei or from the mitotic apparatus, possibly from some as yet unseen centers near the nucleus (whose existence is suggested in other amoebae), cannot be decided at the moment.

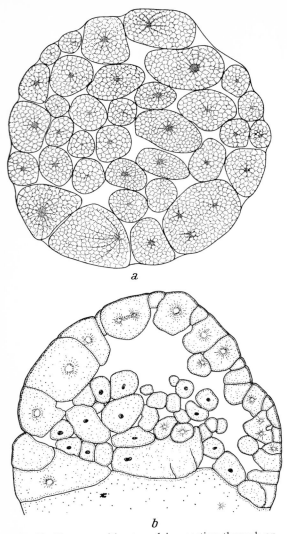

a

b

Fig. 89. Cleavage without nuclei. a section through an embryo from an unfertilized egg of the starfish *Asterias forbesi*, which was activated chemically after removal of the first maturation spindle; b section through a blastula derived from half a *Triturus* egg originally containing a sperm nucleus but no egg nucleus; most cells are anucleate but some contain descendants of the sperm nucleus. (a after McCLENDON, 1908, b after FANKHAUSER, 1929)

The significance of the centrioles and the asters for cytoplasmic division in cleaving cells is shown by the division of anucleate egg fragments, which can proceed impressively far when centrioles are present in the daughter cells and form asters (Figs. 88, 89). The division of anucleate egg regions can be fairly irregular, or it can be just as regular as in normal cleavage, so that, for example, in a case of abnormal nuclear division in an *Ascaris* egg whose ongoing cleavage had been under observation for some time, it was only in the subsequent examination of stained preparations that the ectoderm cells were found to have no nuclei[73, 149].

We can say even less about the events leading to the formation of a separation zone in the cytoplasm than we could about the causes of the cytoplasmic constriction. In fixed and stained preparations of cells with reticulate cytoplasm the equatorial cytoplasm has a coarser mesh during nuclear anaphase and telophase; this zone may also contain a layer of vacuoles (Fig. 90 b, c). Later these vacuoles may merge, forming a cavity separating the two daughter cells. It is difficult to see why the cytoplasmic activities underlying this event should differ from those which underlie constriction; the one process of cell division often cannot be distinguished clearly from the other, or it may be replaced by the other in an otherwise similar cell, as in the cleavage of Echinoderm eggs under various external conditions. We may speculate that the asters whose rays meet in the equatorial zone (Figs. 47, 90) also alter the colloidal condition of the cytoplasm here. During mitosis in some animal cells, moreover, the spindle fibers are thickened at anaphase in the equatorial zone (Fig. 90 b), which recalls the initiation of cell plate formation in plant cells.

The rays of the aster lead to the formation of a wall in a most remarkable way during spore formation in the ascomycetes. The common cytoplasm of the ascus typically divides into eight regions, each surrounding a nucleus during "free cell formation." A pointed process extends from each nucleus, at whose tip lies a centriole fastened to the nuclear membrane (Fig. 91 a). The astral rays emerging from this centriole bend back toward the nucleus like a fountain and surround it (Fig. 91 b, c). In conjunction with an outer fibrous coat a layer of cytoplasm is

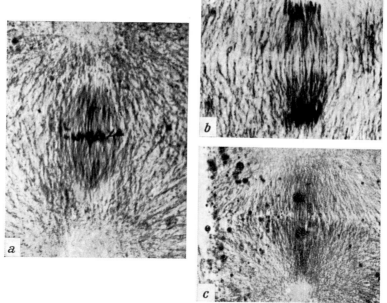

Fig. 90. Nuclear divisions in *Coregonus lavaretus* eggs. a metaphase of the first cleavage; b, c later cleavage divisions, b late anaphase, c telophase. In b and c the cytoplasm is beginning to form a boundary zone. (After Monti, 1933, microphotographs)

delineated. The nucleus, together with the centriole, detaches from this outer layer which now secretes at its outer margin the cell membrane of the spore (Fig. 91 d).

In the cytoplasmic division of most plant cells the spindle provides the substratum on which the new cross wall is laid down: the spindle is transformed into the phragmoplast. It spreads, and its fibers become coarser. They are often visible even in the living cell and appear to undergo substantial chemical changes. The spindle now stains with certain vital dyes which previously did not stain it. In the equator of the spindle there appear thickenings which extend out from the spindle axis, turn into small droplets, and gradually flow together as a cell plate. At the same time, the spindle fibers of the daughter nuclei disappear, beginning near the poles. The cell plate which extends from the center of the spindle out to the old cell wall now becomes the middle lamella on which the layers of cellulose are laid down. We do not know whether the droplets which appear in the equatorial plane of the spindle result from a transformation of the spindle substance or whether they represent a component which has now dissociated from it.

Like the astral rays in free cell formation in the ascus, spindle fiber bundles can also be used independently of mitosis to sequester a region around a nucleus from a common cytoplasmic mass. This has long been known for the endosperm of many plants. When hundreds of free

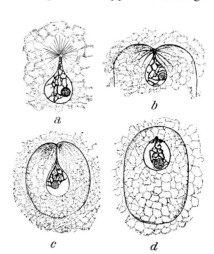

Fig. 91. Delimitation of an ascospore of *Erysiphe* by free cell formation. (After Harper, 1899)

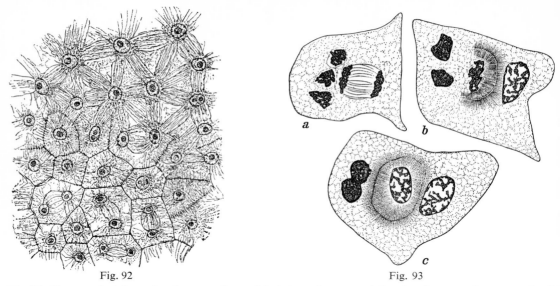

Fig. 92   Fig. 93

Fig. 92. Phragmoplast formation between the nuclei of the embryo sac of *Reseda odorata* at the onset of cell delimitation, which proceeds from below upward. 240×. (After STRASBURGER from "Lehrbuch der Botanik")

Fig. 93. Delimitation of the generative cell in the pollen grain of *Scirpus uriglumis* by free cell formation. a late anaphase of the division of the primary pollen nucleus (on the left are the other three nuclei of the microspore tetrad which degenerate in the Cyperaceae); b extension of the phragmoplast and partial encirclement of the nucleus; c phragmosphere completely enclosing the generative nucleus. 1,340×. (After PIECH, 1928)

nuclei have been formed in the cytoplasm next to the wall of the embryo sac by numerous nuclear divisions without cytoplasmic division, bundles of fibers stretch between nuclei long after the mitotic spindles have disappeared. They do this irrespective of the lineage of the nuclei and accomplish the division of the cytoplasm by forming typical phragmoplasts (Fig. 92).

In *Scirpus* (Cyperaceae) the generative cell in the pollen grain is delineated by means of a most remarkable phragmoplast formation in the cytoplasm (Fig. 93a–c): The spindle of the division of the primary pollen nucleus comes to enclose one daughter nucleus by the addition of new fibers in telophase and thus forms a "phragmosphere" around the generative nucleus. A cell plate arises in the middle of the radially oriented fibers, and on it the cell wall of the generative cell is laid down. Although the geometry is different, the basic process is exactly the same as that seen in the formation of the cell wall after mitosis.

When we survey the diversity of nuclear and cytoplasmic division mechanisms, two standard functional types emerge. in which certain functional elements occur as a package: On the one hand we have the type of mitosis with centrioles, astral rays, and a spindle whose formation and orientation is clearly related to the centrioles and where the astral apparatus also mediates cytoplasmic division. On the other hand, there is a type of mitosis whose spindle lacks centrioles and which leads to phragmoplast formation. BOVERI once said[71, 155]: "Division with centrioles is the most elegant solution to an assignment which can also be carried out in a variety of other ways." "Elegant" here can only mean "easiest for us to follow"! What is most impressive is that the elements of such a "complete" mitotic figure—astral rays, spindle, centrioles—can be used singly and combined in a variety of ways as a means to produce particular results in development of the cell. This "biotechnical" approach is reasonable and necessary, but it is not satisfying; for we wish to learn the physical-chemical nature of the functional elements, as well as the conditions which give rise to them and organize them into a complex mechanism. Here, however, our insights are scant indeed. We have little solid information on the physicochemical changes in the colloidal processes during mitosis, and the information we do have does not explain the formation of the mitotic apparatus, the forces involved in the chromosomal cycle, and the mechanism behind the diverse modes of cytoplasmic division.

Mitosis is apparently accompanied by interconversion between sulfhydryl and disulfide groups in glutathione. Whether these events are directly involved with changes in structural proteins, or whether they relate to the activation or inactivation of enzyme systems is unknown. It appears that cell division always requires oxidative or anaerobic energy metabolism. In explants of mouse epidermis, mitotic frequency depends upon oxygen tension. Cell division stops in the absence of oxygen; in addition, a suitable carbohydrate must be present. Inhibitors of respiration and of glycolysis also inhibit cell division. Like the lack of oxygen, these inhibitors work during the "antephase," before the visible nuclear changes which signal prophase. Once mitosis has begun, it continues unimpeded. This may be because carbohydrate metabolism is required for an "antephasic" process, or because the cell has accumulated sufficient reserves for the whole process before the onset of mitosis[87]. The cleavage of sea urchin eggs evinces a rhythmic rise and fall of oxygen uptake, apparently associated with prophase.

The chromosomal cycle and the cycle of the achromatic apparatus can, as we have seen (pp. 49, 62), run independent courses; but in normal cell division they must be synchronized. This can be mediated only by the physiological condition of the cytoplasm. We can see this particularly clearly in embryonic cells without nuclei but with centrioles. Here, an achromatic figure forms in the normal tempo and leads to cell division. Conversely, when the achromatic apparatus is paralyzed, endomitosis can take place in an egg synchronously with the cleavage divisions of normal eggs. Generally, to be sure, the rhythm gradually deviates from the norm in these incompletely equipped eggs, as in originally synchronized clocks. It is very surprising to learn that the respiratory rhythm and the cortical cycle, an alternation between smooth and wrinkled surface, continue in spite of the arrest of the achromatic cycle and cytoplasmic division by colchicine[665]. Are the metabolic rhythm and the cortical cycle dependent upon the continuing cycle of changes within the chromosomes themselves or on an autonomous cytoplasmic process? Perhaps on the basic clock mechanism itself? That there are autonomous cytoplasmic processes which continue right through cleavage we shall see shortly.

In the sequence of divisions there may be a continuing subdivision of a cell region, as in certain embryonic periods, or multiplication of cells involving an alternation between cell growth and cell division. In this case the mitotic cycle is woven into the overall life cycle of the cell, which runs a rhythmic course under constant environmental conditions. This can be seen in protozoan cultures, in tissue cultures, and in uniformly growing tissues in multicellular organisms.

Naturally, nuclear and cytoplasmic growth require nutrients: building blocks, energy sources, and catalysts. We shall lay these special metabolic factors aside; they cannot tell us why cells never exceed a certain size and what agents cause the onset of division as the limiting size is approached.

From general considerations, the physico-chemical necessity of cell division has been inferred from the nature of the cell as a growing, metabolizing system[529]. One starts with a simplified model: spherical shape, internal homogeneity, external homogeneity, the involvement of a single substance in metabolism, and certain permeability properties of the cell wall. The concentration gradients of the substrate diffusing in and the product diffusing out are calculated, as well as the osmotic pressure and surface tension. Now a "critical radius" can be calculated beyond which further growth leads to instability of the system, so that it will divide in response to the slightest deviation from spherical shape. If, in such a one-phase spherical system, the reaction velocities of the processes occurring in cell respiration are now stipulated, one arrives at values similar to those of the average radii of living cells.

Interesting though these calculations may be, they help us very little with developmental physiology. They demonstrate that there are factors in the functional design of cells which limit their dimensions. But even apart from the fact that cells are anything but homogeneous systems, this excursion into theoretical physical chemistry does not make the actual events comprehensible where the division of the cytoplasm is only a rather spectacular episode in a long and complex continuum of structural change.

Under constant conditions cell growth proceeds from a constant initial size to a constant final size, both of the cytoplasm and of the nucleus.

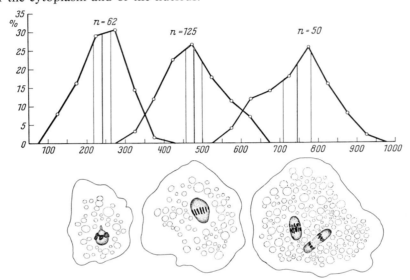

Fig. 94. Size distribution of amoebae (*Naegleria bistadialis*) in agar plate culture; just after division (still with telophase nucleus), with nucleus in mitosis, binucleate amoebae in mitosis. Mean values $\pm$ 3 standard errors are shown. $n =$ number of amoebae measured. (After KÜHN, unpublished)

Figure 94 illustrates the doubling of cytoplasmic size of amoebae in a vigorously multiplying culture. The amoebae have been growing in a thin water film on agar, together with bacteria, and under these conditions they have a flat disc shape (p. 62). The areas of these discs were measured for amoebae which had just emerged from division, as seen by the fact that their nuclei were in telophase, and for amoebae whose nuclei were in earlier mitotic stages. The thickness of the amoebae, which probably did not change significantly because of their flat shape, has been neglected. Each group had a certain variability.

The volume change of interphase nuclei between divisions is seen beautifully in the epidermis of caterpillars from the second to the fourth larval instar (Fig. 95).

In considering the trigger mechanism of cell division, we must first know the relationship between cytoplasmic and nuclear growth. R. HERTWIG (1908) employed the size relationship between nucleus and cytoplasm (N/C, the nucleocytoplasmic ratio) as a possible causal factor in cell division, and on the basis of his observations in protozoans he developed this hypothesis: every cell has a certain characteristic N/C ratio. Immediately after division N/C = K. [In German the letters are different]. Now the cell doubles its volume. The metabolic machinery of growth does not result in a proportional increase of nuclear and cytoplasmic volumes; the nuclear mass increases relatively little ("functional nuclear growth"). This condition of the cell in which N/C is less than K is called "nucleocytoplasmic tension." HERTWIG considered this to be the trigger of cell division. When a critical tension was reached, a "divisional growth" was evoked in the nucleus causing the nuclear mass to double. Now the N/C ratio is restored and division follows immediately.

The N/C ratio is defined as a ratio of masses, and this is the only way it can be a meaningful variable. Only volumes, however, are directly measurable. Even for the quantitative relationship of volume between nucleus and cytoplasm during cell growth and cell division, we have available, unfortunately, only the early experiments on ciliates. According to these studies, for example, cell volume increases fairly steadily in *Paramecium caudatum* between divisions (Fig. 96a). The nucleus grows very slowly at first; it lags behind the cytoplasm in volume.

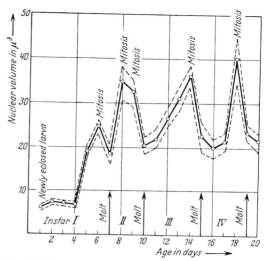

Fig. 95. Mean nuclear volumes (± 3 standard errors) in the epidermis during larval development (first to fourth instar) of *Ptychopoda seriata*. (After RISLER, 1950)

Then, suddenly, a rapid nuclear growth leads to a doubling of its original size (Fig. 96a, N/cT). The critical value of the nucleocytoplasmic tension which triggers "divisional growth" thus is clearly reached here. In Figure 96b this change is represented as the ratio of cytoplasmic and nuclear volumes, which rises from a normal value (set at one) to the critical nucleocytoplasmic tension and then drops sharply back to the norm. At 25° C. with abundant bacteria for food, the

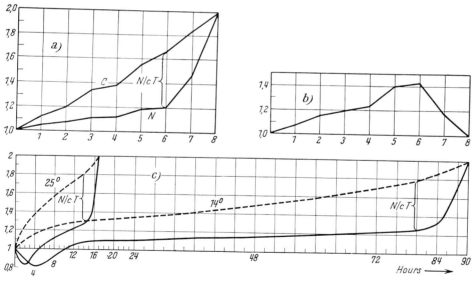

Fig. 96. Volume changes of the macronucleus and the cytoplasm during the growth of Ciliates. Volume immediately after division set equal to one. a, b *Paramecium caudatum*; a increase in volume between divisions; b changes in the nucleocytoplasmic ratio; c *Frontonia leucas* at two temperatures; cytoplasmic growth shown by dotted line. (After POPOFF, 1908, 1909)

interval between divisions of the paramecia averages eight hours, and after six hours the critical N/C ratio is reached. In *Frontonia leucas*, which spends seventeen hours between divisions at 25°C., the nuclear volume first actually drops for a short time; then it rises, and the course of nuclear and cytoplasmic growth corresponds once more to the hypothesis of HERTWIG. After fifteen hours the critical N/C tension value is reached. Growth rate, the absolute dimensions of the *Frontonia* cells, and the N/C norm vary with temperature. The time between divisions rises to an average of ninety hours at 14°C.; both cell and nuclear volumes are greater at 14°C. than at 25°C. But the increases are not of the same proportions in nucleus and cytoplasm. While the normal N/C volume ratio at 25°C. is about 1:67, at 14°C. it is about 1:56. Nevertheless, although the volume ratios differ at the two temperatures, the growth curves have the same form (Fig. 96c).

Unfortunately there are no other examples in ciliate cells which test the hypothesis of HERTWIG. The ciliate macronucleus is, however, atypical in its structure and mode of division: it is really a compound nucleus and its division as a whole is amitotic. Since it has no typical prophase, one cannot define a period with "divisional growth." One might suspect that during this time the doubling of the chromonemata takes place. But this need not be expressed directly in a volume increase of the entire nucleus; and of course a nucleus can increase in volume without multiplication of the chromatids, by the uptake of water in preparation for division. The micronucleus is certainly not without importance for the growth relationships of the cytoplasm in ciliates.

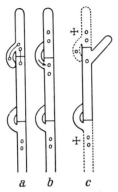

Fig. 97. Diagram of cell division in the Hymenomycetes. a, b hypha with clamp formation; c operative production of a haploid mycelium

That a certain mass ratio between cytoplasm and nucleus is at least one factor involved in the onset of nuclear and subsequent cytoplasmic division is shown by experiments in which the cytoplasmic mass was reduced[259]. Individuals of *Amoeba polypodia* have been kept alive without nuclear division for 130 days by the daily removal of a portion of their cytoplasm, while during the same time the control amoebae, sister cells of the experimental individuals, underwent 65 divisions. (Whether the nucleus suspended its growth during this long experimental time or whether it grew slowly or periodically as new cytoplasm was being formed was unfortunately not recorded.)

Another influence on the final size reached by a cell before the onset of a new mitotic cycle is just as surely the constitution of its chromosomes. Binucleate amoebae grow to approximately twice their cytoplasmic volume before their nuclei start division again (Fig. 94). In the Hymenomycetes growth can be compared in hyphal cells with one nucleus and with two nuclei. From the spores mycelia capable of continuous growth arise with uninucleate cells. If mycelia of opposite mating types meet, each two cells of the opposite sex which touch fuse. The binucleate cells grow into hyphae. The "paired nuclei" divide synchronously. At this time and during the cross wall formation which follows immediately, a remarkable event always takes place: next to the pair of nuclei a hook-shaped outgrowth forms, the "clamp", which is always directed proximally. One daughter nucleus enters the clamp, while the other moves toward the hyphal tip (Fig. 97a). The two daughter nuclei of the second nucleus in the original pair become separated by a cross wall at the level of the clamp. Just behind this wall an opening forms to admit the daughter nucleus from the clamp into the proximal cell (Fig. 97b); now each of the two cells is binucleate. If the terminal cell, the clamp, and the basal cell behind the second cell are destroyed after the formation of the cross wall and before the entry of the clamp nucleus into the surviving second cell (Fig. 97c), as can be done with a fine glass needle in a micromanipulator, a uninucleate cell is left which can grow into a mycelium. Now the diameters of the colonies resulting from single cells can be measured at various intervals. It can be seen that the growth rate as measured by the longitudinal extension of individual cells is about twice as great in the binucleate mycelia as it is in the uninucleate mycelia, whether the latter are derived from the uninucleate spores or from uninucleate mycelia obtained by means of the operation described (Fig. 98).

In the mosses various stages of polyploidy can be obtained easily: if cuttings are taken from the moss capsule, the diploid sporophyte, the cells grow into plants which now are diploid rather than haploid, as normal gametophytes are. If the diploid egg cells are fertilized by sperms from

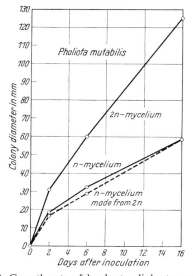

Fig. 98. Growth rate of haplonts, diplonts, and haploid mycelia of *Pholiota mutabilis* obtained by operation. (After the data of HARDER, 1927)

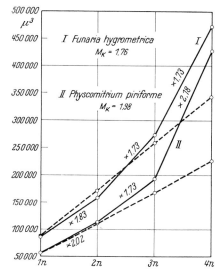

Fig. 99. Volume increase in the leaf cells of moss gametophytes with increase in ploidy. ---- arithmetic progression. (After data of F. VON WETTSTEIN, 1928)

normal plants, triploid sporophytes result. The union of diploid eggs and diploid sperms produces tetraploid sporophytes. The regeneration of polyploid gametophytes from polyploid sporophytes and the subsequent combination of sex cells permits the production of individuals with all sorts of genomes corresponding to exact multiples, odd as well as even, of the wild genome. The volume of the gametophyte "leaf" cells, that is to say, the final size of somatic cells which will not divide further, does not rise in a simple linear fashion with the number of genomes ($V_n : V_{2n} : V_{3n} = 1:2:3$), but rather cell size increases in a geometrical progression depending on a certain index of increase (K), so that $V_{2n} = K \cdot V_n$ and $V_{3n} = K \cdot V_{2n} = K^2 \cdot V_n$, and so on (Fig. 99). The value of K is determined by the genotype of the species or strain and is apparently influenced by individual genes. It also varies from tissue to tissue in a single plant. We do not know whether this "exponential rule" relating number of genotypes to cell size is general for protists, plants, and metazoans, and whether it is general for cells that can divide as well as those that cannot. In those plant cells with large vacuoles, cell volume is obviously a very complex parameter: we have no idea as to the relative contributions that should be assigned to vacuoles and cytoplasm.

In multicellular animals and plants, polyploidy is not infrequently involved in tissue differentiation and the production of cells of diverse size classes. As has been seen in various tissues in the Heteroptera (an order of true bugs) endomitosis takes place (Fig. 100) in which the chromosomes split longitudinally within the interphase nucleus; they do not shorten to the

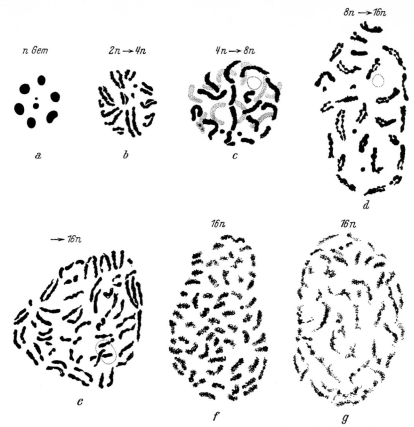

Fig. 100. Nuclei of *Lygaeus saxatilis* (Heteroptera), $2n=12+$ XY. a first meiotic metaphase (six dyads, in the center the X and Y chromosomes in "distant pairing"); b, d–g nuclei of the septa of the testis; b endotelophase of a nucleus in transition from $2n$ to $4n$ (four Y chromosomes); c tetraploid prophase in the fat body; d endo-anaphase in transition from $8n$ to $16n$; e, f endotelophase; g 16-ploid interphase nucleus. (After GEITLER, 1939)

same extent as in normal mitosis. The chromatids separate somewhat but remain joined at their ends for a time. Herein lies a difference from colchicine-induced endomitosis: in both cases the achromatic apparatus is eliminated, but in natural endomitosis the separation of the kinetochores is not inhibited (cf. Fig. 71). In many forms the degree of polyploidy can be determined from the number of chromocenters in the resting nucleus. These correspond to the heterochromatic sex chromosomes (Fig. 101 b–f). Sometimes the polyploid nuclei undergo ordinary mitosis later and show the replication of the genome directly in the mitotic figures (Fig. 101 k, l). Nuclear volume increases with genome number; in high degrees of polyploidy the dimensions become gigantic, and the nuclei may assume a highly branched form (Figs. 101 g, h). In this way the nuclear surface, where the nucleocytoplasmic exchange takes place, increases tremendously.

In polyploid plant nuclei the chromosomes undergo a wide variety of modifications. Giant common chromocenters of the chromosomes containing the nucleolus organizer lie next to the nucleolus. The may be two (Fig. 102a) or, in the triploid endosperm, three (Fig. 102b). The chromocenters of other chromosomes are scattered throughout the nucleus. The chromosomes may appear as a tangle of threads (Fig. 102b), or at times as loose bundles (Fig. 102c). These can be united into "giant chromosomes" (Fig. 102d); indeed, when they are shortened and their

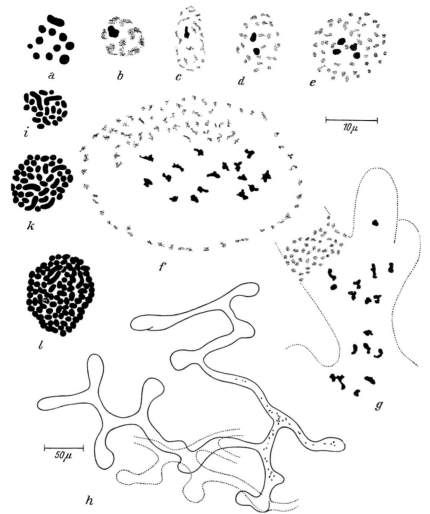

Fig. 101. Equatorial plates and interphase nuclei of the water strider *Gerris lateralis* (Heteroptera). a equatorial plate of the first spermatocyte division (X chromosome above); b spermatid nucleus; c diploid nucleus from the tracheal epithelium; d, e midgut epithelium nuclei, d tetraploid, e octoploid; f 32-ploid nucleus from a malpighian tubule; g small part of a branched salivary gland nucleus; h salivary gland nucleus, slightly enlarged; i–l metaphase plates in the fat body; i diploid; k tetraploid; l octoploid. (After GEITLER, 1938, 1939)

corresponding chromomeres are in register, they resemble the banded chromosomes seen in the higher Diptera (Fig. 102e). On the other hand, the chromosomes resulting from repeated endo-mitosis can also form groups in the diploid number with radiating, fairly helical strands (Fig. 102f).

In the Diptera the chromosomes become polytene in the salivary glands as well as in other tissues. In *Calliphora* the epidermal cells do not divide from the time of hatching to the time of pupation: the entire epidermal growth is attributable to increase in cell volume, and this is associated with an enormous increase in nuclear volume, up to fortyfold. The nuclear growth is discontinuous (Fig. 103). It passes in *Calliphora* from the basic size in the newly hatched larva through four size classes each about 2.5 times as great as the previous class. The chromosomes,

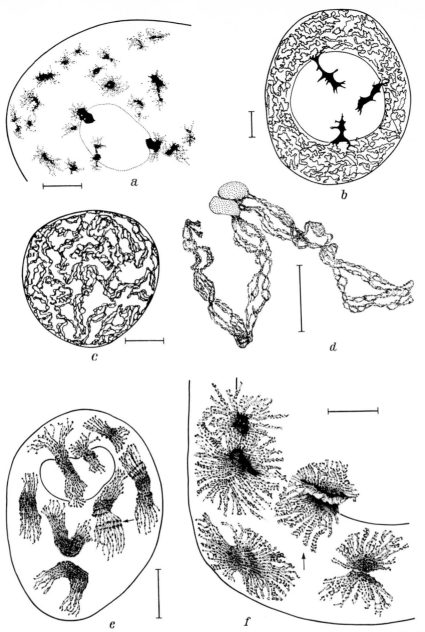

Fig. 102. Polyploidization in plant cell nuclei. a *Lupinus polyphyllus*, polyploid nucleus (apparently 64-ploid) from the suspensor of an embryo, at the nucleolus two dense chromocenters and somewhat less dense chromocenters elsewhere in the nucleus; b *Potamogeton densus* polyploid nucleus from the endosperm with clearly fibrous structures and three large SAT-chromocenters at the nucleolus; c *Alisma lanceolatum*, 16-ploid nucleus of the basal cell of the suspensor with loose chromosomal bundles; d *Allium nutans*, giant "chromosome" from a synergid nucleus; e *Dicentra spectabilis*, apparently 16-ploid nucleus from an antipodal cell. Chromosomes united in eight bundles; each bundle showing transverse structures (arrow); f *Corydalis nobilis*, part of a nucleus of an antipode with (a total of eight) groups of radiating and clearly coiled chromosomes. Line segment=ten μ in each case. (a after GEITLER, 1941; b, c after HASITSCHKA-JENSCHKE, 1959a; d after HÅKANSSON, 1957; e, f after HASITSCHKA-JENSCHKE, 1959b)

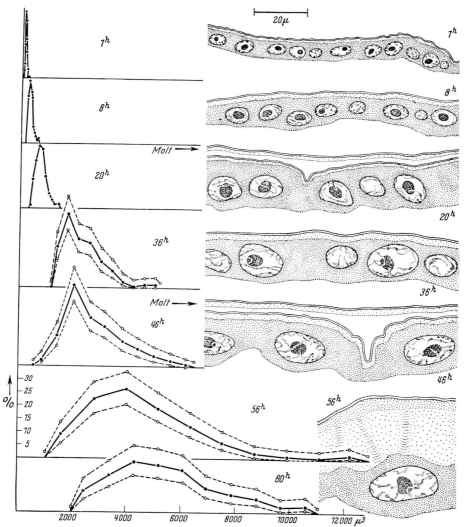

Fig. 103. Growth of the epidermal nuclei of *Calliphora erythrocephala* during larval development. Nuclear volumes separated into ten evenly spaced classes. *n* always equals 200. Dotted lines indicate 95.5 per cent confidence limits. Right, histological appearance of the epidermis at the corresponding stage. h, hours after hatching from the egg. (After WAGNER, 1951)

which are visible in the interphase nuclei, increase in length and thickness. If we assume that each stepwise increase in size corresponds to a chromosomal doubling, a final polyteny sixteen times as great as in the nuclei of the youngest larvae can be reached. In other epithelial nuclei, for example, in the hair-forming cells, much higher degrees of polyteny are reached.

In *Chironomus* the epidermal cells divide, but the nuclei and cells become progressively larger (Fig. 104). It is surprising that the giant chromosomes can become metaphase chromosomes even as the polyteny is increasing, and that they can divide into polynemic chromatids. So the question raised at the end of the second lecture (p. 29) for the bundles of chromofibrils must now be raised for the bundles of chromonemata. Figure 105 illustrates one hypothetical scheme

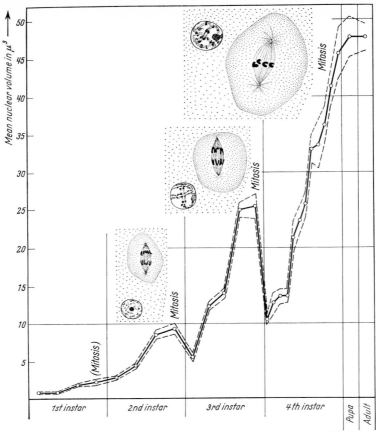

Fig. 104. Growth and multiplication of nuclei in the larval epidermis of *Chironomus*. Mean nuclear volume values ± 3 standard errors; nuclei and mitotic figures are illustrated for each of the last three instars. (After Besserer, 1956)

according to which the stepwise doubling of the chromosomal strands can be reconciled with the maintenance of division surfaces between two groups of strands.

Hand in hand with polyploidy and polyteny, both of which involve the multiplication of chromonemata in the cell nucleus, the size of the cell increases. We can conclude from this that there is a quantitative relationship between the components of chromonemata and cytoplasmic growth. We do not know what sorts of structures and substances may be significant here and whether they exercise their influence directly or indirectly (for example, via the nucleolus whose size increases with chromonema number). We know just as little about the structure, substances, and processes which counterbalance the nuclear effects in the cytoplasm and thus eventually limit cell size.

Since Jakobj[326] established in organs with considerably different nuclear sizes, a variation in nuclear volume concentrated in classes in the ratio of 1:2:4:8, we have always ascribed this sort of rhythmic nuclear growth to an "internal division of the elementary nuclear structure" on the basis of the modern cytology of chromosomes even in cases where this assumption could not be substantiated directly. In some cases cytophotometry of Feulgen positive materials in the nucleus supports the assumption of polyploidy. For example, in mouse and amphibian tissues three classes of nuclei have been found whose DNA contents have the approximate ratio 1:2:4.

76

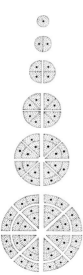

Fig. 105. Hypothetical scheme illustrating the retention of subunits of a strand during replication of its elements. (After SCHLOTE, 1961)

Volume measurements in conjunction with cytochemical studies have shown, however, that this assumption does not always hold. In the pentatomid bug *Arvelius albopunctatus* the spermatocytes in the six compartments of the testis grow to quite different sizes (Fig. 106a). The cell volume of sizes I, II, and III are in the approximate ratio 1:2:8. This ratio holds approximately for cytoplasmic, nuclear, and nucleolar volumes (Fig. 107a). At metaphase in the first meiotic division, cells of all sizes develop identical sets of chromosomes, even though the cytoplasmic volume varies greatly, and in all the cells all the chromosomes are the same size (Fig. 106 e, l). Microphotometric studies indicate that the DNA content, shown by Feulgen staining, is the same in all three sizes of spermatocyte nuclei (Fig. 107b), which vary from 200–1,600 cubic microns (Fig. 106d, k). After both meiotic divisions are over, the DNA content in each spermatid nucleus (Fig. 106f, m) is a quarter of that found in the mature spermatocyte. During the period of maturation division, therefore, no increment in DNA takes place. The protein content of the nucleus, determined after staining with Millon's reagent, corresponds to the nuclear size (Fig. 107b). It is four times as great in size III as in size II. The increase

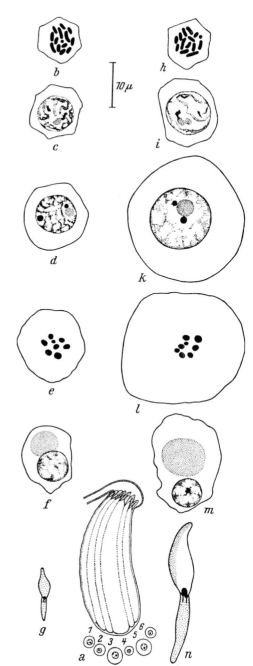

Fig. 106. Cell growth in spermatogenesis of *Arvelius albopunctatus*. a testis, with spermatocyte size in each of its 6 lobes; b–g cells from lobe four with some spermatocytes; h–n cell from lobes three and five with large spermatocytes. (After SCHRADER and LEUCHTENBERGER, 1950)

in RNA content (azure staining, Fig. 107b) in the nucleolus is equally clear. After meiosis, right after the nuclear membrane reforms, the spermatid nuclei are all the same size (Fig. 106f, m). This size is $1/16$ of the largest spermatocyte nucleus size and $1/4$ of the middle spermatocyte nucleus size. Afterwards, each nucleus grows to a size in proportion to its cytoplasm.

In the cytoplasm the protein content increases three- to fourfold and the RNA about three fold. Cytophotometry shows that the protein increase in the nucleus is not accompanied by a corresponding increase in DNA. The parallel increase in protein and RNA in the nucleolus and in the cytoplasm suggests that these processes are closely connected. During cytoplasmic growth, the Golgi material increases as do the mitochondria, which assemble into the "nebenkern" of the spermatid (Fig. 106f, m).

The tremendous nuclear and cytoplasmic growth of which small cells are capable without chromosomal replication is shown by the gigantic growth of the egg cell and its nucleus.

The experiments and the analysis of the diversity of cellular phenomena in nature have shown that the relationships among the components of cells are more complex than had been assumed. On the one hand, nuclear size and cytoplasmic growth have evinced a dependence on genome number; genome replication through polyploidy or polyteny leads to rhythmic cell growth. On the other hand, the differential production of cytoplasmic protein can occur independently of chromosome number, chromosome size, and DNA content in cells. This is related to the increase of RNA in the nucleolus and in the cytoplasm. The surprising phenomenon that even this aspect of cytoplasmic growth proceeds rhythmically without regard to chromosome number suggests the possibility of a rhythmic replication of extrachromosomal structures. The rapid progress being made in this area promises much new insight; and rather than speculate at this stage of the game, it would be well to direct our attention to the results of new experiments as they are published.

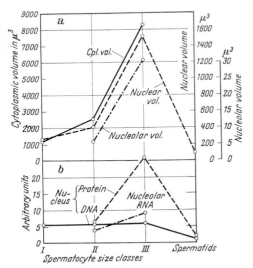

Fig. 107. Volumes of cell, cytoplasm, and nucleus and nucleolus and contents of DNA, RNA and protein in the three size types of spermatocytes of *Arvelius albopunctatus*. (After the mean values of SCHRADER and LEUCHTENBERGER, 1950)

At the beginning of this lecture, the question was raised as to the cause of cell division as a whole. In retrospect, the findings to date, even including the biochemical studies, have yielded rather little. A competent and certainly not pessimistic judge of the problem, Jean Brachet, acknowledged in 1947: "The reason why a cell enters mitosis remains an absolute mystery."[76, 199] And in 1952, in an admirable review, ARTHUR HUGHES stated "that the chemical and physical factors which determine the onset of the process of cell division are completely unknown."[315, 163] It is still true.

Lecture 6

The domain of single-celled organisms embraces a tremendous diversity of forms and modes of development. It illustrates how varied is the differentiation of which single cells are capable and the great extent to which an organism can establish separate functional structures without dividing itself into cells. The course of development in an individual cell and in a sequence of cell generations can be laid out in a strict sequence and thus lead directly to often highly complicated forms when the organism does not have to cope with a variable environment. This is the case for Radiolarians and Foraminiferans, which hardly ever experience any change in the milieu of the ocean, and for Sporozoans and other organisms placed in a rigidly prescribed series of environmental transitions which determine form and method of reproduction.

In other organisms, changes in form (modifications) are the responses to sudden changes in conditions. The amoebae of the family Bistadiidae creep around with plump pseudopodia in dense populations of bacteria, which they eat (Fig. 108a). The amoebae transform into a flagellated individuals as soon as they find themselves in pure water (Fig. 108b, c). At the edge of the nucleus basal bodies arise; the nucleus moves to the cytoplasmic surface, and from the

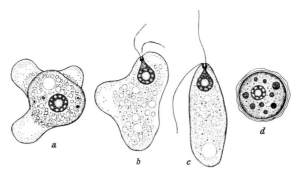

Fig. 108. *Naegleria (Vahlkampfia) bistadialis*. a amoeboid state; b, c transition to flagellated state; d cyst. (Original A. KÜHN)

basal bodies, flagella sprout forth, beating as they grow. Pseudopodium formation stops, the cell rounds up and then elongates; the flagellate now swims away. With this transformation to a swimming form, the fine structure of the cytoplasm changes: the creeping amoeba has a delicate cortical layer, a clearly delineated hyaline ectoplasm which gathers in the pseudopodia, and granular endoplasm sprinkled with vacuoles. The nucleus and the contractile vacuole are moved around by cytoplasmic streaming. With the transition to the swimming form, the distinction between ectoplasm and endoplasm disappears (Fig. 108c), and the elongated cell acquires a fairly constant form. The colloidal condition of the cytoplasm has clearly changed; it becomes stiff and gelled throughout, or else a relatively firm cortical layer arises without the formation of a visible membrane. The nucleus in the flagellated form remains at the anterior end: connected to the basal bodies, it is fixed just under the cell surface and drawn out towards the flagellar attachment. The contractile vacuole lies near the other end. The use of this profound transformation of the individual is clear: When an amoeba is deprived of the bacteria, the prospect of finding a new source of food is more favorable for a fast-swimming form, particularly if it is attracted chemotactically by some metabolic product of the bacteria. After a few hours the swimming form becomes an amoeba again. The nucleus detaches from its site near the flagellar attachment, the flagella are resorbed, mainly dissolving at the base. Pseudopodium formation resumes. Nuclear and cytoplasmic division (Fig. 87a, b) take place only in the amoeba form. Unfavorable environmental conditions, such as the accumulation of metabolic products in decaying medium, drying, and severe cooling cause the amoebae to encyst. The cytoplasm rounds up; within it drops or granules of reserve material appear, and a cuticle is formed at the surface, which is resistant to drying and impermeable to many substances (Fig. 108d).

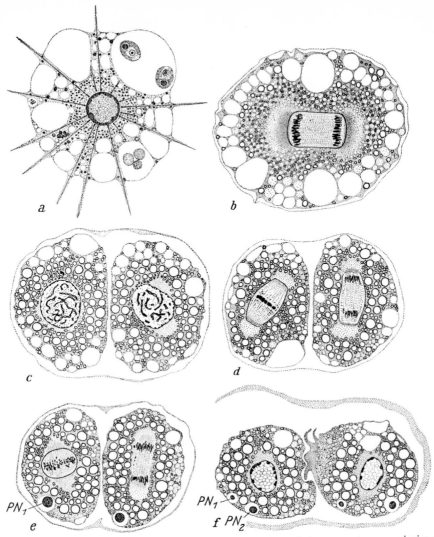

Fig. 109. *Actinophrys sol.* a vegetative individual; b progamic division of the gamont; c–e meiosis; c diplotene; d first maturation division; e second maturation division; f copulation; 1,130×. PN$_1$, PN$_2$ polar nuclei. (After Belar, 1922)

In the category of modifications we can include meiosis and the process of fertilization in some Protista. A species of rhizopod which has been the subject of an excellent study is *Actinophrys sol* (Fig. 109). This heliozoan can be maintained in a vegetative state indefinitely with something like the flagellate *Chlorogonium* as food. In one experiment this "agamic" reproduction was maintained almost three years with vegetative division taking place between one and three times per day. The sexual process, which includes meiosis, gamete formation, and fertilization, can be triggered at any time by particular conditions: If the food supply is removed, gamete formation and fertilization take place simultaneously for all individuals after 24–36 hours. The participation of metabolic products in the medium can be excluded by changing the medium in which the animals are starving every six hours. Fertilization is autogamous, that is, daughters of the

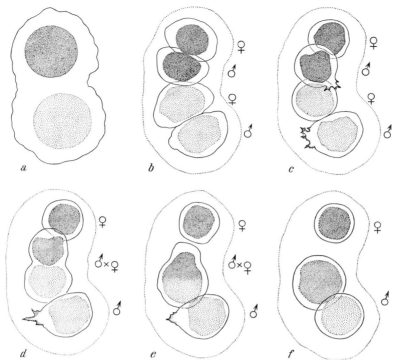

Fig. 110. *Actinophrys sol*. Cross fertilization resulting from the ectoplasmic fusion of two gamonts. The endoplasm of each gamete pair is stippled distinctively. Observation *in vivo*, 425×. (After BELAR, 1922)

same mother cell mate. In the vegetative form, the ectoplasm is pervaded by a stratum of large vacuoles (Fig. 109a). The axial fibers of the pseudopodia extend through the endoplasm to the nucleus. The onset of the events leading to fertilization is signaled by the egestion of food remnants, the retraction of the pseudopodia, and the dissolution of the axial fibers. Now, a thin membrane is laid down on the cell surface, and the nucleus undergoes a division which resembles ordinary vegetative mitosis (Fig. 109b). Cytoplasmic division follows closely. Both daughter cells emerging from this "progamic" division now undergo meiosis. The nuclei pass through typical stages from leptotene to diakinesis (Fig. 109c, cf. Fig. 40). Here, one nucleus always races ahead: while one is still in metaphase, the other one is already completing anaphase (Fig. 109d). Of the two daughter nuclei of this first maturation division, only one survives, undergoing a second maturation division (Fig. 109e), which again results in one mature gamete nucleus and another nucleus which perishes. In meiosis, therefore, two nuclei are extruded, like the polar bodies of metazoan eggs. About 40—50 minutes after the end of the second maturation division one gamete forms an ectoplasmic pseudopodium with thin spikes at a position facing the other gamete (Fig. 109f). When this pseudopodium touches the other gamete, which has until now remained completely passive, the second gamete now responds by forming a shallow groove, whereupon the pseudopodium of the active gamete fuses with the passive gamete, and the cytoplasm unites, as the endoplasm of the active gamete flows into the other one. The zygote rounds up, and the gamete nuclei approach and fuse (Fig. 111d, e).

Thus, the two gametes behave quite differently, in spite of their morphological similarity. One has passed through the maturation divisions faster and is the active partner in mating. We can

81

call the active gamete male and the passive gamete female. Sexual differentiation takes place in the progamic division. Of the two sister cells, one is always female, the other male. This is shown in the exceptional case of a "foreign fertilization": when two *Actinophrys* individuals lie close together during the transition to autogamy, their ectoplasm may fuse (Fig. 110); then a common membrane is laid down, within which the two pairs of gametes separate, each pair including an active male and a passive female gamete. In this case, a male gamete of one pair is as likely to mate with the female gamete of the other pair as with its own sister gamete (Fig. 110c–e). If the remaining gametes are so placed that the male does not reach the female (Fig. 110), the male

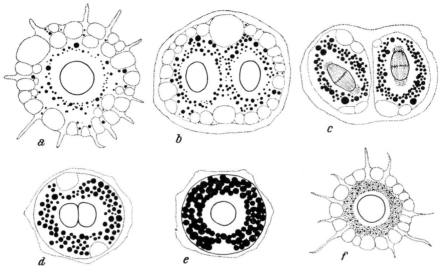

Fig. 111. *Actinophrys sol.* a–d accumulation of fatty reserves during autogamy; e zygote; f consumption of fat during zygote germination; osmic acid fixation. About 750×. (After BELAR, 1922)

stops his futile pseudopodium formation after a while and encysts, as does the lonely female. The mechanism by which this sexual difference arises through the progamic differential cell division is unknown.

The sexual process is closely tied to material changes in the cytoplasm. In the vegetative individuals, hardly any reserve materials accumulate (Fig. 190a). A few little fat droplets are scattered in the outer layer of the endoplasm and in the ectoplasm. As soon as the animals withdraw their pseudopodia and enter the prophase of the progamic division, the fat droplets grow in size and number. Gradually they fill the entire endoplasm, except for a narrow region around the mitotic apparatus, the gamete nuclei and the zygote nuclei (Fig. 111a–e). This lipid in the nonfeeding cells is apparently produced from the metabolism of proteins. The cell volume decreases progressively after the progamic division (Figs. 111, 112a, b). The ectoplasm shrinks, the large vacuoles disappear, and in the mature zygote there is no trace of ectoplasm (Fig. 111e). The contractile vacuoles continue to work undisturbed for a long time (Fig. 111d); their activity ceases only after the fusion of the gamete nuclei. As the cytoplasm draws together, a thick lamellar jelly coat is laid down (Fig. 112). Its outer boundary now has about the same dimensions as the original gamete-forming cell; this observation makes clear the great reduction in volume of the cell that produced the jelly. The mature zygote lays down a tough ectocyst; the cytoplasm now contracts once again and forms a thin endocyst within (Fig. 112c). The germination of the cysts follows regularly three to twenty hours after they are removed from the fluid in which they formed and placed in one of lower osmotic strength. The cytoplasm now swells until the endocyst is closely appressed to the ectocyst. Then both cyst membranes burst and the cell body comes out through the slit. The young free *Actinophrys* individual, whose cytoplasm is at first entirely uniform, gradually acquires a layer of small and later large ectoplasmic vacuoles, after which pseudopodia grow out (Figs. 111f, 112d). The mature young *Actinophrys* individuals, before they have been fed, have the same diameter as the swollen cysts. The accumulations of lipid disappear rapidly as the cyst swells and the endoplasm of the young animal is sprinkled with innumerable tiny fat droplets (Fig. 111f). If fed generously, it grows to maturity in twelve to

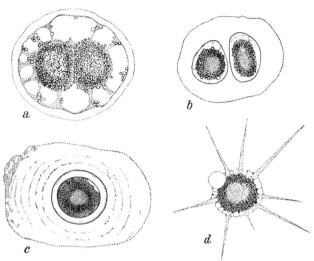

Fig. 112. *Actinophrys sol.* Successive stages of paedogamy. a–c formation of jelly coat and cyst; d newly emerged animal. *In vivo* observation; about 380×. (After BELAR, 1922)

twenty hours. The *Actinophrys*, having emerged from its cyst, can be made to undergo autogamy again after about forty hours if it is subjected to starvation. During this time, no vegetative division has to take place. When *Actinophrys* enters progamic division for gamete formation, it is firmly committed to a developmental path leading all the way to the encystment of the zygote. The series of events may be disturbed but not replaced by any normal alternative, as far as we know. In all other circumstances *Actinophrys* can be caused to choose a new modification from its developmental repertory at any time.

Examples of a great diversity of structure and of modes of reproduction can be seen in the Phytomonads. For the moment we can restrict ourselves to unicellular forms. These can be maintained in a constant sequence of growth and division for hundreds of generations, indeed indefinitely, under constant conditions. The interval between divisions depends upon the environmental conditions including light, food, and, in the case of marine forms, salt concentration. In this way, the flagellates achieve a reproductive balance under a variety of conditions. The range of reproductive rates and the optimal conditions are determined by the genotype (Fig. 113). In certain genera the division rate, the number of cell divisions per unit time, is in general inversely proportional to the cell size of the species or strain. The form and size of the cells can be modified strongly and in various ways by environment factors. Certain modifications of form are retained as "dauermodifications" for a long time after the conditions change and it often takes a large number of generations before a new form, corresponding to the new conditions, is acquired.

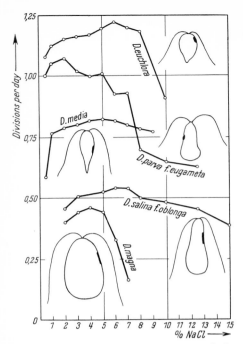

Fig. 113. Division rates in various *Dunaliella* species and strains at different salt concentrations. Cell sketched to common scale. (After LERCHE, 1937)

Many species (Chlamydomonads) can form non-motile "palmella" stages under certain circumstances: the membrane swells and forms a lamellar jelly layer within which the flagellaless cells reproduce.

Normal division takes place in the membraneless Polyblepharids, of which *Dunaliella* is a good example, in the motile state (Fig. 114). First, the nucleus divides, and then the pyrenoid, which is situated in the cup-shaped chromatophore, elongates, and the two flagella move apart. The pyrenoid and the chromatophore divide, and the constriction of the cell body proceeds from the anterior and posterior ends simultaneously. Each of the two daughter cells forms a new flagellum, one daughter cell forms a new eye-spot, and finally the cells separate.

In those unicellular Phytomonads possessing a cellulose outer membrane the division takes place within the membrane. The flagella are cast off and the cell body undergoes two longitudinal divisions which proceed as in *Dunaliella*. The four flagellated daughter cells now hatch out of the membrane of the mother cell.

Some unicellular Phytomonads form morphologically identical gametes (isogametes), while others form two distinct types. In the genus *Chlamydomonas* there is a whole spectrum, from morphological isogamy, through motile gametes differing increasingly in size (anisogamy), to oögamy in which the larger gamete has no flagellum. When the two sexes can be distinguished morphologically the larger is designated as female (gynogamete) and the smaller as male (androgamete).

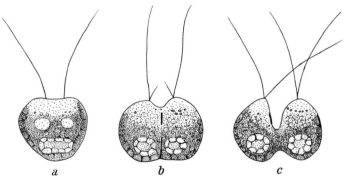

Fig. 114. Vegetative division of *Dunaliella salina*. (After LERCHE, 1937)

In isogamy and anisogamy the process of mating (Fig. 115) begins when the two gametes appose their (anterior) flagellar ends. They become attached by a narrow cytoplasmic bridge, and then gradually they fuse; in forms with a membrane, it is dissolved at the point of contact. The zygote casts off the flagella and surrounds itself with a tough coat. In oögamy the contents of the androgamete pass into the gynogamete.

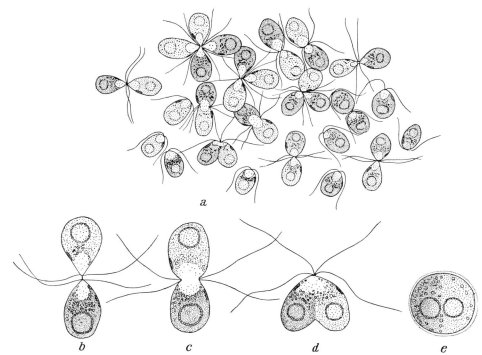

Fig. 115. Copulation of the isogametes of *Dunaliella salina*. a aggregation; b–e course of copulation. The gametes of one clone (and therefore one sex) have turned red. (After LERCHE, 1937)

The germination of the zygote involves meiosis in the Phytomonads; the first two divisions are maturation divisions, resulting in the production of haploid chromosome sets (Fig. 116). The Phytomonads are thus haplonts; only the zygote is diploid. This contrasts with the diplont *Actinophrys*, in which reduction takes place immediately before the formation of the gamete nuclei.

In many Phytomonads the vegetative "swarmers" can be transformed physiologically into gametes by the appropriate environmental influences. In *Dunaliella* sexuality can be induced by lowering salt concentration; other changes in the ambient conditions also work.

Separation into two sexes is completed during meiosis in many species: during the reduction division, chromosomal sex-determining alleles segregate in the two daughter nuclei. Of the four cells which emerge from the two meiotic divisions of the zygote (Fig 116). two belong to one sex and two to the other; these forms have separate sexes and are thus dioecious. If the progeny of one of these

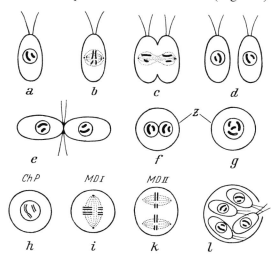

Fig. 116. Diagram of chromosomal behavior in the Phytomonads ($n=2$). *Ch P* chromosomal pairing; *MD I, MD II* first and second maturation division; *Z* zygote

85

four cells are cultured, none of the cells within this clone will mate. The cells of this clone will mate with those of two other clones derived from the same zygote, if the appropriate conditions are established, but not with the fourth clone whose individuals are of the same sex. The fact that when two clones are together, matings are always between an individual of one clone and an individual of the other clone can be shown elegantly in *Dunaliella salina*: When nitrogen and phosphorus are at low levels in the culture medium, carotene accumulates in the chromatophores, making them a deep reddish orange. Now one can take a clone, which is to say one sex, and mark it in this way (Fig. 115). By using cells of the opposite sex which have been raised at higher nitrogen and phosphorus concentrations so that they have the ordinary green color, it can easily be seen in the microscope that it is always one red and one green cell which first attach and later form a half-red and half-green zygote. Thus, the cross mating between clones is a phenomenon at the individual level as well as the population level. This behavior indicates a physiological anisogamy in the absence of any morphological distinction. The isogametes of opposite mating types (sexes) are called + and − gametes.

In addition to dioecious species with genetic sex determination, monoecious species also frequently occur in the same genus, e.g., *Dunaliella*, where sex determination in some species is a phenotypic modification process. In such cases the transition of the cells into gametes also involves the differentiation of sexes. For example, if *Dunaliella media* cells of one clone are placed in conditions favoring copulation, the flagellates swim around randomly at first and then suddenly clump into groups in which numerous individuals are associated, beating their flagella vigorously. Within these groups pairs of cells separate out and mate. Here it is not the case that every cell can fuse with any other; generally a greater or smaller number of cells remain which cannot mate in pairs. These remaining cells are not incapable of mating; they are not asexual "swarmers" which have failed to become gametes, for when one brings together these remaining cells from different cultures of a clone, pairing does take place in certain combinations and proceeds to completion. These remaining cells are really gametes, and they belong to two different types, the two sexes.

This sexual bipolarity in monoecious species is an example of bidirectional modifiability. The cells have in their norm of reaction a bisexual potency, that is, they can respond to the appropriate conditions by developing the physiological properties of either + or − gametes. The change to one or the other phenotype may depend upon very slight differences in conditions just as random differences in the conditions in a particular field can result in the appearance of two phenotypes side by side in an "alternating strain" of a plant, e.g., "normal" and "contorted" in the classical *Dipsacus* example of DE VRIES.

The number of "+" and "−" gametes can be altered by major differences in conditions, involving temperature and pH among other things. For example, at pH 9.5 the "−" gametes are the predominant or sole class, while at pH 4.5 all or almost all of the gametes are "+." Acid solutions promote the formation of "+" gametes, and basic solutions favor "−" gametes. Low and high temperatures have the same respective effects. The fact that conditions favoring gamete formation cause a modification in the cells of the monoecious species does not mean that the establishment of gametes of the opposing sexes is a dauermodification. This can be seen from the further culture of the remaining gametes which did not mate. The progeny of a single gamete, cultured under vegetative conditions and later induced to mate, include both "+" and "−" gametes.

In many monoecious and dioecious species each vegetative cell, even during emergence from the zygote, can be modified directly into a gamete if conditions favoring copulation are established. This is not possible, however, in species whose gametes arise as a result of a special progamic reproduction process. In *Chlorogonium elongatum*, as in *Chlamydomonas*, vegetative repro-

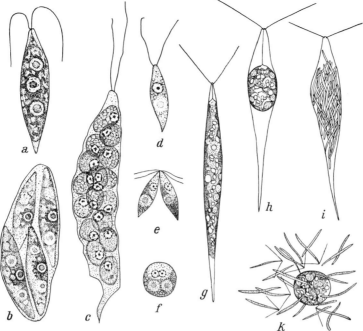

Fig. 117. *Chlorogonium*. a–f *Ch. elongatum;* a vegetative form; b agamic reproduction; c gamete formation (16 gametes); d free gamete; e copulation; f zygote; g–k *Chl. oogamum;* g vegetative form; h gynogamete formation; i androgametes; k androgametes swarming around gynogamete. (a–d after HARTMANN, 1918; e, f after DANGARD, 1899; g–k after PASCHER, 1931; all from HARTMANN, 1953)

duction as a rule involves two divisions within the cell membrane of the mother cell (Fig. 117b). During gamete formation, four or five divisions take place within the membrane, so that instead of four cells, sixteen or thirty-two much smaller ones arise, the gametes of both sexes being morphologically identical (Fig. 117c–e). Here, therefore, the conditions which trigger sexuality act on an earlier cell generation and modify the sequence of cell divisions. In the extremely oögamous *Chlorogonium oogamum* (Fig. 117g–k), the female cell responds to mating conditions by contracting its cytoplasm within the membrane into an "egg" (Fig. 117h), which then emerges from the membrane; it thus transforms as a unit into a gynogamete. The male-determined cell divides several times and produces a large number of tiny needle-shaped androgametes ("sperms," Fig. 117i, k).

In the Foraminifera there is a regular alternation of generations between the sexually reproducing gamonts and the vegetative agamonts. The gamont is haploid; it has half the amount of DNA as the agamont, according to cytochemical studies in the nuclei[756]. The gametes are flagellated or amoeboid. The agamont, which emerges from the zygote, undergoes meiosis, forming a "pseudopodiospore," which grows into a gamont. The development of a culture method by K. G. GRELL has made it possible for the entire course of development of this marine Rhizopod to be followed in detail and be studied experimentally.

In *Patinella corrugata* the uninucleate gamonts (Fig. 118a) are generally considerably smaller than the agamonts (Fig. 118j). As a prelude to sexual reproduction, several gamonts come together in an "aggregate" (gamontogamy, Fig. 118b). Two to fourteen individuals can come together in this way; in most cases, however, they aggregate in threes. As soon as the gametes

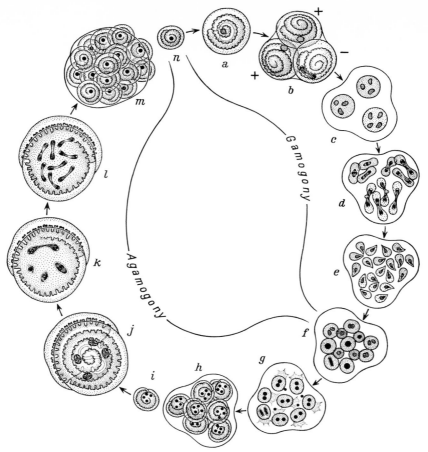

Fig. 118. Diagram of the life cycle of *Patinella corrugata*. a gamont; b aggregate of 3 gamonts (2 "+" and 1 "−"); c nuclear multiplication within the cytoplasm which has emerged from its shell; d final mitosis prior to gamete formation; e gametes; f eight gametes and four remainders; g binucleate agamonts; h eight young (tetranucleate) agamonts; i young; j mature tetranucleate agamont; k, l meiosis; m development of young gamonts. (After GRELL, 1958)

have aggregated, each lays down a delicate organic membrane which holds and binds them together and to the substrate. After several nuclear divisions the cytoplasmic contents emerge from the chorions and round up. Now a body of cytoplasm is delineated around each nucleus and this divides into two gametes (Fig. 118d). These separate and unite in pairs to form zygotes. Although the gamonts forming an aggregate appear uniform morphologically, they are differentiated sexually. Gametes unite only when they come from gamonts of opposing sexes. This can be demonstrated by the use of neutral red staining on the gamonts (Fig. 119). One can designate the gamonts as "+" and "−", just as was done with the clones of the dioecious Chlamydomonads. The "+" and "−" gametes move in an amoeboid fashion until they meet a partner of the opposite sex (Fig. 119 c_1, c_2). When an odd number of gamonts is present in an aggregate, some gametes of one sex must remain unpaired (Fig. 119 b). These unfortunates are later phagocytized by the zygotes. In each zygote two nuclear divisions take place; and in the typical case each agamont that emerges from the zygote membrane has four nuclei and a rudimentary shell

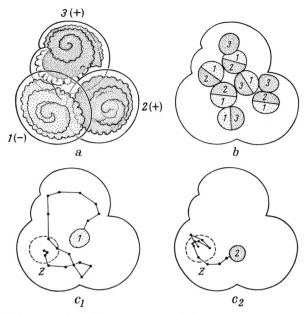

Fig. 119. *Patellina corrugata.* a aggregate of three gamonts consisting of one "−" and two neutral red stained "+" mating types; b composition of the six zygotes and two remainders; c_1, c_2 paths of two gametes on their way to fusion into a zygote (Z) from analysis of a movie. (After GRELL, 1960)

(Fig. 118h, i). The agamonts grow, and when they have reached a certain size meiosis begins (Fig. 118k, l). After these two divisions (and recall the two previous nuclear divisions) sixteen gamonts are formed (Fig. 118m).

In *Rotaliella roscoffensis* autogamy is the sexual process. While *Patinella* is dioecious, in *Rotaliella* the nuclei resulting from the division of one gamont nucleus fuse (Figs. 120a–d). The sexual difference between the gametes is thus a phenotypic alternative modification as in *Actinophrys* (p. 81) and certain Phytomonad clones (p. 85).

In the agamonts of *Rotaliella* and other "heterokaryotic" forms, the nuclei differentiate: three nuclei remain condensed, while the fourth one swells and forms a nucleolus. It moves out of the initial chamber into a succeeding chamber rich in cytoplasm. Here it grows vigorously (Fig. 120g). It is a somatic nucleus like the macronucleus in ciliates, and it disintegrates when the agamont is mature. The three remaining nuclei are generative nuclei which undergo meiosis (Fig. 120h–j) and produce the nuclei of the young gamonts. The somatic nucleus participates in a rudimentary way in the meiosis of the generative nuclei: prophase appears to begin, but the chromosomes do not pair. The metaphase chromosomes are distributed irregularly in an intranuclear spindle; then the nuclear membrane dissolves, and the chromosomes scatter into the cytoplasm and dissolve (Fig. 120h–j). The somatic nucleus thus also reacts, although incompletely, to the influences which trigger the individual events of meiosis.

The physiological significance of this nuclear dimorphism has an elegant explanation, That the somatic nucleus has to do only with the control of metabolism during the development of the individual is shown by a natural experiment: In *Rotaliella*, it frequently happens that one or more of the four original nuclei degenerate. All the agamonts with three or two nuclei form a somatic nucleus from one of them and develop normally. In the rare uninucleate individuals there are no generative nuclei (Fig. 121). The agamont develops, and when it reaches its mature size,

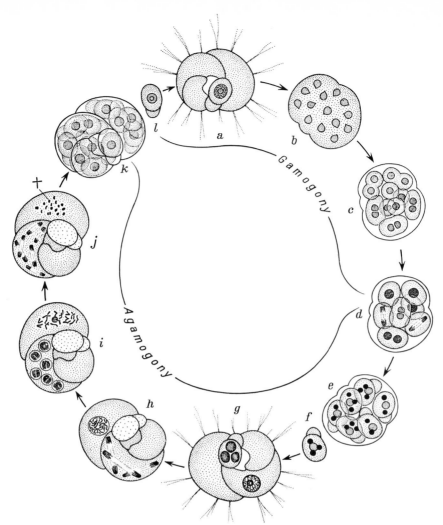

Fig. 120. Diagram of the life cycle of *Rotaliella roscoffensis*. a gamont; b gamete nuclei; c autogamous copulation; d zygotes in first division; e, f, young agamonts (some tri- and some tetranucleate); g mature agamont; h anaphase of the first meiotic division; i metaphase; j anaphase of the second meiotic division; k, l young gamonts emerging. (After GRELL, 1960)

the somatic nucleus degenerates, undergoing exactly the same rudimentary caricature of meiosis that is seen when generative nuclei are present (Fig. 121 b). At this point, the cell dies. As to the determination of nuclear differentiation, the following experiment provides one conclusion: UV irradiation with a device producing a beam only a few microns in diameter was used to induce degeneration of the somatic nuclei in *Rotaliella* agamonts (Fig. 122 a). In most cases, the cells survived, and one of the three presumptive generative nuclei acquired the typical structure and size of a somatic nucleus. From the other two nuclei, eight gamonts emerged after meiosis, while the new somatic nucleus degenerated. The somatic nucleus is thus rigidly determined, and it prevents all other nuclei from becoming somatic, presumably by chemical means. Each generative nucleus thereby retains its capacity to undergo meiosis as long as the presence of the somatic

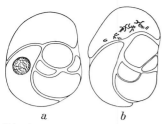

Fig. 121. Agamonts of *Rotaliella* without generative nuclei. a somatic nucleus in the penultimate chamber of the shell; b in rudimentary meiosis. (After GRELL, 1958)

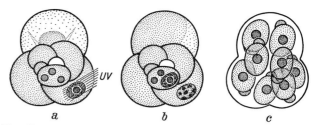

Fig. 122. Inactivation of the somatic nucleus of a living agamont by UV (a) and its further development (b, c); c eight young gamonts. (After GRELL, 1960)

nuclei prevents it from transforming into one. We know nothing of the basis of the choice of a nucleus to assume the somatic role, either in normal nuclear differentiation or in the irradiation experiment. One might imagine that one nucleus has the lowest reaction threshold to a metabolic condition in the cytoplasm during the growth of the young agamont, and that in responding to this influence the nucleus prevents the others from following suit.

Nuclear dimorphism and the course of conjugation in ciliates has long been known, but investigators are still struggling with some consequent problems. The general scheme of conjugation (Fig. 123) is simple. Two individuals come together and become connected by a cytoplasmic bridge. In a region in front of the mouth where the conjugants are in contact, the pellicle dissolves in a region about 10 microns in diameter, and there is cytoplasmic continuity. At the margins of this cytoplasmic bridge, the pellicles of the partners fuse, creating a mechanical bond between the conjugants[603]. In each of the two conjugants, the micronucleus passes through the two progamic divisions of meiosis. Of the four resulting nuclei, three perish, and the fourth

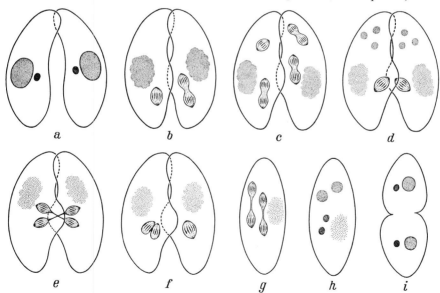

Fig. 123. Diagram of conjugation in *Paramecium bursaria*. a pairing begins at anterior end; b–d progamic nuclear divisions; e exchange of gamete nuclei; f left union of stationary and migratory nuclei, right zygote nucleus (Synkaryon); g second metagamic nuclear division; h differentiation of the macronuclear anlagen; i first cell division of the exconjugants

divides once again forming two gamete nuclei; the stationary nucleus and the migratory nucleus. The migratory nucleus passes through the cytoplasmic bridge and fuses with the stationary nucleus in the other partner, after which the two cells (now exconjugants) separate. The pellicle reforms at the site of the cytoplasmic bridge, beginning at the margin so that the exposed region closes like an iris diaphragm. While the old macronuclei (somatic nuclei) are degenerating in the exconjugants, the zygote nuclei undergo mitosis. After one or more divisions the micronuclei

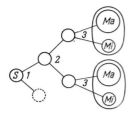

Fig. 124. Sequence of metagamic divisions from synkaryon to the differentiation of micronuclei and macronuclear anlagen in *Paramecium bursaria*

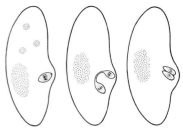

Fig. 125. Diagram of autogamy in *Paramecium*

(germinal nuclei) can be distinguished from the macronuclei (somatic), which now increase in size. Their details, the events of conjugation, and particularly the restitution of the nuclear dimorphism are quite variable. In *Paramecium bursaria* three divisions take place before the formation of a new micronucleus and macronucleus (Fig. 124). In many forms autogamy also occurs: isolated individuals pass through the progamic divisions and then the two ultimate nuclei fuse (Fig. 125).

If the stationary and migratory nuclei are considered to be of opposite sexes, conjugation must be thought of as similar to gamontogamy where dioecious sexually undifferentiated gamonts produce gamete nuclei that are sexually determined by modification as the gamont nuclei in *Rotaliella* are. But it has been established in all the species and varieties so far investigated that normally the individuals of a clone cannot conjugate with one another; rather there are mating

Varietät	Paarungs-Typ	A	B	C	D	E	F	G	H	J	K	L	M	N	O	P	Q	R	S
I	A		+	+	+														
	B	+		+	+														
	C	+	+		+														
	D	+	+	+															
II	E						+	+	+	+	+	+	+					(+)	
	F					+		+	+	+	+	+	+						
	G					+	+		+	+	+	+	+						
	H					+	+	+		+	+	+	+						
	J					+	+	+	+		+	+	+						
	K					+	+	+	+	+		+	+					(+)	
	L					+	+	+	+	+	+		+					(+)	
	M					+	+	+	+	+	+	+						(+)	
III	N														+	+	+		
	O													+		+	+		
	P													+	+		+		
	Q													+	+	+			
IV	R																		+
	S																	+	

Table 3: *Conjugation between mating types of four varieties of Paramecium bursaria. + normal paring, (+) pairing reaction followed by the death of the conjugants.* (According to experiments by JENNINGS and CHEN from SONNEBORN, 1947)

types, two of which are required for the normal union of individuals and subsequent development of the exconjugants. Within the species there are varieties whose mating types cannot conjugate (or cannot produce viable exconjugants) with individuals of any other variety. Some varieties have only two mating types, but in most varieties a system of mating types has been encountered in which every mating type can conjugate successfully with every other mating type except its own. Table 3 illustrates an example from *Paramecium bursaria*, varieties I–IV and mating types A–S. The allelic basis of some mating type distinctions has been established[267a, 629a]. Their expression in the exconjugants of a genetically identical pair is not completely understood. In any event, a modifying force plays a role. Among 139 exconjugant pairs tested for two mating types of *Paramecium aurelia*, 70 exconjugant pairs contained one cell of each type, 35 of one type exclusively and 34 of the other.

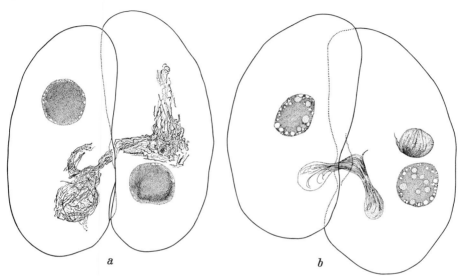

Fig. 126. Colchicine effects on conjugants; a inhibition of the first progamic division; b inhibition after the first progamic division, movement of the chromosomes from the daughter micronucleus of one conjugant into the other conjugant which belongs to an amicronucleate strain. (After EGELHAAF, 1955)

When a variety has only two mating types, they can be designated "+" and "−" and considered as sexually differentiated gamonts as in *Patinella*. But the existence of more than two mating types remains unexplained. If the difference between mating types is considered to be a genetically determined "multiple sexuality," it is necessary to assume a second step in the sexual determination of the gamete nucleus dependent on other factors.

Two developmental periods in the course of conjugation are now of interest. *Paramecium bursaria* may serve as an example. This species is easy to raise in the laboratory: Because of their symbiosis with green algae, the paramecia, and even fragments of cells, can be maintained autotrophically under constant conditions; *P. bursaria* has only one micronucleus and one macronucleus, and there is no fragmentation of the old macronucleus at the time of conjugation. Amicronucleate strains also occur; these are particularly useful for certain experiments.

We shall first consider the movement of the migratory nucleus into the opposite cell in conjugation. The fusion of the conjugants takes place in a region near the mouth where a special bulge, the "paraoral cone," arises during meiosis. Here the pellicle dissolves and the migrating nuclei cross the bridge of cytoplasm. The fourth nucleus, which remains after the second progamic

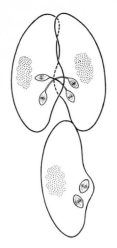

Fig. 127. Diagram of pairing involving three individuals

division, is situated in the region of the paraoral cone, where it produces the gamete nuclei (Fig. 123d, e). During the migration of the nucleus, a push by the spindle may participate in normal cases, but this is no more necessary than the presence of differentiated gamete nuclei. This is shown by experiments in which mitosis before the first and second progamic divisions is inhibited with colchicine. The micronuclear chromosomes form a loose tangle and move irregularly through the cell with various cytoplasmic currents, occasionally arriving at the paraoral cone (Fig. 126a). When this happens, nuclear migration can actually take place (Fig. 126b). Evidently at some time during the course of conjugation, the oral region must be the goal of some cytoplasmic movement whose orientation is not influenced by colchicine and which normally determines the direction of the spindle of the third division.

In autogamy, too, this division takes place in the region of the paraoral cone (Fig. 125). Once in a while, an additional partner of the appropriate mating type can attach to exconjugants (Fig. 127). Then it also forms a paraoral cone, and a second progamic division spindle is attracted to that region.

Again and again we encounter cytoplasmic conditions which cause trigger nuclear processes, as well as cytoplasmic movements, with autonomous rhythms which normally accompany nuclear events and have a decisive effect on them.

In the exconjugants, after the third metagamic division, the differentiation of the micro- and macronuclei begins. The lengthwise orientation of the spindles (Fig. 123g) places two nuclei anteriorly and two posteriorly (Fig. 128a). The macronuclear anlagen appear first in the anterior half of the cell (Fig. 128b). If the exconjugants are cut transversely, extra nuclear divisions are brought about in both pieces; and in 83 cell fragments surviving the operation, about 50 per cent

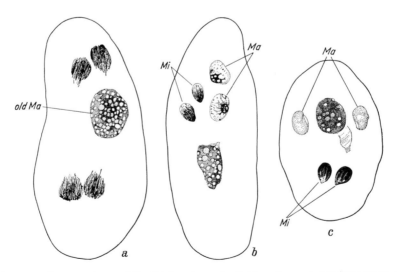

Fig. 128. Exconjugants. a late anaphase of the third metagamic division; b somewhat later stage, micronuclei and macronuclear anlagen; c posterior half from a transverse section experiment. *old Ma* old macronucleus; *Ma* macronuclear anlagen; *Mi* micronuclei. (After EGELHAAF, 1955)

94

of the nuclei in anterior pieces and about 50 per cent in posterior pieces became macronuclear anlagen, which were generally found in the anterior portion of the fragment (Fig. 128c).

These observations prove that the micronuclei are not yet completely determined after the three metagamic divisions; rather it is an anteroposterior polarity of the cytoplasm which appears to determine the nuclei. In fragments containing neither the entire old macronucleus nor a part of it, macronuclear differentiation is delayed or does not take place at all. Perhaps materials from the breakdown of the old macronucleus are required in the construction of the new one.

The development of polyploidy in the macronucleus, which has been seen to consist of clear endomitotic steps in other Ciliophora (Suctoria)[218], has not been observed in *Paramecium*.

Lecture 7

A great diversity of modifications in one individual is made possible by an increase in size of the organism. This is generally achieved by multicellularity; but in nature there are also good-sized organisms which are not organized into cells, but in which a variety of "organs" and reproductive events occur in a single individual. Most often this increase in size is connected with the presence of many nuclei in an undivided cytoplasmic mass; whether the acellularity of such syncytia is phylogenetically primitive or derived from cellular organization does not concern us here.

The Phycomycete *Saprolegnia* is a good example. This fungus grows into a mycelium composed of branched tubular multinuclear hyphae without cross walls (Figs. 129, 130a). Close to the

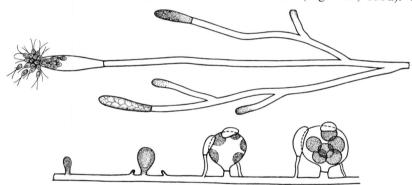

Fig. 129. Diagram of the developmental events of *Saprolegnia*. a Formation of zoosporangia at the hyphal ends; b formation of sexual complexes (oogonia and antheridia) in centrifugal sequence

cellulose wall lies the cytoplasm with fairly uniformly distributed nuclei; the interior of the hypha is filled with fluid (Fig. 130a, b). *Saprolegnia* is an aquatic saprophyte found mainly on dead insects or decaying plant material. Reproduction is both asexual, involving flagellated swarmers or zoospores which arise in club-shaped zoosporangia, and sexual, where eggs formed in oogonia are fertilized by male antheridia. Since the classical experiments of KLEBS beginning in 1896, *Saprolegnia* has served as a model for the control of developmental events by external influences. These influences on the development of individual stages can be determined precisely in pure culture, either on a firm nutritive substratum (2 per cent agar, 2 per cent meat extract) or in water using heat-sterilized ant pupae as food. When a pupa is encountered by zoospores in sterile tap water, a course of development follows which is customary in nature: the mycelium grows out radially into the water from the dead pupa. The duration of vegetative growth is

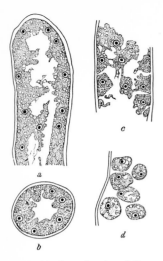

Fig. 130. *Saprolegnia*. a, b Longitudinal and cross sections of a hyphal tip; c, d formation of zoospores. (After Davis from Harder, 1934)

dependent on the concentration of nutrients present, in this case on the size of the ant pupa which the mycelium pervades. When the mycelial diameter reaches 10–15 mm., growth stops. This takes 24 to 30 hours at 15°C. At the ends of the hyphae zoosporangia begin to form (Fig. 129 a): the hypha swells terminally, and a cross wall separates the swollen end from the rest of the hypha. The zoosporangial anlage grows, and its cytoplasm divides into uninucleate regions, which round up (Fig. 130 c, d), sprout two flagella, and become pear-shaped. At the hyphal tip the wall of the zoosporangium dissolves, and the swarmers emerge. From the third day on, while zoospore formation is proceeding, a centrifugal wave of formation of female sex organs begins in the center of the mycelium (Fig. 129 b). Lateral outgrowths of the hyphae arise which first become pear-shaped and later round. When they are full-sized, a wall separates them from the hypha. The cytoplasm of the young oögonium contains many nuclei (Fig. 131 a,b). At first it fills the entire volume, but later a vacuole forms in the interior. In the cytoplasmic coating of the wall, most of the nuclei gradually degenerate (Fig. 132 a); those nuclei that remain after a certain time has elapsed embark on synchronous division (Fig. 132 b). Again most of the daughter nuclei break down. In between the nuclei which do persist the cytoplasm constricts (Figs. 131 c, d; 132 d, e). The centriole clearly plays a role here: the astral spheres of the last division extend out quite a distance from the nuclei, which remain intact; their rays reach the cytoplasmic surface. We have already seen the importance of astral rays in the delimitation of cytoplasmic regions (pp. 61, 64), and one might well suspect that this last nuclear division in the oogonial cytoplasm, most of whose daughter nuclei are eliminated, takes place only in order to activate the

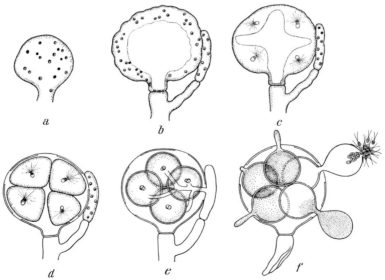

Fig. 131. Development of an oögonium and an antheridium; fertilization; formation and germination of zoosporangia. *Saprolegnia; diagrammatic*

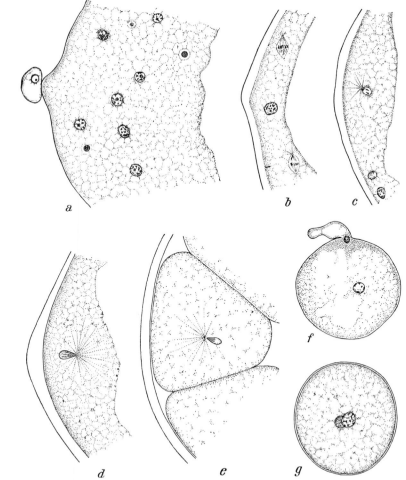

Fig. 132. Development of the egg cell and fertilization in *Saprolegnia monoica*. (After CLAUSSEN, 1907)

centrioles for this task. The mature eggs round up. Their number, depending on the size of the oogonium, varies from two to more than thinty. The antheridia grow from the hypha near the site of the oogonium or else from the oogonial stalk itself (Figs. 129b, 131). The end of this tubular outgrowth is separated from the basal portion by a cross wall. The antheridium becomes apposed to the oogonium. The numerous nuclei in its cytoplasm undergo a division more or less synchronously with those in the oogonium. When the egg cells have formed, the antheridium sends a tube into the oogonium (Fig. 131e). As a rule, especially thin places in the oogonial wall are used as ports of entry. The antheridial tubes branch out more or less strongly according to the number of eggs. In the egg, thick heavily staining cytoplasm accumulates at the point of contact with the tube; then the antheridial tube opens and releases one of its nuclei into the egg (Fig. 132f). During the fusion of the male and female nuclei the zygote surrounds itself with a membrane which is thin at first, but which later thickens substantially (Fig. 132g). Chromosomal reduction takes place during germination of the zygote. *Saprolegnia* is a monoecious haplont.

In most *Saprolegnia* species the encysted zygote passes through a considerable resting period. This diapause may be broken by a single episode of freezing.

97

Particular environmental influences evoke the various developmental processes in *Saprolegnia*: If the zygotes germinate in water, the germinating hypha forms zoosporangia as soon as it has pierced the oogonial membrane (Fig. 131f). On nutrient agar zoospores grow into a star-shaped spreading mycelium which can be maintained permanently in the vegetative state. If the hyphae are moved from the agar into pure water at not too low a temperature, they form zoosporangia at once. Zoosporangium formation can, however, be prevented completely: at low temperatures (3° C.) ant pupae placed in water culture cause the formation of sex organs after about ten days of vegetative growth. If the temperature is raised from a moderate to a higher level (e.g., raised

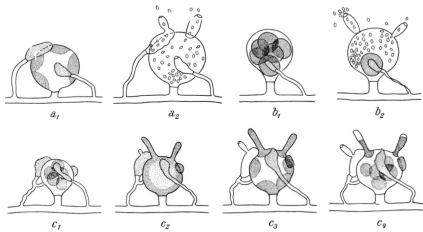

Fig. 133. Modification by temperature change of the sex organs of *Saprolegnia* in liquid culture. a, b Transformation of oogonia and antheridia to zoosporangia after a sudden rise in temperature; c results of successive changes of temperature: c_1, c_2 after warming; c_3, c_4 after cooling. (After SCHLÖSSER, 1929)

from 15° to 22°) after the period of zoosporangium formation is over and the formation of sex organs is in full swing, the development of the oogonia and antheridia terminates even when they are in late stages, and vegetative growth is resumed. Even when the phase of nuclear degeneration has passed and the cytoplasm has accumulated around each of the remaining oogonial nuclei (Fig. 133a_1, cf. Fig. 131c), indeed even after the egg cells have rounded up and are ready for fertilization (Figs. 131b_1, c_1), the cytoplasm in the oogonium flows back into a single mass. Rapid nuclear division follows, and the cytoplasm now divides into zoospores. Tubes pierce the oogonial wall and give rise to zoosporangia, from which swarmers soon emerge (Fig. 133a_2, b_2). The antheridia also develop zoosporangial tubes (Fig. 133a_2). Only when the egg is already in contact with an antheridium is zoospore formation no longer possible for either of them (one egg in Fig. 133b_2). The reversal of oogonia and antheridia back in the direction of vegetative reproduction can also be stopped and even reversed again: in the experiment of Figure 133c the eggs have fused after warming, and the outgrowth of zoosporangial tubes from the oogonium and from an antheridium has begun (Fig. 133c_2); now they are cooled again, and immediately the mass of cytoplasm separates into eggs again (Fig. 133c_3, c_4). Only one antheridium was already too far along; it released swarmers. If individual oogonia are cut away from the hypha and placed on nutrient agar, their egg anlagen also fuse, the egg nuclei multiply, and vegetative hyphae sprout (Fig. 134a_1–a_3). Isolated antheridia also give rise to a vegetative mycelium on nutrient agar (Fig. 134b).

In the oogonial anlagen, therefore, environmental influences affect the behavior of the nuclei, doubtless by means of functional changes in the cytoplasm. It is surprising that even the egg nucleus can be made to divide again. Even more mysterious, however, is the behavior of the nuclei during the normal course of oogonial development. What causes the degeneration of most of the nuclei? And why are some nuclei spared, and why exactly as many as the eggs which will result from the division of the cytoplasmic mass?

Mycelia obtained from oogonia, as well as those from antheridia, can produce both kinds of sex organs. Sexual differentiation is, therefore, only transiently related to organ formation, and the determination of this organ formation is extremely labile; it can be triggered by environmental factors and similarly can be set off in another direction.

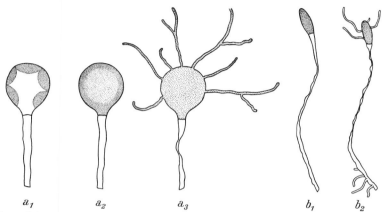

a_1 a_2 a_3 b_1 b_2

Fig. 134. Formation of hyphae from sexual organs after transfer of oogonia (a) and antheridia (b) from liquid culture to nutrient agar. (After SCHLÖSSER, 1929)

It is noteworthy that in the formation of zoosporangia, oogonia, and antheridia, a crosswall always cuts off the remaining syncytial cytoplasm. In this way particular metabolic processes are isolated in the organ anlagen, which are in a labile state of determination.

Between an oogonium and its attendant antheridia exists a direct developmental relationship. The antheridia first arise in the mycelia when the oogonia have reached a certain stage of development, and the antheridial tubes arise only in close proximity to the oogonia. They bend toward the oogonium and make contact with it, ostensibly chemotropically attracted by some stimulus released by the oogonia. Experiments show further that the site of formation and the course of development of the antheridia are also controlled by the oogonium: next to an oogonium the developing antheridia are removed with a fine glass knife. Then the oogonium is brought near a hypha or the same or another mycelium which does not yet bear sex organs (Fig. 135a). Now, antheridia arise on the foreign hypha next to the oogonium, if the mycelium has reached a certain age (Fig. 135a_2, a_3).

The triggering of a developmental process in one part of an organism by another part is what we call induction. The triggering agent is called the inducer, and we say that the resulting reaction is induced. The induced response is one of the array of activities of which the induced part is capable. The inducing stimulus which emanates from the oogonium and exercises its effect at a distance through the water can only be a chemical agent; its nature is still unknown.

The release of this substance, which induces the formation of the antheridia and attracts them, is limited in time; it lasts only until a number of antheridia have attached to the female organ. Then the chemotropic agent disappears. This is shown by antheridia which have not yet become attached and now turn to a younger oogonium of a nearby hypha or degenerate (Fig. 135b). The antheridia reach maturity only if the oogonial stimulus continues to work on them. If one

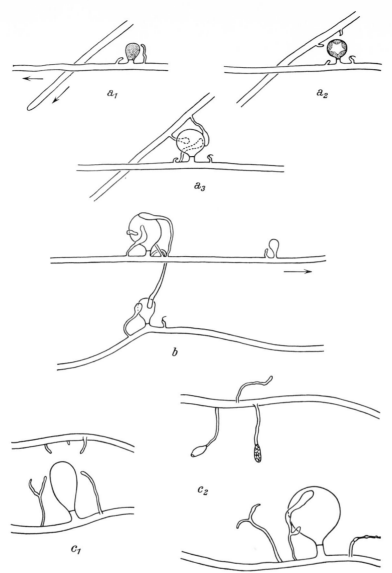

Fig. 135. Induction of antheridium formation and determination of the direction of growth of the antheridial tubes by the oogonium. c_2 Transformation of the antheridia into zoosporangia after cessation of oogonial effect. Arrows point toward hyphal tips. (After Schlösser, 1929)

removes the inducer from the hypha on which it has induced the formation of male organs, the antheridial tubes do not proceed with the formation of a terminal antheridium separated by a crosswall; instead they become vegetative hyphae at low temperatures or change over to zoosporangium formation if warmed (Fig. 135c_2).

In order for the hypha to respond to the inducing stimulus, i.e., in order for antheridium formation to be evoked, the hypha must be in a particular physiological condition. Using growing oogonia whose associated antheridia have been cut off, hyphae of other mycelia of varying ages and in varying developmental states were tested to see whether they could form antheridia. In this way it was found that in mycelia where formation of the sex organs is just beginning in the center, hyphae in the peripheral quarter do not form antheridia; but in those central parts which do not yet have sex organs antheridia can be induced. Also in an area of hypha where oogonia of all stages of maturation and their associated antheridia are already present, one can induce new antheridia in between those already present by bringing up an oogonium on another hypha. This physiological condition, in which a particular developmental reaction is inducible, is called competence.

These experiments show that starting from the growth center of the mold on a nutrient substrate two successive changes in the physiological condition of the cytoplasm proceed centrifugally in the hyphae: first, competence is acquired to form antheridia in response to oogonial induction, and second, the ability to form oogonia at certain intervals along the hypha. The oogonia then complete the dioecious sex organ complex by means of induction.

Vegetative growth, formation of zoosporangia, and formation of sex organs follow in sequence during the course of normal mycelial development after the colonization of a source of food and the subsequent changes in conditions caused by outgrowth of the hyphae into the water and the exhaustion of nutrients. The adaptation to the natural habitat is expressed in these developmental reactions: in water depleted of nutrients, the production of swarmers which can seek out new food supplies; in the established mycelium after exhaustion of the food supply, the formation of encysted zygotes, which can live a long time without food and which are in general very resistant to environmental factors.

Saprolegnia is a thoroughly plastic and open system. Each piece of mycelium can begin a new course of development including all possible modes of reproduction. The individual developmental events are, as in the unicellular organisms described previously, responses to environmental influences. These can be created experimentally and evoke the same sequence of modifications in *Saprolegnia* that is found in nature. Thus, we can demonstrate in simple cases, in answer to our second and fourth questions in Lecture 1 (p. 4), a set of reactive competences and the temporal and spatial arrangement of conditions which bring about a typical course of development in the organism.

In *Saprolegnia* the one-way path of development of this open system is controlled only through the progressive alteration of environmental conditions. In other organisms development proceeds toward a particular final state of a closed system, even under constant conditions, so that a typical form and reproductive process are reached, and the end of an individual life will be brought about through a series of internal conditions. Nevertheless, the pace at which the final condition is approached and the formation of organs in the individual can still depend substantially on external conditions.

Certain highly differentiated algae which are not divided into cells provide some well-studied examples. In *Bryopsis* (Siphonales) erect shoots grow up from creeping filaments, which attach to the substrate as rhizoids. From the main shoot lateral branches grow out on both sides in a single plane (Fig. 136). The whole *Bryopsis* plant contains a large vacuole surrounded by multinucleate cytoplasm along the cell wall. In the main shoot and in its branches are numerous chromatophores; at the base of the stalk they are sparse, and there are none in the rhizoids. When this branched stalk has reached a certain size, and certain temperature and light conditions prevail, the lateral branches transform into gametangia. This process begins with the lowest branches and proceeds toward the tip of the stalk. Walls form at the bases of the branches, and after vigorous nuclear division the cytoplasm is divided up, each portion containing one nucleus and one chromatophore. These units become pear-shaped, form two flagella, and swarm out as gametes. The empty walls of the branches fall off, and finally the whole stalk dies. The zygotes

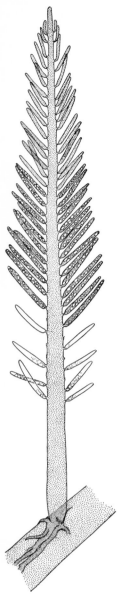

Fig. 136. *Bryopsis cupressoides*. The intermediate branches are gametangia; below are empty gametangia. (After OLTMANNS, 1922)

can germinate at once. They grow out into tubes which crawl along the substratum, branch, and then send up a stalk.

The *Bryopsis* plant thus develops apically differentiating branches behind the growing tip of the stalk and basally differentiating rhizoids below. Thus, we confront the problem of the determination of polarity, the formation of diverse organs at opposite ends of an axis.

Experiments show that the polarity of *Bryopsis* can be inverted. If a *Bryopsis* plant is placed upside down under ordinary light conditions, the downward-pointing tip of the stalk, together with its already formed branches, changes its direction of growth, grows upward, and becomes an erect stalk. But if such a developing plant is inverted in sand, the stalk tip and branchlets turn into rhizoids. At the base of these newly formed rhizoids erect stalks can now form (Fig. 137 a_2) and replace the old stalk. Frequently the old basal pole redifferentiates into an apical one: The old rhizoids, now free in the water, shrivel up, fall off and are replaced by a normal apical branched structure, while rhizoids are growing out of the original apical end (Fig. 137 b_3–b_5). Here it does not matter whether a young plant has already formed branches (Fig. 137 c) or not (Fig. 137 b). The chromatophores are always carried by cytoplasmic streaming from those parts of the plant which redifferentiate into basal portions, and they collect in the new apical regions (Fig. 137 a, b). The direction of growth of the stalk and of the rhizoids is determined by gravity. If a stalk fragment from which the basal and apical ends have been cut is laid flat, growth begins at both cut edges: at both ends a stalk grows upward, and rhizoids grow downward. Apical differentiation is dependent on light. If the upper part of a plant is shielded (e.g., with tinfoil), the apex and upper branches turn into rhizoid-like threads. The chromatophores move down as far as the rhizoid, from which new stalks grow straight up. The stalk region between the apical growth region and the rhizoids cannot form branches even when illuminated. The plant has zones of differentiation at both ends, whose morphological character is alterable.

The movement of the cytoplasm in addition to the chromatophores is shown by the vital staining of cytoplasmic granules. If the apex is stained before inversion, the stain is later found in the original basal end (Fig. 137 c, d). If the rhizoid end is stained before inversion, the stain moves into the original apical part, which is now forming rhizoids (Fig. 137 e_1, e_2). These experiments show that some apical cytoplasm and some basal cytoplasm change take place as the polarity is reversed, suggesting that these localizations of cytoplasm are causally related to organ differentiation. This conclusion is supported by the distribution of the migrating apical cytoplasm into the new apical part of the stalk: the vital stain concentrates at the growing tip and in the regions just behind it which correspond to the sites of formation of branches (Fig. 137 d_2–d_4).

Deeper insight into the determination of polarity and organ formation is provided by the "umbrella" alga *Acetabularia* (Siphonocladiales). Whether one considers *Bryopsis* as a closed system is arbitrary; even though the pinnate stalk dies after gamete formation, the creeping

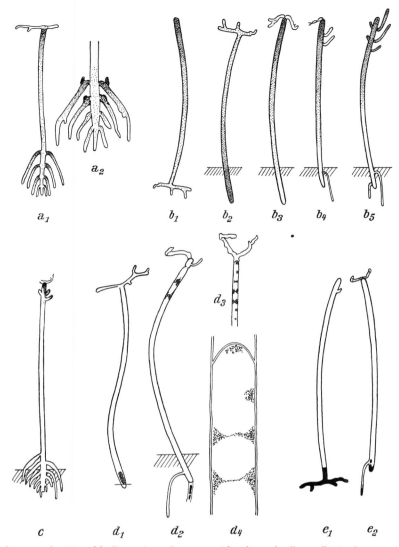

Fig. 137. Inversion experiments with *Bryopsis:* a *B. cupressoides;* b–e stippling reflects chromatophore concentration; c–e vital stain; d_4 upper portion of d_3 enlarged, showing the crosswall sealing off the degenerating rhizoid segment. (After STEINICKE, 1925)

rhizoids survive below, and from them grow new stalks, forming a constantly renewed lawn of algae. One can consider the pinnate stalk as an organ of assimilation and sexual reproduction, and thus simply a part of the entire organism derived from a single zygote; or it can be considered as an individual alga which reproduces asexually via rhizoids. In *Acetabularia* the course of development is unequivocal: from each zygote there arises only one plant with a closed course of development. *Acetabularia* has a remarkably favorable array of properties for developmental investigation: the whole plant is, in contrast to the otherwise multinucleate Siphonocladiales, a uninucleate giant cell. It has great powers of regeneration. The complexity of organ formation in the different species of the genus and the possibility of making heterologous transplants offer

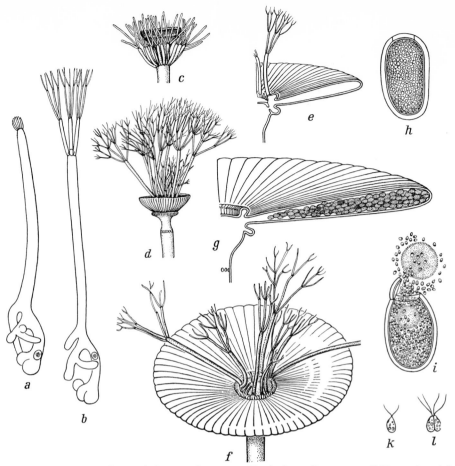

Fig. 138. Development of *Acetabularia mediterranea*. (Partly from Oltmanns, 1922, partly original)

extraordinary points of departure. *Acetabularia* has been the subject of an outstanding series of highly informative experiments by Hämmerling and co-workers.

First we shall follow the normal development of *Acetabularia mediterranea*. The zygote attaches at its flagellar pole, germinates, and forms an upright stalk, and a rhizoid which attaches to the substratum. The primary polarity apparently results from a polarity in the zygote. After the stalk has reached a certain length a tuft (hair tuft = sterile tuft) forms just below the growing point. This tuft's fibers branch dichotomously two or three times (Fig. 138a, b). With the further growth of the stalk, one tuft after another is formed near the tip and spread out. Each tuft lasts only a short time. The old ones shrink and fall off, so that only the uppermost one or two are generally found on a plant. When the alga is fully grown, a fertile tuft is formed out of gametangia (Fig. 138c–g). These grow together as contiguous chambers of the umbrella, or hat. Above the hat still another sterile tuft is formed, which later dies away also. The fibers of this uppermost tuft arise from upward projections at the base of each gametangium. These projections are arranged in a circle, the corona, which corresponds to another wreath of projections, bearing *no* fibers, below the hat (Fig. 138e, g). The zygote nucleus remains in the rhizoid during germination and the growth of the cell (Fig. 138a, b). As the cell becomes larger the nucleus grows

104

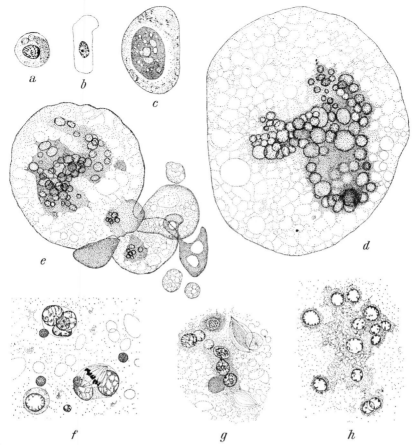

Fig. 139. Nuclear processes in *Acetabularia wettsteinii* (a–e, g, h) and *mediterranea* (*f*). a zygote; b young individual; c, d primary nuclei in the rhizoid; e beginning of fragmentation of the primary nucleus; f, g secondary nuclear divisions in the rhizoid; h secondary nuclei in the stalk. a–d, h 1,600×; e 1,050×; f 1,500×; g 800×. (After Schulze, 1939)

vigorously, forming a substantial nucleolus (Fig. 139a–d); nevertheless, the nucleocytoplasmic ratio decreases progressively, and in the fully grown plant the nucleus is small indeed compared to the entire mass of the cell. In this mature plant, where the hat has reached its full size, the giant nucleus or primary nucleus in the rhizoid region breaks up (Fig. 139e). Its form becomes irregular as it pinches off large and small fragments. Somewhere in the disintegrating nuclear mass chromosomes assemble and undergo a series of rapid divisions, resulting in the production of thousands of secondary nuclei. The first mitotic figures and secondary nuclei often lie within a sharply delineated membrane which is probably derived from the the substance of the primary nucleus (Fig. 139f). Individual spindles without chromosomes are often found (Fig. 139g). They may well stem from extra divisions of the centriole of the primary nucleus. When the rhizoid is filled with secondary nuclei, these begin to be moved upwards in dense packs by cytoplasmic streaming. Chromatophores are also carried up in the cytoplasmic stream: the rhizoid and stalk, deep green up to now, become pale. This upward streaming is quite fast. Finally almost all the cytoplasm with the secondary nuclei and chromatophores enters the gametangial chambers. Here, however, gametes are not formed directly, but first uninucleate "cysts" appear (Fig. 138 g, h);

now several series of further divisions of the cyst nuclei give rise to the very small gametes (Fig. 138 i–l). *Acetabularia* has no asexual reproduction. Its multiplication is assured by sexual reproduction, whose productivity is extraordinarily high: in the hat of *Acetabularia mediterranea* some 15,000 cysts form, with about 1,800 gametes per cyst; thus the formation of 13,500,000 zygotes is theoretically possible (15,000 × 1,800/2), that is, a single unicellular plant could give rise to about 13 million progeny.

The rhizoid, stalk, sterile tuft, hat, and cysts are the morphologically differentiated components of the individual plant. They are formed in a specific sequence. The tuft and hat invariably

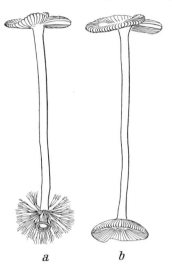

a   *b*

Fig. 140. Regeneration in an anucleate piece of stem about 1 cm long. *Acetabularia mediterranea*, 5 × . (Original drawing by FREIBERG after preparations by HÄMMERLING)

arise at the tip of the stalk, never below. Each formation of tuft and hat therefore follows stalk formation. With the formation of cysts, the development of the richly differentiated cell comes to an end. Cytoplasm and nuclear material enter the cysts and then the gametes. *Acetabularia* is in terms of its cytoplasmic mass the largest known cell. The stalk reaches a length of 5 cm, in laboratory culture; the hat occasionally becomes 1 cm. in diameter. In nature the plants are sometimes substantially larger. If the apical end of an *Acetabularia* is excised at any level, the basal portion almost always regenerates a new apical end, even when a full-sized hat as already been formed. After closure of the wound the stalk grows a bit further and then forms at least one new sterile tuft before a hat arises.

It was very surprising to find that in *Acetabularia* even anucleate pieces of stalk could regenerate; and their life span is astonishing, averaging about three months and reaching six to seven months in extreme cases. During this time the anucleate fragments grow vigorously, even when they are very small (0.25 or 0.1 cm., Fig. 142 a, b), and they are capable of morphogenesis. Thus, the living substance increases and the extensive new surface is covered with a cellulose membrane. After exposure to radioactive CO_2, the incorporation of carbon into protein in anucleate fragments could be demonstrated[78]. As the cytoplasm increases, the chromatophores multiply. This is shown by the regreening of pieces which have lost much cytoplasm with chromatophores after being cut. It can be observed directly also: At the extreme tip of the growing stalk or in tufts and incipient hats, the chromatophores are much smaller than in nongrowing parts. Thus, they are not being supplied from the lower regions. In any event, there is a relatively strong increase in cytoplasmic mass in the regions of morphogenesis. And although slow cytoplasmic streaming, seen in the movement of chromatophores, does take place everywhere, one never sees a steady flow of cytoplasm in the direction opposite to that of the rapid streaming characteristic of the time just before cyst formation. In the cytoplasm of the anucleate cell fragments of *Acetabularia*, therefore, factors are present which support the metabolism necessary for maintenance and growth for a considerable time and even permit a certain amount of morphogenesis.

Anucleate stalk fragments undergo apical differentiation at the undisturbed apical end of the stalk or at the cut apical surface, if there is one; they can also form apical structures at the cut basal surface. The structures produced in this reversal of polarity are called heteromorphoses. On apical stalk fragments at least 1 cm. long, typical hats arise in many cases after the usual preliminary formation of typical sterile tufts (Fig. 140). Often only incomplete apical structures arise, stunted sterile tufts or hats. Morphogenetic ability is dependent on the region and length of the stalk fragment. Figures 141 and 142 summarize the results of a large series of experiments. Apical stalk sections 1.5 cm. long always regenerate apical structures, and 29 per cent do so also at the basal cut surface (Fig. 141). With increasing distance from the apical end, similar-sized pieces lose the ability to differentiate apical structures, first slowly, then sharply. Basal halves regenerate dwarf structures, and even then only in half the cases; relatively few regenerate at the basal surface. Again, all apical pieces 0.5 cm. long regenerate; proceeding basally this ability is rapidly lost (Fig. 142d, e). 0.25-cm. pieces are even less competent (Fig. 142b, c); but even tiny

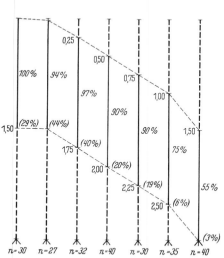

Fig. 141. Summary of a series of regeneration experiments with anucleate stem fragments of *Acetabularia mediterranea*. Diagram of a standard 3 cm. plant; test piece always 1.5 cm. long, at varying distances from the apex. n = number of pieces tested; per cent of regeneration of apical structures; in parentheses, per cent of regenerates at cut basal surface. (After the results of HÄMMERLING, 1934)

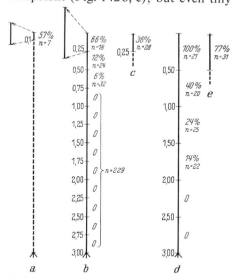

Fig. 142. Summary of regeneration experiments with stem pieces of varying length and at varying distances from the apex (*Acetabularia mediterranea*). Left in a and ab, increase in length of the particular pieces; in c and e, the extreme tip has been cut off; in a, b and d the tip is untouched. n = number of experimental pieces; per cent of apical differentiation is given. (After the figures of HÄMMERLING, 1934)

0.1-cm. pieces from the apex manifest the first steps toward regeneration (Fig. 142a). If the extreme growing point is cut off, morphogenetic ability is strongly reduced (Fig. 142c, e) At the same time, the tendency for apical differentiation is shifted toward the basal surface by amputation of the growing point: in pieces 0.5 cm. long, 21 anteriorly uninjured apical fragments (Fig. 142d) formed one dwarf tuft, sixteen typical tufts, and four dwarf hats apically; basally, only three typical tufts were formed. From 24 of 31 apical pieces from whose growing tip had been amputated seven dwarf tufts, eight typical tufts, and four dwarf hats arose apically, and four dwarf tufts, sixteen typical tufts, and six dwarf hats were formed at the basal end (Table 4;

Table 4: *Apical pieces of* Acetabularia mediterranea *0.5 cm long.* (After HÄMMERLING, 1934)

Regenerates	Intact tip anterior	posterior	Extreme tip amputated anterior	posterior
Dwarf tuft	1	0	7	4
Typical tuft	16	3	8	16
Dwarf hat	4	0	4	6
	21	3	19	26

107

ostensibly some individuals formed structures at both ends and sometimes both a tuft and a hat in succession at the same end).

Rhizoid regeneration has also been observed (Fig. 143). In 1.5-cm. pieces whose lower surface was immediately above the rhizoid, the proportion of rhizoid regeneration was greatest. Over the first 0.5 cm. distance from the base, the percentage dropped sharply, whereas the incidence of rhizoid formation at the upper surface actually rose slightly; 0.25-cm. pieces regenerated rhizoids only in a few cases, and then only when they were derived from the immediate vicinity of the rhizoid.

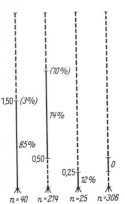

Fig. 143. Summary of rhizoid regeneration in anucleate stem pieces of *Acetabularia mediterranea* of different lengths and varying distance from the base. n = number of experimental pieces; percentage of regenerates; in parentheses, those at apical end. (After the figures of HÄMMERLING, 1934)

These results show that the polarity of morphogenesis in *Acetabularia* is not based on a polar axis which runs through the whole cell. If it did apical differentiation would always have been seen at the apical surface, and rhizoids would have formed only at the basal surface of the cut pieces. The circumstances of the particular morphogenesis observed indicate a gradient: Conditions favoring apical differentiation are maximal apically and decrease toward the base, while those favoring rhizoid formation are distributed in the opposite direction. This slope is expressed in the frequency and degree of morphogenesis at a cut surface. Competence for complete apical morphogenesis is maximal at the growing point, but it extends past the middle of the stalk (Figs. 141, 142). Competence to form rhizoids also extends past the middle of the stalk (Fig. 143). Thus, in one part of the stalk both competences overlap, and this leads to a variety of results: apical or basal differentiation alone, or one type on each cut surface, either in the expected orientation or the reverse.

The morphogenetic ability of a section is not solely dependent on the level of the cut. Morphogenesis on the posterior cut surface of any given section is better than at the anterior cut surface of the next section. Morphogenesis at a cut surface depends therefore on the quantity of morphogenetic factors in the whole piece; these accumulate at the site of morphogenesis, i.e., they migrate to any cut surface and especially to the uninjured growing point. This is shown clearly by the rise in heteromorphoses in pieces of the apical end whose tips have been amputated (Fig. 141, Table 4). Here anterior and posterior cut surfaces appear equally attractive to the morphogenetic substance. These transportable substances are contained in the cytoplasm, but are not identical with the cytoplasm of any particular region. This emerges from the fact that there is no general cytoplasmic movement toward the sites of morphogenesis in *Acetabularia* and from the fact that the cytoplasm increases appreciably in amount during growth and morphogenesis, although the morphogenetic substances clearly do not: the morphogenetic capacity depends on the initial size of the section (Fig. 142).

These experiments suggest that there are at least two kinds of morphogenetic substances: apical and rhizoid substances, which are distributed in growing plants in two opposing concentration gradients. This is diagrammed in Figure 144. The apical substance causes the formation of sterile tufts and hat anlagen, whether the cut surface is anterior or posterior; the rhizoid substance operates similarly in the formation of rhizoids. A large amount of morphogenetic substance results in good morphogenesis, with less results in dwarf or incomplete structures. When both substances are present in a stalk section, regeneration never results in intermediate structures but rather only one of the alternatives, presumably depending on the relative concentrations of the two substances in the section.

108

In an *Acetabularia* in which 0.5-cm. apical pieces showed only slight differentiation and similar anucleate basal pieces showed almost none, transplantation of such a basal piece to the basal cut surface of the apical piece markedly raised the degree of regeneration at the apical end. This suggests that in the basal region, inactive precursors of the active morphogenetic substance are present, which in the combined stalk fragments move to the regenerating point and are there activated[584].

The primary polarity, differentiation into rhizoid and stalk, is dependent on the nucleus: a rhizoid is formed near the nucleus. This is shown by rare cases in which middle or basal pieces

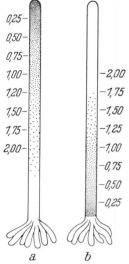

Fig. 144. Diagram of the concentration gradients of morphogenetic substances; left, for apical substance; right for rhizoid substance. Distances in centimeters from apex and base, respectively, are given. (After HÄMMERLING, 1934)

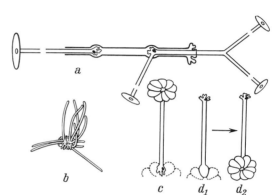

Fig. 145. Formation of rhizoids and apical end from pieces of *Acetabularia* plants with hats. Primary nuclei derived from secondary nuclei (migrating nuclei left in the stalk). (After HÄMMERLING, 1955)

isolated from hat-bearing plants grow into entire new plants. In the stalk some secondary nuclei always remain behind, instead of moving into the hat: New plants arise which develop from a piece of the mother plant in which secondary nuclei transform into primary nuclei instead of forming cysts, gametes and zygotes. The migrating nuclei are therefore not yet completely determined for a particular path of further development. Without regard to the original polarity of the piece, a rhizoid forms near the nucleus, and from this rhizoid grows a stalk (Fig. 145). The new plants separate as the cytoplasm contracts and forms a membrane separating it from the rest of the cytoplasm in the maternal portion. From a multinucleate piece a whole bundle of stalks can spring up (Fig. 145b). Such direct offspring can even emerge from abnormal hat chambers in which cyst formation does not take place; and to be sure, the rhizoid is then formed in a chamber (Fig. 145c) or at the end of an emerging stalk (Fig. 145d), according to the location of the new primary nucleus. Often these direct offspring are multinucleate, but each new plant comes from only one primary nucleus; however, from one basal portion with one primary nucleus, several (often branched) stalks with hats can arise (Fig. 145a). From basal pieces with nuclei isolated from direct offspring, complete plants can once more regenerate.

Underlying the gradient distribution of the morphogenetic substance there must be a structural polarity of the cell which causes the movement of the apical substances to the growing point and the accumulation of the rhizoid substances at the base. In the growing zone the superficial cytoplasm has been seen to have a distinctive structure[720].

The dependence of the morphogenetic substance on the nucleus is already evident and is demonstrated by double regeneration experiments. Anucleate 0.5-cm. basal pieces never form an apical end (Fig. 142). Now one may cut off the piece immediately above such a basal piece and leave the nucleus in the part to be tested; after some time, the rhizoid portion with the nucleus is also cut away. The resulting anucleate piece now shows excellent morphogenetic ability, better than all other 0.5 cm. pieces: of 24 such pieces in one series, all regenerated, producing six dwarf structures, thirty typical tufts, and even five typical hats. The nucleus thus emits morphogenetic substances, although it is unclear whether they act directly on morphogenesis in the cytoplasm or whether they are precursors or catalysts necessary for the formation of substances that act directly. The substances emitted by the nucleus are species-specific, as is shown by transplantation between various species. *Acetabularia mediterranea* differs from *A. wettsteinii* in several morphological characters as well as in size (Fig. 146a, b). The number and shape of the chambers in the hat, the appearance of the sterile tufts, and the presence in *A. wettsteinii* of only one upper and one lower corona are examples of these differences. It has been possible to graft anucleate pieces of *A. mediterranea* onto nucleate rhizoids of *A. wettsteinii* which are attached to a short piece of stalk. In this procedure one stalk must be inserted into the other, so that cytoplasm from each source is brought into close contact. Now the nucleus of one species has an effect on the cytoplasm of the other. In one series of experiments, the first new structures formed were *wettsteinii* hats, corresponding to the nuclei (Fig. 146c). In other transplants *mediterranea*-type structures (typical tufts and, in a few cases, dwarf hats) arose first and then a *wettsteinii* hat. This sequence of events is not surprising, for the *mediteranea* grafts surely still contain some species-specific apical substances. One might now suspect that in such a transplant the cytoplasm foreign to the nucleus is gradually replaced by *wettsteinii* cytoplasm and that only then does a *wettsteinii* hat arise. In addition to the fact that in the days before the beginning of regeneration there was no indication of such a transformation process, the examination of the chromatophores, which fortunately are very different in the two species, showed that the *wettsteinii* differentiation was accomplished with *mediterranea* cytoplasm. The morphogenetic substance produced by the nucleus thus spread into the foreign cytoplasm, in which it directed morphogenesis.

In transplants between *Acetabularia mediterranea* and *Acetabularia crenulata*, as well as between *Acetabularia mediterranea* and *Acicularia schenkii*, hats produced were intermediate between those of the two species. In *Acetabularia crenulata* and in *Acicularia* the hat chambers separate from one another and each bears a single spur at the tip (Fig. 147c). *A. crenulata* forms

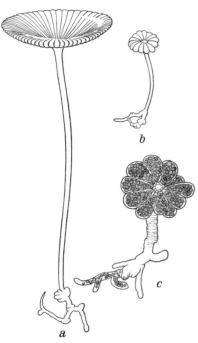

Fig. 146. *Acetabularia*. a *mediterranea*; b *wettsteinii*; c anucleate *mediterranea* stem on nucleated *wettsteinii* rhizoid. a, b about 5×, c 12×. (Original drawing by FREIBERG after preparations by HÄMMERLING)

several hats in line as a rule, and transplants behave this way also. They form at once, or eventually, hats corresponding to the nuclear type; usually intermediate forms of varying degrees appear first (Fig. 147a, b). Transplants with long end-pieces of the species different from that of the nucleus form more intermediate hats than those with short end-pieces (Table 5). Different species-specific morphogenetic substances for apical organs can thus collaborate, the expression of form being determined by the relative concentrations.

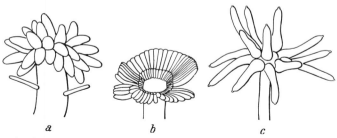

Fig. 147. a, b Hat formation in a graft of an anucleate middle stem piece of *Acetabularia crenulata* on a nucleated rhizoid of *A. mediterranea*. a First hat (intermediate); b second hat (*mediterranea* type); c rather incomplete *crenulata* hat regenerated by the anucleate anterior piece of the donor plant. About 9 ×. (After HÄMMERLING, 1934)

Table 5: *Regeneration in grafts:* Acicularia *stem on nucleated* mediterranea *rhizoid.* (After HÄMMERLING, 1934)

	Intermediate hat %	Almost *Mediterranea* hat %	*Mediterranea* hat %	n
First hat from				
0.8–2.4 cm *Acicularia* pieces	68	32	0	19
0.3–0.7 cm *Acicularia* pieces	18	27	55	11
First hat	50	30	20	30
Second hat	11	17	72	18
Third hat	0	0	100	10

The results of these experiments do not enable us to distinguish whether the developmental series of sterile tuft and hat is evoked as a single process or whether the formation of each depends upon its own specific substance, but the transplantation experiments with *Acicularia* do show that substances in addition to the hat-forming substance must be present in order that apical differentiation may begin at all: *Acicularia schenkii* does not form a hat under culture conditions. The cysts arise in the stalk. An *Acicularia* stalk grafted to a nucleated *mediterranea* rhizoid forms an intermediate hat, so that a substance that determines the form of the hat is clearly stored in the *Acicularia* cytoplasm. But the substance that *triggers* hat formation is either lacking or scarce in cultured *Acicularia* plants[45].

The development of *Acetabularia* cells concludes with the formation of cysts. The processes leading up to cyst formation are initiated by the division of the primary nucleus. What causes this new developmental process which now takes place in the nucleus? The simple regeneration experiment has already shown that the excision of a completed hat which was ready to be supplied with nuclei can defer the division of the primary nucleus until a new hat is formed. This consequence of regeneration can be repeated many times, always with the same result. Further information comes from transplantation experiments between plants of different ages. Division of the primary nucleus is delayed if the stalk of a young plant is placed on the rhizoid of an old one (Fig. 148a). Conversely, when a mature stalk with a hat is united with the rhizoid of a young plant (Fig. 148b), primary nuclear division is hastened. When the hat has reached a certain size, a physiological condition sets in which triggers the division of the primary nucleus. The morphogenetic substance released by the nucleus induces differentiation, which in turn acts back on the nucleus.

If secondary nucleus formation is already in progress or completed, and when cytoplasmic movement has already begun, the multiplication of the secondary nuclei and cytoplasmic movement can both be arrested by the removal of the hat. Such a plant can no longer regenerate.

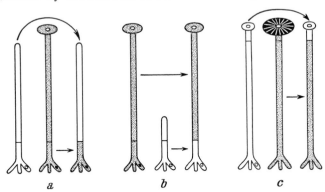

Fig. 148. Diagram of grafts between *A. mediterranea* plants of differing ages. a Young stalk on old rhizoid; b old stalk on young rhizoid; c young hat on the end of a plant whose secondary nuclei have almost all entered the hat chambers. (After HÄMMERLING, 1939, modified)

Obviously there is no more potent apical substance in the cytoplasm, and the primary nucleus which could supply it is gone. If a young hat is grafted onto such a plant, cytoplasmic movement begins again when the hat is fully grown. And now a last experiment: when the movement is ended or almost ended, and only a few secondary nuclei remain in the stalk, the hat now supplied with nuclei is removed, and a young hat is grafted on to the old plant (Fig. 148c). Now secondary nuclear division and cytoplasmic movement begin anew when the hat has reached full size. The nuclei resulting from the additional divisions enter the chambers and cyst formation continues. Thus, an effect emanating from the full-sized hat releases secondary nuclear division and cytoplasmic movement also.

So the experiments on the *Acetabulariaceae* have given us our first insight into the sequence of internal conditions which determine a fixed developmental program in a single-celled system rich in morphogenesis.

Lecture 8

With the organization of cells, simple multicellular systems of varying grades of interdependence have emerged in different ways from the animal and plant Protista. In many cases, it is a matter of definition as to whether one calls something a colony of single-celled individuals or a single multicellular organism. The bases of such a decision may include the ability of isolated cells to survive, the degree of organization into tissues, and the degree of differentiation among the individual cells. What concerns us now is the matter of developmental principles relating to the construction of simple multicellular sytsems. Examples from various groups reveal three diverse principles of organization.

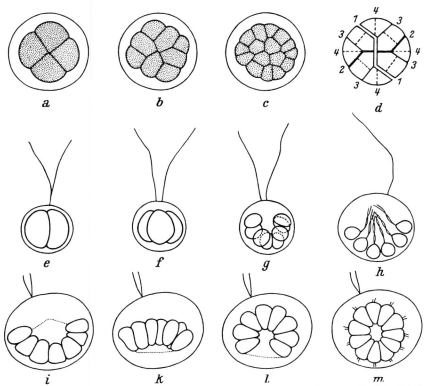

Fig. 149. Development of Phytomonad colony. a–c Division sequence to a sixteen-cell plate (*Eudorina*), surface view; d division scheme; e–h *Gonium*, side view; i–m *Eudorina*, side view. (e–m after HARTMANN, 1924)

The principle of the ordered division of a primary cell into descendants of specific number and arrangement is expressed in the *Volvocales* (Phytomonads) by a remarkable stepwise series which includes several genera: in *Oltmannsiella*, four *Chlamydomonas*-type cells form a row. In *Gonium*, four, eight, or sixteen cells unite as a flat plate in which all flagella emerge on the same side (Fig. 150a). In *Pandorina*, 16 cells form a ball (Fig. 152a). In *Eudorina*, 32 cells are embedded in an ellipsoid of jelly (Fig. 151a). In these forms each cell, according to the environmental circumstances, may either develop asexually into a new colony (Fig. 151b) or produce gametes. The genus *Pleoodrina*, however, shows differentiation of the peripheral cells in a gel ball: In *Pleodorina illinoisensis*, the four anterior cells of the 32, i.e., the ones that remain in front during the rotating movement, are somatic cells which are incapable of reproduction; in *P. californica* (Fig. 156) the anterior half of the ball contains small somatic cells and the posterior half much larger generative cells (Fig. 153). The generative cells have a greater number of contractile vacuoles but no eyespots (stigmata); each of the somatic cells has a very large spherical *Volvox*; only a few generative cells are found scattered among the somatic cells; most of the cells perish after the individual reproduces.

From an asexual cell or from the zygote a new swimming form arises through a specific sequence of divisions and specific morphological changes of the whole anlage. The first series of divisions are always equal (Fig. 149a–d), producing both in the simpler and the more complex genera a saucer-shaped cell plate, whose open end faces the old flagellar pole (Figs. 149e–i). The cell plate then turns inside out, and in *Gonium pectorale* sixteen cells now lie in a characteristic

113

arrangement within a gel plate bowed slightly toward the side with the flagella (Fig. 150a). In *Pandorina* the inverted sixteen-cell plate proceeds to a full closure; the cells are at first densely packed; they then disperse in an enlarged jelly ball in preparation for cell division (Fig. 152a, b). In *Eudorina* a hollow ball is formed by a similar process (Fig. 149i–m). The cell divisions are all longitudinal, as in all Phytomonads (Fig. 114), and the polarity of the cells never changes.

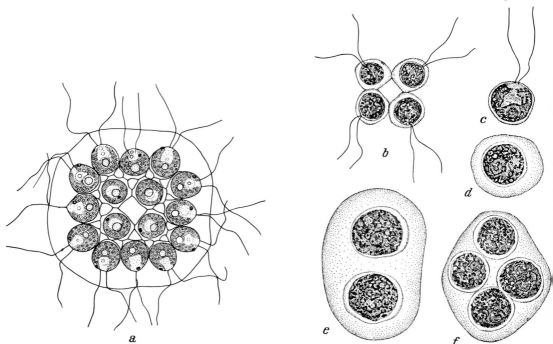

Fig. 150. *Gonium pectorale*. a Typical colony in dilute nutrient solution; b, c Four-celled colony and single cell in concentrated Knop solution; d–f "*Gleocystis* form" on Knop agar. (After HARTMANN, 1924)

The founder cell of a Phytomonad colony begins in a condition of "stress," from which after a certain series of divisions without intervening growth, it passes into an equilibrium condition in which growth begins, until the stress leading to division emerges anew. This condition determining the number of divisions can be altered environmentally. In clonal cultures of *Gonium pectorale* in concentrated Knop solution (0.1 per cent), the normal sixteen-cell forms are replaced by eight- and four-cell forms (Fig. 159b), and ultimately by single cells (Fig. 150c). The number of divisions which gives rise to a connected group of cells is thus reduced until finally the cells separate after each division and live like *Chlamydomonas* cells. In higher concentrations the single cells are as a rule larger than those in the colonies seen at lower concentrations. In the few-celled and single-celled culture forms, which occur also in nature and have occasionally been described as species, we thus see a simple modification which can be reversed by transfer into 0.05 per cent Knop solution, so that the next divisions lead immediately to a sixteen-cell form; dauermodifications in which the alteration is more or less firmly fixed are also seen; these wear off only after several divisions. The four-cell modification of *G. pectorale* resembles the species *G. sociale*, which in clonal culture under normal conditions is always four-celled, and which is distinguished morphologically from *G. pectorale* by the cells' smaller size. Culture in still stronger Knop solution (0.2–0.5 per cent) results in single giant *G. pectorale* cells averaging three or four times the diameter of normal cells. After the transfer of these giant forms into less concentrated nutrient solutions, vigorous division begins; but the daughter colonies produced, in which the cell number can become as high as 32 after one extra division, become spherical rather than flat and "look confusingly like young *Eudorina* colonies."[258]

G. pectorale can be modified in another direction by culturing on Knop agar (1 per cent agar, 1 per cent Knop solution): the flagella are retracted, and in place of the delicate jelly zone found in fluid culture, a massive jelly coat is formed, within which the individual cells multiply (Fig. 150d–f). These forms closely resemble certain species of *Gleocystis*. After return to nutrient

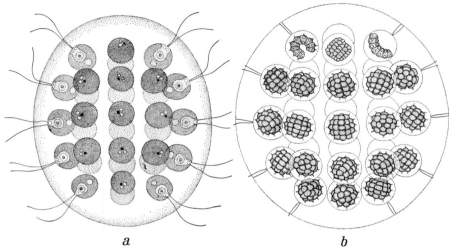

Fig. 151. *Eudorina elegans*. a Mature typical colony; b colony in division. Top, anterior; bottom, posterior. (After HARTMANN, 1921)

solution, flagella are formed once more, the jelly coats are left behind, and sixteen-cell groups are formed once again at the first subsequent division or shortly afterwards.

In *Eudorina elegans* colonies, the 32 cells are arranged in five circles (Fig. 151a, top to bottom) when the cell plate curves in (Fig. 149k–m). The anterior four cells are frequently somewhat smaller than the rest and somewhat behind in their divisions (Fig. 151b). The individual cells are completely isolated in the gel and, in contrast to *Pandorina*, are fairly distant from one another. Under certain culture conditions (nitrate as nitrogen source, strong light), "*Gonium* forms" of *Pandorina* can arise as division stalls at a particular stage: the inversion of the ball-shaped young colony anlage does not proceed until a sphere is formed, but stops with a plate that is bowed slightly upward (as in Fig. 149k). These plates can contain 32 cells; most, however, have only sixteen in the typical *Gonium* arrangement (Fig. 149c), and with further multiplication each cell gives rise to a "*Gonium* form." Frequently, division stops at the eight-cell (Fig. 149b) or the four-cell stage. Single cells also arise.

Thus, the norm of reaction in these Phytomonads includes modifications which resemble the phenotypes of species of other genera: cell numbers and arrangements that arise in one species as modifications under extreme conditions are normal for another under usual conditions. So *Gonium* and *Eudorina* are always different from one another when they grow together in nature, but under exceptional conditions the first species can extend its development to a form similar to the second, and under other circumstances the second species can stop short of its usual form in a stage resembling the first.

115

The typical mature form in the series from *Gonium* to *Pleodorina* has a certain number of cells in a particular array. What happens now when the arrangement is disturbed and the cells are separated from one another?

In *Gonium* the cells can be separated easily with a glass needle. *Pandorina* cells are more firmly bound in the jelly; one can only destroy some of the cells and separate the rest as far as possible. In both cases the isolated cells divide prematurely. The lost cells are not all replaced; rather, each cell undergoes the species-specific sequence of divisions until it forms a new little colony. Often these young *Gonium* colonies swarm out before the first division has begun in the unoperated sixteen-cell control sister colonies derived from the preceding asexual reproduction. With reduced

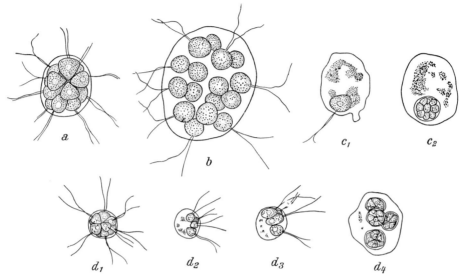

Fig. 152. *Pandorina morum*. a Typical colony; b shortly before the beginning of divisions; c_1 after destruction of all cells but one (1/16); c_2 this cell divided into sixteen, two days after the operation; d_1 young colony; d_2 the same after destruction of twelve cells; d_3 after three days; d_4 after four days. 280×. (After BOCK, 1926)

cell number in the isolated colony fragment, division is accelerated; for example, after division of a colony into three fragments of unequal size (in which one cell was lost), young sixteen-cell colonies were produced on the third day by a 2/16 fragment, on the fourth day by a 4/16 fragment, and on the fifth day by the 9/16 fragment, while the intact sister colony (= 16/16) reached this stage only on the sixth day. In *Pandorina* isolated cells (Fig. 152c), or 2/16 fragments, completed their divisions several days before the sister colonies. If, in very young colonies only a few hours old, all but a few cells are destroyed, cell division begins in the remaining cells after very little growth and produces tiny daughter colonies (Fig. 152d 1–4). The stimulus for division is thus not tied to a certain cell size in *Gonium* and *Pandorina*, but rather to the isolation of the cells.

In *Eudorina* (Fig. 151), whose cells are separated by considerable distances, the individual cells are easily isolated. These isolated cells, however, only produce typical 32-cell colonies just as they would have done if left in place under conditions promoting asexual reproduction. This, of course, contrasts with the behavior of *Gonium* and *Pandorina*. But cells isolated from younger colonies do resemble *Gonium* and *Pandorina* cells in that they begin division prematurely; nevertheless, even here there is the difference that at least one division is omitted. Thus sixteen- or eight-cell colonies form. Their cells can form normal 32-cell daughter colonies again after normal growth. With a reduction in light intensity, the interval between two reproductive periods in-

116

creases greatly, for example, in *Eudorina* from five days to twenty-eight days. Here the premature onset of division in isolated cells is particularly clear. We do not know how the disruption of the colony leads to premature cell division.

In *Pleodorina californica* (Fig. 153) the determination of the organization of the colony into reproductive and somatic cells can be understood in detail. The cell number in a colony is closely grouped around 128, 64, and rarely 32. The ratio between somatic and reproductive cells is an astoundingly constant 3:5, no matter how large the colony is. Thus it is 48:80 with 128 cells, 24:40 with 64 cells, and 12:20 with 32 cells. Since the total cell number corresponds to powers of two, the colony is formed by a series of regular dichotomous divisions of a reproductive cell.

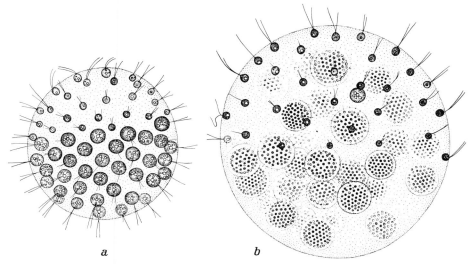

a   b

Fig. 153. *Pleodorina californica*. (After CHATTIN, 1911)

It is fortunate that in *Pleodorina* both flagella of the mother cell remain intact and are passed on to particular cells until the plate is ready to invert; they come to lie on the border of the 32-cell plate (Fig. 154). In *Pleodorina*, after the cell plate curves out (Fig. 149i–m) and closes, the opening that remains is not circular but rather a slit along a seam which distinguishes the mature colony. It is always in the region of the reproductive cells at one pole of the ball. With the help of these marks the future reproductive cells are seen to be arranged at the border of the 32-cell plate and the future somatic cells in the middle (Fig. 154). In each quadrant of the cell plate lie four border cells. In order to produce twenty reproductive cells, one additional neighboring cell must come to each quadrant: whether this is always the same cell is not known. From the cells with uncertain "prospective fate," i.e., future role, each quadrant must also contribute one to the somatic cells.

A count of the cells of a large colony shows that the ratio 3:5 is not always maintained exactly. The greatest deviations found among sixty colonies are listed in Table 6. These cases suggest the omission of individual divisions and the atypical differentiation of some cells. The data also suggest a determinative agent's emanation from the reproductive pole of the colony, with an indistinct border in the equatorial zone. The lack of a sharp border is also expressed in sporadic cases where individual somatic cells divide atypically and give rise to dwarf colonies, and in other cases, predominantly in the border zone, where reproductive cells have a small, pale eyespot.

We may now ask whether the agent emanating from a pole restricts the reproductive tendencies of some of the cells. Or is determination completed in a series of differential divisions, so that certain cells and their derivatives can no longer divide when all the cells are together in a simple colony?

The determination of the somatic cells is final, at any rate: if isolated in nutrient solution, they all die in a short time with only rare exceptions which develop into rudimentary colonies varying in cell number and organization. Isolated reproductive cells produce mainly colonies of typical form.

Young daughter colony stages can be isolated from the mother ball. When they are drawn with their jelly coat into a glass capillary and then blown out again, they become deformed, and their cells are pressed together. This crowding in the four-cell stage destroys the polar distribution of the two cell types and results in the mosaic arrangement of groups of reproductive and somatic cells. A colony, for example, might now possess two somatic poles: an anterior somatic field containing 21 cells would be divided from another somatic group of five cells at the opposite pole by a band of reproductive cells (Fig. 155a). In another case two somatic cells with large eyespots lay side by side in a mature colony among reproductive cells that had already started dividing (Fig. 155b). Thus differences are determined at least as early as the eight-cell stage.

The use of flagella as markers permits one to trace the distribution of presumptive reproductive and somatic cells from the cell plate stage back to the undivided reproductive cell (Fig. 154). Since the flagella of the mother cell always reach the border cells, the border cell material must include the flagellar pole of the founder cell. It follows from this that there is a polar division of reproductive and somatic anlagen in the mature undivided reproductive cell. This cell does have a clear polar differentiation: at the flagellar pole hyaline cytoplasm free of chloroplasts extends to the cell surface. It is embraced by a cup-shaped granular cytoplasmic zone containing the chloroplasts.

The determination of somatic cells has many features: the restriction of growth and of the ability to divide, and the lack of a cytoplasmic arrangement which makes reproductive cells competent to undergo the normal pattern of division. Experiments and observations lead to the conclusion that the pattern of the colony cells in *Pleodorina* rests on the determinative divisions of polar-differentiated reproductive cells.

In *Volvox* modifications of the colony have not been achieved to date, and isolated cells, vegetative as well as reproductive, do not survive isolation.

Table 6: *Extreme deviations from the ratio of three somatic to five reproductive cells in Pleodorina californica* colonies. (After GERISCH, 1959)

Cell numbers total	somatic	reproductive
31	9	22
30	13	17
64	18	46
61	28	33
127	43	84
127	53	74

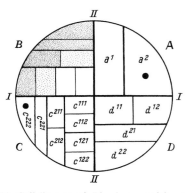

Fig. 154. Cell lineage of *Pleodorina californica* at the 32-cell stage. *I–I, II–II* first two division planes. Flagellar attachment of the mother cell marked by heavy dots. In the *B* quadrant: reproductive cells heavily stippled, somatic cells white, lightly stippled cells of indeterminate fate. (After GERISCH, 1959)

118

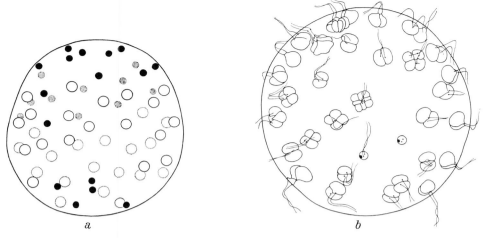

Fig. 155. Colony of *Pleodorina californica* whose cell organization has been disturbed in the eight-cell stage. a black and (on opposite side) stippled: somatic cells; white: reproductive cells; b mature colony, view of reproductive pole, reproductive cells dividing, two somatic cells. (After GERISCH, 1959)

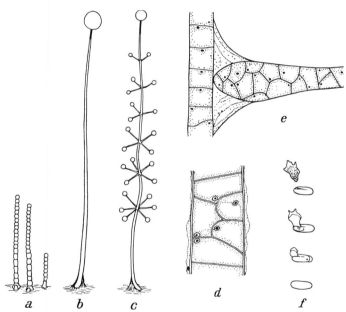

Fig. 156. Acrasiales. a *Acrasis*; b *Dictyostelium*; c *Polysphondylium*; d, e structure of sporophores; d *Dictyostelium*; e *Polysphondylium*; f spore germination. (a–c, f after RAPER, 1940; d, e after OLIVE, 1902)

The Acrasiales show the greatest contrast to the Phytomonads in morphogenetic principles. Their variously formed fruiting bodies (sporangia) arise through the union of amoebae which come together, not to fuse as in the Myxomycetes, but in an aggregate, the pseudoplasmodium, as distinct cells, some of which are somatic and form the supporting structures, and some of which form spores. In the least specialized genus, *Acrasis*, the union of relatively few amoebae results in a single column of cylindrical stalk cells and round spores (Fig. 156a). In *Dictyostelium*,

119

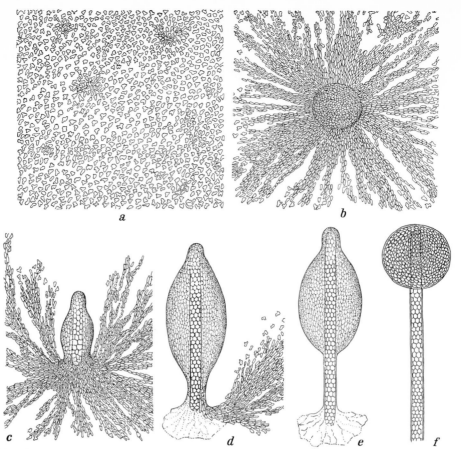

Fig. 157. Development of the sporangium of *Dictyostelium mucoroides*, diagrammatic. a mobilization phase; b colliculus; c cones; d, e clava; f mature spore mass

a longer unbranched stalk, the sporophore, bears a spherical mass of spores (Fig. 156b). In *Polysphondylium* the stalk has lateral branches, generally arranged in more or less regular groups (Fig. 156c). The stalk cells form a somatic tissue reminiscent of the parenchyma of higher plants: these cells are separated by cellulose walls, and they contain large vacuoles which make them turgid and thus stiffen the stalk (Fig. 156d). The lateral branches of *Polysphondylium* (Fig. 156c) are not branches of the main stalk parenchyma, but rather independent smaller stalks which are held on the main stalk by a tent-shaped structure (basal plate) formed by the transformation of amoebae, just as the sporophore is joined to the stalk (Fig. 156e). The development of the Acrasiales begins with the emergence of the amoebae from the spores (Fig. 156f). The amoebae move with amorphous pseudopodia, eat bacteria, and multiply rapidly. The contrast between the Phytomonads and the Acrasiales is thus very sharp: in the Phytomonads morphogenesis emerges from the partitioning of a founder cell; in the Acrasiales, previously independent amoebae derived from one spore, or from many spores, unite for morphogenesis.

The developmental physiology of *Dictyostelium* has been investigated in depth. It can easily be raised on agar with a single bacterial species as food source. Figure 157 illustrates the course of development of *D. mucoroides*. The amoebae wander around, grazing on the bacteria, and if

120

they are continually transferred to new bacterial cultures they remain in this multiplication phase; but when the bacteria are consumed, the amoebae gather into small groups which generally disperse again and reform in another place (Fig. 157a). After some time has been spent in this mobilization phase, which clearly reflects an altered metabolic condition in the amoebae, the aggregation phase begins. A few larger aggregates of amoebae (cumuli) eventually become the final aggregation centers. To them stream the now-elongated amoebae in paths from the surrounding region, the aggregation field.

In the aggregation phase the amoebae in the streams form contacts that reflect a new condition of their cytoplasmic surfaces. The onset of this change is shown by the behavior of cells cultured in fluid. In roller tubes, in which the cells are in constant motion and continually touching one another, the amoebae agglutinate into small groups (up to about twenty cells, Fig. 158a) when the bacteria have been completely consumed. If such agglutination groups are placed on a substratum, the cells wander apart, and aggregation begins only after a few hours. In the roller tubes after the bacteria have been consumed, the agglutination groups become larger (up to a thousand cells, Fig. 158b), and when placed on a substratum the amoebae remain in contact, and the formation of a sporangium begins without any wandering stage.

At the aggregation center a hemispherical pile of amoebae forms (a colliculus, Fig. 157b), which then rises into a cone. Its tip ends in a knob (the capitulum). The formation of the cone is evoked and the polarity determined by the contact with the substrate from which it rises perpendicularly. Agglutinates capable of aggregation form only one cone in liquid if they make contact with a limiting surface. If they are placed on fine mesh gauze, cones and mature sporangia arise from the same amoeba mass on both sides (Fig. 159).

The elaboration of the fruiting body is not always completed in the place where the assembly of the amoebae has resulted in the cumulus; instead the cone generally wanders as a whole over great distances on the agar surface, leaving a trail of amoebae behind. The wandering cone is called a slug, and in its long axis amoebae differentiate into stalk cells. In each cell a large vacuole arises. The inflated cells flatten polyhedrally against one another. On the outside the columns of

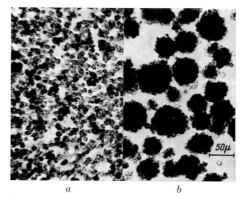

Fig. 158. Agglutination of washed cells of *D. discoideum* after total consumption of bacteria in roller rubes. a immediately after washing; b nine hours later, amoebae ready to aggregate under similar conditions. (After GERISCH, 1961 b)

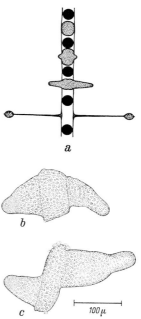

Fig. 159. Polarization in agglutinates ready to aggregate in *D. discoideum* between two limiting surfaces. a diagram of the experimental setup, section through the gauze, which is marked by large dots; b, c cone stages on the gauze; b axes of partial anlagen exactly opposed; c axes opposed, slightly displaced. (After GERISCH, 1960)

cells are separated from the remaining mass of amoebae by a cellulose membrane, and cell walls are built between the stalk cells (Fig. 156 d). The stalk cell column lengthens due to the continuous transformation of amoebae which have migrated to the capitulum (Figs. 157 c–e, 160 d, e). The amoeba mass which has streamed in from the vicinity now moves up the stalk. The sporangial anlage becomes a club-shaped clava (Fig.s 157 d, e; 160 e). At the base of the sporophore a tent-shaped membrane is formed by amoebae which have remained below and become flattened,

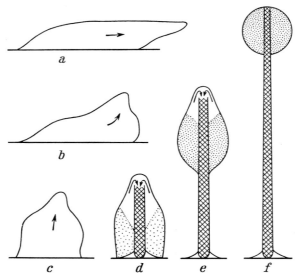

Fig. 160. Scheme of the development of the sporangium of *D. discoideum*. a migrating pseudoplasmodium (slug); b, c end of movement, erection of the pseudoplasmodium; d cones; e clava; f mature sporangium. (After RAPER, 1940, BONNER, 1944)

and this membrane fastens the stalk to the substrate as a basal plate. The stalk becomes free, and those amoebae not incorporated into stalk tissue form a ball at the top. Here the individual amoebae round up and encyst, finally becoming spores. In *D. discoideum* the transformation into rounded "pre-spores" and cysts begins in the basal and outer parts as the amoeba mass rises up the stalk (Fig. 160 d, e). The mass of spores in the sporangium is held together by slime, not by a sporangial coat.

The formation of a multicellular system during the development of *Dictyostelium* thus includes morphogenetic movements of thousands of individual cells and the differentiation of the cell aggregate. Both of these processes are *harmonious* in that they lead to the formation of a typical whole.

The first step toward this whole is aggregation, the streaming of the troops of amoebae toward a certain center. As soon as the center is formed, the streaming of amoebae in cell contact is triggered by a chemotactic substance whose concentration gradient orients the path of the amoebae. This has been demonstrated in several ways, as we shall see.

Aggregation can also take place under water. A dense culture of amoebae is washed from an agar plate and separated from the bacteria by centrifugation. If the suspension of amoebae thus obtained is placed in a glass dish the amoebae fall to the bottom and begin to collect at a center (Fig. 161 a–d). If an aggregation center is pushed laterally from a column of amoebae, this column regularly turns within three to five minutes toward the new site of the center if the distance is not more than 200 μ, i.e., about thirteen amoebae diameters. But at distances even as

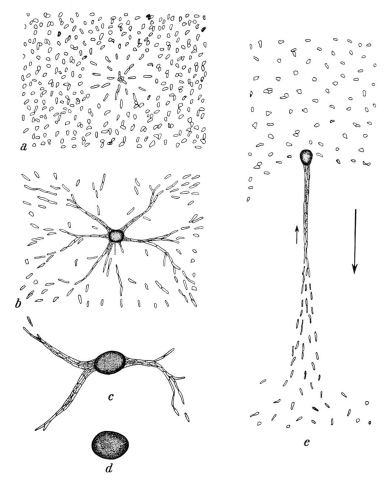

Fig. 161. Mobilization phase and colliculus formation in *Dictyostelium discoideum* under water. a–d Sequence in calm water; e water current in the direction of the large arrow, amoeboid movement in the direction of the smaller arrow. (After BONNER, 1947)

great as 800 μ (some fifty amoeba diameters) a weak attractive influence can be demonstrated. If a gentle stream of water is directed at an area in which aggregation has begun, the upstream troops scatter immediately, while downstream the amoebae move in one column toward the center (Fig. 161 e). A substance which attracts the columns can produce its effect after isolation from the amoebae releasing it. If a piece of agar, upon which a large aggregate has formed, is placed (after careful removal of the colliculus) among amoebae in the mobilization stage, the amoebae from all sides head for the cell-free agar and form columns reaching to the agar, next a cone, and finally a fruiting body (Fig. 162a–c).

Unfortunately, this chemotactic substance, called acrasin, has not been characterized chemically (BONNER, 1947). No doubt it is not a general chemotactic agent but rather one specific for the amoebae of certain species of the Acrasieae. If equal numbers of the red spores of *D. corporeum* and the colorless spores of *D. discoideum* are suspended with bacteria on an agar plate, fruiting bodies of each species arise side by side, with only one kind of spore in the sporangium. In mixed cultures of *D. discoideum* and *Polysphondylium violaceum*, the amoebae separate from one another when the cumuli are formed. The columns of amoebae wandering to the centers of both species, which can be distinguished by their colors, pass over one another in general without mixing or turning away[528].

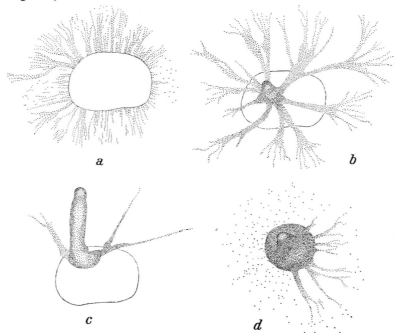

Fig. 162. *Dictyostelium mucoroides*. a–c Migration of amoebae in the sensitive phase to a piece of agar previously under a colliculus; d cones placed on an agar amoeba culture; left, amoeba still in feeding stage; right, in the sensitive stage. (After PFÜTZNER-ECKERT, 1950)

Acrasin works on amoebae only when they are in a particular condition, a sensitive state. If a slug is placed among amoebae in the multiplication stage, the amoebae do not respond, even those closest to the slug; during the transition to the mobilization phase, they orient toward the slug and form columns directed toward it. If *Dictyostelium* spores and relatively few bacteria are plated together on nutrient agar, the mobilization stage generally begins at once. If the bacteria have been streaked and feeding amoebae are now allowed to multiply along the streak, the consumption of the bacteria proceeds, and with it the transition of the amoebae to the mobilization stage takes place along the streak. If a cone is now placed right on the border between the two functional stages, the formation of oriented paths in response is seen only on one side (Fig. 162d).

Clearly it is not only the center which exerts this attraction, for the amoebae do not approach the center in a simple radial fashion, but rather, from the early thin paths running close to one another there arise large streams between which there are few or no amoebae (Figs. 157b, c; 161a–c; 162a, b). Clearly the amoebae which respond to the attractive force also form the attractive substance, and this causes parallel paths of amoebae to converge and fuse.

The liberation of acrasin lasts through the formation of the sporangium. This can be shown by placing a slug or a clava among amoebae in the mobilization phase or near paths which are just forming, and seeing whether and at what distances amoebae are attracted. The strongest attractive force is shown to come from the tip, to which the amoebae continually move to extend the stalk (Fig. 163c, d).

With the differentiation of the amoebae in the sporangial anlage there is a balanced allocation of cells into stalk cells and spore-forming cells in the proportions typical of the whole. The dimensions of the final fruiting body depend on the concentration of the amoebae found in the area of influence of an aggregation center. Their number may range from a few hundred to many

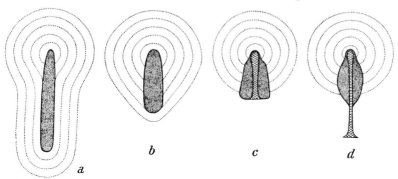

Fig. 163. *D. discoideum*, diagram of the strength of the attraction emanating from various parts of a slug (a, b), a cone (c), and a clava (d). (After BONNER, 1949)

thousand, and so the sporangia may be large or small (Fig. 164). There is also a definite relationship between the length and axial diameter of the stalk and the size of the fruiting body. The number of cells in a cross-section of the stalk rises with its length (Fig. 165a). The total volume of the stalk rises with the volume of the sporangium in a fairly linear fashion (Fig. 165b). When the fruiting body is formed in the dark at 17° C., the average volume of the stalk is 14 per cent and the spore mass 86 per cent of the total sporangium. These proportions may be modified broadly by environmental factors (Fig. 166). Under certain lighting conditions the proportion of stalk cells becomes about 24 per cent. If the plasmodia are moved from the low temperature to a temperature 10° to 15° higher, the number of spore-forming cells rises by a few per cent, and the proportion of stalk cells of course decreases correspondingly (Fig. 166).

Disaggregation experiments on various developmental stages of the sporangium provide insight into the determination of its differentiation. In *D. discoideum* the cone topples over after it has incorporated all the amoebae in its area of influence, and before it begins to differentiate. It now migrates as a slug for one or two hours on the agar surface (Fig. 160a). While in *D. mucoroides* nuclear and cell divisions are restricted to the feeding phase, vigorous mitotic activity is still seen in young *D. discoideum* slugs. The cells in migrating slugs all exhibit amoeboid movement. The mass of amoebae has an explicit polarity: the apical end of the cone is always in front. If a migrating slug is placed among amoebae in the mobilization phase, it will be recalled that a

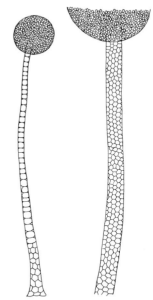

Fig. 164. *Dictyostelium mucoroides*, stalks of different sizes built by different numbers of amoebae. About 200×. (After BREFELD, 1869)

strong attractive influence emanates from the anterior end, and this decreases posteriorly (Fig. 163a, b). Pseudoplasmodia moving in the same direction unite easily when they meet and form a larger single sporangium of typical proportions. This polarity pervades the entire slug: If it is cut in the middle and the posterior half rotated 180°, and if the halves are now rejoined and the amoebae at the junction mixed together with a glass needle, the rear half exhibits its original direction of movement, and separates from the front half; thus two fruiting bodies are formed (Fig. 167a). If two slugs, one vitally stained with Nile blue, are placed side by side but

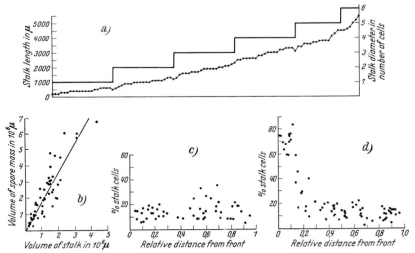

Fig. 165. Relationship between stalk diameter expressed in cell numbers (solid line) and stalk length (dotted line) in *D. mucoroides*. For each diameter, data are arranged in order of increasing stalk length. n = 125. b–d *D. discoideum*; b relationship between volume of the spore mass and volume of the stalk; c, d relationship between proportion of stalk cells and volume of the entire fruiting body formed by pieces of the migrating pseudoplasmodium: c, younger; d, older slug; points are placed over the abscissa value corresponding to the middle of the isolated piece. c Young pieces of pseudoplasmodium which have not begun to move; d from older pseudoplasmodia which have already migrated from 1–4 cm. (a after PFÜTZNER-ECKERT, 1950, b–d after BONNER and SLIFKIN, 1949)

pointing in opposite directions, and the amoebae at the area of contact mixed (Fig. 167b), two masses of amoebae separate from one another. Both contain blue and colorless amoebae; but the fruiting body corresponding to the front end of the blue slug contains more blue cells and the other one more colorless cells. Some of the cells are thus forced, ostensibly by the great mass of their neighbors, to adopt a new polarity of movement. This polarity does not depend on the anterior-posterior acrasin production gradient alone. This is shown by a further experiment: If a slug is bent into a ring and the amoebae mixed at the junction of the front and rear ends, the ring breaks at the junction in about fifteen minutes; the amoebae of the posterior end have not been influenced by the high acrasin production at the front end but retain their old orientation of movement with respect to the entire slug.

In spite of the fact that all of the cells in the wandering pseudoplasmodium are in constant motion, they keep their place in the mass of amoebae. This is shown by experiments with stained amoebae: if *Serratia marcescens* (= *Bacillus prodigiosus*) is used as the food source, *D. discoideum* amoebae store the red bacterial pigment (prodiogosin), which they themselves cannot break down. If pieces of such intensely stained slugs are now combined with colorless pieces, the border remains sharp over the entire wandering period, and the position of individual cell groups can be followed into the mature fruiting body (Fig. 168 b, c). By staining the initial amoeba channels, it can be established even from the beginning of the aggregation stage where the early and late-entering amoebae settle in the wandering slug and later in the sporangium (Fig. 168). Thus, from the beginning of the movement of amoebae toward, a center, the prospective fate of the individual zones of the aggregation field can be established: in concentric rings the presumptive stalk cells are innermost, the presumptive spores next, and on the outside are the presumptive

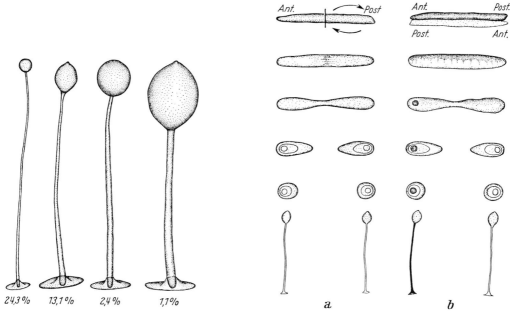

Fig. 166. *Dictyostelium discoideum* fruiting bodies of various proportions: proportion of the stalk expressed as percentage. (After BONNER and SLIFKIN, 1949)

24,3%   13,1%   2,4%   1,1%

Fig. 167. Polarity of the migrating pseudoplasmodium of *D. discoideum*. a Reversal of posterior piece; b parallel placement and mixing of two pseudoplasmodia, one stained vitally with Nile blue. (After BONNER, 1950)

basal plate cells. This topographic fate map does not, however, correspond to a pattern of determination, that is, to a fixed course of differentiation of the cells. This is demonstrated by isolation experiments: if a young slug is transected into pieces, each piece develops into a sporangium with normal proportions, averaging 14 per cent stalk cells (Fig. 165c). A certain *tendency* toward development in the direction of the prospective fate is seen, however, at the front end of older migrating slugs which have already wandered for awhile: The front quarter or third, when isolated, forms a fruiting body with a disproportionately high number of stalk cells, sometimes exceeding 80 per cent of the total, while more posterior pieces form sporangia with the normal proportions (Fig. 165d). A certain *regulation*, that is, a deviation of the prospective fate in the direction of normal development, is seen even in the front end of an older slug: the anterior third ordinarily forms only stalk cells according to the fate map (Fig. 168), but when isolated it does produce spores, even though in low numbers, and basal plate.

Examples of regulation after the loss of parts of the aggregation field are also seen in experiments on *D. mucoroides*. If a large cumulus is reduced in size before the beginning of the differentiation of stalk cells (Fig. 169c), the stalk is small in comparison to the original central aggregation area. This is shown by the distribution of cross-sectional cell counts (Fig. 170), whose mean values are statistically significant. In the aggregation stage (Fig. 157 b), therefore, the entire aggregation field including the center constitutes a single physiological system, in which the center attracts the streams of amoebae, and conversely the accumulating amoebae exercise an effect on the organization of the central mass. When stalk formation has begun, reduction in the size of the cumulus no longer causes a reduction in the diameter of the stalk; the stalk has the same cross-sectional cell number, but its length will be reduced. The emplacement of new amoebae and their transformation into stalk cells ends earlier than when the entire mass of aggregating amoebae is incorporated into the sporangial anlage. Thus, there is also a regulation through which the approximate volume relationship between spore cells and stalk cells (Fig. 165 b) is maintained, although, to be sure, it is not the typical relationship between stalk length and stalk

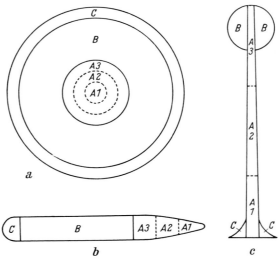

Fig. 168. Diagram of the prospective fates of amoeba zones shown by vital staining. a aggregation stage; b slug; c mature sporangium. (After Bonner, 1944)

cross-section (Fig. 165 a). If a stream of amoebae is separated from a colliculus or cone, the front end of the stream produces a correspondingly small sporangium. Reduced aggregation fields and pieces of young migrating pseudoplasmodia thus organize themselves just as do amoeba masses which had been small from the beginning.

The determination for particular differentiation begins in *D. discoideum* in the young cone (Fig. 160) earlier than in *D. mucoroides*. If the upper half of the cone is killed with a hot needle, the basal cone cells, followed by a trail of amoebae, bypass the dead cell mass and build a new cone and fruiting body (Fig. 171). Thus the cells are not yet determined to become spores. Indeed, even the amoebae of the clava of *D. mucoroides* (Fig. 157 d, e) are still alterable: If they are streaked on agar they come together again, if they have not been separated by too great distances and if their number is not too small, and they form a new fruiting body. If bacteria are present they even enter a new period of multiplication first. The determination for particular differentiation thus begins in *D. mucoroides* only shortly before the completion of differentiation when the amoebae have arrived at a particular place in the whole formed by the morphogenetic movements of the amoeba mass. Until then the individual amoebae are *omnipotent*, that is, they can enter any developmental pathway, and the parts of the amoeba field are *equipotent* and *totipotent*, that is, if they contain enough amoebae, each part when isolated forms a harmonious whole, a fruiting body of normal proportions.

The aggregation field of *D. mucoroides* fits the concept of the harmonious equipotential system of DRIESCH and offers the simplest kind of example of what PAUL WEISS has, with great insight, called a *morphogenetic field*. The field we are concerned with here is a *self-organizing* field (we shall come to know additional morphogenetic fields later). Such a field does not consist of determined parts. Rather, it is dynamically fitted, as a result of particular functional conditions spread over the field region, to organize itself in a particular way. It has a particular axial or radial polarity; a reduced field region becomes organized typically in the same proportions as the undivided whole. If field regions are united with a common polarity, they form a single field which, though larger, is typically proportioned. In a field a pattern of determination arises gradually.

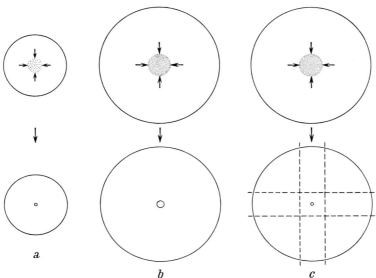

Fig. 169. Diagram of the stalk anlage (cross section) in small (a) and large (b) aggregating field and in reduced aggregating field (c). The outer circles mark the border of the aggregating field; the first heap is stippled; the already formed stalk cross section is shown as a small circle. (After PFÜTZNER-ECKERT, 1950)

We know nothing about the nature of the functional conditions which result in the properties of fields and nothing about the functional alterations which result in determinative organization; thus the essence of even this simple field of *Dictyostelium* remains a mystery. "The field concept is ... but an abbreviated formulation of what we have observed ... Its analytical and explanatory value is nil."[717, 292]. But the field concept is a biological principle, a recurring biological pattern of organization in the collaboration of diverse developmental processes, and it challenges developmental physiological analysis.

The whole course of development from the onset of chemotactic responses in a certain nutritive condition of the amoebae and acrasin formation, through the establishment of the form of the cell aggregate, through the morphogenetic movements of the amoeba mass, to the determination of the individual amoebae reflecting their position in the slug, depends on the specific reactions of the individual amoebae to particular conditions.

The *Dictyostelium* reaction system, mysterious in its totality, nevertheless reveals the independence of individual reactions. Among *D. mucoroides* cultures, perhaps by mutation, a clone appeared that cannot form a fruiting body. The amoebae enter the mobilization phase and they form little aggregates, but no final center. Even streams arise, sometimes of considerable length, but they have no particular goal and do not lead to a colliculus. When the bacteria are consumed at last, irregular cysts are formed on the agar. The amoebae of this nonfruiting strain react to acrasin: they are attracted by a normal slug placed in their culture. Paths lead to the slug from all sides, and the amoebae mix. The abnormal amoebae disrupt further development, however, and the whole structure goes to pieces. The functional condition of the mobilization phase therefore does not determine the chain of succeeding morphogenetic movements; and the amoebae of the mutant strain lack the competence to respond harmoniously to local conditions in the aggregate.[510]

The use of chemicals permits the distinction in *D. discoideum* between cell differentiation and the control of morphogenetic movements (Fig. 172). The cone rises up, and in the presence of

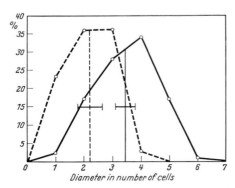

Fig. 170. *Dictyostelium mucoroides*, distributions of axial diameters in normal sporangia (solid line, $n=82$) and after reduction of originally equal sized aggregation fields (broken line, $n=30$). Mean values ± 3 standard errors. (After PFÜTZNER-ECKERT, 1950)

Fig. 171. Killing of the apical half of a cone of *D. mucoroides* and formation of a new sporangial anlage from the basal cone cells and from those recruited from the trail of amoebae. (After PFÜTZNER-ECKERT, 1950)

certain chemicals stalk cells and spores differentiate in a thoroughly atypical arrangement (Fig. 172a, b); other substances inhibit differentiation of both cell types in the club-shaped stage (Fig. 172c).

A morphogenetic principle entirely different from those in the previous two examples is found in certain algae. *Celloniella palensis*, a colonial Chrysomonad occurring in cold, swift brooks, varies tremendously in form according to local conditions. In strong currents, the alga forms wavy, leaf-like colonies up to 2 cm long (Fig. 173a, b). The jelly sheath is held fast to stones by a stalk and spreads into irregularly branched lobes. At the edge of the lobes and in the region of the stalk are numerous cells, each with a yellow-brown, cup-shaped chromatophore. They are arranged mainly in a single layer in the jelly, but at the growing points they are densely packed (Fig. 173b, e). In these apical cell groups the cells divide actively. They do not remain in direct contact; they all produce jelly and thus move apart. In the interior of the lobes and stalk lie a few generally larger, often elongated cells. Where water plunges over a rocky edge, *Celloniella* forms a gel structure entirely different from that in running water. It is a crust consisting of several layers whose margins contain calcium carbonate granules (Fig. 173a, c, f). The layers of crust are formed through the periodic accretion of layers of *Celloniella* cells. Finally, beneath an overhang where the water trickles down and drops away, the Chrysomonad forms sacs (Fig. 173a, d, g). These are filled with a liquid, and the cells lie in the sturdy surface layer, which contains numerous $CaCO_3$ granules. The cells are concentrated on the illuminated side (Fig. 173a, d, g). If stones with pieces of crust are placed in running water, arches arise in the

130

course of one or two days in which cell division is rapid; and in the course of the next four days irregularly cylindrical extensions several millimeters long, pointed at the end and with lateral bulges, begin a transition to the leaflike colonies. If pieces of leaflike, encrusted, or saclike colonies are placed in still water, the formation of motile forms is triggered. After only a few minutes the cells swim out, each with one long flagellum (Fig. 173h). These swarmers can divide and can transform into amoeboid forms, which creep around with blunt pseudopodia (Fig. 173k). In cool water they attach and begin to make jelly. Silicaceous cysts adorned with winding ribbons and a pore with a lid (Fig. 173l, m) reveal the chrysomonad nature of the alga.

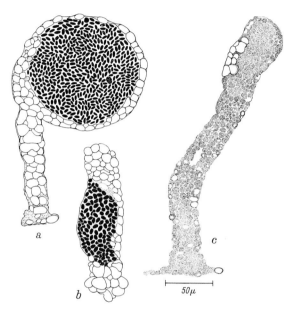

Fig. 172. Disruption of the order of morphogenetic movements of the amoebae and of cell differentiation of *D. discoideum*. a abnormal sporophore brought about by ethyl urethane; b sporophore rudiment after treatment by EDTA; c inhibition of differentiation by mercaptoethanol. (After GERISCH, 1961a)

The particular colony form developed is a result of the mechanical effects of the water. The cells react to these with different degrees of gel production and a different deployment within the gel mass. The wavy leaflike colonies, differentiated into stalk, hold-fast, and lobes, and the encrusted and saclike colonies appear to be adaptations to the conditions which evoke them.

The examples in this lecture illustrate three morphogenetic principles. In the Phytomonads the typical form is determined by ordered division of a founder cell, and morphogenetic movement is carried out only by the entire anlage, as we see in the curvature of a cell plate. External conditions influence the division rate and the number of divisions which lead to a final multicellular whole. In the Acrasiales, the increase in mass takes place through the division of independent cells which then come together for morphogenesis in the morphogenetic field. They bring about a multicellular organism through common morphogenetic movements and differentiation within the aggregate. The size depends upon the number of cells uniting, but the proportions are always the same, External conditions participate only in triggering the aggregation of the individual cells (lack of food), in directing polarity (substrate) and, to a lesser extent, in determining

131

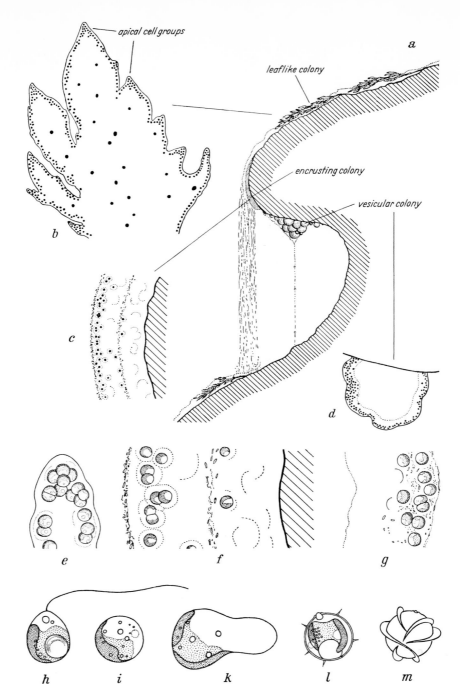

Fig. 173. *Celloniella palensis*. a Diagram of the appearance of jelly colonies in various situations; b–g pieces of the respective colony forms: b leaflike form; e end of a lobe, highly magnified; c,f encrusting colonies; d, g sac colonies; h swarmer cells; i resting cells; k cell in amoeboid movement; l, m cysts; l optical section, m surface view. (After PASCHER, 1929)

the proportions of the parts (temperature, light). Here morphogenesis depends upon the reactions of the individual cells to conditions encountered within the field. In the gel-forming algae a specific repertory of colonial forms is seen in the specific reactions to diverse environmental conditions.

The three morphogenetic principles, ordered division of a founder cell, harmonious organization of a field, and morphogenetic reactions to environmental conditions also find many applications in highly differentiated multicellular organisms in the setting up of the basic organization in embryonic development, in the later development of organ anlagen and in regeneration.

Lecture 9

Fertilization results in the diploid condition. The events that lead to the fusion of the nuclei are very diverse in the plant kingdom. The transfer of a male nucleus to the egg cell is originally the function of free-swimmung microgametes (spermatozoa); it ultimately employs the whole course of development of a reduced generation, the microsporo-phyte, in the pollen grain. In metazoans, fertilization is fairly uniform: Almost invariably the male sex cells, the sperms, are highly specialized, flagellated motile cells which seek out the eggs whether they are in open sea water, fresh water, or in the female reproductive tract.

While in plants the egg cell receives cytoplasm and plastids from the male partner at the time of fertilization (and this takes place to a varying extent, including equal contributions from both parents), it is not certain whether the sperms of animals bring in any self-duplicating cell structures other than the genome and a centriole. Mitochondria are clearly present in all sperms; but their autonomous duplication in the egg cytoplasm is questionable.

Eggs and sperms must be so constructed that fertilization is a near certainty; that the sperm must find the egg and unite with it, and also that multiple fertilization and fertilization by sperms of another species must be prevented.

The sperm moves by means of a tail whose great structural complexity has been revealed by the electron microscope. An energy supply permits this mechanical movement, which can be turned off or turned on and controlled by particular environmental influences. It contains energy rich substances and the appropriate

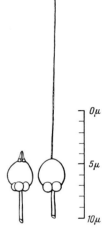

Fig. 174. Sperms of the mussel *Spondylus cruentus*. Right, after the acrosome reaction in egg-water filtrate. (After DAN and WADA, from A. L. & L. H. COLWIN, 1957)

metabolic enzymes. It also carries with it the agents which make possible its union with the egg and evoke certain cytoplasmic changes which activate the egg cell and thus lead to further development.

In the union of the sperm with the egg a structure participates which forms from the acrosome on the tip of the sperm head. On contact with the egg surface a rigid stiletto-like acrosomal filament is everted to a considerable length (Fig. 174) and penetrates the egg surface. The acrosome reaction can be released by various "unnatural" mechanical and chemical stimuli. The normal stimulus is apparently a species-specific material in the egg surface. When the point of the acrosomal filament has reached the egg cytoplasm, it in turn triggers a characteristic reaction. Yolkfree cytoplasm accumulates and forms a fertilization cone, which often bears pseudopodia and which captures the sperm. In the hyaline cytoplasm the sperm, whose acrosomal filament has been released, moves deep into the egg cell (Fig. 175b–d). The free-swimming sperm uses its own supply of fuel, but within the male or female organism it can also take up energy-rich compounds.

In undiluted semen, echinoderm and fish sperms remain immobile. When the semen is diluted with sea water or fresh water they become active, but their activity lasts only a short time, minutes or seconds. Rapid motility and fertilization competence are lost gradually by the sperms, not suddenly (Fig. 176). The short duration of sperm activity rests in every case on the rapid consumption of the small energy supply. The sperm's energy metabolism can be measured by CO_2 production. The rate of energy utilization depends on the speed of movement. Influences

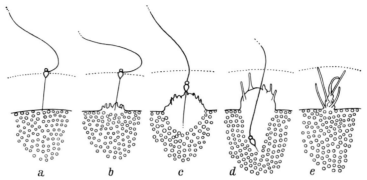

Fig. 175. Entry of the sperm into the egg of *Holothuria atra*, seen in the living egg. (After A. L. & L. H. Colwin, 1957)

which reduce speed without damaging the sperms lengthen their life. To a certain extent, low temperature and low pH reduce sperm activity and increase its duration in this way. In concentrated sperm suspensions, the metabolism of the sperms raises the CO_2 concentration and thus arrests their activity.

Carbohydrates are the main energy source of sperm motility. The addition of glucose prolongs respiration and activity of sperms in many cases, but other substances, particularly albumin, raise the activity of sperms and prolong their ability to fertilize eggs even longer in sea urchin eggs than glucose. It is not clear why this happens. The chemical events of the union of the gametes and activation are still poorly defined. Informative experiments on "fertilization substances" have been performed, particularly in the echinoderms. It is clear that in many animals the sperms themselves release, in addition to CO_2, inhibitory substances that arrest motility and thus conserve energy. What HARTMANN has called Androgamone I can be extracted from dried echinoderm sperm with methanol and is heat-stable. Its chemical nature is unknown. It works in very small concentrations: extracts from sperms of *Arbacia pustulosa* paralyze these sperms in dilutions as great as one to eight thousand. The substance appears neither species- nor genus- nor order-specific. From the eggs of various kinds of animals (echinoderms, molluscs, fish), substances are released into the surrounding water which activate sperms and thus annul the effect of Androgamone I. No common chemical nature has been found for these substances, collectively known as Gynogamone I (G. I), which can be extracted from eggs or from their jelly coats. Experiments on sea urchins and molluscs suggest that G. I is not species- or genus-specific. Inhibition and activation of sperms due to the antagonistic interplay of A. I and G. I are completely reversible and depend only on the concentration of the substances.

After the sperms are attracted chemotactically to the eggs, perhaps through a phobic reaction whose optimum with respect to the concentration of the substance is on the egg surface, fertilization is controlled by further substances given off by the sperms and the eggs, whose nature and function are being investigated vigorously at present. The sperms release a substance which changes the jelly coat of the egg so that the sperms can be drawn to the egg surface. A thermo-labile substance with this effect, a proteolytic enzyme, has been isolated from sea urchin sperms. This is not the same substance that has been included in the Androgamon II (A. II) complex released by the sperms. A specific sperm substance must react with the egg membrane which lies on the surface of the egg. This is a protein membrane which separates from the cytoplasm of the oocyte and can easily be dissolved by trypsin. From the eggs of polychaetes, molluscs, echinoderms, and tunicates, a Gynogamon II (G. II) is given off which has a striking effect on the agglutination of sperms, as experiments with extracts of numerous eggs and jelly coats have shown. The active agent here, "agglutinin," (the "fertilizin" of LILLIE), can be separated from the activating and attractive factor in egg water. In high concentrations it acts on the surface of sperms so that they stick together. In the normal process of fertilization only a small amount of this material is present in the jelly coat and its effect on the sperm which has moved through this coat to the egg surface apparently aids in its entry into the egg cytoplasm (Fig. 175). In the A. II complex a substance has been separated from the factor which dissolves the jelly coat and which acts as an antagonist to Gyno-gamon II and reverses the agglutination of sperms ("antifertilizin").

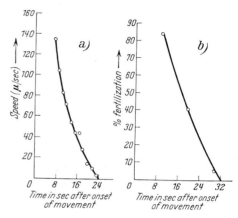

Fig. 176. Trout sperm. a Speed of sperm movement according to cinematographic records of their course; b fertilizing ability of sperms in terms of the percentage of fertilized eggs. (After SCHLENK and KAHMANN, 1935)

The prevention of fertilization by a sperm of the wrong species is effected by a complex system of substances with varying degrees of specificity. The most specific is the species-specific agglutinin reaction. Even between closely related species, sperm agglutination either does not take place or does so to a slight extent and for a short time (a few seconds). This is the case between *Strongylocentrotus droebachiensis* and *Strongylocentrotus pallidus*. In these species male hybrids are found. Their sperms are agglutinated equally well and long by *droebachiensis* egg water and by *pallidus* egg water. This observation shows that the reaction specificity depends equally on the egg secretion and the surface properties of the sperm. Also, the egg membrane forms a somewhat less specific barrier to foreign fertilization: only 5 to 10 per cent of *Psammechinus* eggs are fertilized by *Sphaerechinus* sperms; after removal of the egg membrane, the per cent of fertilized eggs rises to 80 per cent. Therefore the cytoplasmic surface also opposes, though to a lesser extent, the entry of foreign sperms.

For a long time the gynogamones have been characterized more by their biological effect than by their chemical nature. Some of them are certainly not unique but rather common to many kinds of animals, but the experiments show that the process of fertilization is always controlled by a finely tuned system of substances.

As to the processes which proceed after the entry of the sperm head into the egg, we have little precise knowledge except for the echinoderms. The outwardly visible phenomenon which follows fertilization immediately is the rising of the fertilization membrane. It begins as a blister at the point of sperm entry and spreads as a fluid-filled cap covering about a third of the egg in five seconds. After ten to twenty seconds the membrane has lifted off the entire egg surface

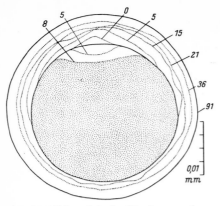

Fig. 177. Lifting of the fertilization membrane in *Psammechinus miliaris* immediately after fertilization. The egg jelly coat is not shown. Scheme derived from analysis of frames of movie film. Numbers: seconds after the first sign of lifting. (After W. & G. KUHL, 1950)

(Fig. 177). It is at first irregularly wrinkled. As it rises further from the egg surface it becomes smooth (Figs. 177, 178). After one or two minutes the whole process is complete. Once the fertilization membrane begins to lift off it is impenetrable to the other sperms which may remain attached to it. These sperms, still moving vigorously, are moved farther and farther away from the egg surface as the membrane lifts away from it. Meanwhile, the egg cell itself undergoes visible changes. In many eggs, even before the fertilization membrane begins to lift off, a series of ripples spreads from the point of sperm entry over the egg surface, and at the point of entry a transient depression is formed (Fig. 177).

The formation of the fertilization membrane is a reaction of the outer cytoplasmic layer of the cortex. The cortex of a mature sea urchin egg is about 1–1.5 μ thick and contains specifically staining granules which are uniformly distributed (Figs. 179a, 180a). The diameter of these cortical granules is about 0.5–1.0 μ and, they number in the neighborhood of 30,000. They are made in the endoplasm and distributed through the entire oocyte; during maturation, cytoplasmic movements carry them into the cortex[455]. The cortex becomes a stiff gel during maturation; during centrifugation neither the cortex nor the distribution of its granules is distorted, while the interior portion of the egg, the endoplasm, becomes stratified in contrast (Fig. 179a). During the cortical reaction which begins immediately after the sperm meets the cytoplasmic surface at its entry point, the cortical granules are singly or severally enclosed in droplets of equal size which adhere to the egg membrane (Figs. 178, 180b). In many species they can be seen for some time in the accumulating perivitelline fluid (Fig. 178). The droplets fuse with the egg mem-

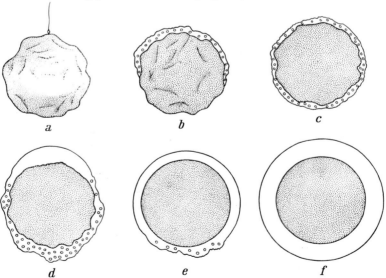

Fig. 178. Formation of the fertilization membrane in *Dendraster excentricus*. Egg jelly coat not drawn. (After CHASE, 1935)

136

brane (Fig. 180c), and this process signals its conversion to the fertilization membrane. With the addition of this cortical material the membrane acquires a negative radial birefringence. If by one means or another the fusion of the cortical granules with the egg membrane is prevented, the granules are converted, evidently due to the effect of a substance released from the egg, from their previous isotropic granular form into birefringent rods in the perivitelline space between the cytoplasmic surface and the egg membrane (Fig. 180d). Normally the substance of these birefringent rods is lodged in the fertilization membrane and also the ground substance in which they are embedded becomes anisotropic. This acquisition of birefringence is reminiscent of the conversion of fibrinogen to fibrin. Thus, through an ordered precipitation, the birefringent, mechanically tough fertilization membrane is formed. The fluid which accumulates between the fertilization membrane and the egg surface (Fig. 178) is obviously highly viscous; for the egg

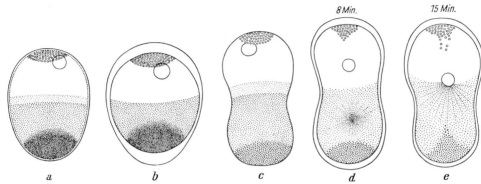

Fig. 179. Centrifuged eggs of *Arbacia punctulata*. a Unfertilized egg; b fertilized egg; c–e eggs fertilized after strong centrifugation. (a, b somewhat diagrammatic sketches after photomicrographs of MOSER, 1939, c–e after HARVEY, 1932)

cell remains concentric with the membrane no matter what its position. The volume of the egg cell is unchanged after the rising of the membrane. The viscous fluid below the membrane has about twice the volume of the egg cell itself. Therefore, as the membrane is raised water must enter from outside to fill the perivitelline space.

With the expulsion of the granules, a change takes place in the fine structure of the cytoplasm. The cortex becomes a firm gel. This is seen in the difficulty with which fertilized eggs are deformed. In contrast, unfertilized eggs can easily be distorted by mechanical pressure and separated into smaller spheres; a fertilized egg resists much stronger pressure, and when it finally bursts the endoplasm flows out without a general breakdown of the cortex. Again, while centrifuged unfertilized eggs are stretched in the centrifugal axis (Fig. 179a), fertilized eggs lengthen little or not at all during centrifugation, even though the endoplasm becomes stratified (Fig. 179b). Unfertilized eggs round up a short time after centrifugation ceases. If they are fertilized immediately, however, they retain their dumbbell shape, and the fertilization membrane follows the surface contour of the egg cortex, which now stiffens in a distorted form (Fig. 179b). One can remove these eggs from their fertilization membranes in a pipette, using strong suction, and even then they retain their form and so begin cleavage.

In addition to the cortex, the endoplasm also undergoes structural change which can be recognized in part in the stability of its various components to centrifugation. When unfertilized *Arbacia* eggs are centrifuged, five layers separate out (Fig. 179a). At the centripetal pole lies an oil droplet layer. Next is a wide layer of clear cytoplasm (hyaloplasm). The egg nucleus lies between the first two layers. Next, about the equator a narrow finely granulated layer forms,

137

and above it the thick yolk layer. The pigment granules of the cortex are displaced sufficiently easily in unfertilized eggs so that they form a cap at the centrifugal pole of the egg (Fig. 179a, c). After fertilization these pigment granules are displaced only by very strong centrifugation. In the stratification the narrow finely granular layer between the hyaloplasm and the yolk layer disappears simultaneously with the spinning down of the cortical granules (Fig. 179b). Experiments on the protein content of unfertilized eggs have revealed changes in solubility and molecular weight which evidently are related to the formation of a cross-linked structure in the colloid of the egg immediately after fertilization.

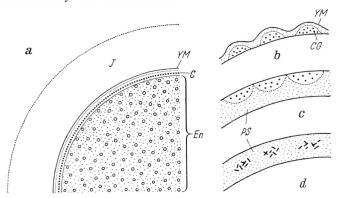

Fig. 180. Diagram of the layers of the cytoplasm of echinoderm eggs (a) and the course of the cortical reaction. *C* cortex; *CG* cortical granules; *YM* yolk membrane; *En* endoplasm; *J* jelly coat; *PS* perivitelline space. (After MOSER, 1939, and RUNNSTRÖM, 1949)

A series of metabolic changes are known to take place immediately after fertilization: liberation of calcium ions, transient formation of an acid whose nature is unknown (but it is not lactic acid), proteolysis, and synthesis of high-molecular-weight carbohydrate and of ATP—but the relationship of these changes to the structural alterations, as well as to the onset of development, is unknown.

The condition of the egg nucleus at the time of fertilization varies even within the echinoderms. The sea urchin egg completes its maturation divisions in the ovary. After the extrusion of the second polar body, the aster at the egg nucleus disappears. No visible egg cytocenter remains, and the egg is released into the sea water in fertilizable condition. Starfish eggs are released in the germinal vesicle stage, and maturation begins upon their release: In *Asterias forbesi* the nucleolus of the oocyte nucleus disappears in eight to nineteen minutes at 16–18°. After 72 to 90 minutes the first polar body is formed and the second one appears after 105 to 119 minutes (Fig. 181). The eggs can be fertilized as early as the onset of nuclear change. Then the sperm centriole divides during the maturation divisions of the oocyte nucleus (Fig. 181a); but only after their completion, at least two to three minutes after the extrusion of the second polar body, does the aster grow around the sperm centrioles. This sperm aster can be seen even in the living egg. Both sets of astral rays now quickly pervade the entire egg and the egg nucleus unites with the sperm nucleus (Fig. 181b). Cytokinesis follows some six to eight minutes after the appearance of the sperm aster (Fig. 182). The period between fertilization and the appearance of the asters takes at least 35 minutes at 16°. If eggs are fertilized during or immediately after the formation of the first polar body, the sperm aster appears shortly after the formation of the second polar body. In eggs fertilized earlier, the time from fertilization to appearance of the sperm aster is lengthened by the time between fertilization and the beginning of formation of

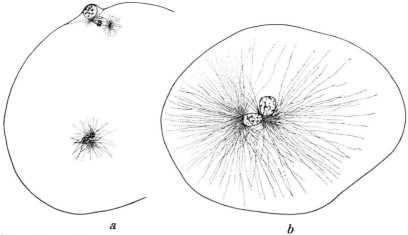

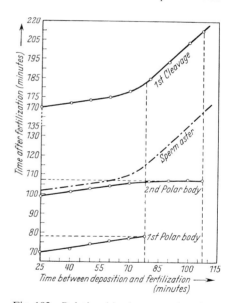

a   *b*

Fig. 181. Maturation and fertilization of the *Asterias forbesi* egg. (After Wilson and Mathews, 1895)

the first polar body. Fertilization hastens the onset of the first maturation division (Fig. 182). The time between the first and second maturation division remains the same. Microsurgical experiments have thrown some light on the relationship between maturation and sperm aster formation. Immature eggs and unfertilized eggs in the process of maturation were divided by a needle into two halves with the aid of a micromanipulator. Both halves could be fertilized and developed, one a diploid and the other a haploid with a sperm nucleus alone. When the eggs were cut up to ten minutes before the time of the first polar body formation and both halves then fertilized, the sperm aster in the half without an egg nucleus formed earlier than in the half with an egg nucleus, and the first cleavage division started considerably earlier in the former half (Fig. 183a). After cutting and fertilizing immediately before or during the formation of the first polar body, the sperm asters appeared simultaneously in both halves two to three minutes after the formation of the second polar body in the fragment with the egg nucleus. In addition, some eggs were fertilized 25 minutes after their release into the sea water, that is, shortly after the breakdown of the germinal vesicle; they were then cut in two at various times until shortly after the formation of the first polar body (Fig. 183b). If the egg and sperm nuclei were separated as a consequence, one half developed a sperm aster while the other half proceeded with maturation. After cutting 35 minutes after release into sea water, the sperm aster appeared shortly after the first polar body in the fragment with the egg nucleus and two to three minutes before the second polar body. In eggs cut fifty minutes after release, the sperm aster appeared at the

Fig. 182. Relationship between the time of fertilization after the release of eggs into sea water (abscissa) and the time of extrusion of the first and second polar bodies, appearance of the sperm aster, and the first cleavage division (ordinate) in *Asterias forbesi* at 16°. The thin dotted horizontal lines mark the time of extrusion of the polar bodies in unfertilized maturing eggs. (After R. & E. L. Chambers, 1949)

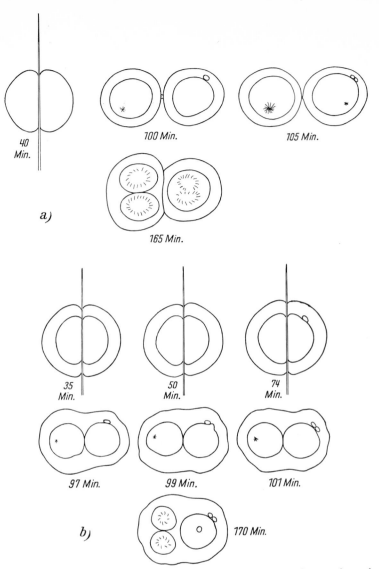

Fig. 183. Cutting experiments with *Asterias forbesi* eggs. a Transection forty minutes after release into sea water, both parts fertilized at once, condition of the egg halves after 105 and 165 minutes; b three eggs fertilized 25 minutes after release and cut ten minutes later (35 min.), 25 minutes later (50 min.), and 49 minutes later (74 min.), just after extrusion of the first polar body. Second row: the same eggs at the appearance of the sperm aster (after 97, 99, 101 minutes). Below: condition of all three eggs after 170 minutes. (After R. & E. L. CHAMBERS, 1949)

same time as the second polar body of the other fragment. In eggs cut after 74 minutes, in this case just after the formation of the first polar body, the sperm aster appears only after the second polar body. Cleavage takes place, as expected, only in the fragment with the sperm nucleus. The egg nucleus buds off the first and second polar bodies at the same time as simultaneously-fertilized, intact control eggs and moves back into the interior where it remains. These experiments show that the egg nucleus and the sperm influence one another reciprocally: maturation is hastened by fertilization; the sperm nucleus evidently releases a chemical agent; it is not necessary for the sperm nucleus to remain in the cytoplasm of the egg. In a reciprocal effect, the growth of the sperm aster is delayed by the presence of the egg nucleus. In this way the sequence of events in the division cycle during maturation and cleavage after early fertilization are regulated. The extension of the sperm aster rays into the cytoplasm is prevented as well as their interference with the chromosomes undergoing meiosis (cf. Fig. 181).

Although in the starfish egg maturation is only hastened by the sperm, in other forms maturation either does not begin, or remains at a certain stage, until a certain functional change is induced by fertilization.

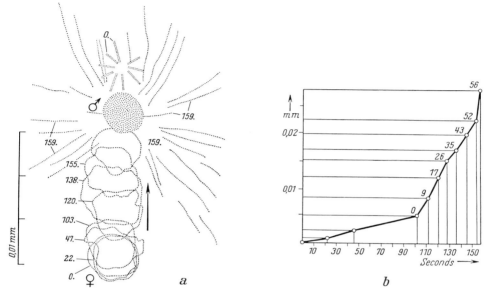

Fig. 184. Movement of the egg nucleus towards the sperm nucleus in *Psammechinus miliaris*. a From the analysis of frames of movie film. Numbers: seconds after the first observation. In the sperm aster are a few long rays which have remained unaltered. b Distance-time curve of the movement of the egg nucleus; at 103 seconds there is a sharp increase in speed which lasts 56 seconds. (After W. & G. KUHL, 1950)

While the formation of the fertilization membrane can take place anaerobically, oxygen appears to be necessary for the movement of the sperm nucleus, its swelling, and the formation of the sperm aster.

They rays of the astral sphere which gradually spread over a great range of the egg cytoplasm are reflected in the living cell by the arrangement of the granular inclusions. In centrifuged eggs they are visible in the yolk zone but not in the hyaloplasm. That they are nevertheless present and indeed actually moving in the cytoplasm can be seen from the displacement of the oil droplets and the pigment granules which stream towards the center of the sphere (Fig. 179e).

In movies one can see that in the sea urchin egg the approximation of the two pronuclei is mainly due to the movement of the egg nucleus. Successive frames show the sudden onset of movement exactly in the direction of the sperm aster. Time lapse movies give the impression that the egg nucleus collides with the rays, at which point it becomes amoeboid. Figure 184 gives an example. The egg nucleus moves slowly at first, then with increasing speed. While it moves 25 μ. in 137 seconds, the sperm nucleus with its aster scarcely covers 10 μ. The time lapse film shows that the rays of the sperm aster remain in a fixed position for a relatively long time. During

141

the advance of the aster center with the nucleus, the rays often undergo considerably curling; the stiff rays are bent in the viscous cytoplasm. After the union of the egg and sperm pronuclei, the zygote nucleus moves with the aster toward the cell center much more slowly than the egg nucleus moved toward the sperm nucleus.

Vigorous cytoplasmic streaming, which the movies reveal through the movement of granules, plays a role in the union of the pronuclei in echinoderm eggs. In other eggs, such as certain nematodes, they carry the egg pronucleus to the sperm pronucleus. While in the echinoderms the first consequence of fertilization is a gelling, especially in the cortex, fertilized nematode eggs

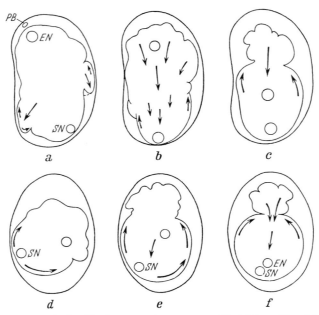

Fig. 185. Cytoplasmic streaming in fertilized nematode eggs. a–c *Rhabditis pellio*; d–f *Diplogaster longicauda*. *EN* egg nucleus; *PB* polar body; *SN* sperm nucleus. (After Spek, 1918)

undergo vigorous amoeboid movement indicating that their cytoplasm is of low viscosity. Cytoplasmic movements resulting in changes of form and streaming can stem from local differences in surface activity affecting viscous droplets of one liquid flowing in a second liquid. At first irregular pseudopod formation takes place over the whole egg surface (Fig. 185a). It stops as soon as the sperm nucleus has swelled to a certain size. Then a bump arises at the place on the cell surface under which the sperm nucleus lies, and toward this point cytoplasm flows in the egg axis. From the tip of this pseudopod the cytoplasm streams superficially away toward the opposite egg pole (Fig. 185b–f). The axial stream leads the egg nucleus to the sperm nucleus. This suggests that the "fountain-streaming" of the cytoplasm is caused by the position of the sperm nucleus near the egg surface and that here, until the union of the pronuclei, a difference between the egg surface and the rest of the egg is established. The cytocenter next to the sperm nucleus forms an aster which can often be seen clearly in the living egg in the stage in which the superficial streaming of the cytoplasm begins from its origin above the sperm nucleus. Whether a chemical or physical effect emanates from it or from the swelling pronucleus and causes a change at the nearby surface, and whether there is a change in surface tension here at a liquid interface cannot be decided.

With the fusion of the pronuclei, fertilization is complete and embryonic development begins. The entry of the sperm into the egg evokes, from the lifting of the fertilization membrane to the first cleavage, a chain of events which is collectively known as the activation of the egg. We know nothing of the causal connections within this chain of events. The phenomenon of natural parthenogenesis and its experimental achievement show that activation can take place without a sperm nucleus and a centriole, but the various means by which parthenogenetic development can be set in motion unfortunately tell us nothing about the critical factors which normally come from the sperm. We see here, as in mitosis and meiosis, an event in a cell whose causality we do not understand.

Lecture 10

The first sign of development in multicellular plants and animals is generally polarity: differentiation along a particular body axis, especially at its ends. Thallus pole and rhizoid pole, shoot and root poles, apical and basal poles, anterior and posterior body poles become recognizable as early landmarks at the ends of the primary axes in the various known types of organization. Even the first division of the initial cell is almost always oriented according to a major axis.

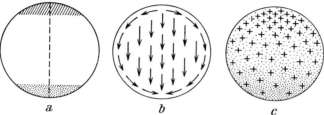

Fig. 186. Diagram of possible kinds of polar differentiation. a Polar field polarity; b structural orientation polarity; c gradient polarity

Theoretically one can represent the invisible polar differentiation of a cell in various ways: (1) one pole or both poles at the end of an axis can be distinguished by particular structures, by the presence of certain substances, or by special physiological properties which endow the cells that later arise there with particular characteristics (Fig. 186a); (2) structures which pervade the entire cell either internally or in the cortex can be oriented in a polar fashion (Fig. 186b), and the polarization will be transmitted to the cells which arise from the egg cell, determining the polarity of each part; (3) a gradient of substances or functions may run along the egg axis, or two opposing gradients can cross one another (Fig. 186c). For short we can term these three types of polarity as polar field polarity, structural orientation polarity, and stratification- or gradient polarity. There may be even more types than this in nature, and the three types described may occur in combinations; but these three categories are worth considering, since the various theoretical possibilities lead us to expect diverse outcomes from experiments.

A gradient polarity in a self-differentiating giant cell was suggested by the experiments on *Acetabularia* (p. 106): the gradients of apical substance and rhizoid substance can be shifted until shortly before cyst formation; the polarity remains labile. A structural contrast (rhizoid region vs. growth region) must account for the distribution of substances along a gradient. In multicellular organisms the polarity of the egg cell or the spore entails a certain property in the daughter cells.

When and how is the polarity of the initial cell determined?

The spores of Bryophytes and Pteridophytes and the eggs of algae which develop in the open can acquire polarity from environmental influences. Usually the direction of light determines the major axis corresponding to the later functional form of the plant. In the spore of the horsetail *Equisetum* the first cleavage wall separates a basal cell, the rhizoid cell, from an apical cell, the prothallus cell. This dividing wall is perpendicular to the incident light rays, and the rhizoid cell is on the side away from the light. The light effect is seen even before cell division in the distribution of the cell contents. Before the spore germinates, its nucleus lies in the center, surrounded by chromatophores (Fig. 187a). The first sign of germination is the displacement of the chromatophores toward the light by cytoplasmic streaming and the movement of the nucleus to the shady

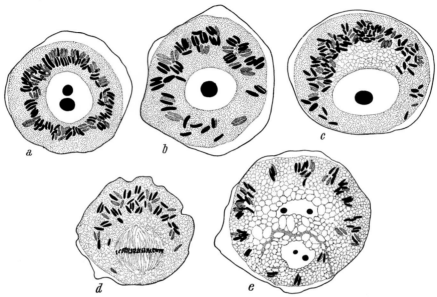

Fig. 187. Sections through *Equisetum spores*, which have been polarized by light. Illumination in 5 per cent gelatin to keep light orientation constant. (After NIENBURG, 1924)

side (Fig. 187b, c). There the nucleus divides, and the spindle forms in the light axis (Fig. 187d). The eccentric position of the spindle results in the separation of a small rhizoid cell from the larger thallus cell by an hourglass-shaped wall (Fig. 187e).

Germination of the eggs of the Fucaceae offers particularly favorable circumstances in which to investigate the determination of polarity by various influences. The eggs are released into the sea water where they are fertilized. After fertilization a jelly coat is formed which fastens the egg to the substratum; if they settle on a glass slide they are easy to manipulate. *Fucus vesiculosus* germinates at 15–17° about seventeen hours after fertilization. In contrast to the *Equisetum* spore, the rhizoid, which indicates the polar axis, grows out even before nuclear division; then the spindle is formed in this axis of extension, and the cleavage wall separates the thallus cell from the rhizoid cell (Fig. 190a). The rhizoid grows away from the light (Fig. 188a) if the egg cell is illuminated for at least two hours during a sensitive period which runs from the tenth to the fifteenth hour after fertilization. If the eggs are illuminated simultaneously with two beams at an angle less than 180°, the polar axis is the resultant of the two light axes (Fig. 188b). The longer the light is on in one direction, the less the polar axis is diverted by the second beam. After fifteen hours the polarity is stable. If the eggs are illuminated by two opposing beams of equal

144

strength and duration (Fig. 188c) the rhizoid forms perpendicular to the two opposing light axes; frequently two rhizoids arise at opposite ends of the cell. (Unfortunately, their further development has never been followed.) In *Cytosira* the polarity becomes fixed much earlier than in *Fucus*. *Cytosira*'s light sensitivity sets in right after fertilization, rises in the first hour and disappears after two to two and a half hours at 18°. During this time illumination for as little as half an hour has a clear effect on polarity. At lower temperatures longer light periods are needed than at higher temperatures.

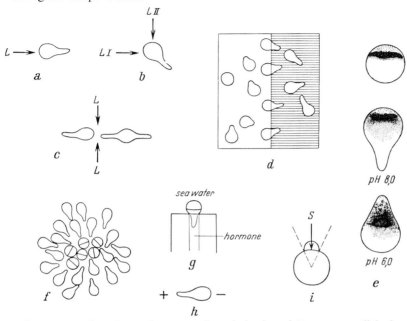

Fig. 188. Schematic representation of experiments on the polarization of *Fucus* eggs. a light from one side only; b light first in direction L I, then L II; c opposing light beams of equal strength; d partial illumination using a half-transparent and half-opaque substratum; e behavior of ultracentrifuged eggs during development in sea water in various pH's; f group effect; g local application of growth hormone; h in an electric field; i polarization effected by sperm entry. [After experiments of KNIEP, 1907 (a–c), NIENBURG, 1928 (d), WHITAKER, 1931, 1939 (e, f), DU BUY and OLSON, 1937 (g), LUND, 1923 (h), KNAPP, 1931 (i)]

The direction of the light does not influence the polarity of the cell directly, but rather indirectly through the differences in light intensity in the parts of the cell. If one illuminates from below *Fucus* eggs or *Equisetum* spores on a plate which is half transparent, half opaque, those cells on the dividing line grow toward the dark part (Fig. 188d); the division wall now runs parallel to the light axis.

If *Cytosira* eggs are centrifuged for fifteen minutes some time during a period from one to three and a half hours after fertilization and kept in the dark both before and after centrifugation, the rhizoid grows out in the direction of the centrifugal force. The chromatophores and the nucleus are centripetally displaced. Clear plasma accumulates at the centrifugal end, and this is where the rhizoid appears. If the eggs are illuminated until an hour after fertilization and then centrifuged, only the centrifugal force affects the axis of polarity; illumination after centrifugation has no effect either, but longer illumination before centrifuging produces a polar axis intermediate between the axes of centrifugal force and light.

The displacement of the chromatophores is, however, not a decisive determinant of polarity. If *Equisetum* spores are centrifuged shortly before nuclear division the unequal cell division will be carried out according to the previously established polarity, even though numerous chloroplasts will now be found in the rhizoid. The polarity is therefore no longer bound to the distribution of those cell components which can be displaced by centrifugation. The polarity must now be structural and lodged in the immovable ectoplasmic layer, the cortex. This stable polarity can be followed by a labile gradient polarity. It can, however, also be abolished immediately by various agents. If sometime during the first twelve hours after fertilization the cell contents of *Fucus* eggs are stratified in an ultracentrifuge at 50,000–200,000 g., rhizoids are formed in the dark at the centrifugal pole, as they are after ordinary centrifugation when the pH of the sea

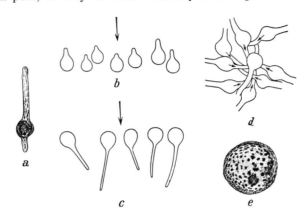

Fig. 189. Germination of *Funaria hygrometrica* spores. a Protonema: above, chloronema; below, rhizoid; b first outgrowth of the chloronemata in strong light; c first outgrowth of the rhizoid in similarly strong light but with addition of β-indolacetic acid to the agar; ↓, light axis; d group effect; the arrows show the direction of outgrowth of the rhizoids; e apolar growth in β-indolacetic acid. All figures to the same scale. (b–d after HEITZ, 1942; a, e after unpublished photographs of HEITZ)

water is about 8.0 (Fig. 188e). Surprisingly, the effect of the stratification is just the reverse when the eggs are placed after ultracentrifugation in acidified sea water at pH 6.0: now almost all the rhizoids grow out at the centripetal pole (Fig. 188e).

In an electric field, germinating *Fucus* eggs form rhizoids on the anodal side. This points to the electrophoretic movement of some substance (Fig. 188h).

Local external chemical differences can also influence polarity: crowded *Fucus* eggs form rhizoids toward the center of the group (Fig. 188f). Ostensibly a substance is released by the cells and establishes a radial concentration gradient. This group effect has been ascribed to the production or consumption of CO_2 or O_2 as a result of the cells' metabolic activities. But no effect of these gases has ever been demonstrated. The group effect, like the stratification effect, depends on the pH of the medium: in unusually alkaline (pH 8.4) sea water it is negative: crowded *Fucus* eggs form their rhizoids toward the periphery. In ordinary sea water the direction of rhizoid outgrowth in crowded ultracentrifuged eggs is a resultant of the two effects. In acidified sea water the group effect is dominant. If the growth hormone β-indolacetic acid is applied locally with a microcapillary to a small part of the *Fucus* zygote surface (Fig. 188g), then the rhizoid forms there. Since the egg cells of *Fucus* actually contain more growth hormone than the vegetative cells of this brown alga, one might surmise that a hormone gradient in the cell plays a role in the determination of polarity.

Spores of *Funaria* germinate at low light intensities, with rhizoids forming on the side away from the light; these contain a few thin chromatophores. Later, broader chloronemata, containing numerous chromatophores, are formed on the light side (Fig. 189a). In bright light the photopositive chloronemata grow out first (Fig. 189b). The *Funaria* spores also show a group effect (Fig. 189d) similar to that in *Fucus* eggs. If growth hormone is added, the photonegative rhizoids sprout first even in bright light (Fig. 189c). This effect is seen in concentrations as low as 0.01 microgram per liter. But when the hormone is added in the dark the polarity is absent: the spores burst their cases and develop into giant spheres which have forty to fifty times the volume of a spore (Fig. 189e). This agrees strikingly with the phenomenon frequently seen in *Fucus* eggs, which often divide without polarity in the center of a dense group (Fig. 188f). The hypertrophy of the apolar *Funaria* spores is true growth, not simply swelling caused by water uptake. The cytoplasmic mass increases, and the chromatophores multiply vigorously. Unfortunately, nothing is known of the further development of this apolar form. Chloral hydrate

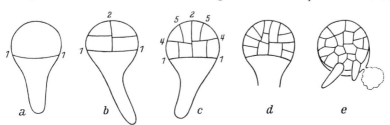

Fig. 190. *Fucus* embryos. a–d The first cleavage walls; a between apical and basal cells, b–d divisions drawn only in the apical cell; e formation of rhizoids from the basal thallus cells after removal of the original basal cell in the two-cell stage. (a–d after NIENBURG, 1928–1933; e after KNIEP, 1907)

does not inhibit the cleavage of *Funaria* spores, but the polarity is removed, and with it the differentiation of the chloronema- and rhizoid-cells; instead an irregular mass of similar round cells arises[97, 721].

In any case, the polarization of the egg cells is neither an instantaneous nor an immutable process; rather, it is labile at first and becomes fixed over a certain period of time, during which the axis of polarity can be moved. This is shown by the fact that it is a resultant among sequential effects of different light axes and between the light axis and the axis of centrifugal force. All these effects on the fertilized egg cell during the sensitive period are imposed on a polar tendency already established by fertilization. At the point of sperm entry a papilla arises on the *Fucus* egg cell, and from it a wave of wrinkles spreads over the egg in about a minute, which is similar to what is observed in animal eggs (p. 136) when the fertilization membrane starts to lift off. Then the egg surface gradually becomes smooth again as the egg contracts. In the microscope, fertilization was followed on a glass slide and the outline of the eggs drawn with special attention to the site of the papilla; then the eggs, firmly fixed to the slides, were placed in the dark. On the following day the site of the rhizoid was observed. The results showed clearly the influence of the sperm entry point on the polar axis: in almost all the eggs the rhizoid grew out to the point of sperm entry or within 45° of it (Fig. 188i). Ostensibly, the entry of the sperm results in an ordered but still labile distribution of material.

The state of determination of the apical cell in the two-cell stage of the *Fucus* embryo can be studied by isolation experiments, which lead to a conclusion on the nature of the polarity of the egg. The removal of the apical cell is technically difficult, but the rhizoid cell can be removed in a simply way. The sea water is diluted by the gradual addition of distilled water to one-third the normal concentration. At this point the rhizoid cell bursts, extrudes part of its contents, and dies. If the embryo is now returned gradually to sea water of normal concentration, the apical cell divides further, and now presumptive thallus cells grow into rhizoids (Fig. 190e). The capacity for rhizoid formation is therefore not restricted to the basal cell, and similarly (in the sense of our diagram, Fig. 186a), the rhizoid-forming substances are not restricted to one place. But only the basal cells among those which can form the apical cell form rhizoids whether the embryo fragment has been placed in the dark immediately after the operation, or lighted from one side, even from the basal side. Therefore, the apical cell and its cell progeny have a fixed polarity and the internal conditions are most favorable for rhizoid formation at the basal end. Whether this polar differentiation is based on an ordered structure or on a concentration gradient is not

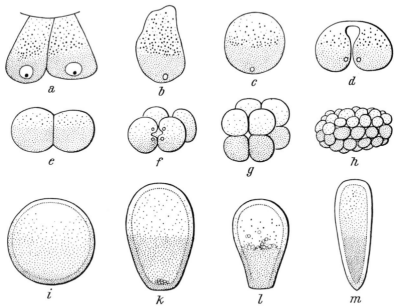

Fig. 191. Development of *Amphisbetia operculata* (Hydrozoa) to planula. a–c Oocytes; a in spadix ectoderm; b, c after release; d–h cleavage stages; i blastula; k, l gastrulation (polar ingression); m planula. Stippling indicates orange pigment in the vegetal half of the egg. cleavage stages and endoderm. 100×. (After TEISSIER, 1931)

known. After this demonstration of the ability of the thallus cells to form rhizoids this question remains: why do the rhizoids form when the connection with the basal body is maintained? One can only suspect that the rhizoid cell exerts an inhibitory influence, but this is a matter relating to the relationship of differentiating cells with one another; it does not concern the primary polarity of the egg cell.

The polarity of animal eggs is often seen during the growth of the oocyte, in the distribution of cytoplasmic inclusions along an animal-vegetal axis, and can often be traced back to the position of the oocyte in the maternal tissue. A good example is seen in the eggs of *Amphisbetia* (*Sertularia*) *operculata*. These begin as an epithelium in the ectoderm of the gastric canal of the free-swimming medusa (Fig. 191a) and are given off individually. The egg cell clearly shows a polar structure: in the outer half of the cell, in this epithelial tissue, is a concentration of orange pigment. The distribution of the components is not altered during release, rounding up, maturation, and fertilization (Fig. 191b, c). The very regular cleavage of the free suspended eggs (Fig. 191d–h) leads to a spherical blastula. Its cells are all similarly shaped, but the pigmentation difference along the egg axis is still clear (Fig. 191i). The blastula stretches in the direction of the egg axis and the embryo transforms into the free-swimming planula larva. The endoderm is formed by ingression of cells at the posterior (vegetal) pole (Fig. 191k, l). They gradually fill

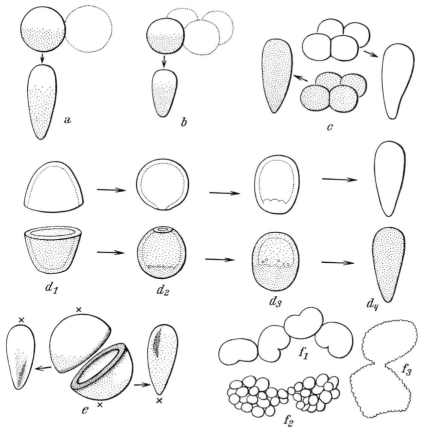

Fig. 192. Isolation experiments von *Amphisbetia operculata* embryos. a, b development of isolated half- and quarter-blastomeres to the planula; c, d separation of animal and vegetal halves in eight-cell- and blastula stages; e oblique transection of the blastula and vital staining of the cut surface, retention of the egg polarity in the planula; × marks the primary pole; f development of an embryo after displacement of the cells in the four-cell stage; f_1 4/8 cell stage (cf. Fig. 191f); f_2, f_3 no differentiation to an individual planula, separation into two cell groups. (After TEISSIER, 1931)

the entire blastula cavity (Fig. 191m). All the pigmented cells become endoderm cells, and so the prospective fate of blastomeres is seen by their pigmentation. This, however, does not mean determination. To consider this question we must turn to isolation experiments: not only half- and quarter-blastomeres (Fig. 192a, b), which contain all the layers in the egg axis, but even the four unpigmented cells and the four pigmented cells of the eight-cell stage develop into normal planulae. Thus, on the one hand presumptive ectoderm alone and on the other hand presumptive endoderm alone give rise to normal larvae (Fig. 192c). Indeed, even single animal cells and single vegetal cells isolated from the eight-cell stage become tiny blastulae and planulae.

The animal (anterior) and vegetal (posterior) halves of the blastula (Fig. 192d) are also still not firmly determined with respect to the formation of germ layers. If these halves are separated (Fig. 192d_1), the equatorial cuts close up, and cells enter each half at its posterior pole until the cavity is filled (Fig. 192d_2–d_4). About half of the cells which normally would have entered at the posterior end now become the ectoderm of the swimming larva that forms from the posterior half: those presumptive endoderm cells from the equatorial region which normally would have

149

moved inside last are now at the anterior pole. This means that they are in the front part of the planula as it moves, which is the part that ultimately fastens to the substrate, whereas they normally would have been in the endoderm in the cap of the growing stalk. In a similar departure from custom, the front half of the cut blastula sends about half ot its presumptive ectoderm cells into the interior as endoderm cells. Even quarter blastulae close over the large, cut edge and produce larvae. The cells during cleavage and the pieces of the blastula wall therefore show a regulatory ability like that of parts of the aggregation field, and of the migrating pseudoplasmodium, of *Dictyostelium*. Nevertheless, the polarity of the egg is maintained. The most vegetal cells always move inside. If a blastula is cut obliquely and the cut surface vitally stained, the group of colored cells which mark the region of wound closure assume antero-lateral positions in some planulae, and postero-lateral in others (Fig. 192e). From the fact that the embryonic cells retain their polarity, we can also see a limit to their ability to regulate and form a harmonious whole: if one disturbs the array of blastomeres (Fig. 192f) so that their individual vegetal poles, recognizable by the pigment, are randomly oriented, the embryonic cells do not organize into a characteristic blastula, but abnormal embryos or cell groups arise instead which separate from one another (Fig. 192f) and then form individual planulae. Embryonic cell groups from *Amphisbetia* therefore behave analogously to halves of migrating pseudoplasmodia of *Dictyostelium discoideum*, which are experimentally disorganized (Fig. 167a). In *Dictyostelium* it is a matter of polarized wandering troops of cells, while in *Amphisbetia* one is dealing with a family of cells which retain, as they divide, the polarity established in the oocyte.

A relationship between egg polarity and the position of the egg in the ovary can frequently be established. In *Ascaris megalocephala* the young oocytes grow on the rachis which runs down the long axis of the egg tube, and they are arranged radially and closely packed. They have the shape of long, narrow wedges whose base is against the wall of the egg tube and whose point is on the rachis. After growth is completed the oocytes separate from the end of the rachis. They then normally round up, are fertilized, and form a close-fitting chorion. The first polar body is then extruded from the highly vacuolated oocyte. Then the contents of the vacuoles are released; the oocyte retracts from the wall and shrinks markedly. At this point, the second polar body is given off and remains fixed to the egg surface. In some *Ascaris* females the eggs do not round up in either oviduct, ostensibly due to a genetically-determined abnormality in the oocyte cytoplasm. The oocytes released from the rachis alter their arrowhead shape only to a pear shape and then form a pear-shaped chorion (Fig. 193a, b).

This abnormality offers the possibility of studying the origin of the polarity of the *Ascaris* egg. The sperm cells can enter the oocyte anywhere at all, at the tip, at the base, or somewhere in between (Fig. 193a). The first polar body is found in most cases at the base of the egg chorion. The second polar body is extruded on the same side (Fig. 193b, c). The spindle of the first cleavage division is formed in the egg axis in *Ascaris* (Fig. 193b). In the formation of the four-cell stage the animal (dorsal) cell divides meridionally while the vegetal (ventral) cell once more forms its spindle in the egg axis. In over 90 per cent of the pear-shaped eggs which develop normally to this characteristic stage, the animal cells lie at the base of the pear-shaped chorion. The axis of polarity of the mature egg therefore corresponds to the long axis of the wedge-shaped growing oocyte, and the blunt end of the oocyte, which has been apposed to the oviduct wall, now becomes the animal pole of the mature egg.

In sea urchins and Holothurians the vegetal pole of the oocyte is attached to the wall of the ovary; the polar bodies are pinched off at the free end of the cell in the so-called micropyle (Fig. 202a)[407].

Since the egg polarity is often expressed in a polar arrangement of materials, one might suspect that polarity is brought about in a growing oocyte by the impact of some immediate environ-

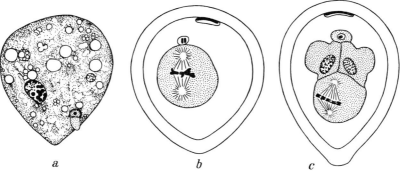

a   *b*   *c*

Fig. 193. Pear-shaped egg and cleavage stages, *Ascaris megalocephala,* from a female whose oocytes did not round up normally

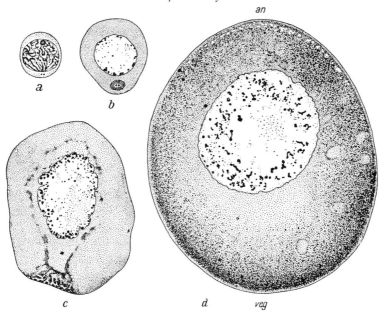

Fig. 194. Growth and development of polarity in amphibian oocytes. a *Amblystoma*; b–d *Rana, an, veg* animal and vegetal poles. a 500×, b–d 200×. (Drawings by FREIBERG after microphotographs by WITSCHI)

mental factor on metabolism; that due to unilateral contact with sea water or extracellular fluid, or through a local contact with blood vessels or nurse cells, an oxygen or nutritional gradient is established, leading to the localization of particular metabolites in the cell. In many cases morphological studies lead to such a notion, even when experimental proof is lacking. However, this assumption is often taken as proved, perhaps due to the preconceived notion that "apolar" oocytes *must* be polarized by an external influence, since it is believed that the axis of polarity cannot be left to "chance." But even in the oocyte an axis is already defined: the axis in which the centriole lies next to the nucleus. Its effect has already been seen in the orientation of leptotene chromosomes in the beginning of meiosis (Figs. 39a, 194a); moreover, in many cases, e.g., in amphibian and fish eggs, it is known that the oocyte nucleus-centriole axis becomes the major axis of the egg (Fig. 194). In the region of the centriole, mitochondria accumulate and multiply,

151

as well as other structures related to yolk formation, which form a so-called yolk nucleus (Fig. 194b). From here the mitochondria and liposomes spread out (Fig. 194c), and on this side of the egg increased yolk protein formation begins, while on the opposite side lipid droplets arise (Fig. 194d). Apparently, in many cases, the position of the centriole with respect to the nucleus autonomously determines an axial difference in metabolic states in the oocyte amidst uniform or not critically asymmetric inflow of oxygen and nutrients. When the polarity of the cell is determined by external influences, they also determine a certain displacement of the centriole with respect to the nucleus, or the direction of movement of daughter centrioles. This is shown by the orientation of spindles in the experiments on *Fucus* eggs and moss spores.

Since in a variety of circumstances the polarity of the egg cell is revealed by the visible layering of substances during oocyte growth, one might ask whether they are responsible for the polar differentiation which determines the polar behavior of embryonic cells. One might hope that through an exaggeration of the layering or through an alteration of its plan, some conclusion might be obtained on the nature of egg polarity. For this reason an abundance of centrifuge experiments have been performed. We shall first discuss only experiments with *Hydractinia echinata* eggs. In these hydroids endoderm formation begins during the late cleavage stages through multipolar ingression (Fig. 195d, e). The spindles are not parallel to the embryo surface as is usual in blastula formation, but some are even radial, and thus daughter cells are given off into the segmentation cavity (Fig. 195d). Once inside, these continue to multiply as endoderm cells at the same rate as those on the outside (Fig. 195e). This multipolar endoderm formation reflects a radial polarity of the egg cell. Gradually the entire segmentation cavity is filled. Then the solid embryo stretches into a planula. Vital staining of the animal or vegetal pole in an early cleavage stage (Fig. 195a–c) shows that in spite of the multipolar endoderm formation, the primary egg axis determines the long axis of the planula just as in *Amphisbetia*: the animal pole of the egg is always at the front end of the planula. Even gentle centrifugation stratifies the egg: at the centripetal pole, oil droplets accumulate; at the centrifugal pole, yolk and mitochondria; and between lies a zone of clear cytoplasm which in the normal egg forms both a coherent layer beneath the egg surface and a network among the granular contents. The nucleus can appear in any one of these strata in the centrifuged egg, and the first cleavage can be at any angle with respect to the strata (Fig. 195f–h). The size of the daughter cells varies greatly according to whether the cells contain yolk or only clear cytoplasm and oil droplets. If the cells are separated at the two-cell stage (Fig. 195f) they undergo normal development, whether one contains only cytoplasm and the other the various other layers, or whether one contains the oil cap and cytoplasm, or cytoplasm and yolk granules. The planulae produced have their normal polar differentiation in spite of their diverse contents and size. The cells of the planula ectoderm and endoderm nevertheless contain different materials (Fig. 195g, h). Therefore, neither the cleavage plane, nor the release of endoderm cells into the interior, nor the lengthening of the planula and the consequent formation of a front end, that is to say, neither the radial polarity nor the apical-basal polarity are controlled by the redistribution of those substances which are stratified by centrifugal force. There must be a submicroscopic structure present pervading the microscopically visible inclusions which move in opposing directions through the cell, and this submicroscopic structure either must not be affected or must be restored to its original condition by the properties of its molecular components after being disoriented. It is difficult to imagine a pervasive cytoskeleton within a cell which is so rigid or whose mesh is so coarse that the movement of so many rather large inclusions does not tear it apart. The polar organization must therefore either be easily repaired, or it must be separated from the flow of granules and oil droplets. The latter alternative would mean that the polar structure lies in the cortex, which is not altered by centrifugal force.

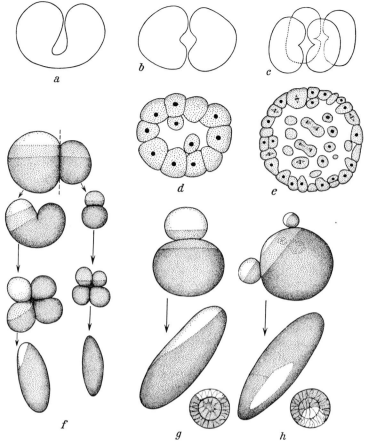

Fig. 195. *Hydractinia echinata*. a–e Normal cleavage, multipolar endoderm formation; d, e optical sections; f, g, h cleavage and planulae formation after stratification of the egg contents by centrifugation; f after separation of the first two blastomeres. (a–e after TEISSIER, 1931; f–h after BECKWITH, 1914)

Experiments on the determination of the events of embryonic differentiation promise further insight into the nature of egg polarity. In higher organisms, in addition to the animal-vegetal axis, differentiation takes place in other axes, which must be similarly laid out in the egg cytoplasm.

Lecture 11

Cleavage is the first morphogenetic process in the embryonic development of metazoans. In this process the egg cell is divided into a number of blastomeres. Cleavage into individual cells can proceed to a final stage without differential determination of the blastomeres taking place; this is the case in the hydroids just described. On the other hand, determination may accompany cleavage. We shall consider, for the moment, only cleavage as such, ignoring determinative events whose expression is seen later in development. In the simplest case, in total equal cleavage, which occurs in a great variety of animal groups, cleavage ends with the arrangement of a number of cells of similar size and form in a hollow sphere, the blastula. Until the blastula is formed, the cell divisions take place more or less synchronously or in waves from one pole to the other. We can find examples of development with a typical blastula among the coelenterates, the lower crustaceans, the echinoderms, and the chordates (*Branchiostoma*); thus the blastula is a typical stage in embryonic development. Even where early development is markedly different, we encounter a corresponding, homologous stage, closely followed by a subsequent stage in which the germ layers become distinct. Large deposits of nutrients (protein granules, fat, glycogen) can

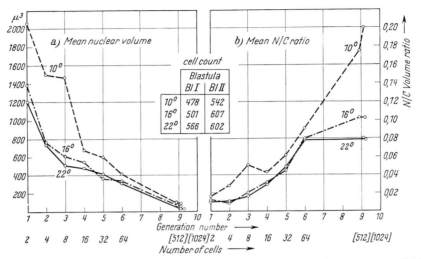

Fig. 196. Changes in nuclear size and nucleocytoplasmic ratio during cleavage of *Paracentrotus lividus* at different temperatures. Theoretical cell numbers after the ninth and tenth cleavages are in brackets. Insert: cell numbers at two blastula stages. *Bl I* hatching; *Bl II* beginning of primary mesenchyme formation. (After KOEHLER, 1912)

change the cleavage into an unequal, partial, or superficial type. But cleavage always runs its course in a certain rhythm of nuclear and cell division set in motion through the activation of the egg by fertilization or parthenogenesis. The cleavage period ends with the formation of the blastula.

The first question which arises is: how is this morphogenetic process terminated?

With cleavage a change in nucleocytoplasmic ratio takes place. With the multiplication of the cell nuclei and a division of the cytoplasm, this ratio rises and reaches a final value in the blastula cells. What we can measure is only the ratio of the volumes (Fig. 196b). The cell volume is no criterion of the quantity of cytoplasm. Even sea urchin eggs, which have very little yolk, show when centrifuged a layering in which the otherwise uniformly distributed cytoplasmic inclusions, fat, yolk granules, and pigment granules take up considerable space (Fig. 179). Nevertheless during oogenesis a great mass of cytoplasm is accumulated for the embryonic cells. Consequently the nucleocytoplasmic ratio in mature eggs and in zygotes is in every case much lower than that found in somatic cells. The reduction in nuclear size during cleavage (Fig. 196a) shows that in the volume of interphase nuclei in this period there is a functional relationship to the current cytoplasmic mass. Whatever materials contained in the cytoplasm reflect the nuclear events, we do not know (cf. p. 77). In any case the synthesis of chromosomal material must accompany the nuclear divisions. The assumption that in every division cycle a doubling of this material *must* take place no longer appears compelling, to be sure, since we are familiar with the highly polytene conditions of all chromosomes and must take into account a possible parcelling out of the elementary strands during a rapid series of divisions, such that the chromosomes would become correspondingly smaller (p. 29). We may expect important conclusions from the cytophotometry of DNA in Feulgen-stained nuclei and from the chemical analysis of cleavage stages. For a long time the experimental evidence has been full of contradictions, however. In *Paracentrotus lividus* the total nuclear DNA measured according to cytophotometric methods increases in proportion to the number of cells. The results of chemical analysis, however, show a much smaller increase. In any event, during the course of cleavage there is a stepwise change in the ratio between active nuclear material and the cytoplasm, and the change proceeds in favor of the chromosomal material.

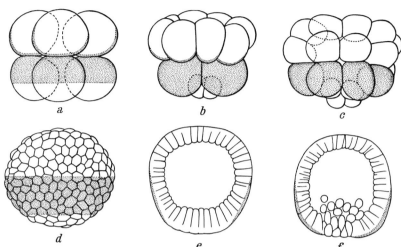

Fig. 197. Cleavage and onset of primary mesenchyme formation in *Paracentrotus lividus*; e, f optical sections (After BOVERI, 1910)

That the extent of the division of the egg cytoplasm during cleavage depends on the amount of chromosomal material, is shown by experiments in which the chromosomal number is changed: In artificial parthenogenesis in sea urchins, there are only one-half as many chromosomes as are present in a fertilized egg with the same amount of cytoplasm; here cell division in the haploid embryos goes one step further, until the cells have one-half the volume of normal diploid cells. Tetraploid fertilized eggs, obtained by interference with the formation of the mitotic apparatus of the first cleavage division (Fig. 201 a), complete cleavage with one-half as many cells, which are twice as large as those in normal embryos. Moreover, one can reduce the amount of cytoplasm: egg fragments with nuclei can be fertilized, resulting in embryos with normal cell size but a correspondingly smaller cell number. The cell divisions during cleavage always lead, therefore, to the same ratio of genome number (and consequently active chromosomal material) to cytoplasmic mass.

In the different regions of the embryo the behavior of the cells in division is uniform. The sea urchin egg is not, in contrast to a number of other echinoderms (e.g., the holothurian *Synapta*), divided into blastomeres of uniform size; rather, and in spite of the uniform distribution of yolk, it undergoes a regular pattern of unequal cleavage (Fig. 197). After two meridional and then one equatorial cleavage the cells of the animal half of the embryo arrange themselves around the meridional furrow in a ring of eight animal cells of equal size. The four vegetal blastomeres are separated by an unequal horizontal furrow into four large macromeres which lie just below the equator and four small micromeres at the vegetal pole. During the next division, as the sixteen-cell stage becomes a 32-cell stage, the eight animal cells are divided by a horizontal furrow into two rings or layers (Fig. 197c). The macromeres divide meridionally, becoming a ring of eight equally-sized cells. The four micromeres give off four more very small micromeres in the direction of the vegetal pole. In the micromeres the nucleus is relatively large in comparison with the cytoplasm. At this point the nucleocytoplasmic ratio has risen to a very high value (Fig. 198). If the ability of the blastomeres to divide decreases with a rising nucleocytoplasmic ratio, one would expect the micromeres to divide less often than the large blastomeres, and this is indeed the case. At the end of the blastula stage (Bl II, Table in Fig. 196) the embryonic cell count ranges from 542 to 607. If each cell had resulted from nine divisions the number would be 512; after ten divisions

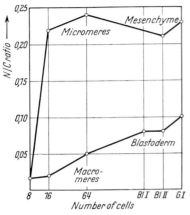

Fig. 198. Nucleocytoplasmic ratio in macromeres, blastoderm cells, micromeres and primary mesenchyme cells of *Paracentrotus lividus*. (After KOEHLER, 1912)

there would be 1024 cells. And if the four micromeres of the sixteen-cell stage had divided as often as the other cells, the 512-cell stage would include 128 micromere derivatives. Now these micromere derivatives remain easy to distinguish from the neighboring blastomeres. They arise (in *Paracentrotus lividus*) from the part of the egg cell below the pigment band and later they move into the interior of the blastula as primary mesenchyme cells; so they can be counted at the end of the blastula stage, and in fact their number never even reaches one-half of 128. When the micromeres are isolated in the sixteen-cell stage, they produce about forty cells. Accordingly, they divide less than one-half as often as the large blastomeres. The nucleocytoplasmic ratio which the micromeres attain after the fourth cleavage (at 22°, 0.23; at 10°, 0.28) is almost exactly the same as the value found when they migrate inward as mesenchyme cells (0.21 at 22°, 0.29 at 10°; Fig. 198).

The curves in Fig. 196b show that the nucleocytoplasmic value at the end of cleavage is temperature-dependent. The nuclear size curve at 10° is clearly higher than that of curves at 16° and 22° in *Paracentrotus lividus*, until the blastula stage (Fig. 196a). The nuclei of all stages are larger at lower temperatures. Similarly the curve for nucleocytoplasmic values is much higher at 10° (Fig. 196b), and the final values are different at each temperature. Even the cell number at which a certain blastula stage is reached is temperature-dependent. At 10° the blastula (Bl I, Fig. 196) emerges from the fertilization membrane with a lower cell number than is the case at 16° and 22°; likewise the inward migration of the primary mesenchyme cells (Bl II, Fig. 196) begins at a stage with a smaller cell number in colder temperatures than at warmer ones.

The morphogenetic process of cleavage thus ends neither with an absolutely determined cell size or cell number, nor with an absolutely determined nucleocytoplasmic ratio. Rather a certain physiological state of the individual cells is evidently reached, which is dependent on temperature and on local conditions in the cytoplasm, in addition to the metabolic conditions of the whole system. This physiological state in which the series of cleavage divisions ends is brought about by the metabolic events which regulate division and the synthesis of nuclear substances. A qualitative alteration of the cytoplasm must be achieved, since the cytoplasm must contribute energy-rich compounds and the components of the nucleic acids and the proteins of the chromosomes. We do not yet know the form in which they are stored in the egg cytoplasm, nor how the synthesis of chromosomal material takes place. That the blastula cells all attain a comparable condition, which is expressed in the nucleocytoplasmic ratio, is shown in a unique way by the development of a Scyphomedusa. In *Chrysaora hyoscella* the eggs are not released into the water,

156

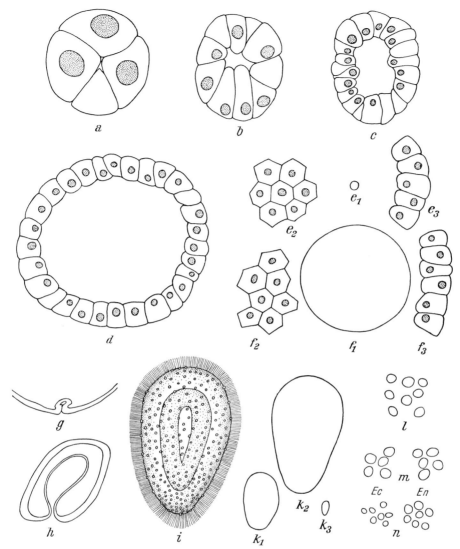

Fig. 199. Development of *Chrysaora hyoscella* (Scyphomedusae) to the planula stage. a four-cell stage; b later cleavage stage; c blastula; e_1, f_1 outlines of blastulae of extremely different sizes; e_2, e_3, f_2, f_3 blastula cells from above and in section; g, h gastrulation; i planula; k planulae of different sizes; l–n nuclei: l blastula, m gastrula, n planula; *Ec* ectoderm nuclei, *En* endoderm nuclei. Magnifications: a–d, e_2, e_3, f_2, f_3 640×; e_1, f_1 56×; g, h 120×; i 200×; k 40×. (After TEISSIER, 1929)

but cleave in the gonads. During cleavage, as in echinoid eggs, the blastomere nuclei get smaller until a stage of about 120 cells (Fig. 199 a–c). At this time the cells have undergone an average of seven divisions and have reached the final cell size and nucleocytoplasmic ratio of the blastula. These blastulae, however, are not released into the water at this stage as other medusae are, but grow with further cell division in the mother medusa, which nourishes them (Fig. 199 d). During this time they can grow remarkably large (Fig. 199 e, f); nevertheless, the cell and nuclear dimensions remain constant (Fig. 199 e_2, e_3, f_2, f_3). After each division each cell grows to its former size before dividing again. The invagination of the endoderm can take place after a long growth period (Fig. 199g) or in a blastula which has grown relatively little (Fig. 199h). From the gastrula arises the organization of the planula, whose size may vary over at least a 300-fold range, according to the size attained during the growth of the blastula (Fig. 199k$_1$–k$_3$). Only during the differentiation of the ectoderm and endoderm of the planula do the cells and nuclei assume new dimensions which are now specific for each tissue (Fig. 199l–n).

During the extreme growth of the blastula of *Chrysaora*, nuclear and cytoplasmic substances are synthesized in amounts fifty to one hundred times those present in the egg, and the blastula cells are therefore in a developmental equilibrium, just as cells in a tissue culture. It is, therefore, not the achievement of a certain cell number and a certain nucleocytoplasmic ratio that lead to the next morphogenetic process, gastrulation, since the availability of further nutrients makes possible the continued synthesis of nuclear and cytoplasmic materials. During further cell division and growth of the blastula, the nucleocytoplasmic ratio remains constant; thus cleavage is an independent component of development distinct from the subsequent phase of embryonic development.

The onset of gastrulation in *Chrysaora* is evidently triggered by the cessation or decrease of the food supply of the growing blastula. This suggests that in all eggs with their own food supply, the continuing quantitative and qualitative changes in this food supply provide the embryonic cells with the internal conditions which cause the formation of germ layers to follow immediately after cleavage.

A second question concerns the determination of cleavage patterns, such as the arrangement of cells through unequal cleavage into blastomeres of diverse sizes. The change from meridional to latitudinal cleavage has been attributed to the usual formation of the spindle in the *long axis of the cell* (O. Hertwig's Rule). Differences in blastomere size are attributed to differences in yolk concentration. In examples chosen for elementary courses these principles seem sufficient to cover the whole course of cleavage, and indeed they may apply frequently to certain periods of cleavage; but in very many cases, if not most, specific additional factors determine the pattern of cleavage.

Cleavage in sea urchins provides a very informative example. Our question can be cast in this way: How are the spindle positions determined through which a sixteen-cell stage comes to comprise eight animal cells, four large vegetal cells (macromeres) and four small cells at the vegetal pole (micromeres) (Fig. 197a–c, 200a)? The three first spindle positions, the first and the second in the equatorial plane of the egg and the third vertical, correspond to the scheme for regular cleavage of an egg with very little yolk. Then follows the disparate behavior of the spindles in the animal and vegetal halves of the embryo: in the animal cells the spindles are again horizontal, but in the vegetal cells they are almost vertical and displaced toward the vegetal pole. If the blastomeres of the two-cell stage are separated from one another they continue to cleave as if they were still together. Figure 200b illustrates this "half-cleavage." Isolated blastomeres of the four-cell stage carry out a similar "quarter cleavage" (Fig. 200c).

The normal change of spindle orientation from horizontal to vertical in successive divisions can be altered experimentally. The first mitosis can be prevented by the inhibition of mitotic apparatus formation (monaster formation) (Fig. 201a). Then only one division is carried out with a horizontal spindle, at the time of the normal second division, and now the second cleavage takes place in both blastomeres with vertical spindles; thus two animal and two vegetal blastomeres are formed. In the third division the spindles are once more horizontal and four animal blastomeres result, while in the vegetal half of the embryo micromeres are produced. The whole egg thus undergoes "half cleavage" (cf. Fig. 201b). Similarly, if the first two nuclear divisions are prevented, the whole egg undergoes "quarter cleavage" (Fig. 201b, cf. 200c); in this case the first

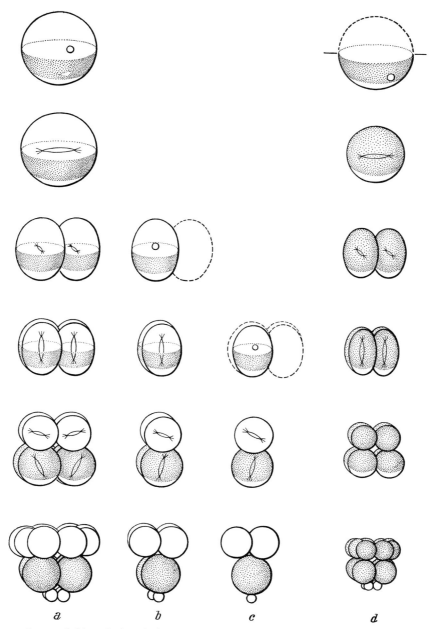

Fig. 200. Scheme of normal (a) and altered cleavage in *Paracentrotus lividus*; b cleavage of a half blastomere; c cleavage of a quarter blastomere; d cleavage of the vegetal hemisphere. (In part after HORSTADIUS, 1928)

spindle permitted to form is vertical. The plane of the spindles in the vegetal half and their eccentric position near the vegetal pole are therefore not determined by the completion of the normally preceding cleavages nor by the resulting arrangement of blastomeres, but rather they result from a condition present at fertilization. From this we can only conclude that an independent

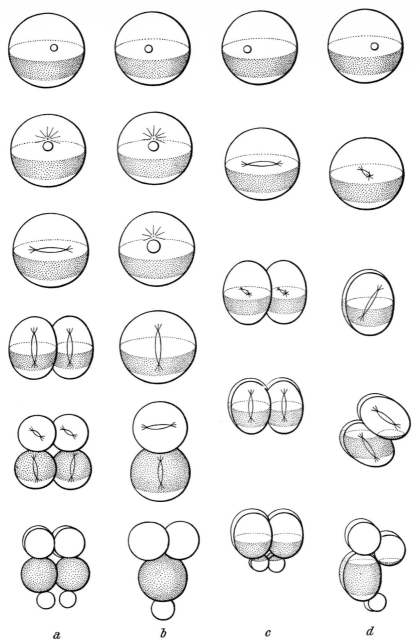

Fig. 201. Scheme of altered cleavage in *Paracentrotus lividus*. a "half cleavage" in a whole egg treated in the first division; b "quarter cleavage" in a whole egg treated during the first and second divisions; c, d abnormal cleavage resulting from the delay of nuclear division in dilute sea water. (After HÖRSTADIUS, 1928)

rhythmic process in the cytoplasm determines the position of the spindles. If the first nuclear division is prevented, then each division which subsequently takes place does so in accordance with the established rhythm under conditions of the cytoplasm which would ordinarily prevail one division later; and when two mitoses are prevented, the first division is carried out at a time and under conditions normally relevant to the third division. These conclusions are confirmed by further experimental alteration of cleavage. In dilute sea water the nuclear division rate gradually slows down, evidently without a corresponding effect on the underlying cytoplasmic process. Thus, the third division can be delayed until the cytoplasm evokes the formation of micromeres (Fig. 201 c). Or cleavage can be moved into a transitional stage of the cytoplasmic process, in which the conditions evoking the horizontal position are just changing to the conditions evoking the vertical position. This transitional stage ordinarily falls during interphase. Now the spindless assume a slanted intermediate position, and the resulting cell tiers, whose cell number corresponds to a "half cleavage," lie at an angle to the animal-vegetal axis (Fig. 201 d, cf. 200 b). The cytoplasmic condition which moves the spindles towards the vegetal pole arises at the eight-cell stage

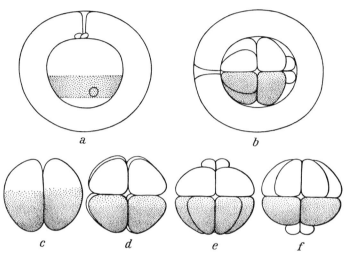

Fig. 202. *Paracentrotus lividus.* a mature egg with pigment band; b–f cleavage in various orientations with respect to the egg axis in stratified eggs. (a after BOVERI, 1901; b after MORGAN and SPOONER, 1909; c–f after MORGAN, 1927)

in normal cleavage of the egg. The time course of this process is evidently already determined in the egg stage, and the determinants of the sites of micromere formation, which become operant during the fourth cleavage, lie in the vegetal half of the egg. If one cuts fertilized or unfertilized eggs in the equatorial plane, the animal half cleaves just as the animal half of a normal embryo does. But the vegetal half mimics the cleavage of an entire egg on a reduced scale (Fig. 200 d). The course of cleavage can therefore be regulated in the vegetal half of the egg, but not in the animal half.

We might well suspect that this difference in behavior between the two halves of the egg is determined by the apportionment of the contents of the egg cell; this has not been confirmed, however. The animal-vegetal axis of the *Paracentrotus* egg can be recognized by the horizontal pigment band and by the marking of the animal pole by a canal (the micropyle) which runs through the jelly coat (Fig. 202 a). The layering of the egg contents by centrifuging (Fig. 179) can take any orientation whatever with respect to the primary egg axis. The first spindle in a centrifuged egg is placed perpendicular to the layering axis; the first cleavage plane goes through the pole with the fat droplets and through the other layers, and it divides the egg into two equal halves (Fig. 202 c). The second and third cleavage planes are perpendicular to the first and to each other, and so there arises an eight-cell stage with four yolk-rich pigmented cells, and four yolk-poor and unpigmented cells (Fig. 202d). At the fourth cleavage the micromeres appear at one pole where two of the three first cleavage furrows intersect; and the four blastomeres at the opposite and cleave meridionally into eight cells of equal size (Fig. 202b, e, f). At this point the further cleavage of the cells assumes the normal pattern and a surprisingly normal sixteen-cell stage results. The site of micromere formation has no particular relationship to the layers of the egg; it can be in the pigment-free zone (Fig. 202e), in the pigment cap (Fig. 202f), or on the border between the pigment zone and the zone of clear cytoplasm (Fig. 202b). The three first spindle positions are therefore determined by the density layers; but then the original polarity of the egg is reasserted: The micromeres are always near the original vegetal pole, which is opposite the micropyle (Fig. 202b), and this happens no matter how the layers are oriented with respect to the vegetal pole. In this way a partitioning of the cytoplasm takes place; the micromeres are almost always unpigmented, even when the cells from which they are formed are pigmented. The cells on the opposite side become a circle of eight cells, just as in the normal embryo (Fig. 202b, e, f; cf. Fig. 197b). A polar morphogenetic event, which through a certain continuing alteration of the cytoplasm determines the spindles' position in the fourth cleavage, is therefore itself not altered by the rearrangement of some of the cytoplasmic contents into layers. A certain compromise must thus take place between the original polarity and those cleavage planes determined by the layering, so that the layering now runs at an angle to the primary egg axis. Here, as in the polarity of the *Hydractinia* egg, there may remain as carrier of the determining cytoplasmic pattern only an exceedingly reparable egg skeleton, or more likely, the egg cortex, which is not disturbed by centrifugation.

From this course of cleavage it can be concluded that a continuing cytoplasmic process independent of nuclear divisions underlies the cleavage of the sea urchin egg. In other cases such a process is seen directly in autonomous movements of the egg cytoplasm.

In a few annelids (*Chaetopterus*, *Myzostoma*) and many molluscs (*Dentalium*, *Ilyanassa*, and others) a cytoplasmic lobe forms in the region of the vegetal pole during certain of the early divisions, and these polar lobes are separated from the rest of the egg to a greater or less extent; they may be pear-shaped and attached only by a thin stalk. They are usually formed shortly before the first cleavage (Fig. 203a). While the remainder of the egg divides equally (Fig. 203b), the polar lobe remains attached to one of the first two blastomeres and is resorbed after the first division (Fig. 203c). At the second division the polar lobe is again pinched off (Fig. 203d, e), after which it rejoins one of the four blastomeres (the D blastomere) (Fig. 203f). In *Dentalium* still another polar lobe is formed when blastomere D divides (Fig. 392h); then it rejoins the daughter cell nearer the vegetal pole, which is called 1D (Fig. 392i). If a fertilized *Dentalium* egg is cut to produce an anucleate vegetal fragment, this fragment does not cleave, but it forms three polar lobes in succession at the same time that the nucleated animal fragment undergoes each of its first three cleavages. The polar lobes thus formed are of the same size as those seen in a normal egg. Indeed, while normal eggs form polar lobes only during the first three divisions, the anucleate vegetal fragment now goes on to produce a fourth polar lobe again just when the animal fragment is dividing for the fourth time. The nucleated animal portion cleaves like related eggs which do not make polar lobes. *Ilyanassa* forms and resorbes a polar lobe four times; during the ejection of the first and second polar bodies and during the first and second cleavages. If the polar lobe is cut off just before the first cleavage, it changes shape synchronously with the cleavage of the egg (Fig. 204). It becomes pear-shaped, nearly amoeboid, and returns to a spherical form. It can also send out its own little polar lobe in a way closely resembling normal polar lobe formation in a normal egg. If an isolated polar lobe is cut in two, both parts follow the alternating courses of activity and rest.

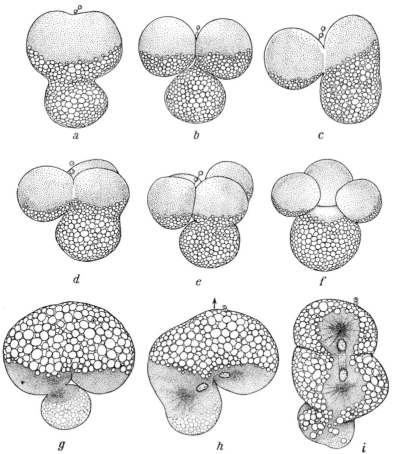

Fig. 203. Polar lobe formation in the snail *Ilyanassa obsoleta*. a–f normal cleavage: a–c formation of the polar lobe during the first division and its resorption into one blastomere (*CD*); d–f polar lobe in the second division and its resorption into a blastomere (*D*); g–i formation of the polar lobe at the first division opposite the animal pole (recognizable by the polar bodies), after inverted centrifugation. (After MORGAN, 1927, 1933)

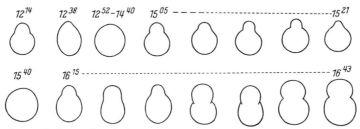

Fig. 204. Rhythmic changes in form of an isolated polar lobe of *Ilyanassa*; polar lobe isolated in a young two-cell stage (as in Fig. 203b) at 12:14 p.m. Two cycles of change corresponding, with a slight delay, to the second and third divisions of the whole egg

Normally, polar lobes contain yolk in eggs where the yolk is concentrated at one pole (Fig. 203 a–f), but centrifuging experiments show that their eversion is independent of the material they contain. Eggs embedded in gelatin, with a particular orientation, can be centrifuged to produce a layering of their movable contents in an inverted or oblique orientation. Thus, the yolk can be collected at the animal pole, which is still recognizable by the attached polar bodies; in this case an oil cap and clear plasma are seen in the vegetal half. Nevertheless, the polar lobe always emerges in the region of the vegetal pole, whether it contains only clear cytoplasm, or oil, or yolk on one side; similarly, the first cleavage is usually meridional (Fig. 203g, h). When centrifuging does displace the spindle, so that the first cleavage is not meridional but rather perpendicular to the egg axis, the polar lobe nevertheless arises on the opposite side of the egg from the polar bodies (Fig. 203i).

Here also, consequently, there must be in the unfertilized egg both a polar morphogenetic pattern, which reveals itself in the polar partitioning of the oocyte contents (but is not influenced by them), and also an autonomous rhythmic process, determined in the cytoplasm. This determination is certainly related to the vegetal cortex, which is not disturbed by centrifuging.

The importance of the cortex for the cleavage pattern is also seen in eggs of very high yolk content. In the oligochaete *Tubifex* a relatively stable pattern in the cortex appears to direct the streaming of the plasma, so that the spindles are brought to the right place.

Most surprising of all is the normal cleavage pattern of the very yolky Cladoceran eggs, which is independent of the partitioning of the cytoplasm and those cytoplasmic contents which are displaced by centrifuging. The parthenogenetic *Daphnia* egg contains a great oil droplet and numerous large yolk granules (Fig. 205a). A honeycomb of cytoplasm pervades the yolk granules and spreads over the upper surface of the egg, especially at the animal pole, where the polar bodies are ejected, and also at the vegetal pole where materials are received from the nurse cells. Since the oil droplet lies near the vegetal pole, the egg is oriented in the brood pouch and in the fluid within its chorion, with the vegetal pole up (Fig. 205a). The eccentric nucleus lies in the animal hemisphere in a pool of cytoplasm, from which the astral rays of the first cleavage reach out between the yolk granules through the cytoplasmic network. The first two nuclear divisions take place in the animal hemisphere inside the egg, with the spindles almost horizontal, but displaced somewhat toward the long axis (Fig. 205b, c). Then the four nuclear regions move to the upper surface of the egg and the spindles orient meridionally for the third division (Fig. 205d). During this division the cytoplasm cleaves simultaneously, separating each of the regions of the first four nuclei in the vegetal half and each of the four in the animal half; and the subsequent cytoplasmic cleavages display the characteristic pattern seen also in the yolk-poor Cladoceran embryos (*Polyphemus*).

In the centrifuge the lighter, oil-droplet-containing vegetal pole assumes a centripetal position, and the yolk is tightly packed in the animal hemisphere (Fig. 205f). The clear cytoplasm flows from the superficial regions and from the reticulum between the yolk granules, up into the vegetative half, where it now surrounds the oil droplet. The nucleus with the aster, whose extent is more restricted than in the normal egg, lies next to the oil droplet. Here the first two divisions take place in the vegetal hemisphere, rather than in the animal hemisphere, as is normally the case (Fig. 205g, h). Now, from the cytoplasmic cap four superficial tongues of plasma stream toward the equator meridionally, and they are followed by the nuclei with their astral regions (Fig. 205h, i). Some 120 minutes after the eggs have been laid these cytoplasmic streams come to rest in the centrifuged egg. The nuclei then lie more or less in the equatorial plane of the egg, while the streams of cytoplasm have now reached the animal pole. These cytoplasmic processes now begin to spread out, joining one another; they make a thin plasma layer which surrounds the cap of yolk. Thus, the cytoplasmic and nuclear divisions approach the normal condition. This regulatory streaming which spreads the cytoplasm over the yolk cap and blazes a trail for the migrating nuclei is seen not only when the egg is centrifuged in the direction of its polar axis: even when the layers are formed at an angle to the egg axis (Fig. 205k) (once more this is achieved

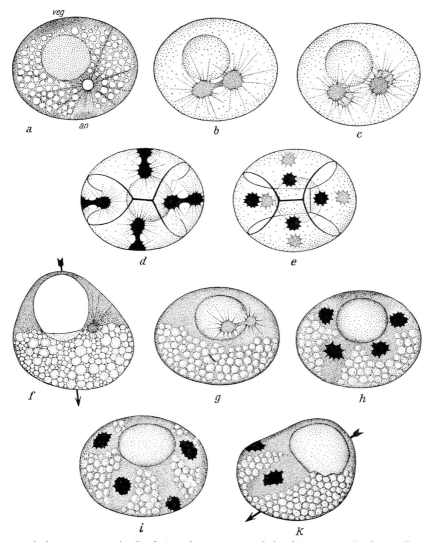

Fig. 205. Eggs and cleavage stages in *Daphnia pulex*. a–e normal development at the four-cell stage; f–k after centrifugation (f–i in the direction of the egg axis; k oblique to the egg axis). a, f longitudinal sections; b, c, g side view; d, e animal pole view; h, i oblique view of the vegetal pole. The arrow points in the direction of centrifugation

by centrifuging eggs embedded in gelatin), the same astounding events take place. The cytoplasm always plays its restorative role: it leads the nuclei back and determines the place and orientation of division. In exceptional cases, especially after diagonal layering, tongues of cytoplasm can also extend meridionally without bringing nuclei along. If the centrifugation has not lasted too long, these regulatory events lead to an embryo with a typical arrangement of organs, although the normal cytoplasmic order has been completely destroyed by centrifuging. Contrary to our expectations, neither the animal nor the vegetal cytoplasmic contents are critical for the cleavage pattern and thus for further development. Only a structure in the stable cortex can evoke and direct the cytoplasmic streaming in these eggs and thereby determine the pattern of cleavage.

Lecture 12

In many animals cleavage leads to a structurally uniform blastula, a hollow ball whose wall consists of a simple layer of similar cells, the blastoderm. We call such a layer of morphologically still undifferentiated cells a blastema (F. E. LEHMANN). It is characteristic for many phyla that, in the development of the germ layers and organs, such cell layers, rather than individual cells, play the main role. From the primitive blastema of the blastoderm, the components of the basic body plan arise stepwise in primitive development through morphogenetic movements and segregation; thus, a spatial ordering of the organ anlagen of the typical body plan or of the organs of a larva is completed. If the larva is really different from the mature form, as, for example, in echinoderms and amphibia, the end of a developmental stage is clearly delineated; in other cases the termination of early development is less sharp.

A first step is the formation of the component blastemas of the embryo, the germ layers ectoderm and endoderm, and in most phyla also mesoderm and mesenchyme. From these blastemas differentiate the organ anlagen which in a later phase of development undergo their histological differentiation. In some hydroids, for example *Amphisbetia* (Figs. 191, 192), the whole blastoderm has the character of a morphogenetic field, a self-organizing field similar to the aggregation field of *Dictyostelium* (p. 129): all parts of the blastula of sufficient size develop regulatively into typical proportions of ectoderm and endoderm, manifesting the polarity determined in the egg (p. 148). Such complete regulation is rare in metazoans.

In this and in the next lecture we shall be concerned with the determination of developmental processes leading to the formation of the echinoid larva. It is (especially through the experiments of RUNNSTRÖM, HÖRSTADIUS, LINDAHL, and VON UBISCH) one of the best known and most conclusive studies in all of developmental physiology.

First, we shall follow normal development to the pluteus larva (see pp. 178f., Fig. 197). In the 64-cell stage (Fig. 206a) the animal cells are no longer clearly arranged in a circle, but on the whole they correspond to the two animal tiers an_1 and an_2 of the 32-cell stage (Fig. 197c). From the eight macromere derivatives two tiers of eight cells each have arisen. The upper one near the equator is called veg_1, the lower one veg_2. With further divisions the size differences among the cells disappear. The segmentation cavity is filled with a gel whose first traces have already appeared in the space between the first two blastomeres. This blastocoel gel increases as the segmentation cavity enlarges. It consists mainly of a polysaccharide-protein complex. Its composition resembles the thin gel layer which, as the hyaline coat, surrounds the embryo and apparently also covers the individual blastomeres and restricts their fields of movement.

The round blastula (Fig. 197d) is soon covered with active cilia (Fig. 206b); the blastula revolves at first within the egg membrane, then it hatches and swims freely in the water. At the animal pole much longer apical cilia grow out (Fig. 206c), and the region of the vegetal pole becomes flattened (Fig. 197e, 206c). From here the primary mesenchyme now wanders out into the blastocoel (Figs. 197f., 206d); it is derived from the micromeres of the sixteen-cell stage (Figs. 197b, c; 206a). Then the primitive gut invaginates from the vegetal pole (Fig. 206e). At this time the blastocoel gel is redissolved, evidently by enzymes of the endoderm cells, and a gel is now secreted into the lumen of the primitive gut. The primary mesenchyme now builds around the primitive gut a ring with two large accumulations on the right and left sides of the future ventral portion of the embryo (Fig. 206e, f). Thus, the gastrula now shows bilateral symmetry. In the paired aggregations of primary mesenchyme cells, two triprongs are formed as the rudiments of the larval skeleton (Fig. 206e, f). From the top of the primitive gut secondary mesenchyme cells migrate out. Laterally the mesodermal coelomic pouches are pinched off. The primitive gut differentiates into esophagus, midgut, and hindgut. The gut now bends

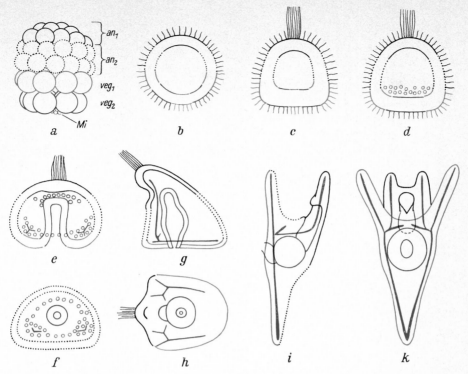

Fig. 206. Normal development in *Paracentrotus* (*Strongylocentrotus*). an_1 black lines; an_2 black dots; veg_1 yellow; veg_2 green; micromeres (Mi) red; a 64-cell stage; b young blastula; c older blastula with tuft of flagella; d blastula after formation of primary mesenchyme; a gastrula with secondary mesenchyme; f same gastrula as e in cross-section; g prism stage, side view, with oral depression; h prism stage, anal view; i, k pluteus larva: i from the left side, k anal view. (After Hörstadius, 1955)

toward the future ventral surface, where the animal part flattens into an oral field (Figs. 206 g, 207 a).

The dorsal side of the ectoderm grows faster than the ventral side and is drawn to a point (Figs. 206 g, 207 a). From the ventral ectoderm an oral depression forms, which anastomoses with the top of the primitive gut. The first part of the esophagus is therefore an ectodermal derivative. The ectodermal portion of the columnar epithelium of the blastula and young gastrula changes into a cuboidal or squamous epithelium, with the exception of a ciliated band which surrounds the oral field (Fig. 207 b). The larva continues to grow. It appears triangular from the side (prism stage, Fig. 207 b). The larval axis, which corresponds to the primary egg axis, goes in the radially symmetrical gastrula through the apical tuft, the top of the gut, and the blastopore. During the transition to bilateral symmetry, this axis is bent (Fig. 207 a). The dorsal line runs from the apical tuft to the blastopore, which becomes the anus. The dorsal side continues to grow faster than the ventral side. It will be drawn out into the apical arm of the larva. The ventral side includes in the pluteus larva the oral field and also the region of the "anal side" as far as the anus. To the dorsal side belongs the entire "abanal side," together with that part of the anal side from the anus to the apex (Figs. 206, 207 a). At the edge of the oral field grow the oral and anal arms (Figs. 206 i, k; 210 a). These arms are stiffened through skeletal struts, branches of the primary triprongs (Fig. 210).

167

The prospective importance of the blastomeres and, further back, the various ooplasmic regions, frequently can be found with the aid of the pigment band in the cortex, which rings the mature egg below the equator (Figs. 197, 200, 202). This pigment can, however, also extend a greater or lesser distance toward the animal pole from the equator. The vegetal polar region of the egg remains pigment-free. Thus the pigment is later found essentially in the veg_1 and veg_2 cells. To a very slight extent the an_2 cells can also be pigmented. The primary mesenchyme cells are almost always pigment-free. At the time of gastrulation, a considerable number of the pigmented cells remain in the ectoderm in the region of the blastopore, while the primitive gut is always fully pigmented. Transplantation of vitally stained cell groups of the 16-, 32-, and 64-cell stages

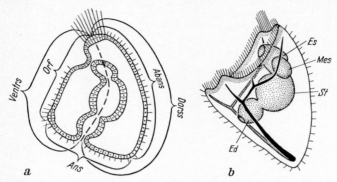

Fig. 207. Prism stage of Paracentrotus, diagrammatic. a Sketch of the body regions: dotted line, egg axis; b somewhat oblique view of the left side with apical tuft and flagellar band; *Abans* abanal side; *Ans* anal side; *Dorss* dorsal side; *Ed* hindgut; *St* stomach; *Mes* mesodermal pouch; *Es* esophagus; *Orf* oral field; *Ventrs* ventral side

to an unstained complementary part of an embryo permits the prospective fate of the blastomeres to be defined exactly. Staining of an_1 in the sixteen-cell stage shows that this cell layer produces a clearly defined part of the ectoderm (Fig. 208 a). The transplantation of veg_2 and the attached micromeres from a 64-cell stage to an unstained complementary portion of a larvae leads to the resulting primitive gut being completely colored (Fig. 208 b). Similar staining of the macromeres of the 32-cell stage (the whole vegetative half) and subsequent transplantation results also in a great part of the ectoderm being colored, the anal side from apex to the tips of the anal arms (Fig. 208 c). The ectoderm-endoderm border reflects therefore the furrow between veg_1 and veg_2. From the comparison of the experiments in Fig. 208 a and c, one can find the ectoderm region which is produced by an_2. In Fig. 206 are the regions of the embryonic stages and of the larva which are derived from the cell layers of the 32- and 64-cell stages on the basis of vital staining. Not half the blastula, as was formerly believed, but only a quarter is invaginated.

After the departure of the primary mesenchyme cells the presumptive endoderm, which, as vital staining shows, stems from the veg_2 material, forms a thick plate. Invagination is accompanied from the beginning in the whole plate by the epiboly of the vegetative blastula wall, which thus becomes thinner. During gastrulation the ectoderm stretches, and the volume of the larva increases. One is tempted to trace the mechanism of invagination back to certain visible factors of normal development: the blastula is covered by an elastic hyaline coat. Inside it is filled with gel; one might suspect that the invagination of the endodermal cell plate is an essentially passive consequence whereby its form or the colloidal matrix of its cytoplasm responds to a meridional pressure of the upper blastula wall in the elastic coat or yields to a negative pressure in the blastocoel, where the gel is being resorbed. In both cases gastrulation would depend on the integrity of the entire blastula. But gastrulation is not arrested when a part of the blastula wall is cut away: it proceeds, and frequently the wound is closed and a smaller gastrula than normal results. If the vegetal plate is isolated, the wall folds under and forms a tiny hollow ball, into which the central part of the endodermal plate invaginates (Fig. 209 a–e). Invagination is therefore not dependent on the integrity of the embryo. In the first decisive step of morphogenesis, the morphogenetic movement of the blastoderm, the endodermal plate therefore displays autonomy but also regulation, since here the plate does not reflect its prospective fate and change entirely

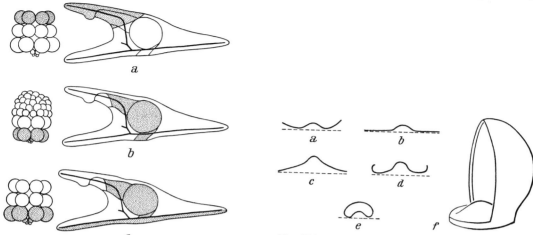

Fig. 208. Vital staining of embryonic layers to show their prospective fates. a, c 32-cell stage; b 64-cell stage. (After Hörstadius, 1935)

Fig. 209. a–e Behavior of excised endodermal plate in the hour following its isolation; f diagram of an egg at the beginning of gastrulation; left part of ectoderm cut away. (After Moore and Burt, 1939)

into an endodermal tube; rather a considerable part becomes the outer surface of the miniature gastrula.

How far morphogenesis can proceed in isolated parts of sea urchin cleavage stages is a question which was posed in the early years of developmental physiology. That isolated cells from the two- and four-cell stages can develop into well proportioned but correspondingly small pluteus larvae has long been known. This discovery of Driesch in the early 1890's was a major point of departure in the study of developmental physiology. Further cleavage in such isolated blastomeres generally proceeds until a cell number is reached which corresponds to the 32-cell stage of the whole embryo (Fig. 200). Then the cell plate, whose normal form would be bowed out, closes over into a blastula. One-eighth embryos can be obtained by cutting the 32-cell stages into eight equal meridional parts like sections of an orange (the micromere derivatives must be left out because they are not easily separated). Pluteus larvae often arise in this way from one blastomere each from an_1, an_2, and a macromere (see Fig. 197c). They can develop into normally proportioned miniature larvae (Fig. 210). A contrast to the regulating ability of isolated meridional embryo fractions is seen if one separates the animal half from the vegetal half in an eight- or sixteen-cell stage. Blastulae arise from both halves, but only the vegetal halves gastrulate and form larvae with a gut divided into three sections. The organization of these larvae varies within a certain range. The rarest, most complete result is a pluteus larva with four arms and a disproportionately large gut (Fig. 211e). A progressively reduced amount of morphogenesis in vegetal halves is seen in the increased lack of those morphogenetic structures which normally arise from the animal parts of the embryo. First seen is the oral arm (Fig. 211f.; cf. Fig. 208), then arms in general (Fig. 211g); in extreme cases, the oral depression is missing, as well as the flagellar band

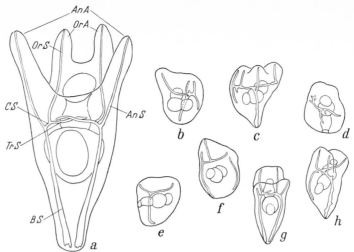

Fig. 210. a Normal pluteus larva of *Paracentrotus*, all examples drawn to same magnification; *AnA* anal arm; *AnS* anal skeleton; *BS* body skeleton; *OrA* oral arm; *OrS* oral skeleton; *TrS* transverse skeleton; *CS* connecting skeleton; b–h seven larvae derived from equal meridional eighths of an embryo (one was lost): g complete in body form and skeleton. 180×. (After HÖRSTADIUS and WOLSKY, 1936)

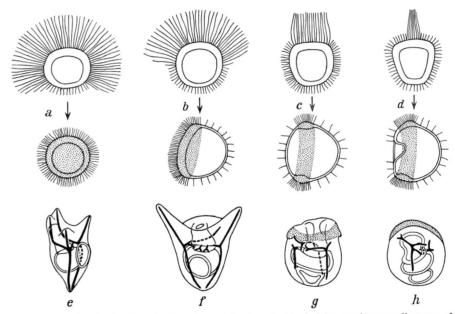

Fig. 211. Development of animal halves (a–d) and vegetal halves (e–h) of eight- or sixteen-cell stages of sea urchin embryos. (After HÖRSTADIUS, 1939)

bordering the oral field, and a field of flagella remains as the only animal differentiation at the most animal part of the larva derived from a vegetal half.

Blastulae from animal halves rarely make a normal apical tuft, which ordinarily includes about an eighth of the entire upper surface at the animal pole (Fig. 211d; cf. Fig. 206c). Generally, the region of the long stiff setae extends over the greater part of the blastoderm (Fig. 211a–c). A few days after fertilization the stiff setae disappear and are replaced by short, motile cilia. In the final stage of development attained by the animal halves, even the differentiation of the epithelium varies. The most nearly normal blastulae form an oral field fringed by a ring of flagella (Fig. 211 d). These blastulae with an extensive distribution of stiff setae become globular dauerblastulae, whose wall always consists of columnar epithelium with motile cilia (Fig. 211a), so that only the epithelium of the flagellar band undergoes normal histological differentiation. Between these extremes are forms in which the oral depression is lacking (Fig. 211c) or in which a flagellar field is formed on one side (Fig. 211b). Thus, at best the animal blastulae possess all of the ectodermal derivatives found in a normal young prism stage (Figs. 211d, 207a). The ability to

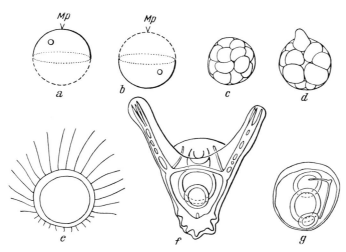

Fig. 212. Cleavage and differentiation of animal and vegetal halves of sea urchins (*Arbacia punctulata*) after the artificial division of an unfertilized egg and subsequent fertilization. a, b orientation by means of the micropyle (*Mp*); c equal cleavage of the animal half; d cleavage of the vegetal half, micromeres at the vegetal pole; a cell at the edge of the section protrudes through a gap in the fertilization membrane; e differentiation of an animal half, blastula with abnormally large apical tuft; f, g differentiation of vegetal halves to pluteus (f) or to an ovoid larva (g). (After HÖRSTADIUS, 1937)

form the primitive gut, skeleton, and arms is always lacking. In the early animal blastula, there-fore, a region of excessive animal tendency is seen in the extent of the tuft of setae together with a corresponding animal histological differentiation which spreads over the whole oral field in its final stages and indeed may spread over the entire blastula.

In the behavior of animal and vegetal embryonic halves there is therefore a polarity which does not exist in the development of hydroids (page 148f., Fig. 192). This animal-vegetal difference is already determined in echinoid eggs before fertilization. Eggs can be cut with a fine glass needle. The nucleus can lie close to the vegetative pole or to the animal pole, which can be recognized as the site of the micropyle (Fig. 212a, b; cf. Fig. 202a). Whether the egg fragments are diploid or haploid, blastulae arise from the animal halves with an enlarged tuft of setae after equal division (Fig. 212e). From the vegetal halves, after more or less regular cleavage during which micromeres are made (Fig. 212d; cf. Fig. 200d), normal looking plutei (Fig. 212f) or ovoid larvae (Figs. 212g, 211g, h) arise. Polar determination in early development thus stems from the polarity of the oocyte in echinoids.

The entire isolated presumptive ectoderm of the 64-cell stage (Fig. 213 a; cf. Fig. 206) generally results in nothing more than the best cases derived from the animal half of the embryo (Fig. 211 d): a normal blastula with apical tuft and then an ectodermal larva with a band of flagella around an oral field containing an oral depression (Fig. 213).

The vegetal half contains material which becomes primary mesenchyme, endoderm, and a part of the ectoderm during normal development (Fig. 206). During the development of more or less complete larvae, during "vegetal anormogenesis" (LEHMANN) (Fig. 211 e, f), some of the cells derived from veg_2 give rise to ectoderm. But the primitive gut is made from veg_2 material in vegetal anormogenesis as well as normally, and likewise the skeleton is derived from the micromeres. This course of events is not necessary, however, as we see from experiments involving the abnormal combination of embryonic parts: if one combines an animal half from the

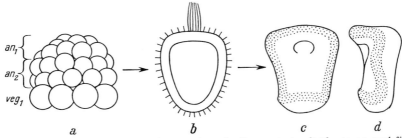

Fig. 213. Development of the presumptive ectoderm in *Paracentrotus lividus* to an oral field larva. (After HÖRSTADIUS, 1935)

32-cell stage with a (vitally stained) veg_2 cell layer, a normal larval skeleton is produced in addition to the primitive gut (Fig. 214a). Similarly, if one combines the animal half, veg_1, and four micromeres from the sixteen-cell stage, the primitive gut will be made from veg_1 material, which normally becomes ectoderm (Fig. 214b). In the combination of an animal half with four micromeres, a primitive gut arises from the presumptive ectoderm of an_2 (Fig. 214c). The individual cell layers are therefore not confined within the limits of their prospective fate in "self-differentiation," differentiation not influenced by the other parts of the larva.

These experiments show on one hand (as also the development of the one-eighth larva in Fig. 210 shows) that in the absence of micromeres a larval skeleton can arise from veg_2. On the other hand they show that micromeres attached to presumptive ectoderm, which ordinarily never forms a primitive gut, work a vegetalizing effect, in which they cause veg_2 or animal cells to invaginate as endoderm.

The isolation of individual cell layers leads to further conclusions. The isolated micromeres divide until they reach the approximate number of the primary mesenchyme cells (p. 156). Then they separate from one another as ciliated blastula cells. The other cell layers complete their usual activities: an_1 and an_2 become ciliated blastulae (Fig. 215); veg_1 can produce a small larva with an oral field and a small invagination of the gut; veg_2 becomes an ovoid larva. This is naturally much smaller, but otherwise almost as differentiated as the usual minimal product of the entire vegetal half (Fig. 211 h). In following the tiers of cells from the animal to the vegetal pole, a gradient is revealed in which an animal tendency declines and a vegetal tendency increases. In the differences in morphogenesis of the early blastula between an_1 and an_2, the difference between these cell layers becomes clear (Fig. 215); in the an_1 blastula the long stiff setae extend over the entire surface. In an an_2 blastula they are absent from one region, as in the poorest developed blastulae from entire animal halves (Fig. 211). The differentiation of isolated an_1, an_2, and veg_1 tiers, therefore, also shows a polarity within the cell layers of the presumptive ectoderm.

When micromeres are attached to the individual cell layers of the 32-cell stage (Fig. 215) a graded vegetalization appears. When one micromere is combined with an_1 the apical tuft is restricted in the early blastula, but development still ends with a dauerblastula. The addition of two micromeres leads to a rudimentary skeleton and introduces in the ectoderm the differentiation of a flagellar field. The combination of four micromeres and an_1 results in a small normal larva.

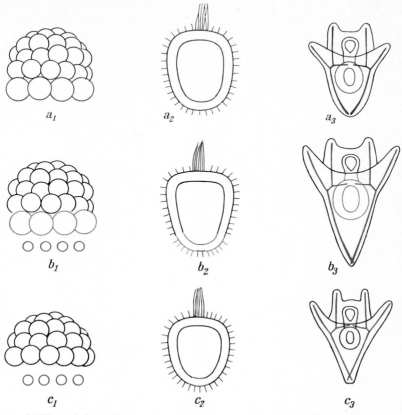

Fig. 214. Diagram of differentiation of the gut and skeleton in sea urchin embryos in which the animal half is combined with veg_2 in a, with veg_1 and four micromeres in b, and only with four micromeres in c; animal half black; veg_1 yellow; veg_2 green; micromeres red (cf. Fig. 206). (After HÖRSTADIUS, 1935)

With an_2, two micromeres suffice for this result, and four micromeres make the larva "too vegetal"; the oral arms are not formed (cf. Figs. 211f., 208). The addition of a micromere to veg_1 improves its developmental ability, but the result still remains in the range seen for vegetal halves (Fig. 211); additional micromeres restrict development further. Development of veg_2 is much too vegetal; too much of the blastula wall becomes endoderm, and the primitive gut bulges outward; it becomes an exogastrula, and the more micromeres one adds to veg_2, the smaller the ectodermal sac. The typical development of a pluteus larva can therefore be achieved through a certain combination of animal and vegetal material.

These experimental results are clarified by the assumption of two opposite gradients, which emanate from the animal and vegetal poles, respectively (RUNNSTRÖM's gradient hypothesis). The highest concentration of the animal tendency is found in the uppermost cell layer, the highest

173

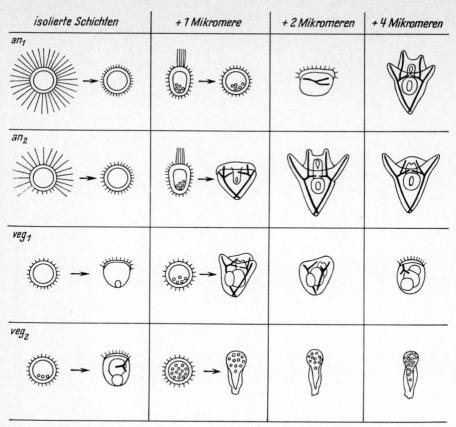

isolierte Schichten	+ 1 Mikromere	+ 2 Mikromeren	+ 4 Mikromeren
an_1			
an_2			
veg_1			
veg_2			

Fig. 215. Diagram of development of layers an_1, an_2, veg_1, veg_2 from sea urchin embryos after isolation and subsequent attachment of one, two or four micromeres. (After HÖRSTADIUS, 1935)

concentration of the vegetal tendency in the micromeres (Fig. 216a is a rough diagram). A certain balance between the two tendencies in a layer or a combination of layers, which corresponds to the balance of tendencies in the whole egg, leads to typical development. This balance obtains when the middle layers are isolated (Fig. 216b): an_2 and the macromeres of the 32-cell stage (Fig. 197c); it can likewise be produced through the joining of certain distant layers (Figs. 214, 215, 216c: An and Mi of the 16-cell stage). The variability of morphogenesis in animal and vegetal halves (Fig. 211) can be explained on the basis of variation in the relative steepness of the two component gradients (Fig. 216d, e): a relative strengthening of the vegetal tendency improves the morphogenetic competence of the animal half and weakens that of the vegetal half; and vice versa.

That the vegetal tendency extends into the animal half is seen in a temperature experiment: in animal halves, low temperatures (11–15°) promote differentiation in the animal direction, high temperatures (21–26°) in the vegetal direction.

In the abnormal development in Figures 213–216, the polar orientation of the gradients was maintained by "orthopolar" transplantation; the most animal and the most vegetal regions maintained their relative positions. But the regulative development of meridional halves (p. 169) shows us something new if one asks how the closure is effected in a half-blastula coming from an

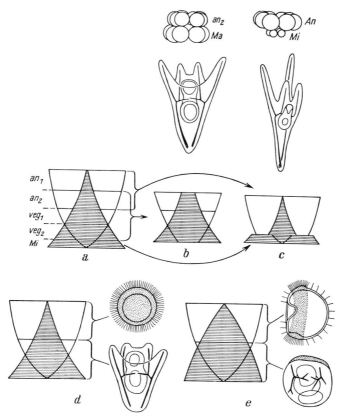

Fig. 216. Diagram of the vegetal and animal gradients in the sea urchin embryo. In the entire embryo (a) and in embryonic fragments; b an_2 + macromeres; c An (= an_1 + an_2) + micromeres; d, e isolated animal and vegetal halves, two variants (cf. Figure 211) with different balance between animal and vegetal gradients. (a in Appendix, von Ubisch, 1936; plutei after Hörstadius, 1936)

isolated blastomere of the two-cell stage. Only if its edges come together on the side do the polarity and the cell layers remain undistorted (Fig. 217d, Äq.). Vital staining of the most animal part of the four an_1 cells of a half-32-cell stage (Fig. 217a) shows that the half-blastula actually closes over its entire free circumference (according to the arrows in Fig. 217d).

The most animal material is brought together with the most vegetal; in the gastrula, vitally stained cells are seen along one side of the embryo up to the edge of the blastopore (Fig. 217b, c). At gastrulation, invagination begins at the original most vegetal point (Fig. 217f, g VEG); the neighboring animal material does not invaginate, but gastrulation proceeds through a constant epiboly of vegetal material (Fig. 217g). After gastrulation, the mouth region consists partially of what was originally the most vegetal material and partly of the most animal (polar) presumptive ectoderm material (AN), which has been pushed back. The tip of the gut (veg) corresponds now to the medial region of the primitive gut of a normal gastrula (Fig. 217e). At the new animal pole (an) of the closed blastula, the long stiff setae arise. As a consequence of the kind of closure which the half-blastula undergoes, the individual regions of both ectoderm and endoderm are used atypically throughout, although the border between ectoderm and endoderm is never crossed. The presumptive ectoderm and the presumptive endoderm develop harmoniously in

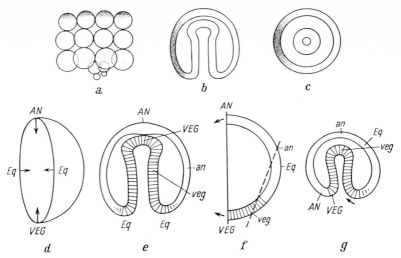

Fig. 217. Diagram of the apportionment of material and of the polarity in a half-embryo of *Paracentrotus lividus*. a half 32-cell stage from a half-blastomere in which the most animal region of an_1 is vitally stained; b, c resultant gastrula in longitudinal and cross optical sections; d diagram of closure of the half-blastula; e section through a normal gastrula; f half-blastula; g resultant gastrula, presumptive and invaginated endoderm; hatched. *AN*, *VEG* sites of the animal and vegetal poles of a whole embryo; *an*, *veg* same in half-gastrula; *Eq* equator. (After HÖRSTADIUS, 1928, 1935)

typical proportions. In the meridional embryonic fragments the gradients are rearranged along the new embryonic axis, as further development into a typical pluteus larva shows.

In heteropolar transplantation embryonic parts are brought together in unusual orientations. When a vitally stained animal half is joined to a meridional half of the sixteen-cell stage (Fig. 218a), a normal larva is produced in accordance with the polarity of the meridional half. A considerable part of the presumptive ectoderm becomes endoderm (Fig. 218 b, c). A colored streak on one side of the primitive gut often extends to the tip (Fig. 218b, c), and even secondary mesenchyme can be formed from vegetalized presumptive ectoderm, for among the colorless mesenchyme cells there are some colored ones (Fig. 218b). The plutei from such combinations appear completely normal. Thus, the animal material has been incorporated harmoniously in morphogenesis.

If the vegetal half of a 16- or 32-cell stage is inverted (Fig. 219a), embryonic halves with opposing polarities are united. Both halves fold over as they would have done in the normal situation. They thus retain their radial polarity. The animal half sits like a cap on the little vegetative blastula (Fig. 219b), but then an opening appears in the wall between the two blastocoels (Fig. 219c), and the embryo assumes the form of a single blastula (Fig. 219d). Here the originally most vegetal material lies at the equator, and this is where gastrulation begins (Fig. 219e). A primi-

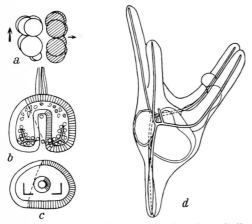

Fig. 218. Fusion of a meridional and a vitally stained animal half in the sixteen-cell stage of *Paracentrotus lividus* (a); b the resulting gastrula; c same, optical cross-section; d resulting pluteus

tive gut invaginates into the animal half, and from its tip a little ectodermal pouch is suspended by a stalk. The stalk either invaginates into the little ectodermal pouch or makes the free primitive gut. In this type of abnormal gastrula the most vegetal part of the larval material now lies near the blastopore (Fig. 219f). The little ectodermal pouch derives from a part of the presumptive ectoderm of veg_1 (cf. Fig. 206); the endoderm comes from the original equatorial parts of the inverted vegetal half (veg_1 with the nearby regions from veg_2). This system retains its polarity. However, in the major portion of the veg_2 material, which has been joined to the animal half and in which the primitive gut is formed, the polarity is reversed. In the mature pluteus, an

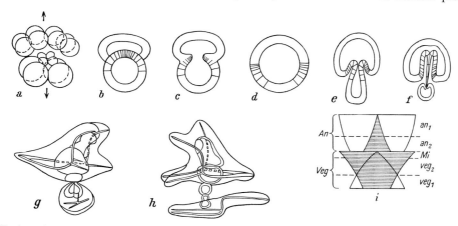

Fig. 219. Fusion of an inverted vegetal half with an animal half in the sixteen-cell stage of *Paracentrotus lividus* (a); b–d formation of a single blastula; e, f course of gastrulation; g, h pluteus stages; i gradient scheme. (a–h after Hörstadius, 1928, 1935)

additional incomplete dwarf larva sits below the anus with skeletal elements and either an internal or an exogastrulated gut (Fig. 219g, h). Figure 219i illustrates one interpretation involving a gradient scheme.

The vegetalizing force of the micromeres placed in their normal orientation (orthopolar) is shown by the experiments in Figure 214 and 215. But micromeres transplanted in a heteropolar fashion to the animal half migrate inward as primary mesenchyme cells and build mostly extra skeletal elements, in addition to having an endodermalizing effect. When they are joined to the side of an entire embryo, a little primitive gut arises from the nearby animal material (Fig. 220). Even when placed right in the animal pole they can induce a little primitive gut in *Paracentrotus* (Fig. 221). The vegetal tendency of the micromeres appears much stronger when they are implanted in the animal pole of an isolated animal half (Fig. 222a). When the micromeres are introduced into the blastoderm in later cleavage stages (Fig. 222b) they move into the blastocoel and collect partly beneath the site of implantation and partly on the opposite (most vegetal) side of the blastula, from which an attractive force evidently emanates (Fig. 222c). The larvae which result from such transplantation experiments are very diverse. Sometimes irregular structures arise at both poles, but no gut. In other embryos a gut invaginates from the most vegetal side, and in still others from both poles (Fig. 222d, e, f). The tuft of the blastula is displaced toward the equator. At times two tufts arise opposite one another (Fig. 222d, e). These phenomena strongly suggest the opposing actions of the original polarity and a new one set up by the micromeres (Fig. 222i). Even a complete reversal of polarity can occur. The apical tuft appears at the original vegetal side, which is still recognizable by its red egg pigment; the primitive gut invaginates

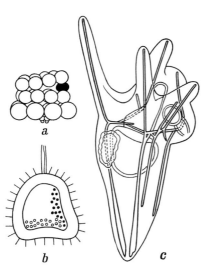

Fig. 220. Implantation of four micromeres between an_1 and an_2 in a 32-cell stage of *Paracentrotus lividus* (a); b the skeletal anlage from the implant arranges itself on one side of the blastocoel; c pluteus with extra little gut and extra skeletal elements. (After HÖRSTADIUS, 1935)

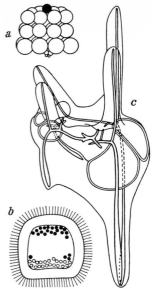

Fig. 221. Implantation of four micromeres in the animal pole of a 32-cell stage of *Paracentrotus lividus* (a); b arrangement of the primary mesenchyme in the blastula; c pluteus with extra two-part "animal" gut and two extra skeletal elements. (After HÖRSTADIUS, 1935)

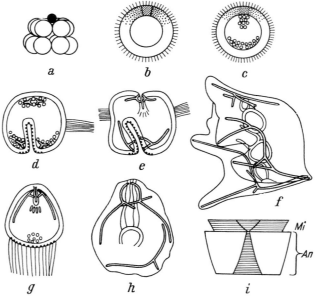

Fig. 222. Implantation of four micromeres in the animal pole of an animal half of *Paracentrotus lividus* (a); b, c position of the micromeres and the primary mesenchyme cells in the blastula, animal pole vitally stained (dotted); d, e formation of skeleton and primitive gut at the vegetal pole and later also at the animal pole; f pluteus with vegetal and "animal" primitive gut and extra skeleton; g, h reversal of polarity; i gradient scheme. (a–h after HÖRSTADIUS, 1935)

from the site of implantation to the animal poles, and here the mesenchyme cells form a large skeleton while forming a smaller one on the opposite side (Fig. 222 g, h).

These experiments show that the differentiation of embryonic parts is not determined by the quantities of animal and vegetal tendencies already present in the individual layers. Where the total composition of the tendencies is sufficiently balanced, the polar role is assumed by the most extreme part of the gradient (Figs. 214a, 216b). This *principle of polar dominance* (LINDAHL) appears in HÖRSTADIUS' gradient hypothesis as the *reconcentration of the poles*. Gradients split

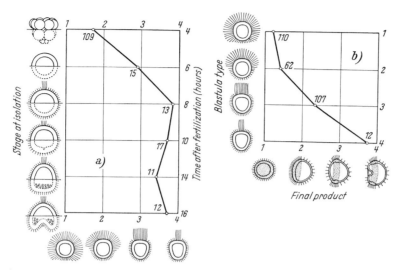

Fig. 223. Dependence of differentiation of the animal half on the age of the embryo in *Paracentrotus lividus* in terms of the time of isolation (hours) after fertilization at 20°: 4 hours, 16- or 32-cell stage; 6 hours, blastula without cilia; 8 hours, blastula with cilia and apical tuft; 10 hours, vegetal wall of the blastula thickened with a depression at the vegetal pole; 14 hours, formation of primary mesenchyme; 16 hours, gastrulation in progress. a Dependence on the size of the area of long straight cilia upon the larval age at which the animal half is isolated; b final differentiation of the animal half ordered according to the size of the apical tuft in the blastula stage. Ordinates: average values for each class; in a, grades of extent of the apical tuft: *1*, more than half the blastula; *2*, about half the blastula; *3*, one-third to one-half; *4*, less than one-third of the blastula. In b, grades of animal differentiation (cf. Fig. 211 a–d). For each point the number of individuals is given. (Diagram based on data of HÖRSTADIUS, 1935, 1936)

by the fusion of differently oriented larval parts (Figs. 215, 216e) re-form, resulting in a continuity of new gradients, which often in turn results in the typical organization of the whole larva or of a larval part. The micromeres, which by themselves can only form the skeleton, act as inducers of a vegetal organ complex upon heteropolar implantation; following orthopolar implantation they act as centers from which a new vegetal gradient emanates, and thus strengthen an existing gradient or weaken the animal gradient.

The contribution of the gradients, which in normal development leads to the determination of the larval parts at a certain time, can be demonstrated experimentally. If the animal and vegetal halves are separated at different embryonic ages (Fig. 223a), the animal halves will develop progressively smaller (and therefore more typical) apical tufts, the later the separation is made, that is, the longer the two halves have remained united. A corresponding pattern is seen in the final differentiation of the animal half (Fig. 223b). Classification of the individuals according to the age of isolation (Fig. 224) shows that the maximal frequency of each degree of differentiation shifts considerably: ciliated blastulae predominate where isolation has taken place

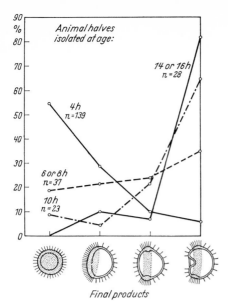

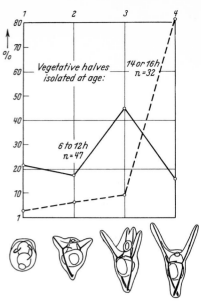

Fig. 224. Classification of various degrees of animal differentiation after isolation of the animal halves at different ages in *Paracentrotus lividus*. (After data of HÖRSTADIUS, 1936)

Fig. 225. Classification of degrees of vegetal differentiation in vegetal halves after isolation 6–12 (—) and 14–16 (···) hours. Abscissa: stages of differentiation (cf. Fig. 211). (After data of HÖRSTADIUS, 1936)

four hours after fertilization; in the ten-hour class, blastulae with flagellar band and oral depression are most frequent. In the six- and eight-hour age groups both extremes are seen. As for vegetal halves, those isolated at six to twelve hours produce a variety of results (cf. Fig. 211), the most frequent of which are plutei, generally with an oversized gut (Fig. 225). After fourteen and sixteen hours plutei without oral arms are predominant (Fig. 195). After the ingression of the mesenchyme cells determination has proceeded so far that isolated vegetal halves now produce mostly a type of larva whose most animal part, which arises from an_1 (cf. Figs. 206, 208a), is lacking, as if it had been cut away (Figs. 225, 226). Determination has therefore taken place, which is to say that regulatory ability has greatly diminished.

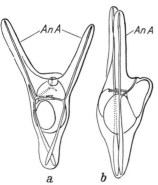

Fig. 226. Plutei without oral arms from vegetal halves of *Paracentrotus lividus*. (After HÖRSTADIUS, 1936)

The time course of determination, the compromise reached by the opposing gradients, is also seen in the reaction of animal halves to implanted micromeres. Embryos can be divided equatorially in the 16- or 32-cell stage, and the animal halves can then be fitted out with four micromeres at various times four to sixteen hours after fertilization (Fig. 227). The effects of the micromeres on differentiation depend here very strongly on whether the micromeres have just been removed from a 16-cell stage or have lain isolated so long that they have begun their own morphogenetic movements. After the implantation of micromeres at four hours, the results are mostly typical plutei or slightly animal plutei with a somewhat small gut; at six hours the second group predominates. The implanted micromeres form a skeletal anlage and induce gut formation. Only two hours later (at eight hours) a gut is formed only rarely. By far the most frequent result is now, as also at ten hours, an oral field embryo with oral depression, skeleton and rudimentary arms. At twelve hours, when in the intact larva micromeres move inward (see the time cycle in Fig. 223a), the implanted micromeres do build a fairly bilaterally symmetrical skeleton, but induce an

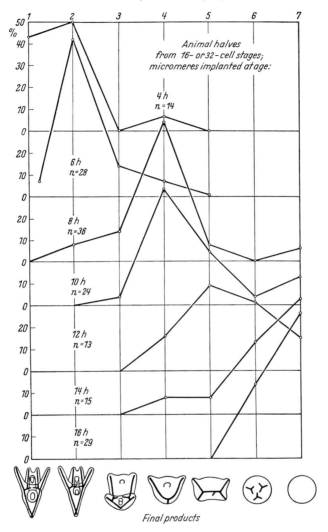

Fig. 227. Differentiation of anomalous halves of *Paracentrotus lividus* in which micromeres were implanted after the halves had been isolated in the 16- or 32-cell stage at different ages (*h*, hours after fertilization). (After data of HÖRSTADIUS, 1937; his fourth and fifth classes of differentiation have been combined)

oral depression only rarely. Now ciliated blastulae are numerous, with triradiate skeletal elements of irregular number and position. After implantation at fourteen or sixteen hours the micromeres have almost no effect on the ectoderm and generally cannot even form skeletal elements. Thus, at four to six hours the micromeres can set up a nearly normal gradient system in the blastula. At eight to ten hours this is rarely possible, but, as opposed to animal halves

181

which have been completely isolated for an equal length of time (cf. Fig. 224), a greater degree of differentiation is nevertheless achieved. At the age when the micromeres move inward in the intact embryo, the responsiveness of the animal blastoderm to implanted micromeres has vanished. The disappearance of the ability to form gut in the blastula stage may rest in part on the fact that the micromeres placed in the blastocoel no longer maintain contact with the blastoderm. The ectodermal responsiveness varies with the form used. In *Echinocyamus*, animal halves isolated in the eight-cell stage do not form a gut when four micromeres are implanted as they reach the blastula stage, but they produce otherwise fully differentiated plutei (Fig. 228).

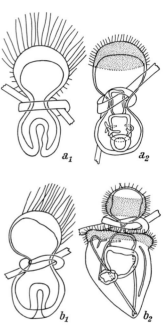

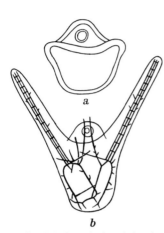

Fig. 228. Animal halves isolated in the eight-cell stage of *Echinocyamus pusillus*. a final stage of animal anormogenesis; b gutless pluteus after four micromeres had been implanted into an animal half in the blastula stage. (After von UBISCH, 1933)

Fig. 229. Differentiation of larvae of *Psammechinus miliaris* after equatorial ligation in the 8- or 16-cell stage. a_1, b_1 blastula stage; a_2, b_2 resulting larval stages. (After HÖRSTADIUS, 1938)

The influence of the vegetal material on the differentiation of the animal material is also seen in equatorial ligation of cleavage stages during the period before the expression of the gradient system. The connection between the animal and the vegetal material is narrowed, and the distance between the poles is increased. In this way any chemical effect emanating from the vegetal pole will obviously be hampered: the apical tuft becomes abnormally large (Fig. 229 a_1, b_1) and then a flagellar field develops (Fig. 229 a_2); a discrete band of flagella appears only rarely (Fig. 229 b_2), and an oral depression never. The vegetal half during ligation undergoes a course of development similar to that after complete separation (cf. Fig. 211 g, h); it develops a more or less disordered skeleton and often also the flagellar field or flagellar band characteristic of ovoid larvae (Fig. 229 b_2).

The gradient hypothesis is consistent with all of the experimental results we have described.

Lecture 13

The basis of the animal and vegetal tendencies, whose existence is concluded from the defect and transplantation experiments, must be the specific localization either of materials or of metabolic processes. For the experimental approach to this problem the discovery has been of great importance that early development of echinoids can be shifted in a vegetal or in an animal direction by exposure to certain chemical substances.

Lithium ions have a vegetalizing effect. LiCl, LiBr, LiI, $LiNO_3$, and Li_2SO_4 have been tested. LiCl is especially effective and has been used almost exclusively; lithium restricts animal differentiation and causes endodermal hypertrophy. The first sign of vegetalization is a movement toward the animal pole of the ring of primary mesenchyme cells (Fig. 230). The hypertrophy of the primitive gut prevents it from taking up its normal position in the larva: it is not invaginated but rather bulges out. This exogastrulation produced by relatively low concentrations of lithium results in the fairly normal formation of a larva without a gut. As endoderm formation is enhanced, the differentiation of the rest of the larva is reduced; attached to a partitioned or unpartitioned gut whose ciliated surface is on the outside, one finds a differentiated ectodermal bag (Figs. 231, 232). If this is only slightly reduced, the primary mesenchyme cells make an irregular skeletal structure (Fig. 232a); if the ectoderm is further reduced, the mesenchyme cells scatter in the blastocoel without forming skeleton (Fig. 232b). The degree of lithium effect is dependent upon its concentration, temperature, the time of onset of treatment, and its duration (Fig. 231). Higher temperatures promote vegetal differentiation here, as they do in animal halves (p. 201). The isolation of animal and vegetal halves at different stages in embryonic development (Figs. 224, 225), and the implantation of micromeres into animal halves of different ages (Fig. 227), has showed that between twelve and sixteen hours after the onset of development, the joint action of the gradients has a decisive effect on the further development of the larval parts; and as one may expect, the lithium treatment must begin before this time in order to be effective.

When embryos are put in LiCl solution immediately after fertilization and left for twenty-four hours, the highest degree of vegetalization is achieved: extreme exogastrulae with very little ectoderm and no skeleton. Six-hour embryos become mostly ovoid larvae; after ten hours the lithium sensitivity of the embryos decreases. If the embryos are treated with lithium for a

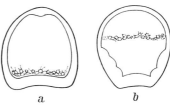

Fig. 230. Placement of the mesenchyme cells in a normal (a) und a LiCl-treated vegetalized (b) blastula of *Paracentrotus lividus*. (After Czihak, 1962)

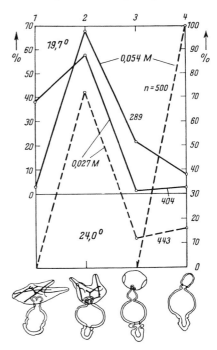

Fig. 231. Differentiation of *Paracentrotus lividus* embryos in LiCl solutions of various concentrations at various temperatures. Left ordinate: percentages of embryo types (illustrated below) at 19.7° (solid lines); right ordinate: percentages at 24° (dashed lines). Total number of embryos for each experiment is placed near the appropriate line; molarities as indicated. (After Lindahl, 1938)

duration of only ten hours, the effects are considerably weaker; larvae treated as early as six hours after fertilization can still produce good plutei. Even when treatment begins immediately after fertilization, a pronounced vegetalization requires at least a twelve-hour exposure period.

The clearest picture of the effect of lithium during the course of determination is obtained when animal halves of different ages are placed in lithium solution (Fig. 233). The very young animal halves show the highest sensitivity. Right after fertilization, their tendency to extreme animal development (Figs. 211, 223) can be completely reversed: at the highest concentrations, ovoid larvae are produced, as is the case with untreated vegetal halves. In addition, there may even be exogastrulae, some normally proportioned plutei, and a number of flagellar band blastulae with or without an oral depression and — in contrast to the development of untreated animal

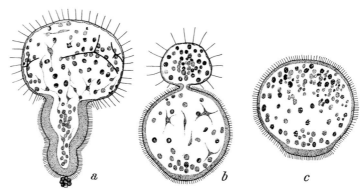

Fig. 232. a–c Lithium larvae of *Sphaerechinus*; varying degrees of enlargement of the endodermal area. (After HERBST, 1893)

halves (Fig. 211)—with a greater or lesser degree of gut invagination (Fig. 233). This last result predominates when treatment begins at the six-hour stage (Figs. 233, 234 b–d). After eight hours this result is rare (Figs. 233, 234e). Now almost all the larvae are of distinctly animal types (cf. Fig. 211): most frequently blastulae with flagellar band but without an oral depression (Fig. 234f); next, those with an oral depression; and there are also some flagellated blastulae (cf. Fig. 211). When later stages are used, the lithium effect is further reduced, and after sixteen hours the development of treated embryos is the same as those not exposed to lithium (Fig. 224). It appears that the lithium treatment is especially effective in restricting mouth formation, for flagellar band larvae without an oral depression, which are the most frequent result when the onset of treatment is between six and ten hours, are relatively rare among untreated isolated animal halves (Fig. 224).

In the whole embryo as well as in animal halves, therefore, lithium strengthens the vegetal tendency during the determination phase; in terms of the gradient hypothesis, it strengthens the vegetal gradient. But the forms which result from lithium treatment do not lie on a continuous line from the most vegetal anormogenesis to the most animal in the sense of a displacement of the gradient along the major axis (corresponding to Figs. 211, 213, 215). A certain anarchy of determining forces is revealed in that particularly after the exposure of six-hour embryos to lithium solution, several primitive gut invaginations occur together (Fig. 234c, d). Moreover, especially just before the disappearance of lithium sensitivity, and long after the ability to form a gut has been lost, blastulae appear which have neither flagellar band nor flagellar field, nor even a complete columnar flagellar epithelium (as in Fig. 211a); instead, their wall is composed of irregularly distributed flat epithelium (Figs. 233, 234g). These are most common when treatment

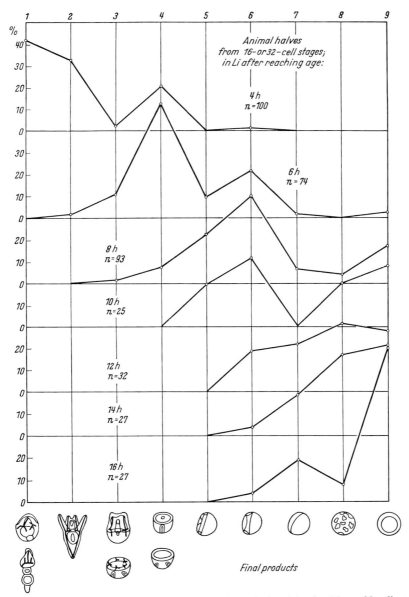

Fig. 233. Differentiation of animal halves of *Psammechinus miliaris* isolated in the 16- or 32-cell stage and placed at various ages in LiCl solution. (After data of HÖRSTADIUS, 1937; his differentiation type *e*, which appeared only once in this experimental series, has been left out)

begins at twelve hours and are still very numerous if it begins at fourteen hours (Fig. 233). Accordingly, lithium appears to act at different places in the blastoderm during certain sensitive periods. In the vegetal parts of the embryo the most vegetal determination, invagination of the primitive gut, is easily disturbed; later, in the entire blastoderm the normal course of animal epithelial differentiation is disrupted.

That lithium can induce a complete vegetal gradient in isolated animal halves is shown by a really elegant experiment: the vitally stained vegetal part of an animal blastula which has developed in lithium solution (Fig. 235a, d) is joined to an untreated animal half of the 32-cell stage (Fig. 235f, g, h). Untreated animal halves of the same stage normally result in blastulae with an extended apical tuft field and later in ciliated blastulae (Fig. 235b, c); if they are left in lithium solution they are somewhat overvegetalized and produce ovoid larvae (Fig. 235e). If an_2 from an animal half exposed to lithium is joined to a normal animal half (Fig. 235g, h), the apical tuft is restricted (Fig. 235i); and an_2 produces primary mesenchyme and the tip of the primitive gut

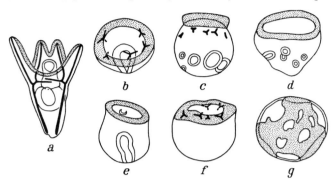

Fig. 234. a–g Lithium effect on animal halves of *Paracentrotus lividus* which were isolated in the 16- or 32-cell stage and placed in lithium solution at various ages. a After four hours; b, c, d after six hours; e, f after eight hours; g after twelve hours. (After HÖRSTADIUS, 1937)

(Fig. 235k). Finally, a normal pluteus results (Fig. 235l). On the other hand, if the animal cap of a comparable Li^+ blastula (Fig. 235f) is vitally stained and joined to the vegetal side of an untreated animal half (Fig. 235m, n) this animal half produces no effect; it is incorporated into the blastoderm and the development of the embryo (Fig. 235o, p) resembles that in controls (Fig. 235b, c). The controls referred to here are, of course, not normal embryos but simply untreated animal halves. The most vegetal cells from a lithium animal half, therefore, act in a young embryo as the center of vegetal influence; they behave just as micromeres would behave when implanted in the same stage (cf. Fig. 227). The lithium effect imparts to an_2 material the full vegetal tendency of micromeres.

An *animalization* of the embryo can be achieved through a variety of chemical effects on the *unfertilized* egg. Thiocyanate ions (e.g., NaSCN) are most effective, then iodide (NaI), and to a still lesser extent, trypsin, among other things. The weakest effects result in types of differentiation resembling those resulting from combinations of embryonic layers (Fig. 215). In the lowest grade the apical tuft of the blastula is greatly enlarged, but a pluteus can still result, although generally with a reduced gut (Fig. 236a$_1$, a$_2$). Then, passing over intermediate grades (Fig. 236b–d) one arrives at forms which have undergone essentially no vegetal differentiation, although there is still some inward migration of the primary mesenchyme. Still further animalization leads to a new and very surprising phenomenon at the vegetal pole; the blastula wall becomes somewhat thickened here, and another apical tuft appears (Fig. 236e). In general, with still stronger treatments this adventitious apical tuft spreads over the vegetal surface (Fig. 236f); in the equatorial zone the long setae are not quite as thick, and they are mingled with short, beating flagella. In these embryos the regression of the long setae is followed by the establishment of a band of flagella, and an oral depression can appear, or several oral structures can arise. A kind of oral field thus becomes clear in that around the oral depression the long setae disappear (Fig. 236g). When

186

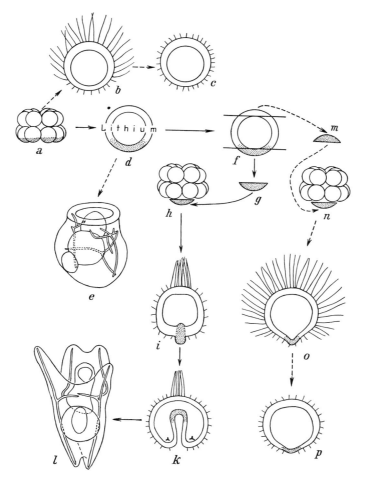

Fig. 235. a–p Diagram of the effect of a lithium-treated larva's most vegetal part, which acts as a secondary vegetal center in transplantation experiments. a–c First control experiments: development of the animal halves to a blastula with enlarged apical tuft and then to a blastula with setae; a–d, e second control experiment: development of the animal half in lithium solution to an ovoid larva; a–d–f, g, h–l main experiment; m–p third control experiment. (After HÖRSTADIUS, 1937)

the SCN⁻ effect is maximal, blastulae result which have their entire outer surface covered with long stiff setae (Fig. 236h). The animal differentiation of the whole embryo is therefore even more extreme than that in isolated animal halves (Fig. 211a). These embryos become dauerblastulae without any further differentiation (as in Fig. 211a).

Animalization caused by SCN⁻ ions before fertilization, and the paradoxical appearance of an apical tuft at the vegetal pole, reveal an effect on the egg cytoplasm which leads to a change in polarity and therefore to an entirely abnormal distribution of animal and vegetal tendencies.

This reversal of polarity in the presumptive vegetal part becomes even clearer when animalizing and vegetalizing effects are combined. Treatment with SCN before fertilization moves the vegetal tendency toward the equator and away from the vegetal pole (Fig. 236e, f). This displaced vegetal focus can now be strengthened greatly by the use of LiCl during larval development. The blastula elongates in the direction of the egg axis, and primary mesenchyme cells move inward in the region

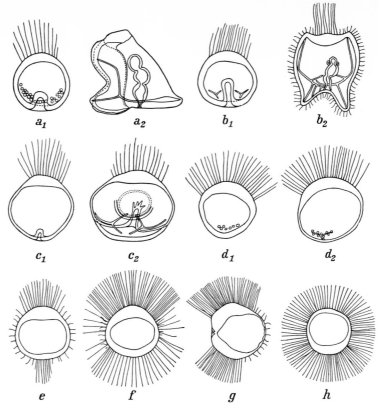

Fig. 236. a–h Animalization of eggs of *Paracentrotus lividus* (a–g) and *Arbacia pustulosa* by treatment before fertilization with SCN⁻ in sea water (15–36 hours). a_1 24 hours old; a_2 4 days old; b_1 2 days old; b_2 4 days old; c_1 2 days old; c_2 3 days old; d_1 24 hours old; d_2 4 days old; e 24 hours old; f, g 32 hours old; h 36 hours old. (After LINDAHL, 1936)

of red pigment, i.e., in the region corresponding to veg_1 and veg_2 (Fig. 237a). The embryo constricts, and the mesenchyme cells are parcelled out to the two spherical extremes (Fig. 237b), which will become ectoderm, while the region in between becomes endoderm (Fig. 237c). The ectodermal spheres can develop further to varying degrees. In some cases twinning results (Fig. 237d, e), so that when differentiation is completed, the products are very similar to what one sees after the inversion of the vegetal half of a sixteen-cell embryo (Fig. 219g, h). In both cases, each in a different way, the maximal vegetal tendency is placed in the equatorial zone of the embryo. The endodermal bridge is typically divided into two primitive guts connected by the esophageal parts (Fig. 237d). In some cases the endodermal invagination occurs on only one side, and thus only one primitive gut is formed (Fig. 237f–h). The animalization of the vegetal pole leads to embryos with two flagellar bands (Fig. 237g) and two skeletal pairs with corresponding arms (Fig. 237h). In some cases the tendency for invagination in the equatorial zone is chaotic: several primitive gut anlagen arise (Fig. 237i).

The results of this chemical vegetalization and animalization can be interpreted in line with the gradient hypothesis (Fig. 238), assuming that the relative displacement of the two tendencies can come about through a weakening of one gradient, a strengthening of the other, or both. That lithium increases the vegetal tendency in an absolute sense is shown by the fact that the vegetal-

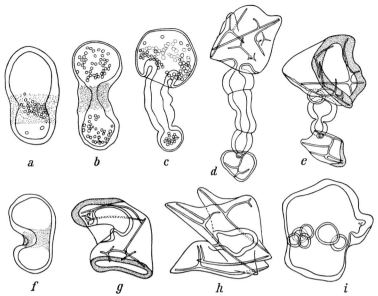

Fig. 237. a–i Larva from *Paracentrotus lividus* eggs treated before fertilization (14–24 hours) with NaSCN in sea water (sometimes calcium-free sea water) and after fertilization (for about 24 hours) with LiCl in sea water. a 1 day; b $1^1/_2$ days; c 2 days; d 3 days; e $3^1/_2$ days; f $1^1/_2$ days; g–i $3^1/_2$ days old. (After LINDAHL, 1936)

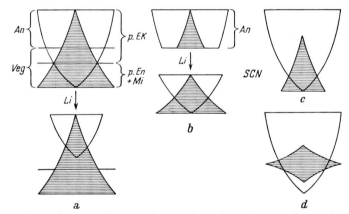

Fig. 238. a–d Gradient scheme for vegetalization of an entire embryo (a) and of an animal half (b) by lithium; animalization by SCN⁻ (c); and combination of both effects (d)

izing part of a lithium blastula has the same effect on an untreated animal half as do micromeres (Fig. 235). Moreover, the other results involving quantitative alterations of the original polar gradients lead to the conclusion that the reversal of polarity (Figs. 236e–g, 237) is caused by a displacement of the center of the vegetal gradient (Fig. 238d).

Novel distorted pluteus larvae describable neither as animalized nor as vegetalized arise when eggs are put in sugar solution after fertilization and are returned to sea water after about twenty hours, having reached the gastrula stage. Hexose monosaccharides and disaccharides, and also the pentoses xylose and arabinose, all produce the same effect. The "sugar larvae" are always much smaller than normal (Fig. 239a). The lowest effective dose prevents formation of the oral arms (Fig. 239b; cf. Fig. 210a). The oral lobe characteristic of this kind of experiment is enlarged

189

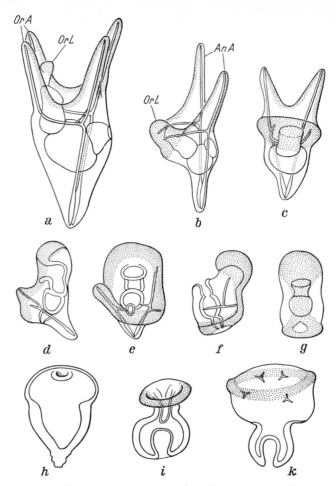

Fig. 239. a–k Effect of sugar solutions on the development of *Psammechinus miliaris*. a Normal larva from the same batch as the experimental animals; b–g increasingly strong effects; b, d, f side view; c, e, g dorsal view; h–i individuals after combined treatment with sugar and LiCl; k control larva treated only with LiCl; *AnA* anal arms; *OrA* oral arms; *OrL* oral lobe. (After HÖRSTADIUS, 1959)

and covered with a broad flagellar band (Fig. 239 b, c). Stronger treatments prevent the formation of the anal arms also; the entire oral field is covered with a greatly enlarged flagellar band. The mouth and gut are very large in proportion to the small larva. The entire aboral side, normally a long, pointed structure (cf. Figs. 206, 207), remains undeveloped. In the most extreme cases the entire larva is reduced to an oral field covered by a broad flagellar band, together with a tiny anal field including an anus (Fig. 239 f, g). The sugar larva resembles none of the reduced forms resulting from animal or vegetal halves (Fig. 211) or from presumptive ectoderm (Fig. 213). It is an oral field larva, and yet it has a large gut. The nearest approximation to this form is seen in larvae derived from animal halves which have been placed at different ages in LiCl (Figs. 233, 234 b, c), but these abnormal forms are at the end of their development, while the sugar larvae move on in the direction of normal structure: a pluteus does arise, though admittedly a misshapen one. A compromise of a still unclear nature evidently takes place between the animalizing and

190

vegetalizing tendencies. If the sugar treatment is combined with lithium, the endoderm enlarges still further. From the tip of the usual gastrula an oral depression is formed, the oral field is reduced, and the skeleton disappears or consists only of irregularly placed three-rayed elements. The resulting larvae (Fig. 239h, i) resemble, except for the closely restricted oral depression, larvae treated simply with lithium under the same conditions (Fig. 239k).

The question naturally arises as to which conditions or processes underlie the opposing polar gradients, their regulatory polar concentrations, and their determining effects.

From the experiments discussed thus far we only know that different processes can be attributed to the animal and vegetal tendencies which do not correspond in their temperature dependence (pp. 174, 183), and that transplanted micromeres, evidently through the release of a substance, effect a maximum vegetal tendency.

Echinoid eggs appear to be an especially good theater in which to elucidate the bases of polarity, and consequently the ordering in gradients, from as far back as the oocyte stage, of substances or processes, and their chemical consequences for determination. This is so, since eggs and embryos can be studied alive in great numbers in comparable developmental stages with biochemical methods. Both their metabolism and the effects of particular chemicals on development can thus be studied. To solve the riddle of determination, however, an insight into the chemical composition and proces-

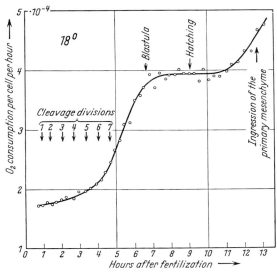

Fig. 240. Respiratory rate during the first hours after fertilization of *Psammechinus miliaris*. (After BOREI, 1948, from RUNNSTRÖM, 1949)

ses in differently determined embryonic parts is necessary, or at least in the animal and vegetal halves. To obtain such partial embryos in the required numbers in a particular stage is not so easy, for surgery on little embryos is difficult and time-consuming. One may, therefore, substitute whole vegetalized or animalized embryos of a particular grade.

First of all, we observe changes in the metabolism of whole embryos correlated with morphogenetic stages. Since the sea urchin egg is clearly an aerobic organism, its respiratory rate should be a good expression of its general metabolic level. Oxygen consumption changes in the course of development. During cleavage up to the blastula stage it rises exponentially; the it remains constant until shortly before the ingression of the primary mesenchyme (Fig. 240). The end of this first developmental period is marked by a sudden rise in respiratory rate. During the second period of development, in which oxygen consumption increases rather linearly (Fig. 241), the ingression of the primary mesenchyme and gastrulation are completed. During the third period of development, which follows, respiration continues to increase slowly.

The breakthrough of the mouth signals the end of this developmental period; now feeding begins (well before this time, however, when skeletal formation begins fifteen to twenty hours after fertilization, the uptake of calcium from the sea water is initiated). During early development (cleavage, blastula, ingression of the primary mesenchyme, gastrulation, and the concomitant stages of determination) a change takes place in the breakdown and synthesis of specific immunochemically analyzable proteins (Fig. 242).

Ostensibly, the animal-vegetal double gradient system influences the gradients of specific proteins; but the results of experiments designed to reveal the chemical events underlying determination do not present a clear picture.

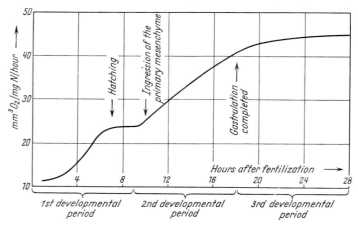

Fig. 241. Respiratory rate during early development of *Paracentrotus lividus* at 22.0°. (After LINDAHL, 1940)

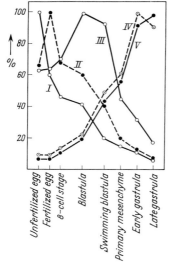

Fig. 242. Concentration changes for specific proteins *Pseudocentrotus depressus* embryos. (After ISHIDA in and YASUMASU, 1957)

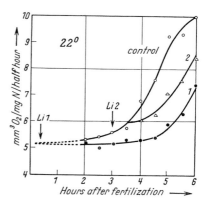

Fig. 243. Respiratory rate in the first six hours in controls and two experimental series placed in *Li* solution at different times (*Li*₁, *Li*₂). (After LINDAHL and ÖHMANN, 1938)

Vegetalization by lithium is accompanied by a lower-than-normal rise in respiratory rate during the first developmental period (Fig. 243). The effect of lithium treatment on respiration, just as on morphogenesis, is less the later it begins. The vegetalizing effect of lithium is enhanced by inhibitors of respiration such as CO and KCN; similarly the respiration-promoting compound pyocyanin enhances the animalizing influence of NaSCN. And yet a higher respiratory rate cannot be interpreted as the specific for animal determination; for measurements have revealed no difference in respiratory rate between the animal and vegetal halves of early blastulae; and, furthermore, the animalizing NaSCN reduces the respiratory rate, just as LiCl does. Accordingly, LINDAHL concluded that different forms of metabolism are involved in the animal and vegetal tendencies respectively. In a cell homogenate from echinoid eggs, with hexose monophosphate added, lithium delays the reduction of methylene blue considerably; this suggests that carbohydrate catabolism is the lithium-sensitive specific animal component of respiration. With respect to specific vegetal metabolism the following phenomenon is of interest: in the blastula stage a sulfate deficiency, apart from general deleterious effects, inhibits vegetal differentiation. It has been suspected that sulfuric acid serves to detoxify aromatic metabolites, as in the vertebrates, and that a sensitive protein metabolism is characteristic of vegetal determination and

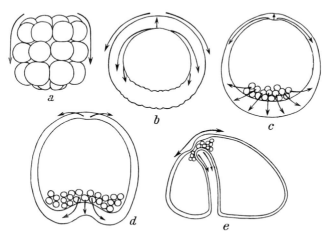

Fig. 244. Reduction gradients in an cehinoid embryo stained with methylene blue. (After CHILD, 1937)

differentiation. This idea receives additional support from the fact that sulfate deficiency damages vegetal halves more than entire embryos and has no effect on animal halves nor on whole embryos animalized with SCN. It is further known that developing sea urchin embryos releases a substance which inhibits the development of other embryos; this substance, which, to be sure, has never been isolated, is released in higher concentration when there is a sulfate deficiency. Finally, a phenol sulfatase has been demonstrated in sea urchin embryos. However, it is still doubtful whether the fall in oxygen consumption is a necessary component of the vegetalizing process: between 20 and 24° both inhibition of respiration and the degree of vegetalization increase with temperature; at 26° the morphogenetic effect is especially strong, but the respiratory inhibition disappears; and finally, known inhibitors of glycolysis do not cause vegetalization; thus, the hypothesis of a specific animal carbohydrate metabolism is open to question.

A suggestion of metabolic gradients in the embryo is provided by experiments using vital staining in oxygen-free or oxygen-depleted sea water. Redox indicators which cause little or no damage are used, such as Janus green, methylene blue, brilliant cresyl violet, and others. During cleavage to blastula the reduction of the dye proceeds from the animal pole (Fig. 244a, b). With the ingression of the primary mesenchyme, a second strong center of reducing activity appears at the vegetal pole, from which the bleaching of the vital stain spreads in the direction of the animal pole (Figs. 244, 245a). The mesenchyme cells are the most powerful reducing cells in the embryo at this time. In the late mesenchyme blastula and during gastrulation, the reduction gradients extend in opposite directions from the two poles. (Figs. 244c, d; 245a, b). In isolated animal halves the reduction does not begin from as restricted a center at the animal pole as in

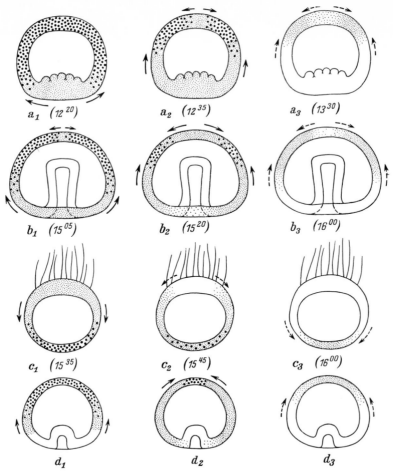

Fig. 245. Reductive bleaching of Janus green via diethyl safranin (which is a reddish intermediate) in embryonic stages of *Paracentrotus lividus*. a_1, b_1, c_1, d_1: time after fertilization when embryos were placed in the stain. Follow each row from left to right. a Late blastula; b gastrula; c animal half of an embryo; d vegetal half of the same embryo. Large dots: blue; small dots: red. (After Hörstadius, 1952)

normal embryos (Fig. 245 a_2, b_1), but rather begins simultaneously in the whole ectodermal region of the enlarged apical tuft and spreads quickly to the opposite pole (Fig. 245 c), while in whole embryos this animal wave of reduction remains restricted to the blastoderm region above the equator (Fig. 245 a, b). In vegetal halves, destaining begins in the mesenchyme cells and spreads during gastrulation to the opposite pole (Fig. 245 d). In animalized embryos (Fig. 246 a) destaining takes place quickly and uniformly; the gradient that appears in untreated animal halves is nowhere to be seen. In vegetalized embryos the reduction spreads out toward the animal pole, as in vegetal halves (Fig. 246 b).

Experiments previously described showed that micromeres implanted in heteropolar fashion in cleavage stages worked as secondary vegetal centers and triggered the formation of a secondary vegetal organ complex (Figs. 220–222). This effect corresponds in the blastula and gastrulation period to the induction of a secondary reduction gradient (Fig. 247). From the pole in which the implanted cells are lodged in the blastoderm, reduction spreads out on all sides. It unites in

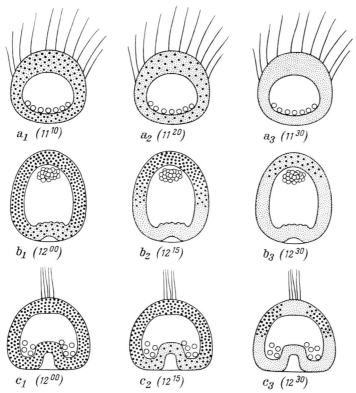

Fig. 246. a embryo animalized with trypsin; b embryo vegetalized with LiCl; c control. All embryos stained with Janus green in the late blastula stage; color goes from blue to red (cf. Fig. 245). (After HÖRSTADIUS, 1955)

its half of the embryo with the wave of destaining emanating from the host micromeres and continues over the animal pole. In isolated animal halves, implanted micromeres, which restrict the region of the apical tuft, also restrict the spread of reduction coming from the animal pole. These experiments show that a metabolic process emanates from the micromeres which expresses itself in the reduction of vital stains; and this vegetal process restricts the reduction gradient coming from the animal pole. From this one may conclude that there is a difference between the animal and vegetal components of metabolism.

During the course of gastrulation the reducing ability of endodermal cells increases; the tip of the primitive gut thereafter maintains a high level of reduction (Fig. 244). In the ectoderm, the

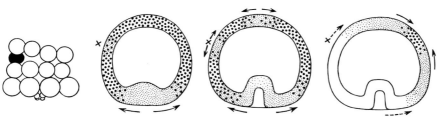

Fig. 247. Implantation of micromeres between an_1 and an_2 of a 32-cell stage. X site of implantation (cf. Fig. 229a, b). (After HÖRSTADIUS, 1952)

center of reducing ability moves toward the ventral side in the region of the oral depression (Fig. 244e). Later an increase in reduction appears at the places where the oral and anal arms will arise. These experiments show the new appearance of reducing activity and therefore of a quantitative or qualitative change in metabolism in places where a morphogenetic event, a consequence of determination, originates. Even in echinoid oocytes a reduction gradient exists, both before and after maturation. Its maximum is at the animal pole, which agrees with the orientation of the reduction gradient during cleavage until the blastula stage.

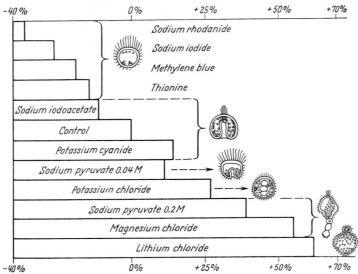

Fig. 248. Comparison of the specific viscosity, using an Ostwald viscometer, of protein solutions (euglobulin solutions a and b from *Arbacia lixula* embryos) treated with different chemicals; and larval types produced by the same chemical treatments. Bars represent specific viscosities, measured as differences from the control. (After RANZI, 1949)

Thus, the formal idea of opposing gradients, which was concluded on the basis of isolation and transplantation experiments, acquires substance. But however useful the biochemical experiments to date on echinoid development may be, we still have great hopes for them; the nature of the vegetal and animal gradients still remains a mystery, and no chemical property nor any metabolic process permits us to elucidate even a single morphogenetic step, not even the simple process of gastrulation.

That the colloid structure of lithium embryos is altered in some way can be seen from a variety of observations. While the cells of a normal blastula round up as soon as they are separated from one another, cells from a lithium blastula retain their extended form for a long time. Moreover, dark-field observations reveal a peculiarity of the colloid structure of the egg cortex in the vegetal region: after fertilization, a yellowish-orange ring, apparently due to a lipid layer, appears around echinoid eggs in the vicinity of the future veg$_2$ tier. In lithium embryos it extends much further in the animal direction; in an exogastrula, the yellow-orange layer is spread over the evaginated endoderm[566].

Experiments with protein fractionation of echinoid embryos shows a parallel between the effect of different substances on the viscosity of the protein solutions and the animalizing or vegetalizing effect of these substances. The ranking of the effects of various substances (Fig. 248) shows that a reduction of more than 14 per cent in the specific viscosity of the protein solution accompanies an animalizing effect, while an increase in viscosity of more than 28 per cent progressively accompanies vegetalization. Surely, however, the viscosity of the proteins is not the only determining factor here: sodium pyruvate at one concentration (Fig. 248, 0.2 M) increases viscosity and vegetalizes; in weaker concentrations (0.04–0.1 M; see Fig. 248, 0.04 M) it still increases the viscosity slightly but has an animalizing effect. Still, these experimental results show that the condition of the cytoplasmic proteins — linear or globular, monomeric or highly polymeric — is closely tied to the determination of echinoid embryos. This conclusion is reinforced by the fact that protein fractions from animal and from vegetal halves or from animalized or vegetalized embryos show

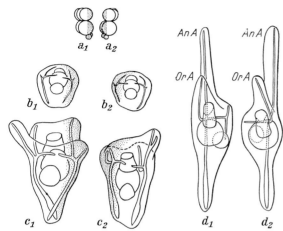

Fig. 249. Left and right halves of *Paracentrotus* larvae: a–c after meridional cutting in the sixteen-cell stage; d after meridional cutting ten hours after fertilization. *AnA* anal arm; *OrA* oral arm. (After Hörstadius and Wolsky, 1936)

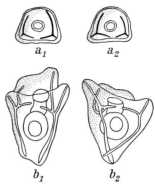

Fig. 250. Dorsally stained meridional halves of *Paracentrotus* separated in the sixteen-cell stage, one of which shows inverted dorso-ventral polarity. (After Hörstadius and Wolsky, 1936)

corresponding viscosity differences. Nevertheless, the direct or indirect effects on the determination which stem from the condition of these proteins remain obscure.

Besides animal-vegetal polarity, bilaterality, and dorso-ventrality also arise during sea urchin development. The right-left axis is first recognizable in the aggregation pattern of the skeleton-forming mesenchyme cells in the gastrula (Fig. 206e, f); later the dorsoventral axis is seen in the transition of the prism larva through the formation of the oral field, the bending of the primitive gut, and the relatively rapid growth of the dorsal side (Fig. 206g). The question as to when these axes are formally determined can be answered by the isolation of meridional halves of young embryos which are stained on the cut surface. Naturally, one does not know in advance whether one has cut the embryo in the sagittal plane or at right angles to it or in any intermediate plane.

After cutting in the sixteen-cell stage, both halves become complete pluteus larvae. If both halves are stained at the same time, when the cut turns out to have corresponded to one in the sagittal plane (Fig. 249a, b) most of the differentiation on the cut side proceeds more slowly than elsewhere (Fig. 249c). If a meridional cut is made in the blastula stage and proves to be a cut in the presumptive sagittal plane, half larvae develop from the half embryos which are incomplete on the cut side and which may even resemble exactly half pluteus larvae (Fig. 249d₁, d₂). These blastula halves develop with respect to the ectoderm and skeleton exactly as they would have done in a whole embryo; that is, through self-differentiation. Bilateral symmetry is therefore already formally established at this stage, and regulation is no longer possible. Each of the incomplete larvae has a three-part gut in spite of the division of the endodermal region.

197

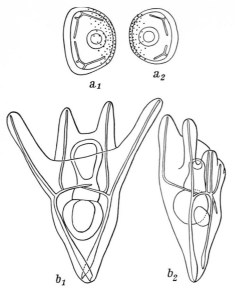

Fig. 251. Ventral (a_1, b_1) and dorsal (a_2, b_2) halves of embryos and of larvae of *Paracentrotus* after meridional cutting ten hours after fertilization. (After Hörstadius, 1936)

When one of the meridional halves of a sixteen-cell stage is stained dorsally (Fig. 251 a_1, a_2) or dorsolaterally (Fig. 250 a_1, a_2), the other half always shows staining in the same place: therefore in one embryo the dorsoventral axis must become inverted; and it is quite clear that the inversion must take place in the dorsal partner, for one embryo, obviously the one with a ventral tendency already present in the cleavage stage, develops its ventral side with the oral field and skeleton quicker and to a fuller extent (Fig. 250). Even when the isolation is not made until the blastula stage, the ventral halves can still develop into complete pluteus larvae, while the dorsal half always develops somewhat abnormally (Fig. 251 b_2); often two ventral sides develop and compete with one another. Finally, embryos can even arise after separation at the beginning of the ingression of the mesenchyme cells: these embryos correspond fairly exactly to the ventral and dorsal halves of a pluteus, respectively.

If eggs are ligated meridionally in the sixteen-cell stage, mirror image embryos can arise with two central portions (Fig. 252), so that even a lessening of the contact between the two embryonic halves causes a reversal of the dorsoventral polarity in one.

Bilaterality and dorsoventrality are closely related. The position of the groups of mesenchyme cells (Fig. 206e) is influenced by the position of the lateral ectoderm, which is in turn distinguished by its thickened epithelium. If micromeres of another species are implanted in animal halves, the

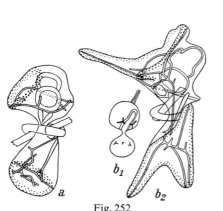

Fig. 252

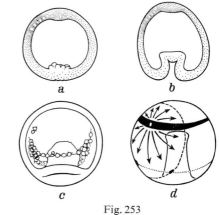

Fig. 253

Fig. 252. *Psammechinus miliaris* larvae with two ventral sides after ligation at the sixteen-cell stage. (After Hörstadius, 1938)

Fig. 253. a, b *Paracentrotus lividus* vitally stained with Janus green, side view: a blastula at onset of ingression of the primary mesenchyme cells; b early gastrula; c *Psammechinus miliaris* ventral view, oral field circumscribed by mesenchyme cells; d schematic representation of the differentiation of the oral field through the spread of a gradient from the oral field center and determination of dorsoventrality. (After Czihak, 1962)

resulting mesenchyme cells reassemble beneath the paired ectodermal thickenings of the host. If these ectoderm cells are killed by UV irradiation, dead cells are pushed into the interior, where they degenerate together with the mesenchyme cells already there. After a few hours new ectodermal thickenings arise and beneath them new cells assemble to form a skeleton. In the blastula or early gastrula there is a place between the apical plate and the equator where the mitochondria (shown by Janus green staining) are densest (Fig. 253a, b). The appearance of this mitochondrial field can be followed back to the eight-cell stage: one of the four cells of the animal tier shows the strongest indophenol blue reaction,[147] which is characteristic for mitochondria. From this center emanates the oral field determination. Wandering mesenchyme cells are brought together by the spreading oral field so that the oral field margin, which later forms a flagellar band, is marked by a circle of mesenchyme cells (Fig. 253c). The lateral ectodermal thickenings appear where the presumptive oral field border meets the ring of mesenchyme cells (Fig. 253d).

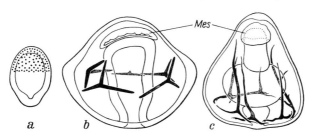

Fig. 254. a, b *Psammechinus miliaris*: a destruction of the animal half of the blastula with UV ten hours after fertilization; b Twenty-seven hours after irradiation; weak expression of gut differentiation, secondary mesoderm separated as an unpaired pouch (five radially arranged skeletal elements); c *Paracentrotus lividus* radially symmetrical larvae after treatment with eight-chloroxanthine. (After CZIHAK, 1962)

If the blastula stage is placed in a toxic gradient emanating from one side, an oral field center appears at the same distance between the poles as usual, but where the toxicity is lowest. This must correspond to the most active point of a gradient (Fig. 253d). If the animal cap, together with the presumptive oral field, is eliminated by UV irradiation, bilateral symmetry never arises, but instead a radically symmetrical, generally five-rayed larva appears (Fig. 254a, b). The same result can also be obtained by the use of various chemicals which vegetalize and inhibit the formation of the oral field, such as LiCl, sodium lauryl sulfate, and eight-chloroxanthine (Fig. 254c).

The dorsoventral axis, like the animal-vegetal axis, is apparently already laid out in the unfertilized egg. But it is still movable in the egg stage: if the eggs are drawn into a capillary tube which is narrower than the egg diameter, the eggs are stretched. Before it rounds up again, the front end of the sausage-shaped egg protruding from the capillary can be stained with Nile blue sulfate. If the axis of stretch is approximately perpendicular to the animal-vegetal axis (as judged by the position of the pigment band), either the stained end will become the ventral side and the opposite end the dorsal side, or else ventral sides will be formed at both ends. Passage through a narrow capillary not only stretches the egg but also forces material from the surface at the leading pole to move backwards; the egg cortex is made thinner; indeed, the eggs sometimes burst at this site. Clearly, then, the determination of the ventral side is related to the cortical structure.

The expression of the dorsoventral differentiation of the echinoderm embryo can be influenced by strong centrifugation of unfertilized eggs. After fertilization of centrifuged *Dendraster* eggs, blastulae develop with an uncleaved centripetal lobe in which a part of the layered material is sequestered and not incorporated into the egg. When it develops a ventral side with fairly well-differentiated arms, the lobe lies in this area. From this we may conclude that the cortex. which is not displaced by centrifugation, contains a region with a ventral tendency decreasing gradually with distance from its center, and that for the expression of this tendency, a material is necessary which *is* displaced by centrifugation. Thus, ventral differentiation is a resultant of the concentration of this material and the position of the inducing region.

The mechanism of the reversal of dorsoventral polarity in dorsal halves of sectioned or ligated embryos has not been made clear. In addition to its importance in the determination of the cleavage pattern, another critical role of the egg cortex in early development is shown by further experiments in which cytoplasm is removed from the interior of a sea urchin egg with a micropipette. When a considerable part of the endoplasm is sucked out, the egg surface wrinkles at first but becomes smooth again in the course of an hour. Even eggs reduced in size by the loss of over half their cytoplasm (Fig. 255) develop into normally proportioned pluteus larvae. Without a doubt, materials and structures (for example, granules and enzyme complexes) are present in the endoplasm which are necessary for the realization of the determined plan; but centrifugation experiments as well as the removal of endoplasm show that any structural pattern is very unlikely to exist throughout the interior of the egg.

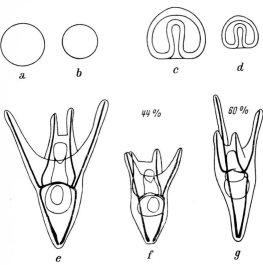

Fig. 255. a Normal egg; b egg reduced in size by removal of endoplasm; c, d gastrulae from normal and reduced eggs; e normal pluteus of *Psammechinus miliaris*. Plutei from reduced eggs with mass stated relative to that of the original eggs. (After HÖRSTADIUS, LORCH, and DANIELLI, 1950)

A number of chemical events have been revealed which accompany the early processes of differentiation; often these provide the first clearcut signs of the completion of a particular determinative event[117, 673] But as of now no primary determining factors have been defined chemically. The nature of the bipolar and dorsoventral cortical pattern and its collaboration with cytoplasmic materials whose synergism and antagonism are expressed in definite gradients remain a mystery.

Lecture 14

The amphibians offer unusual advantages for developmental study, their eggs are easy to obtain. Many species can be raised in the laboratory and caused to reproduce at will. The embryos are quite accessible in all developmental stages. Thanks to the yolk supply which every cell contains, embryonic parts are nutritionally independent until the differentiation of the larval organs is completed, in contrast to fish and reptile embryos, which draw their nourishment from the yolk sac through the circulatory system, and to mammalian embryos, which are nourished from the mother's blood. Amphibian embryos are hardy and large enough to be operated upon easily, so they can be cut apart in many ways, and relatively small parts can be removed and kept isolated and placed in a great variety of combinations by transplantation. The diversity of organ formation is greater than that in the experimentally accessible stages of echinoderms. For this reason, even more developmental principles are subject to investigation in the amphibians.

And a developmental pathway leads, without so radical a transformation as in the echinoderms, through early development to the adult organ, which gives us an easily grasped example of the general pattern of vertebrate development. Here questions can be posed that must be set aside in fish, reptiles, and mammals, which are more difficult to approach experimentally. No wonder that the amphibians have become the classic objects of the study of the causality of form. Even before the experiment on echinoid embryos, ROUX's experiments on frog eggs had begun, and since 1900, beginning with the work of R. G. HARRISON in America and HANS SPEMANN in Germany, the developmental physiology of amphibians has been revealed to be so diverse and so informative that its literature could fill a massive library.

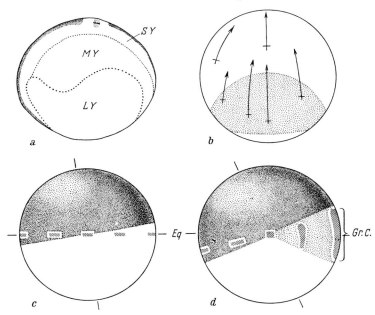

Fig. 256. Egg structure and formation of the gray crescent. a diagram of an oocyte of *Rana temporaria* in the first maturation division, longitudinal section: *L Y* large-granule yolk, *S Y* small-granule yolk, *M Y* intermediate yolk with small and medium-sized granules; b displacement of cortical marks on the side opposite to sperm entry during the formation of the gray crescent (stippled) in a frog egg; c, d movement of material during the formation of the gray crescent (*Gr. C*), seen by dye marks made on the equator (*Eq*) of the axolotl egg. (a after WITTEK, 1952, b after ANCEL and VINTEMBERGER, 1948, c, d after BANKI, 1929)

The morphogenetic movements leading to the gastrula, together with their determination, will be the subject of this lecture. They are much more complicated in the amphibians than in the echinoderms. The innovation of marking with vital stains by Vogt (1923) first made it possible to follow precisely the various movements of the embryonic parts.

Amphibian eggs are rich in yolk. The animal-vegetal yolk gradient is established in the primary egg axis in the growing oocyte. In mature frog eggs, three layers can be distinguished according to the size of their yolk granules, although the separation of these three layers is neither sharp nor smooth. In the vegetal half large yolk granules are packed relatively densely. In the animal half they are very small and sparse. In between, in a central mass whose extent depends upon the species, are small and middle-sized yolk granules, which are also densely packed (Fig. 256a). Among the yolk granules in the ground cytoplasm are various particles, some of which can be distinguished by their cytochemical reactions; mitochondria, lipochondria, RNA-containing

201

granules, and phosphatide-containing granules. The animal half contains many more of these inclusions than does the vegetal half. An intermediate marginal zone contains a certain mixture of these granules and the yolk granules. In the beginning, yolk formation and distribution has one axis of symmetry. Later, in many eggs, a bilaterally symmetric a arrangement of the yolk begins during the growth of the oocyte (Fig. 256). In the egg cortex a dark pigment marks the upper surface. It is fairly thick in the animal half and extends a greater or lesser distance past the equator (Fig. 256). The cytoplasm of the less yolky part of the egg can be more or less lightly pigmented ("brown yolk"), in contrast to the "white yolk" of the vegetal half. In the axolotl a ring of dark cytoplasm, the so-called marginal plasma, lies above the equator in the zone border-

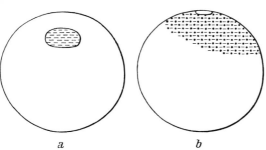

ing the white yolk, and it is wider on one side of the egg. In several forms bilateral symmetry is expressed also in the oblique position of the pigment cap (Fig. 256c). During the maturation divisions there is a striking bilateral distribution of the sulfhydryl proteins which stain red in the nitroprusside reaction. In the ovarian egg (Fig. 257a) only the nucleus stains "like a red pearl in color- less cytoplasm." When the nucleus enters the maturation divisions and the nuclear mem- brane dissolves, the nitroprusside reaction spreads out from the animal pole (Fig. 257b). After fertilization it extends as a band toward the equator. In the same region, RNA granules

a   *b*

Fig. 257. Distribution of sulfhydryl-containing pro- tein (—) and RNA-containing granules in the mature oocyte and during maturation division. (After BRACHET, 1938)

are concentrated. With fertilization, again in several of the forms discussed so far, the pigment cap of the egg changes shape and now appears as a gray or brownish crescent-shaped field, the so- called gray crescent. If before its formation a row of marks is made with a vital stain around the equator of the unfertilized egg, these marks lengthen greatly in the region of the gray crescent (Fig. 256d). Some marks made ionophoretically with an electrode on the egg cortex show that the pigmented cortex contracts toward the pole (Fig. 256b), taking with it a thin layer of large yolk granules (Fig. 258a). In this way the symmetry of the egg is laid out. The sagittal plane of the embryo passes through the widest part of the gray crescent, and the gray crescent itself is the mark of the dorsal side.

If a bilaterally symmetrical distribution of material is already set up in the oocyte, it is never- theless not firmly fixed; as a rule the plane of bilateral symmetry is determined by the entry of the sperm. If the egg membrane of *Rana fusca* is fastened to a surface and sperms are placed at a particular point on the equator with a fine needle, the gray crescent always appears on the oppo- site side of the egg. Apparently the sperm aster, as it develops, has an effect on the symmetrical disposition of cytoplasmic substances (Fig. 258a). During the formation of the gray crescent the yolk moves into the vegetal half of the egg and thereby moves the egg's center of gravity; because of this, the egg rotates. As a result of this fertilization-rotation the primary egg axis becomes oblique (Fig. 258a) with respect to gravity.

The site of the gray crescent is influenced by gravity. If newly laid frog or salamander eggs, which cannot rotate before the fertilization membrane is raised, are fastened to a substratum so that the vegetal pole is about 45° from the top (Fig. 258b), the egg rotates about 135° in the fluid, which arises, after fertilization, between its surface and the fertilization membrane, so that the egg assumes its normal orientation with the vegetal pole downwards (Fig. 258c). The gray crescent then is formed almost always in the area that was uppermost while it was held fixed, and it forms symmetrically about the plane of the aforementioned rotation which reoriented the egg (Fig. 258 d, arrow). If the movement of substances leading to the appearance of the gray crescent has not yet begun, the effect of the first artificial orientation can be replaced by a second. The direction of reorientational rotation from a fixed position can also obliterate the orienting effect of fertilization. It appears that, at the beginning of the movement of the yolk along the egg cortex, a thin layer of white yolk remains at the cortex (Fig. 258 d) and causes a distribution of material which would ordinarily have been influenced by the sperm (Fig. 258 a). The direction of cortical contraction apparently is a consequence of this distribution.

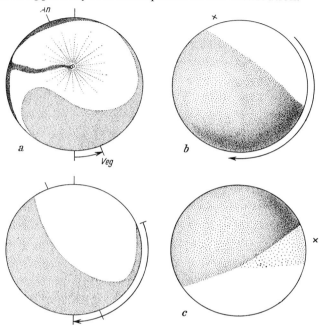

Fig. 258. Formation of the gray crescent in the frog egg. a yolk distribution after fertilization; the arrow shows the movement in the animal-vegetal axis during the fertilization rotation; b egg prevented from rotating; c after orientation rotation corresponding to the arrow in b; d yolk distribution after orientation rotation. a and d schematic medial section, b and c surface view. (a, d in Appendix, PASTEELS, 1951; b, c after PASTEELS from LEHMANN, 1945)

In any event, in the cytoplasm of the oocyte of many species a bilaterally symmetrical arrangement of substances takes place which is nevertheless still labile and which can be rotated under the influence of sperm entry or gravity. This reminiscent of the polarizability of algal eggs and moss spores, brought about by entirely different environmental influences (p. 144f.).

Cleavage (Fig. 259) has no relation in most amphibians to the bilateral symmetry evinced by the structure of the egg cell. The first division plane can be sagittal (medial), dividing the gray crescent; it may be perpendicular to that (frontal), or at some angle to these major axes of the future embryo (Fig. 279). In a few species the sagittal or frontal planes seem somewhat preferred for the first cleavage. Mitoses in *Triturus* are highly synchronous in the first five cleavages; in healthy embryos the nuclei are always in the same phase. In the 32-cell stage the degree of synchrony is reduced, and in the many-celled blastula it is absent. At first one finds all the nuclei in the animal hemisphere continuing to undergo mitosis, while in the vegetal region the nuclei are resting (Fig. 261 a); they remain about one cleavage behind; finally synchrony disappears even within a given hemisphere.

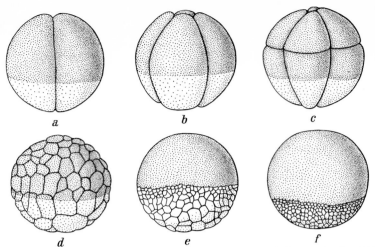

Fig. 259. Cleavage, large-celled (early) and small-celled (late) blastula of *Rana sylvatica*. (After POLLISTER and MOORE, 1937)

The surface of the egg is covered by a thin, though, and elastic gel layer which also extends over the surface of the embryo and over the cleavage spaces (Fig. 260). This layer is mainly responsible for the permeability properties of the embryo. The deeper-lying cells are not closely apposed to one another until the early blastula. Rather there are gaps among them which are often bridged by thin or knobby outgrowths (Fig. 260). Later the blastomeres in the animal half come together more closely and form a several-layered blastoderm. Even before the beginning of gastrulation the upper, small-celled part of the blastula becomes distinct, and its margin moves over the large yolk cells (Fig. 259f). The small-celled blastoderm remains separated from the yolk cells at this time by a cleft. During this stretching of the blastula roof the cells divide vigorously, particularly in the animal cap (Table 7). At the same time, cells deep in the several-layered blastoderm push toward the surface, so that the roof becomes thinner and spreads out on all sides. The externally visible yolk region now draws together in this pregastrulation phase, and the floor of the segmentation cavity is raised (Fig. 261b).

Gastrulation begins in the region of the dorsal lip, within the yolk region but near its upper dorsal border. The cells begin to stretch into the interior in a latitudinal or arched strip. The

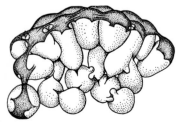

Fig. 260. Living cells from the wall of a large-celled blastula from *Amblystoma punctatum*. (After HOLTFRETER, 1943)

Table 7: *Mitotic frequency in the blastula and early gastrula of* Bufo cognatus *(n = number of cells in the area examined).* (After BRAGG, 1938)

Blastula	Total cell number	Cells in Mitosis, %
Animal half	1,540	64.8
Vegetal half	364	44.5
Early gastrula with crescent blastopore:		
Animal pole region	872	81.1
Blastopore region	642	58.4
Remainder of animal half	7,809	56.0
Yolk region	687	12.7

cells become pear-shaped (Figs. 261 b, 262 b), and where they are connected to the surface only by a thin stalk, the dark surface layer draws together (Fig. 262 a). Soon a crescent-shaped invagination (Figs. 261 c, 262 c) forms. The surface cells nearby stretch toward it (Fig. 262 d). The cells pushing inward still hang together via the sticky surface layer (Fig. 262 e). The margin of the dorsal lip extends ventrally and laterally; it becomes semicircular and gradually surrounds the yolk region, which moves inward as endoderm (Figs. 263 a, 264 b, c).

Stain marks placed equatorially (Figs. 264, 265) or meridionally (Fig. 266) show the extent of the small-celled blastoderm which has invaginated. The region of the blastula between the yolk

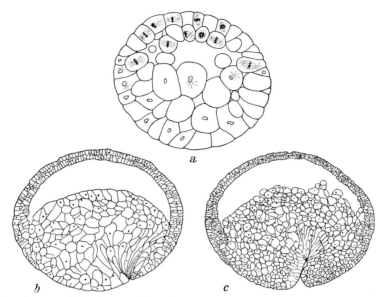

Fig. 261. Urodele embryos. a blastula, transition between synchronous and asynchronous division; b, c early gastrulation; b onset, c somewhat later. (a after Schönmann, 1938; b, c after Daniel and Yarwood, 1939)

field and the invaginating edge on the side of the future ectoderm is called the marginal zone (Fig. 263 b). The stain marks reveal through their movements and distribution the movement of material during gastrulation. After the part of the yolk region which lies dorsal to the dorsal lip (Fig. 261 b, c) is drawn in, the dorsal and most animal part of the marginal zone follows (Figs. 264, 266). On the inside the invaginated pocket pushes forward underneath the animal cap of the embryo (Fig. 267 a) with the lateral and ventral extension of the lip of the blastopore. The mass of yolk cells moves inside and makes the floor of the archenteron (Fig. 267 b–d). The true invagination is confined to the dorsal median part of the yolk region. When the blastopore lip spreads laterally, the marginal zone material separates from the mass of yolk cells and turns inward. This involuted material moves along the inner surface of the blastoderm, which is still outside, and the inner and outer layers are now moving in opposite directions (Fig. 267). The endoderm of the archenteric floor moves into the archenteric roof only at the front end of the invaginated pocket. This pocket becomes the pharynx. Behind it, the entire involuted border zone forms a single germ layer, the chordamesoderm, which forms archenteric roof dorsally and surrounds the endoderm laterally (Figs. 265 d, 268). The endoderm of the archenteric floor curves laterally upward along the mesodermal covering (Fig. 268 a). With the extension of the edge of the blastopore ventrally, new mesoderm continues to move inward around the lip. Once inside,

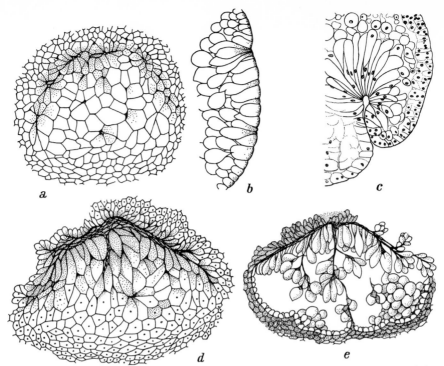

Fig. 262. Gastrulation: a, b, d, e *Amblystoma punctatum*; c *Triturus cristatus*. a, b onset: a view of the yolk region; b longitudinal section; c, d, e invagination in the dorsal lip; c longitudinal section; d surface; e after removal of the region of ingression ventral to the dorsal lip. (a, b, d, e after HOLTFRETER, 1943; c after VOGT, 1929)

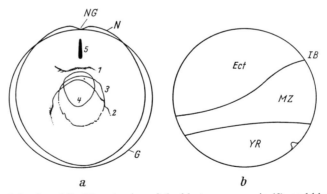

Fig. 263. a appearance of the dorsal lip (*1*), extension of the blastopore margin (*2*), and blastopore closure (*3, 4, 5*) in *Triturus torosus*. G, N outlines of the gastrula and neurula; *Rr* neural groove. (After DANIEL and YARWOOD, 1939.) b diagram of the borders between presumptive ectoderm (*Ect*), marginal zone (*MZ*) and yolk region (*YR*) at the beginning of gastrulation. *IB* involution border. Diagram derived from stain marking experiments

it moves laterally and dorsally (Fig. 267). The distortion of the stain marks shows the direction of these movements from the time they pass the blastopore lip (Figs. 264, 165), and one subsequently finds marks that had been made ventrolaterally (Fig. 265a: *5, 6*) now up in the mesoderm (Fig. 265d). With the involution of the marginal zone the closure of the blastopore gradually

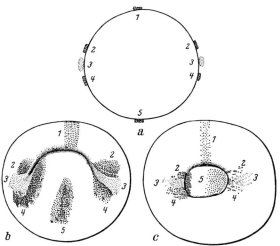

Fig. 264. Stain marking in the equatorial region of a small-celled blastula of *Amblystoma*. a arrangement of the marks; b, c entry of the yolk region and the border zone. (Modified after Vogt, 1929)

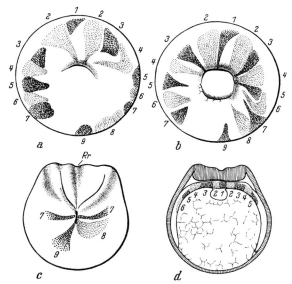

Fig. 265. Gastrulation in *Triturus alpestris* after marking the animal margin of the marginal zone (Nile blue sulfate and neutral red). a dorsal lip; b completed blastopore; c closure almost complete; d cross section through the living embryo. (After Vogt, 1923/24)

takes place (Figs. 263, 265). Its margin becomes an ever-smaller circle from which the yolk plug protrudes for a time. Finally, only a furrow remains perpendicular to the original dorsal lip (Fig. 263a: 5). The dorsal convergence of the mesoderm and the closure of the blastopore are diagrammed in Figure 269. The numbered points indicate the presumptive blastopore lip in the marginal zone at the beginning of gastrulation. The directions of movement on the upper surface up to the time of the half-closed blastopore are shown with thick lines. The thin lines illustrate the movement in the sectors of the involuted border material on the inside of the embryo. These

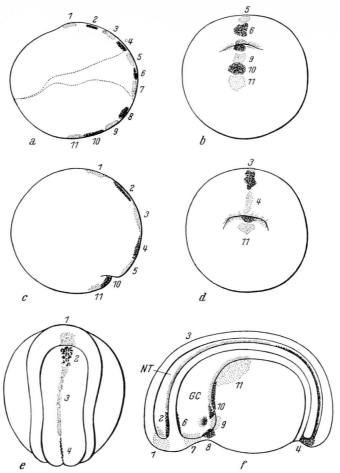

Fig. 266. *Amblystoma*. a–d gastrulation after staining in the dorsal meridian; e neurula; f embryo with closed neural tube, sagittal section. (After VOGT, 1929)

movements were made clear by the use in living embryos of stains which showed through to the outside. The inner lines show also the degree of stretching of the region of the chordamesoderm during gastrulation; this stretching is most pronounced dorsally (Fig. 269 b). The mesoderm which has converged dorsally turns forward beside the lengthening pharynx and ventrally up to an anteroventral region, which remains free of mesoderm. The ectoderm, which covers the whole embryo once the yolk cells and marginal zone have moved inside, undergoes extraordinary stretching during gastrulation and becomes a single layer which is especially thin on the ventral side (Fig. 267). In the midline of the archenteric roof the primitive notochord now differentiates (Fig. 270 b). It is connected in front to the roof of the pharynx by the prechordal plate (Figs. 270 a, 272 c). The position assumed on the inside after gastrulation, by the stain marks placed on the blastula surface, permits one to map in the marginal zone the presumptive regions of the notochord, mesoderm, and the pharynx, all of which lie in the neighborhood of the dorsal lip (Fig. 271). In Figure 272 these regions are sketched in a sagittal section. Above the archenteric roof the differentiation of the neural plate begins in the ectoderm, and thus the gastrula becomes a neurula.

In the process of gastrulation many morphogenetic movements work harmoniously in the various parts of the blastoderm: the stretching of the marginal zone on the surface of the embryo and the stretching of the ectoderm, the invagination of the dorsal part of the yolk region as the archenteron below the dorsal lip of the blastopore, the involution of the marginal zone material

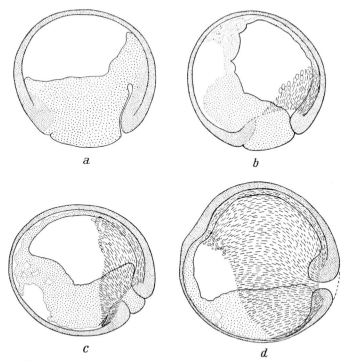

Fig. 267. Gastrulation. a *Bombinator*; b–d *Pleurodeles*. Median sections with reconstructed mesodermal covering. (After VOGT, 1929)

around the edge of the blastopore as chordamesoderm, the stretching and dorsal convergence of the invaginating mesoderm, and the constriction of the blastopore margin leading to closure.

In order to elucidate the determination of these morphogenetic movements, one can begin by testing the behavior of the embryonic parts in isolation experiments. Large or small pieces of presumptive ectoderm stretch greatly when explanted. Explants from an early gastrula, including the dorsal lip, show diverse behavior according to their size and location. In one set of experiments the following explants were made: 1) square pieces cut just below the dorsal lip and extending dorsally (Fig. 273 a, b_1); 2) a larger piece with more of the dorsal border zone (Fig. 273 c_1); 3) a piece including a section of the yolk region below the dorsal lip (Fig. 273 d_1). All of the explants began by rounding up; after four to six hours a bulge appeared, which in the next six to eight hours grew into a vermiform appendage (Fig. 273 b_2, c_2, d_2). The blastopore groove remained below the outgrowth in all explants and even deepened in some cases. The endoderm cells of the d_1 explant remained in a clump and were partially covered over by dark pigmented animal material. Explants from the lateral marginal zone and extending into the yolk region (Fig. 273 e, f, g) flattened out and formed a blastopore groove, and the marginal zone material moved over the endoderm somewhat (Fig. 273 f_2, g_2). Both parts stretched in a dorsoventral direction, the upper part more than the part beneath.

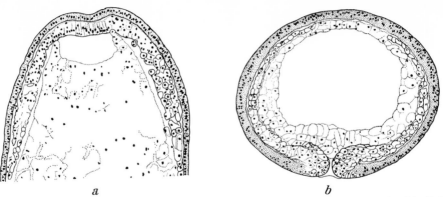

Fig. 268. Gastrulae: a, of *Triturus cristatus*; b, *Pleurodeles*. a frontal section through a gastrula with slit-like blastopore, archenteric roof seen somewhat in front of the dorsal lip, chordamesodermal borders of the archenteric floor
curled dorsally; b gastrula with blastopore near closure, frontal section, the lateral blastopore lips have met;
envelopment of the endoderm by the mesoderm, in front (in the illustration above) of the prechordal plate. (After
VOGT, 1929)

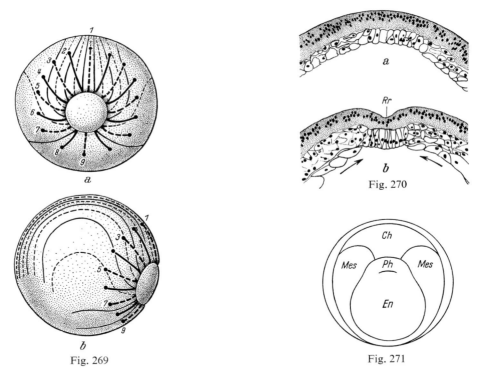

Fig. 269

Fig. 270

Fig. 271

Fig. 269. Diagram of the movements of material of the marginal zone during gastrulation. (After VOGT, 1929)

Fig. 270. Gastrula of *Pleurodeles*. a anterior region, prechordal plate; b posterior region, onset of chorda differentiation, neural groove *(Rr)* above, first sign of neural plate formation. (After VOGT, 1929)

Fig. 271. Diagram of the regions of the material of the presumptive chorda (*Ch*), presumptive mesoderm (*Mes*),
endoderm (*En*) and pharynx (*Ph*)

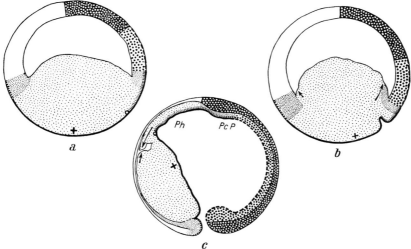

Fig. 272. Schematic median section through the blastula, beginning gastrula and beginning neurula of *Pleurodeles*, neural region dark, chorda region heavily dotted, mesoderm and prechordal plate lightly dotted, endoderm sparsely dotted, ectoderm open. + site of the vegetal pole. The diagrams illustrate the stretching of various sections in the dorsal meridian and the stretching of the ectoderm. *Ph* pharynx; *PcP* prechordal plate. (After VOGT, 1929)

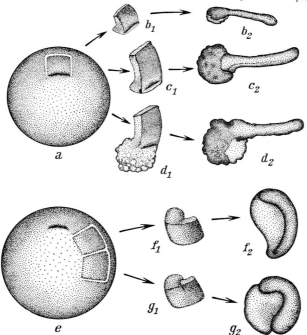

Fig. 273. Behavior of explants from a young gastrula (when the blastopore is a shallow, narrow groove) of *Hyla regilla*. a–d explants with dorsal lip; e–g explants from the lateral marginal zone. (After SCHECHTMANN, 1942)

The tendencies of the cells of the invaginating region are also shown by the following experiment. A rounded-up group of explanted cells of the invaginating area of the dorsal yolk region is vitally stained and placed on naked endoderm. If the ball of cells is still partially covered by its elastic coat, it constricts and sinks, together with the free cell surface, into the endoderm in one to two hours and forms a groove. At the bottom of this groove the cells stretch inward into the endoderm, assuming a flask shape, as at the beginning of normal invagination. Then they begin to spread (Fig. 274a–c).

Such explantation experiments, which can be done in many ways, show that the individual cells of the blastoderm already possess the tendencies for morphogenetic movements at the onset of gastrulation. These tendencies correspond to what the part will do in the intact embryo, and the whole complex array of morphogenetic movements *appears* to rest on the self-differentiation

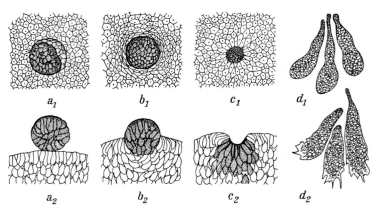

Fig. 274. Behavior of cells from the endodermal invagination area. a–c cell clusters, partially covered by the elastic surface layer, placed on naked endoderm; a_1–c_1 surface; a_2–c_2 section (semi-diagrammatic); d isolated blastopore cells undergoing amoeboid movement. (After HOLTFRETER, 1934)

of the parts, which has been previously determined in the blastula and works autonomously from then on. This, however, turns out to be not entirely true, as we can see from experiments involving the removal of certain parts. The embryo from which the blastopore explants of the first experiment (Fig. 273a) were taken does not carry out normal gastrulation but develops into a so-called "ring embryo" (Fig. 275b, c). The ectoderm and marginal zone stretch, but the embryo flattens out as this takes place. Around the mass of yolk cells a blastopore groove is formed, and through it the marginal zone material moves between the ectoderm and the endoderm. The edge of the blastopore, however, does not grow over the endoderm; this part remains free. A stain mark placed in the equatorial region reveals the movement of the marginal zone material. In a further experiment the dorsal lip is not cut out, but merely separated from the lateral marginal zone by the implantation of two strips of ventral ectoderm (Fig. 276a, b). Now the region dorsal to the blastopore behaves as if it had been explanted: a vermiform extension grows out, and the remainder of the embryo becomes a ring embryo (Fig. 276c), just as if the invagination region had been removed. The cross section (Fig. 276d) shows how the mesodermal mass has rolled in through the blastopore groove below the endoderm.

If the lateral border zone is divided further ventrally by implanted strips of endoderm (Fig. 276e) the dorsal part completes gastrulation normally; in the ventral part, however, the blastopore fails to close (Fig. 276f). These experiments, to which one might also add numerous variations which have been carried out, show that a mosaic of tendencies for particular movements is certainly spread over the blastoderm, but that the completion of gastrulation is not a mosaic maneuver; the invagination of the presumptive notochord region is closely related to the dorso-lateral mesoderm, and the dorsal convergence of the mesoderm and the contraction of the blasto-

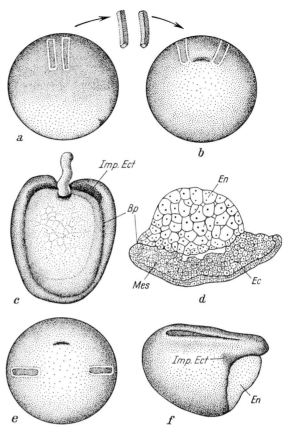

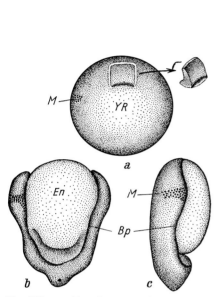

Fig. 275. a–c Development of an embryo of *Hyla regilla* after removal of the dorsal lip (donor of Figure 273b$_1$); b, c ring embryo; b top view, c side view. *YR* yolk region; *En* endoderm; *M* stain marks; *Bp* blastopore groove. (After SCHECHTMANN, 1942)

Fig. 276. Interruption of the connection between dorsal lip and lateral marginal zone (a, b) and within the lateral marginal zone (e) by the implantation of strips of ventral ectoderm; d, c "ring embryo" from b; d cross section; f neurula from e, with partially uncovered endoderm, view of the left side. *Ec* ectoderm; *En* endoderm; *Imp Ect* implanted ectoderm; *Mes* mesoderm; *Bp* blastopore margin. (After SCHECHTMANN, 1942)

pore margin, resulting in closure, fail when the connection between the marginal zone and the dorsal lip is broken. The autonomous movement tendencies of individual pieces of blastoderm must therefore be regulated into a harmonious whole. These morphogenetic movements become a collaborative event which serves as an excellent example for the working together of self-differentiation and induced differentiation.

What forces then unify these individual tendencies? The temporal relations during normogenesis illustrated by the movement and distortion of stain marks permit one interpretation: gastrulation begins with the invagination of the material of the presumptive pharynx. When it has invaginated, the presumptive notochord has reached the dorsal lip. At this time the nearby dorsolateral parts of the marginal zone, with their strong tendency to involution, are already entering the interior through the nearby blastopore margins (Figs. 264–267). Apparently they pull the material of the chorda that lies between them, and thus help it over the dorsal lip into the interior, as well as orienting its great tendency to stretch. Without this collaboration, the connection with the presumptive pharyngeal endoderm is broken, and the presumptive chorda material sticks out free (Figs. 273, 276). The rapid forward growth of the presumptive chorda material underneath the animal roof of the gastrula, accompanied by a narrowing of the broad entering band due to stretching, now evidently draws the contiguous presumptive mesoderm dorsally and causes its entry into the single chordamesodermal plate of the archenteric roof (Figs. 267, 269), and consequently dorsal convergence. In this way the direction of movements into the interior is determined according to the autonomous involution, and the blastopore contracts, since no further material converges from its vicinity, where the mesoderm adheres firmly to the ectoderm until closure is complete. The blastopore is drawn together as if by purse strings.

The following questions naturally remain: Why do the germ layers cohere so tightly? And what force is responsible for the movements? During invagination or epiboly of the epithelial layers, the connection can be maintained by the sticky elastic coat, to whose stringy extensions even the invaginated cells remain attached (Fig. 262e). As long as this connection exists in the germ layer a mechanical force can orient the stretching, but certainly the germ layer cannot be compared with a rubber band. That the cells push against one another is shown by the graduated changes during stretching of the presumptive ectoderm, in which the several-layered epithelium becomes single-layered (Fig. 261). Moreover, there is no doubt that during invagination and involution, the deeper-lying cells separate from one another, more or less, and wander as isolated cells in a specific medium (Fig. 274d). The inward streaming of the cells is reminiscent of aggregation in the Acrasiales (p. 120f.). Should the wandering cells also stick together because of the adhesive properties of their surfaces, the force could only be one of orientation to which the cells of a streaming layer respond as *Dictyostelium* amoebae respond to acrasin. Unfortunately, we are just on the edge of understanding here; but it should be possible to devise experiments that can lead us further.

Lecture 15

The experiments just described reveal the determination, in the various regions of the blastula, of the morphogenetic movements that collaborate in gastrulation. When and how is the unifying pattern of specific movement tendencies determined, and what brings about the typical organization of the blastemas?

In the echinoderms the first morphogenetic processes are governed by the animal-vegetal polarity of the egg cell. In the amphibia the dorsoventral polarity of the egg is first expressed in the separation of the germ layers. This axis is set up before cleavage in the distribution of cytoplasmic material manifested in the gray crescent, and it seems likely that this arrangement of substances is critical for the normal course of gastrulation, and thus for further typical development.

If the four animal blastomeres are separated in the eight-cell stage from the four vegetal blastomeres (cf. Fig. 259c), the animal 4/8, which according to vital staining experiments becomes presumptive ectoderm only, is not capable of gastrulation. The upper part of the blastula becomes a hollow wrinkled sphere, set off more or less sharply from a basal compact part (Fig. 277a). In section one finds a thin stretched epithelium and a loose inner cell mass that is not divided into mesoderm and endoderm. The basal portion consists of a group of rather yolky cells (Fig. 174b). The half embryo from the vegetal 4/8 blastomeres forms a broad dorsal lip. The incorporation of the disproportionately large yolk mass creates problems (Fig. 277c); nevertheless, an almost typical arrangement of germ layers finally comes about, resulting in a well-proportioned neurula (Fig. 277d).

214

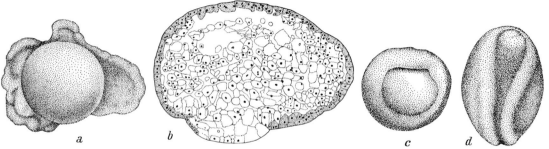

Fig. 277. Embryonic fragments of *Triturus taeniatus*. a, b from the four animal cells of an eight-cell stage; c, d from the four vegetal cells. (After Ruud, 1925)

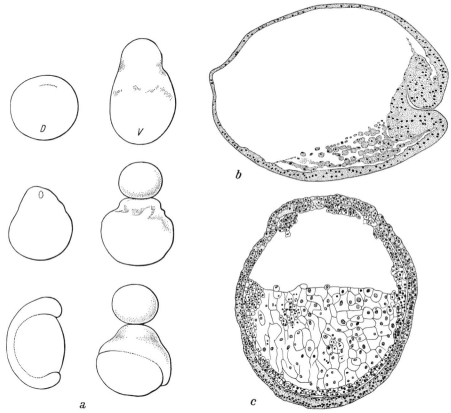

Fig. 278. Separation of the first two blastomeres of *Triturus taeniatus*. a dorsal-ventral twins (*D*, *V*); b sagittal section through a central embryo which has everted the yolk plug (as in a *V*); c cross section through a ventral embryo in which the yolk mass has been incorporated. (a, b after Ruud, 1921; c after Spemann, 1902)

Isolated half blastomeres have been worked with for a long time. They undergo total cleavage, then they can develop further in various ways. Most become either two symmetrical twins, or else one complete embryo and one "ventral fragment" containing ectoderm, endoderm, and mesoderm, but without further differentiation or organ rudiments (Fig. 278). This result is consistent with the idea that the dorsal region, which governs the formation of the dorsal lip and the involution of the chordamesoderm, is laid out in the egg and can be so divided that both blasto-meres are competent to form the normal germ layers; in other cases the first cleavage leaves this important region mainly in one blastomere. Recall that the first cleavage plane may be sagittal, frontal, or intermediate (p. 203); the gray crescent can be transected in a great variety of ways (Fig. 279). Accordingly, regulation must be possible in the region of the future dorsal marginal zone before gastrulation, permitting symmetrical development. If a sufficient amount of the dorsal region of the gray crescent is allocated to each half, then orderly gastrulation can proceed,

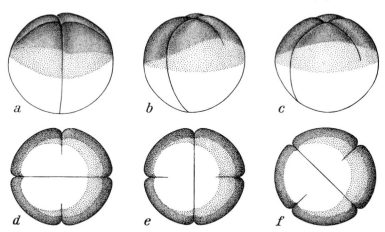

Fig. 279. Division of the gray crescent by the first and second cleavage furrows. a–c frog embryos, dorsal view: a & c exactly dorsal, b oblique. (After Morgan and Boring, 1903.) d–f salamander embryos, vegetal view, dia-grammatic: first cleavage sagittal in a and d, frontal in b and e, intermediate in c and f

and the section shows the typical form of the germ layers (Fig. 280a). The region of the gray crescent thus appears to be the anlage of a morphogenetic field which centers itself if divided and works as a unit. For short we call it the dorsal field. Even when one blastomere gets the whole dorsal region of the gray crescent (Fig. 279b) regulation is necessary in the partial embryo to bring about a typical distribution of cell materials to the germ layers; less material must be invaginated in order to maintain the ectodermal covering of the embryo.

The "ventral" blastomere includes only parts of the "horns" of the gray crescent (Fig. 279b, e); it cannot form normal germ layers. No dorsal lip is formed on its round blastula; between the white yolk region and the dark ectoderm appears a constriction, a blastopore groove, into which the mesoderm moves that spreads over the inner surface (Fig. 278b). Either the ball of yolk cells is pinched off as the blastopore begins to close (Fig. 278a, V) or it is enveloped, taking a position on the floor of the embryo (Fig. 278c). The particular degree to which gastrulation is inhibited in the ventral piece may be related to the plane of cleavage of the egg cell between the frontal and sagittal planes. At times the dorsal embryos are dissimilar in the derivatives of the germ layers, and this has the same cause.

The first four blastomeres can also develop after being separated. They initially undergo total cleavage, then at most two normal neurulae result (Fig. 281), and the other two become ventral pieces which frequently evert the yolk plug. In some cases three ventral pieces and one neurula arise, and in still others the neurulae may not be well proportioned, which again can be under-stood in terms of the placement of the first two cleavage planes (Fig. 279). In the best dorsal quarters the arrangement of the germ layers is normal (Fig. 280b); chorda and mesoderm sepa-rate, and over them a neural rudiment is formed. But apparently there is insufficient regulative

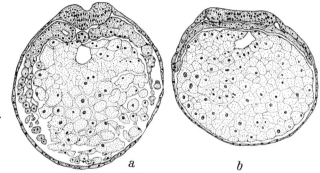

Fig. 280. Cross sections through dorsal embryos arising from isolated *Triturus* blastomeres. a half embryo; b quarter embryo, same magnification. (After Ruud, 1925)

ability; it is no longer as exact as in the half blastomeres; the neural tube appears disproportionately large (Fig. 281).

Still another field property of the dorsal field, in addition to its divisibility, is shown by an elegant experiment: two fields may fuse into a larger single field. By cutting at the first cleavage furrow the egg is freed from its surface gel layer and placed in a special salt solution which dissolves membranes. Then the first two blastomeres are separated from one another to a greater extent than during normal cleavage, and together they form a dumbbell. Two such two-cell stages can be placed together to form a cross; they fuse into one embryo which resembles a normal four-cell stage. This union can be made between two embryos of the same species (homoplastic, Fig. 282a) or between two embryos of different species (heteroplastic, Fig. 282b). In the first stages of their subsequent development these fused embryos can scarcely be distinguished from normal. Cleavage leads to a round blastula which is somewhat flattened due to the lack of a gel layer. The animal-vegetal orientation is maintained during the fusion, and so without exception a common blastopore is formed from all the sectors on the vegetal side. Around it entirely disparate dorsal fields organs can arise according to whether the first cleavage in each partner was in the saggital plane, the frontal plane, or an intermediate plane (cf. Fig. 279). In Figure 283 several of the possible combinations are diagrammed. With frontal plane cleavage in both embryos (Fig. 283, A), the two undivided gray crescents are placed side by side (*I*). If the first cleavage in both partners divides the gray crescent medially (Fig. 283, B), two halves arise with their median cut surfaces together while the other halves of the gray crescent remain separated (*IV*). With frontal cleavage in one and sagittal in the other, there are two possibilities: either the divided crescents are arranged with their medial regions right next to the undivided crescent (Fig. 283, *II*); or the undivided and divided crescents are separated from one another (Fig. 283, *III*). Intermediate first cleavage planes in one or the other partner or in both give still further

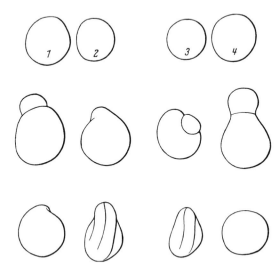

Fig. 281. Development in *Triturus taeniatus* of the first four blastomeres. (After Ruud, 1925)

217

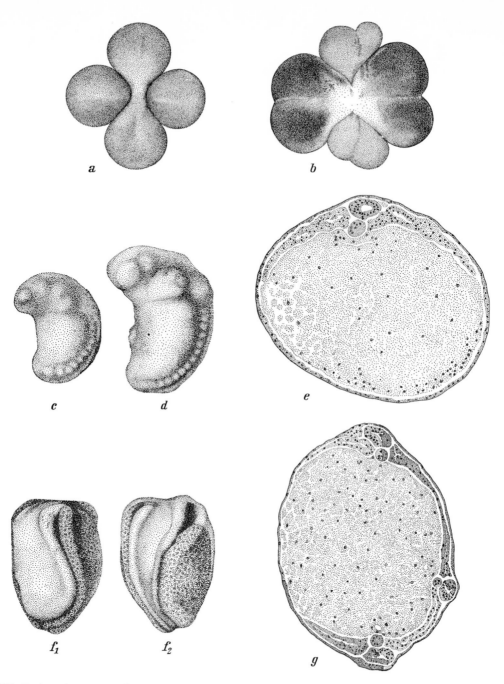

Fig. 282. Fusion of two two-cell stages; a *Triturus taeniatus*; b *T. taeniatus* with *T. alpestris*; in each cell the second cleavage division is underway; c normal embryo; d reintegration after homoplastic fusion; e cross section through a reintegrated embryo; f, g triple form from union of *T. taeniatus* and *T. alpestris* as in b (dark material *alpestris*); f₁, f₂: neurula seen from opposite sides; g cross section through this embryo in a later stage. (After MANGOLD and SEIDEL, 1927)

combinations on fusion, and just two of these are diagrammed in Figure 283, *V* and *VI*. From such fused embryos arise either single harmonious giant larvae (Fig. 282d, e; cf. normal embryo in c) or embryos with two, three (Fig. 282f, g), or four axial systems. These various forms can be attributed to the respective types of fusion combinations which have been diagrammed. In most cases it can be established during gastrulation whether a simple, double, or triple dorsal lip has been formed. If the two starting embryos are stained different colors, it is possible to trace the origin of the dorsal lips of the individual sectors of the fused gastrula. If several neural tubes are built, it can be decided from their relative positions whether the individual dorsal lips were close or distant; and when the neural tube is stained entirely or partially, it can be related to particular sectors (Fig. 282f). In the sections (Fig. 282g) the numbers and positions of chorda and archenteron lead to conclusions about the original anlagen. In addition to complete axial systems abortive formations without chorda but with mesoderm and weak neural tube formation occur; these often lie at the borders between material supplied by the two partners and may have included half or smaller lateral pieces of the gray crescent (Fig. 283 *III–VI*). Well-developed axial systems are formed from whole dorsal fields (Fig. 283 *III, V*), from dorsal fields arising from two halves (Fig. 283, *IV*), or double-sized dorsal fields resulting from the fusion (Fig. 283 *I, II*). The observed criteria suggest that the individual giant larvae can be ascribed to case *I*, where two entire dorsal fields are closely approximated, and *II*, where an entire gray crescent is flanked on each side by a half-crescent. In embryos with axial systems which lie close together, neighboring fields have evidently remained separate (Fig. 283 *I, II*, and, in part, *IV*). Triple formations can be expected from many fusion combinations (Fig. 283 *III–VI*). Rare embryos with four axes in which two of the neural

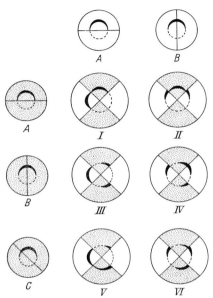

Fig. 283. Scheme of various fusion combinations. (After MANGOLD and SEIDEL, 1927)

tubes are always close together and partially fused can only be ascribed to combination *IV*, in which neighboring half-fields of different partners are not completely fused.

These experimental results show that the causes of germ layer formation are actually present in the unfertilized egg in the arrangement of cytoplasmic substances in the region of the gray crescent.

Thus, it is very surprising that the polarity in uncleaved amphibian eggs can still be altered fundamentally after grey crescent formation and in the two-cell stage by the effects of gravity. Yet long ago (1895) it was found that eggs and two-cell stages inverted in the animal-vegetal axis rearranged their contents, and that in such embryos a good deal of twinning occurs. This "inversion experiment of Schultz" has often been repeated since. The eggs are fertilized, and either immediately or after the beginning of the first division they are pressed between two glass slides so that they cannot turn; then they are kept with the vegetal pole upward. Now the heavy white yolk falls slowly from the vegetal pole to the animal pole below, and the cytoplasm with the light "brown" yolk and its granular inclusions rises to the vegetal pole (Fig. 284a–c). This restratification proceeds until a new equatorial border is established. When the eggs are freed they round up and cleave further. The path of the sinking yolk varies. It may fall through the center of the egg or peripherally, and in one band or in several streams. The light material is carried upward in correspondingly disparate ways. The result, however, is always constant; in the original animal half there is now white yolk and in the original vegetal half, brown yolk (Fig. 284). The extent to which the white yolk reaches the surface of the animal half and the extent to which it remains stuck to the vegetal surface can vary a great deal. In these inverted embryos, two or even more gut invaginations may arise, and these will always be in places where regions containing white yolk and dark regions meet (Figs. 284–286). This may happen on the original animal side, on the original vegetal side (Figs. 284, 286), or on both sides (Figs. 285, 287).

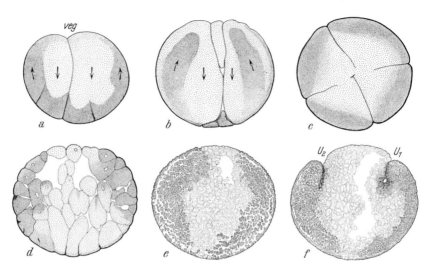

Fig. 284. Sections through embryos of *Rana fusca*, turned 180° in the two-cell stage. *veg* vegetal pole; U_1, U_2 blastopore invagination. (After Penners and Schleip, 1927)

The place where the dorsal lip arises is marked by the border of a secondary yolk region, and the direction of invagination follows the local yolk gradient: it always goes in the direction from yolky cells to yolk-poor cells (Fig. 284). Yolk-poor or yolk-free dark cell material involutes into a chordamesodermal plate. The original animal-vegetal polarity plays a further role. The blastopore margin extends to the border of the secondary yolk region, and closes over it (Figs. 285, 286a_1, b_1). Many different kinds of disorganized malformations appear, particularly when there are more than two gut invaginations, but surprisingly normal twinning can also arise, particularly when the newly created dorsal lips are far apart and where the chorda mesoderm anlagen can develop without coming into conflict (Figs. 286b, c; 287). Especially in *Triturus*, these twin forms can reach the stage of swimming larvae (Fig. 287e). According to the position of the blastopore invagination the involution and axial orientation of the chorda and the neural plate formed above it can be on the original dorsal side, ventral side, or both (Figs. 286, 287). What has been recognized anew after the apparently chaotic restratification of the egg contents is not only the determination of an invagination site on the blastula, but also the whole field structure which determines the contribution of individual events in the invagination of the foregut, the involution of the chordamesoderm, and the closure of the blastopore. These events can proceed in new places, stemming from entirely new distributions of substances, which in some cases are just the opposite of the distributions set up in the growing oocyte and completed after fertilization during the formation of the gray crescent. After the reorganization of the egg cytosplam, a blastopore can form anywhere on the surface of the embryo, but a quantitative distinction does remain:

220

the site of the gray crescent is favored. If in the vicinity of a yolk region only one dorsal lip arises, it appears regularly on the presumptive dorsal side of the embryo where the gray crescent had appeared (Figs. 285, 286a, b). Often two blastopore invaginations can arise in the same yolk region (Fig. 286c). These are always opposite one another and, as a rule, the one closer to the dorsal side develops earlier. Of the two partners of a twin form, the one that progresses further is also the one that has developed nearest the original dorsal side or that remains nearer (Fig. 286b, c). A very illuminating hypothesis of Dalcq and Pasteels ascribes this phenomenon

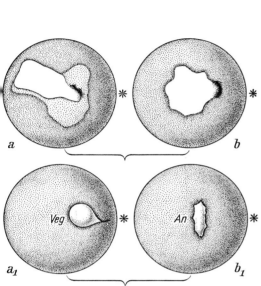

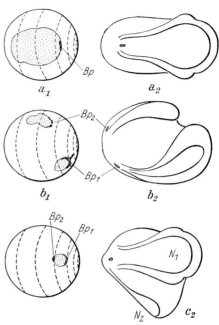

Fig. 285. Gastrulation on the vegetal (*Veg*) and animal (*An*) sides of a frog embryo turned 180° in the egg stage. a, b appearance of dorsal lip; a_1, b_1: blastopore closure; * site of center of gray crescent. (After Pasteels, 1938)

Fig. 286. Three cases of gastrulation and neural plate formation in embryos from inverted frog eggs (diagrammatic). The secondary yolk regions are stippled; in a_1, b_1, c_1 symbolic gradient lines define assumed cortical fields. Bp_1, Bp_2 first and second blastopore; N_1, N_2 the corresponding neural plates. (After Dalcq, 1941)

to a gradient field in the entire egg cortex related to the symmetry established after fertilization (Fig. 286a_1–c_1). The center of this gradient lies in the center of the gray crescent. It is not involved with the position of the yolk. The closer a border between yolky and yolk-free blastoderm is to the center of this dorso-ventral cortical field, the earlier and more vigorous is the invagination of the side facing this center (Fig. 286). At the border of the yolk region the arrangement of the yolk granules and cytoplasmic particles which normally distinguishes the region of the gray crescent must be reestablished. If an inverted egg is moderately centrifuged, before the beginning of cleavage, until the entire coarse white yolk has collected at the original animal pole, a normal embryo with completely inverted polarity can arise; and this can be understood if one assumes that the marginal zone between the white yolk and the yolk-poor particle-containing region acquires a structure which is normal except that it is inverted, and that the place where invagination occurs in the marginal zone is determined by the center of the cortical field[499]. In contrast

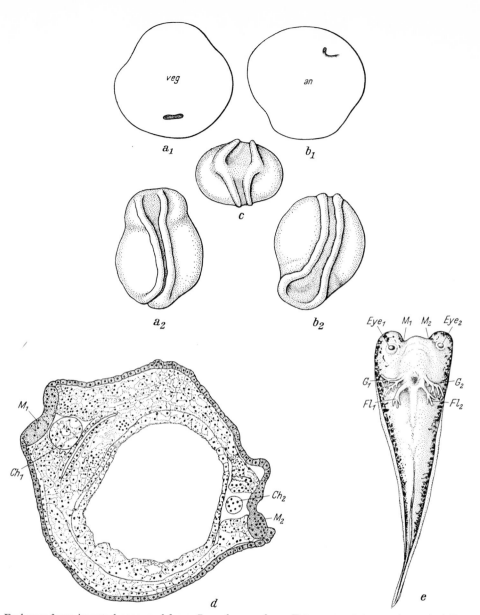

Fig. 287. Embryos from inverted eggs, a–d from *Rana fusca*, e from *Triturus taeniatus*. a_1, b_1 gastrulation on the vegetal (*veg*) and animal (*an*) sides; a_2, b_2 the corresponding neural plates; c anterior view; d cross section through a partial twin; Ch_1, Ch_2 chorda anlagen; M_1, M_2 medullary plates; e final stage of a Schultzian twin; Eye_1, Eye_2 eyes; G_1, G_2 gills; M_1, M_2 mouth openings; Fl_1, Fl_2 front legs of the two partners. (a–c after PENNERS, 1929; d, e after WITTMANN, 1929)

to this hypothesis, which deals with a gradient field in the cortex and a particular mix of cytoplasmic constituents merely as quantitative factors, LEHMANN has proposed another sound hypothesis. With the disorganized displacement of the yolk mass, the brown yolk is forced upwards (Fig. 284), and if the "marginal cytoplasm" (p. 202) plays a qualitative role in the formation of the gray crescent (perhaps providing the "morphogenetic substance of the dorsal lip"), it is possible that "through the displacement of the egg contents, the blastopore arises where the cranio-dorsal marginal cytoplasm meets the white material." As to deciding between these two possibilities, we still know too little about the cytoplasmic structure and its alteration during the restratification of amphibian eggs. At any rate, the conditions for the determination of a new blastopore are not restricted to the dorsal half of the egg; for even after total frontal cleavage (Fig. 279 b, e) both cells of an inverted embryo can gastrulate and develop a good deal further.

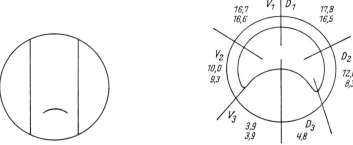

Fig. 288. Diagram of glycogen distribution in six sectors of a median ring of a *Triturus* blastula (upper numbers) and a young gastrula (lower numbers); D_1, D_2, D_3 dorsal, V_1, V_2, V_3 ventral sectors of the ring. (After HEATLEY and LINDAHL, 1937)

A reorganization of the cytoplasmic constituents can therefore lead to the formation of a "dorsal field," even in the ventral cells, which after isolation produce a ventral piece, although, to be sure, this is quite rare[506]. The following experiment speaks for a decisive role of the cortex: A fragment of the gray crescent of a fertilized but uncleaved egg is implanted into the opposite side of another egg; now this egg undergoes double gastrulation, giving rise to two neural anlagen[145a]. So there remains, in spite of the abundant experimental work of outstanding investigators, a great gap in our knowledge of the early determination of germ layer formation.

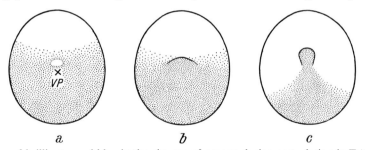

Fig. 289. Reduction of brilliant cresyl blue in the absence of oxygen during gastrulation in *Triturus cristatus*; ×VP vegetal pole. (After PIEPHO, 1938)

Much effort has been spent in studying the metabolic events which take place before the beginning of gastrulation and during the morphogenetic movements.

It is surprising that the DNA and RNA content of the embryo increases little or not at all until the onset of gastrulation[118]. The need for chromosomal DNA in rapidly multiplying nuclei must be met from a reserve in the cytoplasm laid down during the growth of the oocyte, unless the increase in chromosome number is achieved by the parcelling out of a great number of elementary fibrils from the chromosomes of the zygote nucleus (p. 29). At gastrulation the concentration of both DNA and RNA rises sharply, and purines and mononucleotides are consumed.

It was important to find out whether particular regions of the embryo differ quantitatively or qualitatively with respect to their chemical constituents and processes, and whether the local tendencies to movement and differentiation are related to the differentiation of the egg cell. Along the egg axis the yolk protein increases in concentration, and glycogen decreases in the

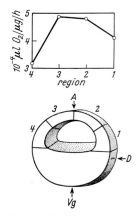

Fig. 290. Respiratory rate in parts of an animal half of the beginning gastrula of *Rana pipiens*. *A* animal pole; *D* dorsal groove (blastopore); *Vg* vegetal pole. (After SZE, 1953, from GREGG, 1957)

direction of the vegetal pole. In the blastoderm there appears to be more glycogen dorsally than ventrally. Histochemical and direct analyses of median blastula and gastrula sections have shown that in the region which moves over the dorsal lip the glycogen content decreases far more than in other regions when the dorsal lip is formed (Fig. 288). When young gastrulae are treated with reduced vital dyes, the color disappears quickest in the absence of O_2 during gastrulation on the dorsal side of the embryo over the involuting region of the marginal zone (Fig. 289), suggesting particularly high oxidation-reduction activity in this region. Oxygen consumption (Fig. 290) in the young gastrula is highest in the animal region in which the preparation of the blastoderm for subsequent stretching is proceeding by means of rapid cell division (p. 204); next highest is the region moving over the dorsal lip. The R.Q. in the dorsal lip approaches the value of one, while it is lower in the ventral ectoderm. This in turn leads to conclusions on the utilization of various energy sources. Protease activity is highest in the presumptive chordamesoderm and in the presumptive neural plate (Fig. 291). This indicates extensive breakdown of protein reserves and the synthesis of new proteins. All these findings point to qualitative differences in the metabolism of different parts of the embryo, not just differences in intensity.

The content of sulfhydryl proteins at the beginning of gastrulation is highest in the dorsal lip and in the animal hemisphere (Fig. 292). RNA-containing granules, which stain intensely with pyronin, show the same distribution. Both have already been noted in the dorsal region of the egg (p. 202, Fig. 257). We are in no position to decide whether these local differences in the blastoderm have any immediate relation to the pattern of movements in gastrulation, whether they are an expression of its determination or links in the causal chain of its completion, or whether they play a role in the determination of future patterns after the morphogenetic movements have been completed. In these biochemical relationships we are still at the question — or hypothesis — stage, as in the developmental physiology of the echinoderms.

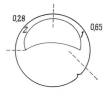

Fig. 291. Protease activity in two regions of the beginning gastrula of *Discoglossus pictus*. Activity in micromoles of tyrosine per liberated milligram nitrogen per two hours at 30° C. Average values for six measurements. (After D'AMELIO and CEAS, 1957)

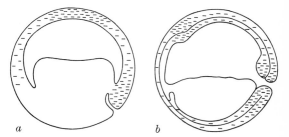

a   *b*

Fig. 292. Distribution of sulfhydryl-containing proteins in the blastula and gastrula of *Triturus*. (Diagram after BRACHET from DALCQ and PASTEELS, 1938)

Lecture 16

After the formation of the germ layers the separation of the first organ rudiments begins. The long strip of chorda rudiment develops a ventral groove and then closes into a solid rod. In the mesodermal layer, the coelomic cavity arises as a cleft, and the rudimentary muscle segments (somites or myotome) arise from the unsegmented lateral plates (Fig. 295). The upward-curving borders of the archenteric floor unite below the chorda and thus close the gut cavity. Above the chorda anlage and the row of somites, the margins of the neural plate rise to form a neural groove and then close together to form a neural tube (Fig. 293). During the involution of the dorsal border zone over the dorsal lip, the major part of the presumptive neural plate has undergone extensive stretching; this is shown by the longitudinal distortion of stain marks in the dorsal meridian (3 and 4 in Fig. 266e, f). The marks at the animal pole (Fig. 266 1, 2) remain in place, are taken up in the anterior (cranial) fold, and come to lie at the anterior end of the neural tube (Fig. 266f). With the closure of the neural tube, as stain marks show once again, there is a gathering of the ectoderm from two sides which entails another strong stretching of the rest of the ectoderm, the epidermis.

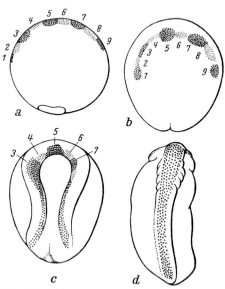

Fig. 293. The appearance of the neural folds and closure of the neural tube in *Triturus alpestris*. Rows of stained cells moved from the region of the animal pole down to the equator. a Yolk plug stage; b after closure of the blastopore; c neural fold; d tailbud stage.
(After GOERTTLER, 1925)

The chorda, the bilateral series of somites, and the neural tube are collectively known as the axial system of the embryo.

The region of contact between somites and lateral plates (nephrotome) gives rise in one region of the embryo to the first visible embryonic organ, the pronephros (Figs. 294, 295). Ventrally, blood cells and vessel anlagen differentiate from the mesoderm (Fig. 294). The embryo stretches longitudinally. In the foregut the gill pouches grow out (Fig. 299a). Posteriorly, the tailbud grows out. The cell materials of all the organ rudiments can be followed back into the blastula with the help of vital staining (Figs. 264–266), beginning with the first sign of axial organization, the appearance of the dorsal lip. Thus, a fate map can be laid out on the blastula surface which predicts the differentiation of the various regions of the blastoderm (Fig. 296).

We know that the regions of the blastula have already acquired certain tendencies for movements which harmoniously bring about the typical organization of the germ layers. Do the parts of the blastoderm have similar tendencies for differentiation, or even rigid determination?

We can answer these questions through isolation and rearrangement experiments.

The whole ectoderm can be isolated from the rest of the embryo by an elegant procedure: the induction of exogastrulation. If a dechorionated axolotl blastula is placed in hypertonic salt solution the endoderm and chordamesoderm will not invaginate; they pouch outward instead. A furrow rings the border between the yolk region and the marginal zone. It corresponds to the blastopore margin. The marginal zone, however, does not involute; rather, its cell material moves over the furrow and onto the expanded endoderm (Fig. 297a, b). In this way it expands as in normal gastrulation, and particularly, the dorsal border zone material corresponds to the

225

archenteric roof. The advancing tongue of tissue moves over the exposed yolky material, and the lateral regions of the mesoderm converge medially on the endoderm, as vital staining shows. Thus, the various parts of the border zone carry out the same morphogenetic movements in exogastrulation as they do in normal gastrulation (Figs. 267, 298a–c); but they execute these movements in an abnormal direction, and from this an entirely abnormal embryonic topography arises (Figs. 297c, 298d, e): The chorda anlage is superficial, and the convergence of the mesoderm results in a compact mass between chorda and endoderm. The endodermal mass bends

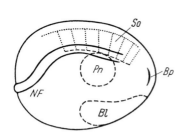

Fig. 294. Position of the somites (*So*), blood cells (*Bl*) and pronephros anlage (*Pn*) in the neurula; oblique dorso-lateral. *NF* neural fold; *Bp* blastopore. (After Dalcq and Pasteels, 1938)

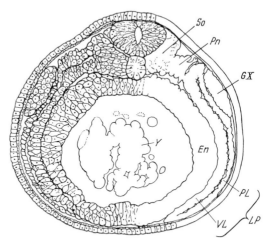

Fig. 295. Cross section, in the region of the pronephric anlage (*Pn*), through an embryo of *Discoglossus pictus* with 5 or 6 somites; *En* endoderm, *GX* anlage of the vagus ganglion; *Y* nutritive yolk; *So* somite; *LP* lateral plates; *PL* parietal layer (somatopleure); *VL* visceral layer (splanchnopleure). (After Dalcq, 1941)

upward with the borders of the mesoderm and gradually covers over the chorda also. This extension of the endoderm maintains its contact with mesoderm and chorda in a totally abnormal topography (Fig. 298c, e).

The ectoderm expands all by itself; it becomes a large, emtpy, folded sac (Figs. 297, 298). Gradually the ectodermal epithelial layers fuse, and a disorganized spongy mass of cuboidal epithelial cells is formed (Fig. 297c). Neither neural anlage nor the typical epidermis with cuticle surface layer, germinal layer, and glands are formed. Only ciliated cells and mucous cells, of the kinds seen in the normal development of the epidermis of early embryonic stages from the neurula on, appear. The blastoderm thus has some tendency to differentiate, but it retains a thoroughly juvenile character. Between presumptive neural plate and presumptive epidermis there is no difference in this respect. Therefore, in the ectoderm neither the morphogenetic movements of the neural anlage nor the histological differentiation of neural plate or epidermis are determined.

The rest of the embryo behaves quite differently in spite of the abnormal arrangement of its germ layers. The chorda undergoes its complete histological differentiation. Mesoderm divides into somites, in which groups of striated muscle cells differentiate (Fig. 297c). In the head region sets of striated muscles and cartilage elements appear (Figs. 297c, 299b). Posteriorly, coelomic sacs arise and pronephric ducts also are formed (Fig. 297c). Blood cells and vascular elements are always seen, and sometimes a beating heart in addition. Most surprising is the differentiation, over the course of ten to fifteen days, of the endoderm, which is now on the outside (Figs. 297c,

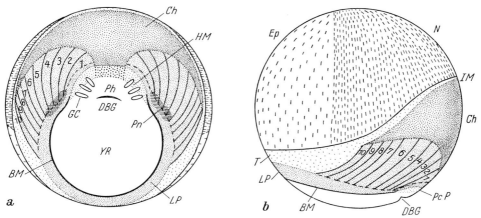

Fig. 296. Fate map; diagram of the arrangement of the presumptive organ anlagen in the urodele embryo at the beginning of gastrulation. a blastopore view; b left view. *Ch* chorda; *YR* yolk region; *IM* invagination margin; *Ep* epidermis; *Ph* pharynx; *GC* gill clefts; *HM* head mesoderm; *N* neural plate; *PcP* prechordal plate; *T* tailbud material; *LP* lateral plates; *DBG* dorsal blastopore groove; *BM* blastopore margin; *1–10* somites. (After Vogt, 1929)

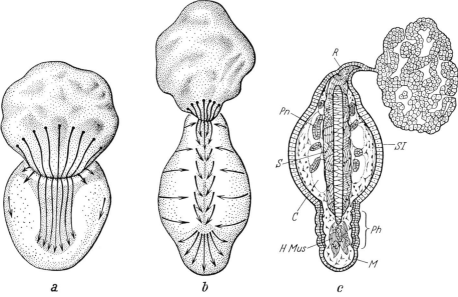

Fig. 297. Exogastrulation in the axolotl embryo. a, b Diagram of the morphogenetic movements according to stain spots; c diagrammatic frontal section through the older exogastrulated embryo. *C* coelom; *SI* small intestine; *R* rectum; *Ph* pharynx; *H Mus* head muscles; *M* oral cavity; *S* somites; *Pn* pronephros. (After Holtfreter, 1933 b, a)

299 b). The head section is closed off by a cap of flat epithelium, just as the anteriormost part of the mouth cavity is covered in the normal animal. Then follows an epithelium with two or more layers and taste buds like those found in the normal mouth. Gill pouches are formed (Fig. 299 b), but these are directed inwardly rather than toward the outside, as in the normal embryo (Fig. 299a). At the pharyngeal end the lungs invaginate as a forked tube. Next comes a short ciliated esophageal section, then a gastric mucous membrane with glandular crypts extending into the

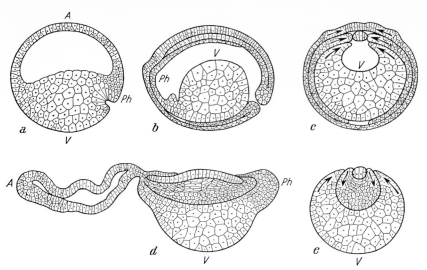

Fig. 298. Exogastrulation in the axolotl embryo, diagrammatic. a–c Sections through normal young gastrula (a) and neurula (b, c) for comparison; d, e sections through exo-embryo. a, b, d longitudinal sections; c, e cross sections. *A* animal pole; *Ph* pharynx; *V* vegetal pole. (After HOLTFRETER, 1933 b)

mesenchyme, then a small-intestinal epithelium with many mucous cells, and finally ciliated rectal epithelium (Fig. 297c). Liver and pancreas rudiments are evaginated from the forward region of the small-intestinal epithelium into the interior of the embryo and acquire a typical structure. Smooth muscle cells are present in the gut epithelium, and rhythmic contractions begin, in spite of the complete absence of nerves in the exogastrula. Although the gut epithelium now lies on the surface, it nevertheless has differentiated into a typical craniocaudal series of tissue types.

The exogastrula explant shows that in the axolotl embryo the ectoderm is not competent for self-differentiation. Whether the extensive typical differentiation of the endoderm and the chorda-mesoderm of the exogastrula stems from self-differentiation of its parts or from the interactions among the spatially arranged germ layers has not been decided by this experiment.

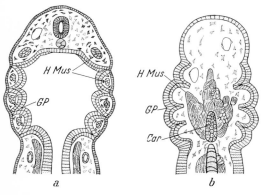

Fig. 299. Diagrammatic frontal sections: a through a normal embryo; b through an axolotl exo-embryo. *GP* gill pouches; *Car* cartilage; *H Mus* head muscula-ture. (After HOLTFRETER, 1933 b)

To make this distinction, we turn now to explants from very young gastrulae. These experiments were made possible by the development of a physiological saline solution permitting the long survival of embryonic parts, and providing an unspecific environment to permit the autonomous differentiation of the isolated parts. In such a sterile standard salt solution (Holtfreter's solution) a great variety of parts of very early blastulae of urodeles and anurans were systematically isolated. From the results of the exogastrula experiment it was expected that the parts of the presumptive ectoderm would show no differentiation beyond the atypical juvenile epidermis seen before. The explants of anuran ectoderm were surprising, however, in that each isolated piece, whether it was taken from the region of the presumptive epidermis or the presumptive neural plate, differentiated into an adhesive gland, which liberated a great stream of mucus. It is noteworthy that only one adhesive gland was formed, whether the piece of ectoderm was small or large. This glandular epithelium contained a mass of fairly disorganized ciliated cells, which made up anywhere from a small portion to 90 per cent of the whole explant. Since in the normal

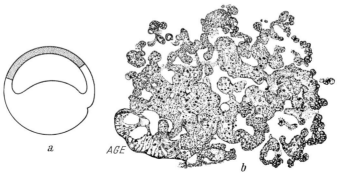

Fig. 300. Explant of presumptive ectoderm from *Rana fusca*. a Explanted region stippled; b loose mass of atypical apidermis with adhesive gland epithelium (*AGE*). (After HOLTFRETER, 1938 b)

embryo only two paired adhesive glands arise on the head (from presumptive epidermis whose normal position on the fate map can be seen in Fig. 306 b), their appearance in all the explants is clearly not the self-differentiation of one particular blastoderm region; for most of the explants would have developed into something else, had they been left in place. So here we see a case in which a portion of larval tissue expresses a morphogenetic (histogenetic) competence in an "indifferent" medium which is permitted only in a specific place in a normal embryo. The absence of typical differentiation in the ectoderm and the appearance of heterotypic differentiation shows that the events of normogenesis are prevented in the exogastrula and explant by the lack of some causal factor.

The behavior of similar explants of presumptive endoderm contrasts strongly with that of the ectoderm explants. The part of the endoderm which remains outside the longest, the yolk plug (Fig. 267), is destined in normogenesis for early death; these cells are pushed into the midgut lumen as nutrient material (Fig. 295) after the end of gastrulation and are digested by the differentiating gut epithelium. These nutrient yolk cells die in the explant at the same time as their career in control animals ends in the midgut; their life span is therefore temporally

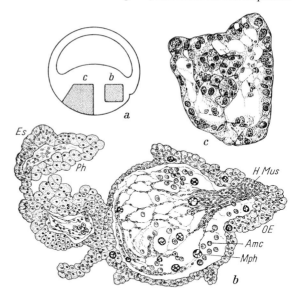

Fig. 301. Explants of presumptive endoderm. a Explanted regions stippled; b dorso-lateral material near the blastopore, border between endoderm and inner marginal zone of *Rana fusca*; c intestine of *Triturus*. *Amc* amoebocytes; *Ph* pharynx; *HMus* head muscles; *OE* oral epithelium; *Mph* macrophages; *Es* esophagus epithelium. (After HOLTFRETER, 1938 a, b)

229

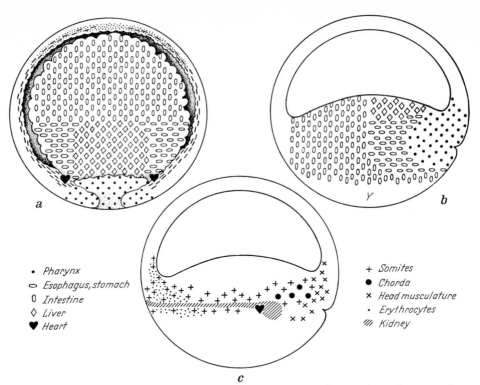

+ Somites

•   *Pharynx*   + *Somites*
○   *Esophagus, stomach*   ● *Chorda*
0   *Intestine*   × *Head musculature*
◊   *Liver*   · *Erythrocytes*
♥   *Heart*   ⫽ *Kidney*

Fig. 302. Diagram of the autonomous differentiation patterns in explants from regions of presumptive endoderm and the ventral marginal zone of the beginning gastrula. a Archenteric floor and lower marginal zone after the raising of the roof of the young gastrula; b longitudinal section; c projection of the regions of the inner marginal zone on the sketch of the longitudinal section. (After HOLTFRETER, 1938a, b)

determined, and they are not simply exterminated by surrounding tissues. Explants from the rest of the yolk cell region give rise to particular histologically differentiated gut epithelium which corresponds to their prospective site in the gut after gastrulation. A region immediately next to and within the dorsal lip produces mouth and gill epithelium; esophagus and stomach epithelium arise from lateral wings of this region (Fig. 301a, b). Liver and pancreas cells come from medial endoderm cells on the inside. The ventral part of the yolk region becomes mainly hindgut material (small intestine) (Figs. 301a, c; 302a, b).

If pieces of presumptive endoderm are inverted in the early gastrula, they also differentiate in the embryo into their expected portion of the gut. The histological differentiation of endodermal components of the gut therefore expresses mosaically the distribution of endodermal anlage material along the craniocaudal and dorsoventral axes (Fig. 302). Just when, on the way to the beginning of gastrulation, the presumptive endoderm cells acquire this determination which makes them competent to self-differentiate we do not know. In ventral pieces from isolated half-or quarter-blastomeres or from ligated blastulae (Fig. 278) the endoderm cells remain large and yolky and show no differentiation whatever. Some of them disintegrate in the segmentation cavity.

The greatest diversity of structures which may be formed autonomously is found in the marginal zone. In the interior lateral and ventral regions, which produce lateral plate mesoderm (Fig. 296), topographic determination similar to that in presumptive endoderm (Fig. 302a, c) predominates:

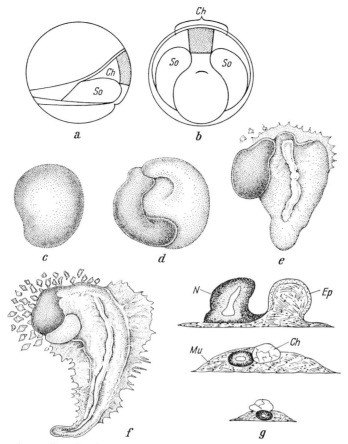

Fig. 303. Explant from the median chorda region of the axolotl. a, b Operative schemes; c–f morphogenesis over the next six days; g cross section. *Ch* chorda; *Ep* epidermis; *Mu* musculature; *N* neural tube; *So* somites. (After HOLTFRETER, 1939)

beating hearts arise in explants from a pair of regions separated laterally from the blastopore groove; pronephric ducts appear in pieces from a longer region of the ventral and interior border zone; the cells of the presumptive ventral blood islands, which in the beginning blastula lie opposite the blastopore, differentiate into red blood cells when explanted. Amoeboid blood cells (macrophages) arise in the dorsal and lateral parts of the border zone of the blastula as well as in the ventral regions near the endoderm.

The upper parts of the marginal zone, which contain presumptive chorda and somites, produce in every explant a great variety of tissues according to their prospective fate. In explants from the dorsal border zone, which includes only presumptive chorda, myotome always appears in addition to chorda; moreover, there is also some differentiation characteristic of other germ layers: epidermis and neural tissue (frequently a closed neural tube or a brainlike part). Figure 303 gives an example. First, the isolated cell material rounds up, then it lifts away from the bottom of the culture vessels, and a dark part separates from a lighter part. It stretches longitudinally, and a chorda becomes visible on the upper surface (Fig. 303 e). The border of the structure spreads out over the underlying tissue; from its anterior border, at which a large neural mass forms, yolky giant mesenchyme cells migrate out. Cross sections (Fig. 303 g) show that the major portion consists of muscle cells. These are arranged in series and thus show the tendency for somite formation. In front they are rolled together into a clump which is covered by epidermis. A neural tube sticks out free in front in a large swelling and, covered by mesoderm, runs tapering backwards near or under the chorda into a pointed tail, in which the differentiated explant terminates (Fig. 303 f, g). Explants of pieces which extend back to the site of invagination also contain foregut epithelium, as one would expect (cf. Fig. 296). Figure 304 shows a product of this region. On the outside it is covered by endoderm; inside are a large chordal element next to a large neural sac. Some mesenchymal connective tissue, amoebocytes, and macrophages are scattered

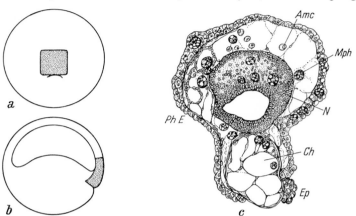

Fig. 304. Explant from the dorsal lip of *Rana esculenta*. a, b Operation diagrams; c sac of pharyngeal epithelium (*PhE*), some epidermis (*Ep*), chorda (*Ch*), a neural vesicle (*N*), connective tissue with amoebocytes (*Amc*) and macrophages (*Mph*). (After HOLTFRETER, 1938 b)

in the lumen of the endodermal sac. A little piece of epidermis lies next to the end of the chorda, which projects free from the endodermal sac. From lateral parts of the marginal zone, even when they are far from the presumptive chorda region (Fig. 305 a, cf. Fig. 296), long chordal elements and neural tissue arise in urodeles, in addition to the expected somite musculature. The explant in Figure 305 consists mainly of muscle tissue. The chorda is curled up in the front part of the explant, and it stretches backward within the musculature all the way to the tail. In the front part a great neural mass lies on one side. Endoderm covers this explant, which extends somewhat into the yolk region (Fig. 305 a). Even explants which contain extremely ventral pieces of the region running along the tips of the presumptive somites near the presumptive lateral plates, presumptive ectoderm and yolk cells, produce, in addition to the expected structures (muscle, pronephric duct, some endoderm and epidermis), also chorda (Fig. 306 a).

The autonomous formation of the chorda-somite zone is therefore not self-differentiation: in contrast to its prospective fate, each piece of the urodele border zone can also produce chorda, somite musculature, and, in addition, even epidermis and neural structures; that is to say, even structures whose presumptive regions lie outside of the marginal zone. The behavior of the marginal zone explants has the character of the self-organization of a morphogenetic field, which we know in its simplest model, the aggregation field of *Dictyostelium*. The development of the isolated part of an embryo runs a course aimed in the direction of a harmonious organ complex; it is regulatory to an extent reminiscent of the regulation which leads the vegetal halves of sea urchin embryos a greater or lesser distance toward the form of a typical whole. To be sure, this regulation has its limits. Endodermal differentiation and that involving the lateral plate region (pronephros, heart) do not result in prospective chorda mesoderm. In the autonomous organization of border zone explants a *de novo* orientation of the structure and relationships of the tissues takes place: the regulatively formed tissues are always clearly separated and have normal form; and they are not chaotically distributed, but assume a particular spatial relationship to one another. Chorda and neural tube are often attenuated or in a pad of muscle. A bilaterally symmetrical form characteristic of the general tail region often arises with an anteroposterior and a dorsal-ventral axis. These regulatively formed tissues frequently have a certain mass relationship

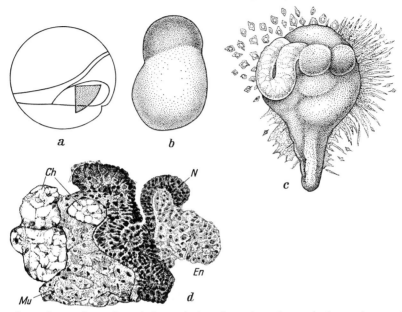

Fig. 305. Explant from the somite region of the axolotl. a Operative scheme; b–d morphogenesis; b after one; c after four; d after five days; d cross section. *Ch* chorda; *En* endoderm; *Mu* musculature; *N* neural tissue. (After HOLTFRETER, 1938a)

to one another. In explants from the chorda-somite region, the ratio of chorda to somite muscle to nerve tissue to epidermis approximates 2:4:2:1. The range of variation is great, as it is in the regulation of vegetal halves of sea urchin embryos (Fig. 211). Chorda without somites, and muscle without chorda, are rare in urodeles. The greater the distance from the blastopore, the less is the regulative ability of the ectoderm (Figs. 304, 306e).

There is a notable difference between urodeles and anurans (Fig. 306c): in anurans the ability to form chorda regulatively does not extend as far ventrally as in *Amblystoma* and *Triturus*. While in the latter two forms chorda is formed from ventral-lateral derivatives to the same extent as somites, chorda formation in frog embryo explants never goes as far as one would expect from the fate map. Moreover, somite formation in the dorsal presumptive chorda region is scant. In the anurans, therefore, a dorsoventral gradient of the differentiation tendencies for chorda and somite structures prevails in the marginal zone. Figure 306 gives the comparison of the autonomous differentiation of explants (c) with the presumptive organ regions of the fate map (b) in anurans.

During normal development the dorsal and dorsolateral marginal region becomes archenteric roof during gastrulation, and above this a neural anlage develops in the ectoderm, which can undergo no neural differentiation if isolated. This analysis of the interactions between archenteric

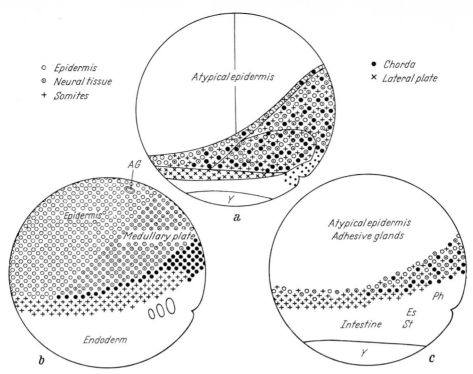

Fig. 306. a Diagram of the autonomous differentiation patterns of the regions of the beginning urodele gastrula (cf. fate map, Fig. 296); b, c beginning anuran gastrula; b fate map; c diagram of the autonomous differentiation patterns of the regions. *Ph* pharynx; *Y* nutritive yolk; *Es* esophagus; *St* stomach. (After HOLTFRETER, 1938b, modified)

roof and neural anlage, first undertaken by SPEMANN (1918), was an extremely important finding for developmental physiology, and some of the questions this discovery raises have still not been answered.

Transplantation experiments lead to some elementary conclusions. These can be carried out in various ways: (1) the dorsal lip of a young gastrula is excised and implanted in the ventral or lateral presumptive ectoderm of an embryo of similar age (Fig. 307a, b); then the implant invaginates autonomously if it is large enough and spreads out on the underside of the ectoderm. (2) A piece of marginal zone is implanted in the region of the ventral marginal zone of a young gastrula; then the implant accompanies the neighboring tissue inside during gastrulation (Fig. 310a, b). (3) The piece of dorsal marginal zone is placed in the blastocoel through a slit in a young gastrula (Fig. 307a, c); the implant then becomes apposed to the ectoderm as gastrulation continues (Fig. 307d).

When, after one of these procedures has been used, a piece of the dorsal marginal zone containing presumptive chorda material comes to lie under the ectoderm, it stretches. The extent of the stretching depends on the size of the implant and on the site of origin of the implant in the dorsal marginal zone. An implant from a very young gastrula extends further than a similarly sized piece from a later gastrula; this corresponds to the greater stretching during normal development of the material which moves inside first (Figs. 267, 269). The direction of stretching is the property of the implant and can proceed at an angle to the long axis of the host embryo. The

234

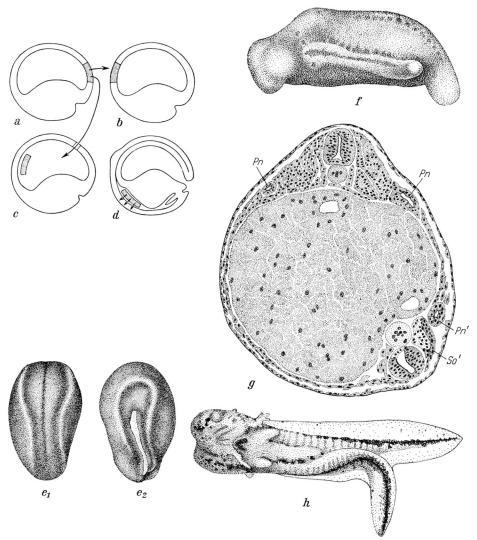

Fig. 307. Transplantation of the dorsal lip; a–b into ventral ectoderm; a–c into blastocoel; d position of c implant next to ectoderm, inductive effect; e–h *Triturus taeniatus* embryos with secondary embryonic anlagen; e neurula; e_1 dorsal view of primary neural plate, e_2 ventral view of secondary neural plate showing pale longitudinally stretched implanted strip; f secondary embryonic anlage with otic vesicles at the front of the neural tube, somite rows and tailbud; g cross section through f; lower right, secondary embryonic anlage, implanted tissue is lighter and includes chorda, a part of the induced somite, and neural tube; induced pronephric duct is entirely host tissue; secondary gut cavity; h embryo with extremely well-developed secondary twin. (e–g after SPEMANN and H. MANGOLD 1924; h after HOLTFRETER from SPEMANN, 1936)

implant organizes itself as do explants from the same region, into chorda and somites. The formation of these anlagen in the host embryo is not restricted, however, to the material of the implant, but instead the completion of a secondary embryonic anlage may utilize some of the host mesoderm which gastrulation has brought into the region of the implant. This can be studied in cases where host and implant are of different species (heteroplastic transplantation), particularly where there is a distinction in cell pigmentation, as between *Triturus cristatus* versus *Triturus taeniatus* and *Triturus alpestris*, or where the implant is vitally stained. In many cases, the individual anlagen, even the chorda and somites, are composed chimerically of both implant and host cells (Fig. 307 g). The mechanism is extremely peculiar: It is not the case that the implanted material first differentiates according to its prospective fate and then is completed by the addition of cells from the neighboring regions of the host, as a crystal grows; rather, the implant incorporates the not yet formally determined surrounding material into its field of influence before differentiation, and then organization in this field proceeds throughout the whole region without regard to the source of the material. This process reminds us of the formation and

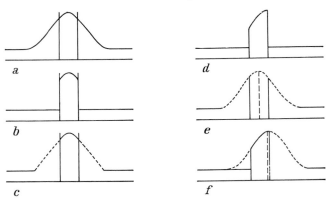

Fig. 308. Diagram of a gradient during the organization of the morphogenetic field in implants of organizer parts. (a–c after PASTEELS, 1951)

organization of an aggregation field in *Dictyostelium*. But there is a great difference: in *Dictyostelium* the cells accumulate in a centripetal morphogenetic movement leading to the incipient sporophore and become determined gradually along the way, while in our amphibian case a pattern of determination extends over the whole field. This field organization can be understood in terms of a morphogenetic gradient as Dalcq and Pasteels have proposed: its top leads to the determination of chorda, a lower level to somites (Fig. 308 a–c). This interpretation agrees with the distribution of differentiation frequencies in the marginal zone of anurans (p. 233, Fig. 306 c), which can also be interpreted in terms of a gradient from chordal to somite tendencies in the dorsoventral direction. Above the secondary chorda anlage so formed and the accompanying row of somites, a neural anlage is induced: it consists of a neural plate with raised edges, which closes into a tube whose orientation is the same as that of the chorda below. This secondary neural tube can also be a chimera if during the course of the first experiment (Fig. 307 a, b) a part of the implant is not invaginated and thus remains in the region of the secondary neural plate (Fig. 307 e_2) and finally in the wall of the neural tube (Fig. 307 g). The secondary axis system also has an inducing effect on the endoderm: a secondary gut cavity (Fig. 307 g) is formed. The secondary embryonic anlage can develop further to such an extent and draw so much embryonic material to it that twinning actually results (Fig. 307 h).

If a lateral half of the dorsal lip is implanted (Fig. 309 a), it regulates into a symmetrical form. Such an implant contains presumptive chorda material at its medial margin. If it comes from a young gastrula, the chorda tissue arises mostly in the middle of the implant, i.e., far from its margin; it has thus been formed from presumptive somite material; the field has formed a new axis of symmetry (Fig. 309 d, e). In implants from older donors the chorda forms at the edge of the implant (Fig. 309 f); the material from the lateral-dorsal marginal zone has thus organized

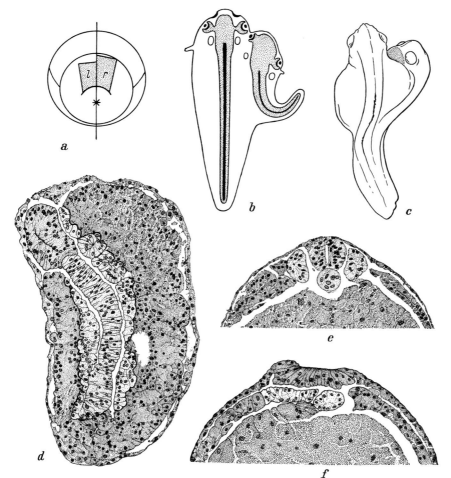

Fig. 309. Induction after implantation of lateral halves of the dorsal lip in *Triturus* embryos. a operative scheme; b sketch of the primary and secondary embryonic anlagen; c view of the primary embryonic anlage: the second is seen on the right side of the host; d frontal section secondary chordamesoderm, cut practically frontally, implant of *taeniatus* (light) into *T. cristatus*; e cross section through the center of an embryo like d; chorda, left and right somites formed entirely from implant; f cross section in the anterior half of an embryo with implant from an older gastrula; chorda and part of the left somite are formed by the implant. (a, c–f after B. MAYER, 1935; b after EKMAN, 1936)

itself according to its presumptive symmetry and incorporated host material unilaterally. During the involution of the marginal zone a determinative organization or segregation thus takes place: an invisible differentiation of the dorsal marginal zone into chorda anlage and somite region. The ability to reorganize into a whole remains, however. Even in this case the organization of the field results in the formation of a whole, not after the differentiation of the implant, but obviously earlier, in the course of the secondary gastrulation; after the differentiation of the implant, the host material itself is organized and determined to such a degree that it will no longer participate in the organization of a foreign structure. The regulative field determination of a half dorsal lip can also be thought of in terms of a gradient scheme (Fig. 308 d–f). The secondary embryonic anlage induced by a half dorsal lip can develop as completely and massively in the host (Fig. 309b, c) as that evoked by a symmetrical piece of dorsal marginal zone in the most favorable cases (cf. Fig. 307g, h).

The competence for self-organization and completion of a secondary embryonic anlage is not restricted to the dorsal lip. Implants from the marginal zone which contain only presumptive somites and no presumptive chorda (Fig. 310a) form, in conjunction with incorporated host mesoderm, chorda and somites (Fig. 310c_2) or only bilaterally-symmetrically-arranged somites (Fig. 310d_2), according, we can assume, to the level of the gradient system in the implant. The chorda-free row of somites also induces a neural tube.

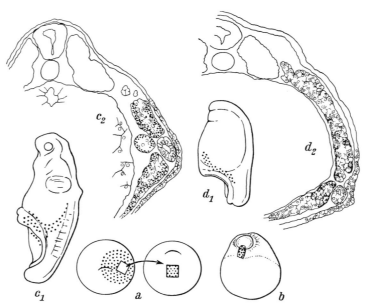

Fig. 310. Transplantation of presumptive somite material into the ventral lip of axolotl embryos. a transplantation scheme; b the vitally-stained implant beginning to move over the ventral lip; c, d embryos with secondary embryonic anlagen; c_1, d_1 entire embryos, induced neural tube ventral, stained implant tissue seen through the outer wall; c_2, d_2 cross sections. (After MINGANTI, 1949)

In the *Triturus* embryo the region has been precisely circumscribed which, when implanted, can induce a neural plate in presumptive epidermis. From beginning gastrulae flat pieces were cut out and then cut into smaller pieces. From these experiments the congruence of the inducing region and the presumptive chorda-somite region is shown. In anurans the tendency for chorda formation both in explants (Fig. 306) and implants is restricted to a smaller region of the marginal zone than in urodeles.

The material of the dorsal and dorsolateral marginal zone can also give rise to a chimeric organ complex through self- and inductive organization. The induction process is completed in two stages, each underlain by diverse processes: first, the host material is brought into the chorda-mesodermal field of organization, through *assimilating induction*; and then, in another blastema, ectoderm, a structure is evoked by *complementary induction*, completing the organ complex. The field of self-organization thus works also as an induction- or determination field which directs spatially-related differentiation in another blastema.

At the same time as a new axial system is set up, vigorous growth and cell division is triggered in the nascent organs. Even the endoderm cells participate in this increase in division with the formation of a second gut cavity under the secondary axial system. This remarkable competence to organize a secondary embryonic anlage in the large surrounding theater of the chorda somite field has caused SPEMANN to designate the dorsal lip as an *organizer* or a center of organization.

In the formation of the secondary embryonic anlage the implant is nevertheless dependent upon its environment: while during the self-organization of explanted pieces of border zone epidermis and neural tissue arise (Figs. 303, 306a), implants lying under the ectoderm form only chorda and somites from the same regions of the border zone (Figs. 307, 309, 310). We can ascribe this only to an ectodermal influence which is lacking in the explants and which is itself evoked by the implanted organizer as it induces the completion of the organ complex. Clearly the blastema, which has experienced induction, now replies with an effect on the inducer.

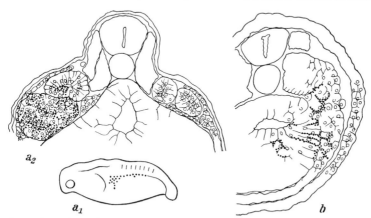

Fig. 311. Implants from transplantations as in Fig. 310a, b. a implant cells differentiated into secondary pronephros; b implant cells, some lodged in the endoderm, others migrating into the somites and lateral plate. (After MINGANTI, 1949)

The implant succeeds as an organizer in the host only when its cells remain together. In the anlage complex of the secondary chorda and somites the implant cells are not scattered among the host cells but rather remain together, even when the implant includes two distinct regions (Figs. 307g, 309d–f). When they are prevented from staying together or are implanted anywhere but a "neutral" place, i.e., near as-yet-undetermined material, the implanted cells come under the influence of the host. Implants which move in over the ventral lip (Fig. 310a, b) often do not move under the ectoderm but migrate upward with the mesoderm, which is being gathered dorsally. Vitally stained implants can be seen from the outside and followed throughout their careers (Fig. 311a_1). They can be seen to be incorporated, in groups or completely scattered, through the assimilating induction emanating from the host, and according to their site they become parts of somites or extra pronephric ducts (Fig. 311a). They doubtless migrate actively, another indication of a chemical attraction exerted by the dorsal parts of the chordamesoderm (cf. p. 214): If a piece from the marginal zone is implanted into the yolk region, implanted cells are seen to wander out of the endoderm and into the mesoderm after gastrulation (Fig. 311b). Whether the cells remaining in the endoderm can also take part in the differentiation specific to that locale is not known.

It is surprising that even large pieces taken from the organizer region can lose their character completely in particular surroundings. This phenomenon occurs after the inversion of a medial strip from the blastopore to the animal pole in the beginning gastrula (Fig. 312a). Here presumptive chorda and, when the blastopore lip is included as far as the site of invagination, also

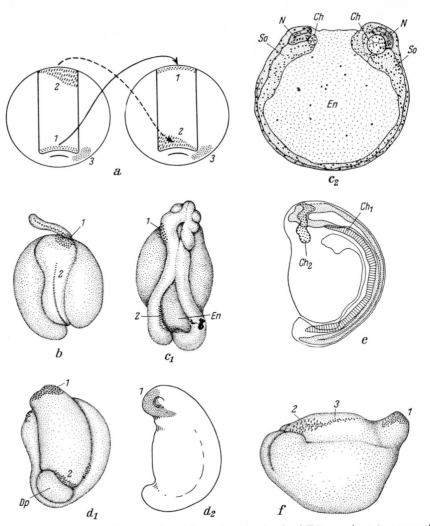

Fig. 312. Inversion of a strip running from the dorsal lip to the animal pole of *Triturus alpestris*. a operative scheme; b neurula with hornlike outgrowth of presumptive chorda; c_1, c_2 embryo with incompletely closed blastopore; c_2 cross section through the posterior part of the embryo; d_1, d_2 early and later stage of an embryo with complete incorporation of the border zone material into the anterior part of the brain; e schematic sagittal reconstruction of an embryo with secondary chorda (*Ch*); f lateral view of an embryo in which dorsolateral material has previously involuted (*3*) followed by presumptive ectoderm (*2*). *Ch* chorda; *Dp* yolk plug; *N* neural tube; *So* somites; *1 & 3* blue marks; *2* red mark. (After TÖNDURY, 1936)

presumptive prechordal plate (foregut roof) are exchanged with presumptive epidermis (cf. Fig. 296). The ends of the inverted strip and the marginal zone near the dorsal lip in the host are marked with different vital stains. In some of the cases a horn-shaped appendage grows out of the material implanted at the animal pole (Fig. 312b), just like those that explants from the dorsal lip produce (Fig. 273a–d). Some of the vitally-stained material of the transplant is incorporated and sits with a broad base on the anterior neural region in the neural plate stage. Later, this appendage disintegrates without histological differentiation and is cast off. The rest of the transplanted presumptive chorda is incorporated into the neural plate, whose medullary rims can close. Often, however, the stretching tendency of the presumptive chorda is entirely inhibited: only a stubby outgrowth appears at the animal pole from the end of the inverted strip (Fig. 312d) and is soon reincorporated; the whole presumptive chorda is incorporated into the neural anlage. Then one can see the vitally-stained cells in part of the brain from the outside (Fig. 312c_1, d_2). Only very rarely does a part of the piece of dorsal lip invaginate at the animal pole and form a small chorda (Fig. 312e); and it never exercises an inductive effect. This ectodermalization of the dorsal marginal zone material is dependent on the age of the gastrula and

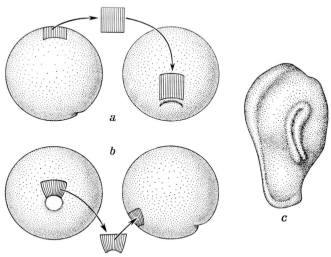

Fig. 313. Inductive organization of presumptive ectoderm in the organizer. a Implantation of a piece of ectoderm in the dorsal marginal zone of an earlier gastrula; b retransplantation of the implant, after a 24-hour stay in the first host, now into ventral ectoderm of a second host in the early gastrula stage; c secondary embryonic anlage induced by an implant of *Amblystoma*. (After RAVEN, 1938)

on the place reached by the piece of dorsal lip: if the inversion of the embryonic roof extends as far as the site of the blastopore, the blastopore material is incorporated smoothly into the neural anlage. That the dorsal border zone has ectodermal competence has already been demonstrated in explants; but what limits the tendency for chorda and somite formation, and what is responsible for the spatial organization of ectoderm? Only one property of the ectoderm of the animal pole suggests itself. If a strip is cut out passing the animal pole obliquely, or if a longer strip extending over the animal pole is cut out and inverted so that the region of the dorsal lip is now in the belly, it behaves essentially as a transplantation into the side or the belly region of another embryo. It develops according to its presumptive fate and forms a secondary embryonic anlage.

These experiments show that the chordamesoderm region of the blastula is still not formally determined as a field. Only during the course of gastrulation does the determination become fixed so that a portion of the border zone assumes the role of organizer with respect to a particular region of the embryo.

Experiments at the other end of the inverted strip are also informative, where the presumptive ectoderm now lies dorsally in the border zone. Generally, gastrulation is more or less inhibited by this foreign material, which is understandable in view of the results of experiments described earlier (p. 212ff., Fig. 276). When a small bridge remains over the blastopore invagination

(Fig. 312a), this inhibition is generally overcome, however, and the presumptive ectoderm involutes and stretches. Stain marks (*2* in Fig. 312a, b) show through the outer layer and are seen to be stretched longitudinally in the archenteric roof beneath the neural plate. In the archenteric roof the incorporated ectoderm is determined to become chorda and somites. Often, however, the stained ectodermal border (*2*) of the strip remains outside at the edge of the blastopore above the yolk plug (Fig. 312d$_1$, *2*). The dorsolateral parts of the marginal zone, which are presumptive somites, of course, have closed together and now form the archenteric roof. Mark *3* in the embryo of Figure 312d$_1$ has disappeared inside. At times even the presumptive ectoderm of the inverted strip is later drawn inside the embryo: mark *3*, which indicates presumptive somite material, is seen up in the archenteric roof, and the ectoderm mark (*2*) is behind it (Fig. 312f). Sometimes blastopore closure is also inhibited by the displaced ectoderm, and part of the endoderm remains exposed (Fig. 312c). The mesoderm involutes over the lateral blastopore margins (as in the experiments on ring embryos, Fig. 275). When the embryo stretches, the still-open blastopore is stretched lengthwise; the presumptive mesoderm organizes at the site of involution into somites and chorda, over which a neural tube is induced (Fig. 312e). The ectoderm remaining on the surface (mark *2*) is also stretched to the lateral blastopore margin and is incorporated into a neural tube (Fig. 312c$_1$). The anterior part of the neural tube, which lies over the archenteric roof and so is differentiating into a brain, must therefore split toward the rear, just as the corresponding chorda does. This experiment is informative from many points of view. It shows that the ectoderm which comes to involute flanked by material from the dorsolateral marginal zone differentiates chorda just as the dorsal marginal zone does. The ectoderm participating in this morphogenetic movement is therefore incorporated into the chorda-mesodermal field due to the assimilating induction exercised by its immediate surroundings. This response is similar to that of the host mesoderm to an implanted organizer. The question now arises as to whether this organization of the ectoderm is accompanied by the induction of competence for autonomous development corresponding to that of the marginal zone material, and whether organizer properties are also induced. The following experiment provides the answer. From an early gastrula, ectoderm from the animal pole region is marked with vital stain and used to replace an excised piece above the dorsal lip in another early gastrula (Fig. 313a). A small margin above the dorsal lip is left in order to assure smooth invagination. In the course of a day the implant advances to the blastopore margin and then begins to move inside. The part of the implant still remaining outside is now excised, and taking great pains to avoid the inclusion of host cells in the presumptive epidermis, it is placed in a third embryo (Fig. 313b). Here this twice-transplanted piece behaves like a transplanted dorsal lip and can form a secondary blastopore, involute, differentiate into chorda and somites, and it can even develop into a more or less complete secondary embryonic anlage. Neural differentiation can take place either in host material or in that part of the implant remaining on the surface. Thus the ectoderm transplanted into the dorsal border zone among presumptive somites acquires both the tendency to invaginate and a competence for self-organization as a part of a chorda mesodermal field, as well as competence to induce neural differentiation.

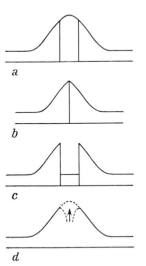

a

b

c

d

Fig. 314. Gradient scheme for the experiments of Fig. 313 (see text)

242

The replacement of the material of the presumptive chorda by the dorsolateral parts of the marginal zone is consistent with the gradient scheme (Fig. 314a, b): the open sides are quantitatively similar, and the highest level becomes chorda. But this simple geometric picture is not sufficient to explain the raising of a piece of presumptive ectoderm placed between presumptive somites (Fig. 314c, d) to the level of chorda determination. We have to know what "declines" or flows in the gradient and evokes this alteration of the tendency of the assimilated ectoderm.

Lecture 17

The induced nervous system is a highly complicated organ system, and it raises the question of how the organization and histological differentiation of the neural anlage is effected. Is the neural plate a self-differentiating field, which is induced in its entirely at the end of gastrulation, or does the substratum still play a role in the longitudinal organization of the neural anlage into brain and its associated regions, and into spinal cord with its differentiation along its length?

First of all let us consider the normal pattern of development: the ectoderm of the late urodele gastrula is a single epithelial layer. Before the neural plate is clearly demarcated, the neural groove sinks in the midline (Fig. 270b), then the border of the neural plate can be recognized by a stripe of heavily pigmented cells (Fig. 315a). Here the cell bodies are flask-shaped; their thick ends point inwards, and their narrow, pigmented ends extend to the surface of the epithelium. As neural ridges rise, the surface of the neural anlage narrows (Fig. 293), and the cells of the neural plate are crowded together and stretched in their long axes (Fig. 315b), particularly the lateral cells. The presumptive epidermis, which is drawn into a cap over each ridge and which then closes over the completed neural tube, becomes a thin epithelium (Fig. 315c). In the anterior part of the neural plate, which must form the brain, the ectoderm is originally thickest, and here it consists of several layers (Fig. 327b). In the closed spinal cord, the neural tube is bilaterally symmetrical, consisting of two thick lateral plates bordering the laterally compressed central canal which runs sagitally. The ventral and dorsal regions are thin (Figs. 315c, 316c, d). A band of cells at the roof of the neural tube (Fig. 316c) and originating in a region from the curled edge of the neural ridge into the ectoderm (Fig. 316a, b) forms the neural crest (Fig. 316d). Its cells grow large and migrate laterally and ventrally away from the neural tube (Fig. 316e, f). This ectodermally derived cell material is a truly remarkable primordium in the vertebrate body. It becomes brain ganglia, spinal ganglia, sympathetic ganglia, neuroglia, and "ectomesenchyme" – cells which spread far and wide: head mesenchyme for parts of the chondrocranium (Figs. 349, 350a), dentine-forming cells in the teeth, the pigment cells found throughout the body, and other cell types, too.

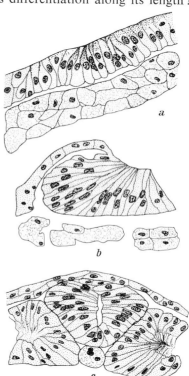

Fig. 315. Formation of the neural plate and the neural tube in *Triturus*. a embryo shortly after the delimitation of the neural plate; b neural ridges approximated. Section through the middle of the trunk (compare Fig. 293); c closed neural tubes, section through the middle of the trunk of an embryo with primary optic vesicles. (After LEHMANN, 1928)

243

From the brain region of the neural anlage come (Fig. 317): the telencephalon, which is divided into two hemispheres; the diencephalon; the mesencephalon; the metencephalon (which is quite rudimentary in the amphibia); and the myelencephalon or rhombencephalon. The telencephalon and diencephalon together are called the archencephalon; the metencephalon, the myelencephalon, and generally the mesencephalon are called collectively the deuterencephalon. In the neural plate stage the border between the two main regions of the presumptive brain is marked

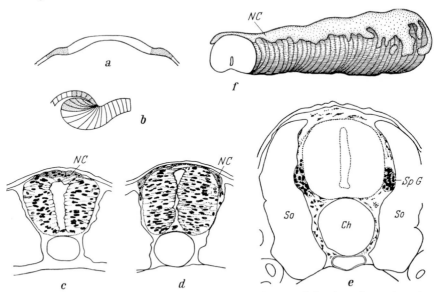

Fig. 316. Development of the neural crest of *Amblystoma*. a, b diagram of the sites of neural crest material during neurulation; c–e cross sections through later embryonic stages; f reconstruction of the neural tube and migrating neural crest material from the second to the seventh spinal segment. *Ch* chorda; *NC* neural crest; *So* somites; *SpG* spinal ganglion anlage. (a after RAVEN, b after FAUTREZ, c–f after DETWILER from HÖRSTADIUS, 1950)

by the front end of the chorda. The spinal cord and deuterencephalon lie over the chorda anlage, while beneath the archencephalon is the prechordal plate (Figs. 272c, 317, 327b). The extent to which the chorda underlies the anlage of the mesencephalon seems to vary with the species. In about fourteen-day-old *Triturus* larvae the brain, together with its closely related large sense organs (nose, eyes, otic vesicles), already has a complicated structure, as is seen in Figure 317c–f. The diencephalon assumes a central role at this time, since the eyes form as lateral evaginations from it, and dorsally the epiphysis pouches out, while ventrally the infundibulum and neurohypophysis are formed (Fig. 317).

To understand the determination of the organization of the neural tube, we may consider first of all some transplantation experiments, in which presumptive ectoderm from various regions of the archenteric roof was placed in a host in the early gastrula stage. Four successive sections of the archenteric roof were cut from young neurulae after the curling of the neural plate (Figs. 318a–d, 319) and placed in the blastocoel of early gastrulae. The implants induced a neural plate, from which a knob and then generally an organ complex arose, containing various organs according to the origin of the implant, and in many cases corresponding to a well-formed body region. In the best cases head, trunk, or tail resulted (Fig. 318). Weaker differentiation resulted merely in individual organs, such as parts of the brain, eyes, otic vesicles, balancers, and olfactory pits. Figure 319 shows the frequency of each organ formed in a large implantation experiment,

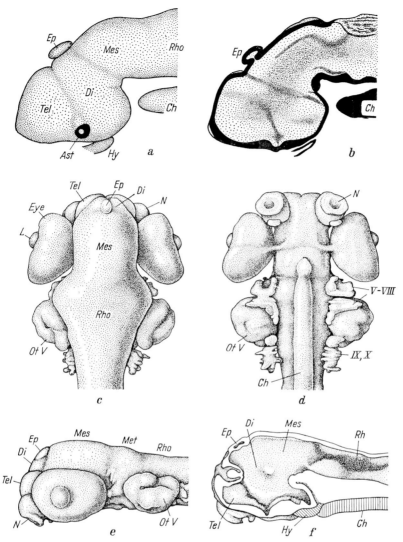

Fig. 317. Amphibian brain according to plastic reconstructions. a, b *Pelobates fuscus*, 6 mm. embryo; a surface; b median section; c–f *Triturus alpestris* larva with two toe buds on the front limb; c dorsal view; d ventral view; e left view; f median section. *Eye* eye; *Ast* eye stalk; *Ch* Chorda; *Di* diencephalon; *Ep* epiphysis; *Hy* hypophysis; *L* lens; *Mes* Mesencephalon; *Met* Metencephalon; *N* nose; *OtV* otic vesicle; *Rho* rhombencephalon; *Tel* telencephalon; *V–X* cerebral ganglia. (a, b after BERGQUIST, 1952; c–f after MANGOLD from MANGOLD and VON WOELL-WARTH, 1950)

employing the four archenteric roof regions. The frequency curves suggest three distinct induction tendencies: the two anterior quarters of the archenteric roof induce anterior head organs mainly, then brain parts and nose and suckers (Fig. 318a_2, b_2); the third quarter induces organs of the ear region predominantly (Fig. 318c_2), and the very different effect of the hindmost quarter is the formation of somites, pronephros, spinal cord, and entire forward-pointing tails (Fig. 318d_2). Thus, we can distinguish two forms of head induction, anterior or archencephalic induction, and posterior or deuterencephalic induction. Moreover, behind these is trunk-and-tail or spinocaudal induction. Whether these regional specificities depend on quantitative or qualitative differences among the inducing tissues is not indicated by these results.

Such complete development of secondary embryonic anlagen, of organ complexes and body parts, as seen in these transplantation experiments, can be aided by their development in an intact host. This possibility has been ruled out by a further, elegant series of experiments: the

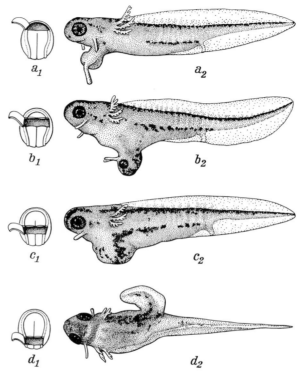

Fig. 318. Regional specific induction caused by implantation of regions of the archenteric roof of the early *Triturus* neurula, (a$_1$–d$_1$) into early gastrulae; a$_2$ head with balancers; b$_2$ head with eyes and forebrain; c$_2$ posterior part of head with deuterencephalon and otic vesicles; d$_2$ trunk-tail induction. (After MANGOLD, 1933)

parts of the organizer whose effects are to be tested, are explanted into ectoderm alone, which still has no tendency to differentiate (except to generalized or atypical epidermis, p. 226). Around the explanted inducer is placed the largest possible piece of presumptive ectoderm from an early gastrula. The edges of the ectoderm are brought close together; they soon unite; and the explant is enclosed in an ectodermal cover. This "sandwich" experiment of Holtfreter even permits the study of combinations of explants. To begin with a simple example, one can ask which organs are built when the ectoderm encloses head inducer or trunk inducer. For head inducers pieces of dorsal lip were taken from young gastrulae (Fig. 320a, cf. Fig. 296); for trunk inducers pieces of the archenteric roof from the posterior trunk region of a young neurula (Fig. 320d) were used. The distinction between explant and ectoderm in the resultant tissues was made easily in many cases by the use of inducer and ectoderm from two different species, even from an anuran and a urodele. In xenoplastic combinations of *Triturus* and *Bombinator* the tissues derived from the respective embryonic components are distinguishable on the basis of differences in yolk content, pigmentation, and cell size.

246

The dorsal lip of the young gastrula evokes a neural plate in the surrounding ectoderm, and this neural plate sinks in and is overgrown. These complex explants are at first rounded—later they assume various shapes. Most swim around in the medium, driven and spun by their cilia. In *Bombinator* ectoderm two secreting adhesive glands arise (Fig. 320b), in *Triturus* ectoderm balancers arise in several places. Often one can cause muscle contractions by prodding the complex explant gently with a glass needle at the time when the control embryos are just beginning swimming movements. After about five days melanophores always appear, and these are characteristic

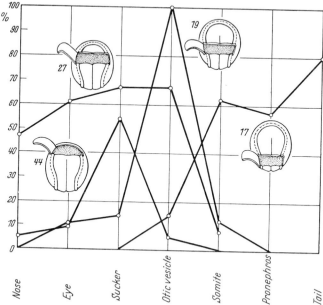

Fig. 319. Distribution of the frequencies of formation of organs after implantation of various regions of the archenteric roof into *Triturus* embryos. Numbers stand for cases of each operative procedure. (Graphical representation of data of MANGOLD, 1933)

of the ectoderm's species. Sectioning shows substantial head organ formation (Fig. 320c). From the material of the dorsal lip itself come fragments of pharynx, chorda, musculature, frequently even nervous tissue, and rarely cartilage. The ectoderm originating in a different amphibian order produces parts of the brain, eyes, olfactory pits, and sometimes even otic vesicles!

The posterior region of the archenteric roof differentiates into the ectodermal covering of the chorda and the musculature. The enveloping ectoderm, which is often arranged in packets like somites, forms an attenuated neural tube and frequently also contributes to musculature and chorda. These tissues grow, as a rule, into long tails edged with a fin (Fig. 320e, f). Often the ectoderm also forms a hindgut (Fig. 320f).

The two organizer regions tested thus work in explanted ectoderm as head and tail organizers respectively and evoke the specific regional organ complexes as in a host.

By what means does the process of induction take place? That it is a chemical mechanism is seen from the discovery that even dead tissues can evoke some of the same reactions brought about by living inducers. Indeed, these effects come not only from killed organizers, but can be brought about by pieces of entirely diverse dead organs of vertebrates, invertebrates, and even plants, placed under the gastrula ectoderm. This shows that pieces of certain organs are target-specific, that is, they evoke only the organs of a particular body region. In this way liver and

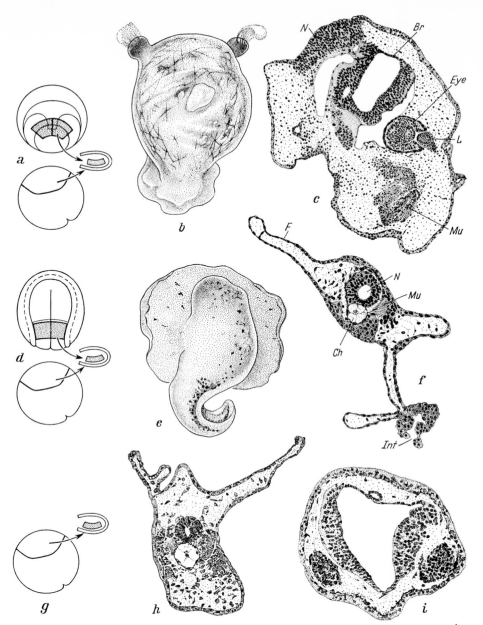

Fig. 320. Sandwich experiments: the inducing material to be covered is enclosed in presumptive ectoderm. a–f pieces of organizer; a, d operative procedures; b, c results after a; b *Triturus* head organizer in *Bombinator* ectoderm; c *Bombinator* head organizer in *Bombinator* ectoderm; e, f results after d: *Bombinator* trunk organizer in *Triturus* ectoderm; g–i abnormal inducers in *Triturus* ectoderm; h liver; i kidney; c, f, h, i sections. *Int* intestine; *Eye* eye cup; *Ch* chorda; *F* fin; *Br* brain; *L* lens; *Mu* muscle; *N* neural tube; *N* olfactory pit epithelium. (a–f after HOLTFRETER, 1936; h, i after CHUANG, 1938)

kidney from perch, snake, and guinea pig have been tested, as well as the kidney of a jay, among others[678]. Fresh living tissues were cut into small pieces and placed in 70 per cent alcohol for a time; then the alcohol was washed out in sterile Holtfreter's solution and the small pieces implanted into early gastrulae. These were transported to the ventral ectoderm by the invaginated endoderm (the same method used with living implants). Over these implanted pieces of dead tissue arose knobs of varying size, containing induced structures. One group of such abnormal inducers, perch, snake, and guinea pig liver, induced a high proportion of almost exclusively anterior head structures (archencephalic structures, p. 245): the first two parts of the brain (telencephalon and diencephalon), olfactory pits, eyes, and suckers. Implants of jay and perch kidney induced predominantly posterior head structures (deuterencephalic structures, p. 245): mesencephalon, rhombencephalon, otic vesicles, and also, in very small amounts, trunk and tail structures (spinal cord, muscle segments, chorda, tail fin). Guinea pig kidney induced mostly tail structures. The distribution of the effects of these abnormal inducers shows a clear similarity with the distinction in the effects between the anterior and posterior serial regions of the archenteric roof (Fig. 319).

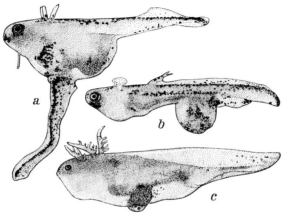

Fig. 321. Regional specific induction caused by application of various fractions of chick embryo extract to early gastrulae. a, b from *Triturus alpestris*; c from axolotl; a trunk-tail induction; b posterior head induction; c anterior head induction. (From MANGOLD, 1958 after the original from H. and H. TIEDEMANN)

Sandwich experiments with abnormal inducers (Fig. 320g) lead one step further with this surprising result. The abnormal inducer, which does not participate in the natural formation of the organs, nevertheless induces organ complexes which are completely typical in form. Fish kidney induces exclusively head organs in the surrounding *Triturus* ectoderm: always parts of the brain, and eyes in about half the cases; frequently olfactory pits and suckers are formed. Often the products are beautifully bilaterally symmetrical: Figure 320i shows in section a forebrain with an epiphysis and, laterally, two olfactory pits; the epidermis is typically formed and underlain by pigment cells. Fresh adult salamander liver has manifold effects; it induces trunk structures, muscle, pronephros, chorda, and well-defined tail (Fig. 320h), but also neural structures and sense organs. If the liver is cooked, however, the ability to induce trunk and tail structures disappears after only two seconds, while the ability to induce head organs remains even after fifteen minutes. From this we can conclude that there are various inducing factors and of these the trunk and tail (or spinal) factor is more easily destroyed by heat than the head factor.

Extracts from homogenized eight- to ten-day old chick embryos and from adult chicken liver have proven especially favorable starting material for the isolation and chemical characterization of the inducing substances[674, 677]. As fractional centrifugation shows, the activity appears in the microsomal fraction, which contains the ribosomes, of course. The ribosomes induce, directly or bound in agar, deuterencephalic and spinocaudal structures in young gastrulae. The supernatant, which after long centrifugation contains little nucleic acid, induces a high proportion of trunk and tail structures; the remaining ribosomal fraction induces posterior head structures strongly and anterior head structures weakly. Further purification has produced a predominantly mesoderm-inducing protein fraction. For the induction of a complete trunk and tail, a neural-inducing factor is needed in addition to the mesodermal factor. Further extraction releases an archencephalic-inducing substance which is more stable to hydrolysis and to heat than the other inducing factors. Figure 321 illustrates some region-specific induction. The mesodermal factor, which is easiest to purify, is a protein or polypeptide whose molecular weight is less than 50,000. A protein component is also important in the anterior head substance and in the posterior head substance, because these inducing abilities are destroyed by proteolytic enzymes (pepsin and trypsin); ribonuclease does not destroy them. The original association of the inducing substances with the ribosomes may well be due to the bound messenger RNA which determines the primary structure of the proteins in question.

The combination of different concentrations of specific abnormal inducers[589, 589a, 680] and of highly purified fractions [677] suggests that the region-specific induction depends on the collaboration of various inducing factors. Apparently the interaction of only two factors, a neuralizing and a mesodermalizing factor, may be involved, as shown in the following scheme[676]:

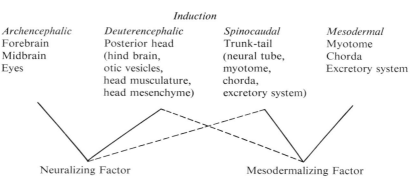

In Figure 322 the two factors are diagrammed as opposing gradients which spread from the dorsal side of the embryo.

As to the primary chemical mechanism of the inductive process we know nothing. The transmission of the inducing substances from the inducer into the other tissues has received little attention. Transplants into whose tissue radioactive amino acids have been incorporated indicate that not only these but the proteins or polypeptides containing them move from the implant into the responding tissue[631]. Direct cell contact is not necessary for the induction of neural differentiation in competent ectoderm, as induction through filters has shown[589a]. The role of the inducing substances contained in a great variety of tissues and liberated by the use of abnormal inductors is not entirely clear. Perhaps they share only certain chemical properties with the natural inducing substances and act as skeleton keys.

Leading closer to the nature of inductive processes is the question of the temporal sequence of neural induction. Presumptive brain, or parts of it, from gastrulae and neurulae of various ages were explanted into a piece of ectoderm; they were thus removed after exposures of various durations to the underlying archenteric roof and then kept for ten to fifteen days in Holtfreter's solution. Ectoderm of the presumptive brain region removed from the earliest gastrula (Fig. 261 b, c) forms lobes of atypical epithelium (as in Fig. 297). Similar isolates from mid-gastrulae up to the small yolk plug stage (Figs. 264c, 265b, 267b, c) also often yield only these undifferentiated epithelial lobes. A certain proportion of the explants do whow the first signs of histological differentiation, however. In the simplest case part of the epithelium forms a vesicle in which loose mesenchyme cells, compact aggregates of neural cells, and pigment cells are found. These minimal inductive effects correspond to structures which normally are formed from the neural crest (ganglia, "ectomesenchyme," pigment cells); but there also arise the first brain structures next to a lot of undifferentiated epidermis in explants from embryos with middlesized-to-small yolk plugs. These are not chaotic or un-specialized structures but resemble individual parts

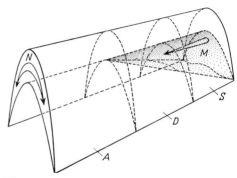

Fig. 322. Diagram of the neuralizing and meso-dermalizing induction gradients believed to operate during normal development. *A* arch-encephalic; *D* deuterencephalic; *S* spinal region. (After Toivonen, Saxen, and Vaino, 1963)

of the brain, even if in very primitive form. From here to the appearance of the neural groove, the frequency of distinct individual brain components in the explant rises sharply, together with the extent of morphogenesis and of histological differentiation. The forebrain and the hind brain are mapped out even after a very short contact with the archenteric roof. An unpaired eye rudi-ment arises even before the midbrain becomes distinct; and the eye, of all the brain components, approaches its final form most closely, with pigment epithelium, retina, and a rudimentary lens. The remaining parts of the midbrain lag behind; they still appear as irregular vesicles or thin-walled tubes when forebrain and hindbrain have proceeded some distance in morphogenesis.

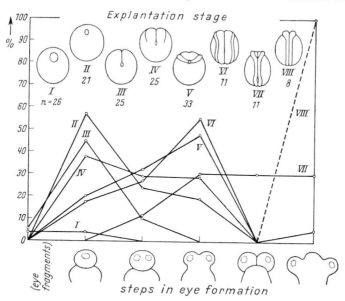

Fig. 323. Advancing determination of bilateral eyes during gastrulation and neurulation in explants of the presump-tive brain region with ectoderm, *Triturus alpestris*. Above, diagrams of stages in which the isolation was made. Abscissa, sequence of stages from fragments to complete cyclopia to complete separation of the eyes. (Graphical representation of the data of von Woellwarth, 1952)

During the neural groove stage and the first appearance of the neural plate the midbrain rudiments acquire the competence for normal morphogenesis in the explant. After the development of the neural ridges this process is essentially complete. Only the doubling of the forebrain and eye remains.

The eye rudiments provide a standard during the entire course of determination, for the determination of bilaterality (Fig. 323). While an eye rudiment generally appears even in explants from early gastrulae, it is almost always "cyclopic," a single optic vesicle (Fig. 324b). In early neurula stages also, cyclopic or partially fused eyes are still frequent. With the closure of the neural tube the invariable normal bilateral pairing of the eyes and of the forebrain appears abruptly.

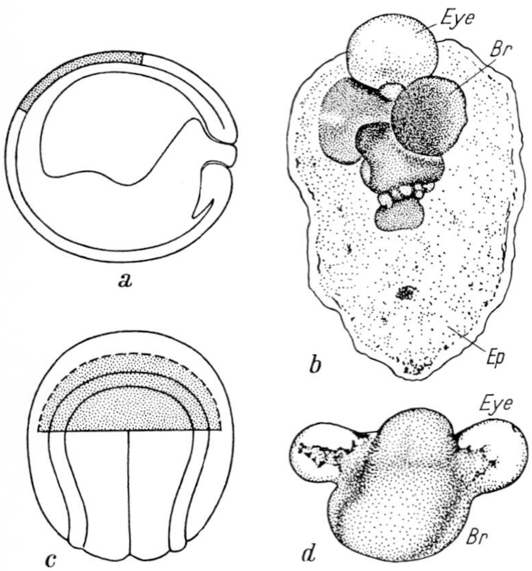

These same cyclopic or incompletely separated eyes are also seen when the prechordal plate (Fig. 327a, b) is removed from young neurulae[2].

The organization of the brain and the differentiation of the brain regions thus improve with longer exposure to the influence of the archenteric roof. The determination of this organization continues steadily from its early beginnings until the explant reaches a developmental stage corresponding to that of the brain of a normal embryo with pigmented eyes and simple limb buds.

At the beginning of induction the forming neural organ rudiments are very small, while a much greater part of the piece becomes undifferentiated epithelium than its prospective fate would predict. Even when the entire presumptive brain rudiment is isolated with plenty of presumptive epidermis, the eye, for example, is formed from only a small fraction of the cell number present in the optic vesicle of a normal embryo just when it is pinching off from the neural tube. The determined rudiment must therefore have been very small, even at the outset. As induction proceeds the size of the rudiment increases at the expense of its still undifferentiated vicinity;

Fig. 324. Autonomous differentiation in the brain; explantation of the anterior quarter of the presumptive archencephalon with a strip of presumptive epidermis. *Triturus alpestris*. a, c sketches of regions removed; a in small yolk plug stage; b from a, several brain complexes, an optic vesicle, manylayered epidermis with strong basal layer adhering to the glass substratum; c medullary plate stage; d from c, brain portion with paired abnormally formed eyes. *Eye* eyes; *Ep* epidermis; *Br* Brain.
(After Mangold and von Woellwarth, 1950)

the amount of material increases which is taken by a rudiment for its morphogenetic task. If an explant is used, the parts of whose midbrain have already acquired their outlines, there is often a disharmony in the size and the degree of differentiation of the individual parts, in which the anterior or posterior part of the brain (in the forebrain, the eye or the anteriormost part of the brain) may be relatively favored or restricted. Thus it appears that one part of an explant can outrace another and have relatively more cell material, as if there is competition among the rudiment regions in the explant.

The continuing determination of the organization of the brain is expressed also in defect experiments: in the donor embryos of the experiments just described, from which pieces of presumptive brain were removed, explanted, and tested. Young donors whose explant had not yet shown any neural differentiation developed into entirely normal larvae; the material defect was completely compensated. Older donors of ectoderm which in the explant formed plentiful neural tissue and an eye (Fig. 324b) developed a head shortened in front, with no anterior brain parts and no eyes. Thus, the regions of the individual parts of the brain are not completely determined even in the young neurula: unilateral defects can be regulated. Vital staining of the remaining regions of the neural plate demonstrate the movement of materials involved in the regulation. Material from the opposite side, but also from the brain regions on the same side in front of or behind the defect, can also replace the missing parts. A rudiment can thus organize a new symmetry and incorporate as yet undetermined material. Separation of left and right is

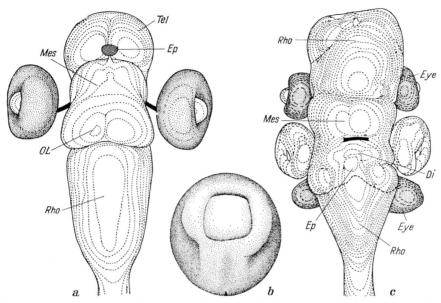

Fig. 325. *Rana esculenta*. a and c graphic reconstructions of brains from sections. a normal larvae; b neurula with rotated portion of neural plate two hours after the operation; c larvae after an operation as in b. *Eye* eyes from pieces of eye rudiment; *Di* diencephalon; *Ep* epiphysis; *OL* optic lobe; *Mes* mesencephalon; *Rho* rhombencephalon; *Tel* telencephalon. (After SPEMANN, 1912)

determined earliest in the neurula, in that part of the neural plate giving rise to paired organs like the eyes. If from a neurula with raised neural ridges the site of its presumptive eye rudiment (Fig. 339a) is cut out, the eye cup is missing on the operated side, and with the removal of a larger piece, the diencephalon also develops unilaterally (Fig. 342).

These results obtained through transplantation experiments show that after a certain stage parts are determined and can no longer be steered away from their normal developmental course by the activities of surrounding regions. While pieces from the presumptive ectoderm of an early gastrula, transplanted into the region of another presumptive germ layer, become identical with the host tissue, transplants of young neurula ectoderm self-differentiate: a piece of neural plate becomes a part of the nervous system, even in a strange place. If the presumptive eye rudiment of another embryo of the same age is transplanted into the as yet not visibly differentiated lateral ectoderm of the body, the implant is not incorporated into the epidermis but is overgrown and sinks below. There it develops further into a two-walled eye cup with retina and pigment epithelium. If a piece of neural plate is turned around, the parts of the brain developing from it self-differentiate in a now reversed cephalocaudal axis (Fig. 325b, c). If a piece of eye rudiment

253

is cut during such an experiment and placed in the back, it develops there into an eye cup with retina and pigment epithelium. The size of the anterior and posterior eyes varies with the mass of the respective portions of the eye rudiment.

The region-specific inductive effect of the archenteric roof raises questions about the origin and nature of patterning within the inducer. When does the organizer acquire its regions? At the beginning of gastrulation there is certainly no firmly established regional structure, as we can see from the explant and transplant experiments just described; but an invisible differentiation sets in with the involution over the dorsal lip (Fig. 261): here a sensitive period begins in which

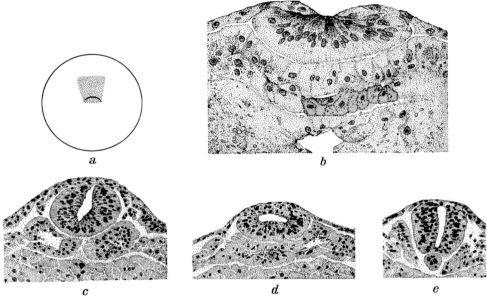

Fig. 326. Mesodermalization of the presumptive chorda material of *Triturus taeniatus* by LiCl. a vital staining before Li$^+$ treatment; b the vitally stained cells in the mesoderm, cross section through the posterior trunk region of the mid-neurula; c–e cross section through the center trunk region of older embryos; c and d from the same embryo; c somites separated medially, neural tube with sagittal cleft; d somites fused, their cavity persisting medially, neural tube with thick basal mass; e normal embryo. (After LEHMANN, 1937, 1938)

the marginal zone material responds to certain chemical influences, with changes in the later differentiation of the archenteric roof. When young gastrulae are treated with LiCl, the subsequent differentiation of chorda and somites in the archenteric roof can be prevented: the presumptive chorda cells become mesoderm. They do not die: cell material of the middle marginal zone, vitally stained before lithium treatment (Fig. 326a), appears later in the mesoderm (Fig. 326b); the somites can be squeezed together in the midline (Fig. 326c) or united (Fig. 326b, d). Anteriorly the prechordal plate, which lies beneath the anterior most region of the neural plate, the presumptive archencephalon (Fig. 327a, b), merges with the adjoining endoderm of the foregut. When the somites of both sides join, their centers of mass move medially, and the total mass of the somites is clearly reduced in comparison to that of the lateral plates. A quantitative reduction in the differentiation of the chorda mesoderm therefore results progressively in the dorso-ventral direction (p. 223).

The sensitivity of the organization material to LiCl was found in experiments on *Triton* to be phase-specific (though not corresponding to the order of invagination) as well as region-specific at the same time. The sequential gastrula stages were each placed for six hours in 0.3–0.5 per cent LiCl solution. Gastrulae whose blastopores had just become visible as little grooves did not respond. When treatment started with the stage in which the blastopore is a transverse groove (Figs. 263a, *1*; 265b), there were almost always defects in the archenteric roof beneath the posterior brain region and anterior trunk region, that is, in the anterior part of the chorda

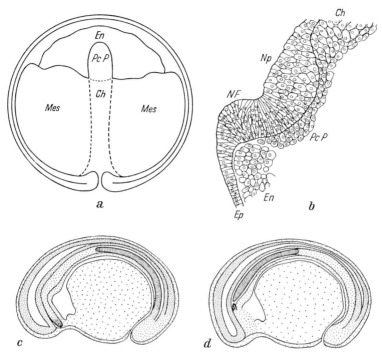

Fig. 327. a Diagram of the archenteric roof of the young neurula of *Triturus taeniatus*; b median section through the cranial part of a young neurula of *Xenopus laevis*; c, d diagrams of the phase-specific defect types after LiCl treatment of *Triturus taeniatus*; c after Li$^+$ treatment during the early gastrula stage; d during the middle gastrula stage; normal parts of the archenteric roof are dark. *Np* neural plate; *NF* neural fold; *Pc P* prechordal plate (as modified after VOGT, 1929; b after PASTEELS, 1949; c, d after LEHMANN, 1938)

(Figs. 327c; 328, *1*). With the further involution of the dorsal marginal zone, the sensitivity extends to the posterior part of the chorda region. When the blastopore is sickle- or U-shaped (Figs. 263a, *2*; 264b) the sensitivity of the posterior head region and anterior trunk region falls while increasing in the posterior trunk region (Fig. 328, II). At the same time the anterior head region, the prechordal plate, becomes more responsive; it reaches its maximum when the blasto-pore contracts around the yolk plug (Figs. 263a, *3, 4*; 265b; 328, III). In this stage the posterior end of the presumptive trunk chorda is sensitive. Now the scheme of defect localization is exactly the reverse of that at the beginning of the sensitive period (Fig. 327d). When treatment begins while the blastopore is a sagittal slit and the neural groove has appeared (Fig. 263a, *5*), defects are seen in the anterior head region while the chorda remains normal even in the posterior trunk region and only the tail is often defective (Fig. 328, *IV*). This distribution of sensitivity to lithium is another indication of the regional organization of the inducer.

The lithium-sensitivity of chordamesoderm differentiation in amphibians is remarkably re-miniscent of the Li$^+$ effect in echinoderms in that it too is enhanced by potassium ions and by

255

high temperature. Even more remarkably, there is also a correspondence in the antagonistic effects of NaSCN: If frog or axolotl blastulae are placed in a 0.5–1.0 per cent NaSCN solution for twelve to twenty-four hours, the chorda hypertrophies greatly at the expense of the somites (Fig. 329). This effect is increased by pyocyanin as in the echinoderms. Naturally, we cannot simply equate the mesodermalization of the chorda by Li$^+$ with Li$^+$ vegetalization in echinoderm embryos, nor can we equate the favoring of amphibian chorda by NaSCN with NaSCN animalization in echinoderms; but the striking parallel does suggest that in amphibian determination certain opposing metabolic processes are important which resemble the ones in the echinoderms that we have characterized on the basis of similar chemical sensitivity. Of course, beyond this, we know as little about one set as the other.

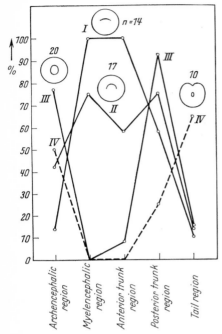

Fig. 328. Regional defects of the archenteric roof after Li$^+$ treatment in a sequence of stages of *Triturus taeniatus* (I–IV) from early gastrula to the early neurula (with neural groove. (Graphic representation from the data of LEHMANN, 1938)

certain opposing metabolic processes are important which resemble the ones in the echinoderms that we have characterized on the basis of similar chemical sensitivity. Of course, beyond this, we know as little about one set as the other.

The reduction of the chorda by Li$^+$ also provides an explanation for the influence of the substratum on the transverse organization of the neural tube. Even in the absence of the chorda, a normal neural tube can arise with a vertical central crevice if the somites remain apart (Fig. 326c), and the neural tube comes to lie between the lateral rows of muscle segments. If the mesoderm forms into a single plate, however, the neural tube forms a basal plate (Fig. 326d) rather than two thick lateral walls. The central canal is now dorsoventrally flattened and displaced upwards. These changes in neural tube formation are not the direct effect of Li$^+$; they can be produced just as well by the operative removal of median chorda mesoderm material (Fig. 330). The somite material thus orients the neural epithelial cells: these arrange themselves and stretch so that they are perpendicular to the somites.

The differentiation of the neural crest is also induced by the underlying tissue. In *Amblystoma* and *Triturus* the inductive effects of medial and lateral parts of the archenteric roof of the young neurula were tested in implantation experiments (Fig. 331). A characteristic distinction appeared: the medial parts induced neural tube and neural crest, while the lateral parts induced little but neural crest derivatives. This clear quantitative difference has as its simplest explanation a mediolateral gradient in the induction field: neural plate and neural crest are induced by a single substance whose concentration is maximal in the medial part of the archenteric roof and falls off ventrolaterally. The ectoderm would react differently to different concentrations of the substance. Thus we see an additional indication of a dorsoventral gradient in the chorda mesoderm.

The lengthwise differentiation of the neural crest depends upon the somites. At first the neural crest is a single mass of cells which extends down between the neural tube and the somites on each side (Fig. 316d, f). Later some of the cells move farther and form sympathetic ganglia, visceral skeleton in the head, pigment cells, and other things (p. 243). That part of the cell mass remaining next to the neural tube forms brain and spinal ganglia (Fig. 316e). If a row of somites is taken out, the neural crest in the somiteless region behaves quite irregularly (Fig. 332a). The organiza-

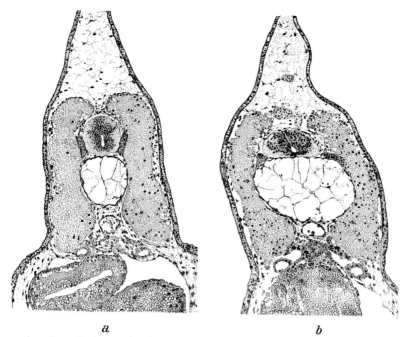

a　　　　　　　　　b

Fig. 329. Cross section through the trunk of two twelve-day-old larvae of *Rana esculenta*. a control; b treated 48 hours beginning in the blastula stage with a 0.5 per cent solution of NaSCN. (After RANZI and TAMINI, 1939)

tion of the dorsal mesoderm into somites begins anteriorly and proceeds to the rear of the embryo. When they are first formed the somites are short; later they grow in length. Thus one can replace a group of already elongated somites of the anterior head region by a greater number of short ones from the posterior region. The number of spinal ganglia formed now corresponds to the number of somites: one ganglion beside each somite (Fig. 332b).

Fig. 330. Cross section through the body of a *Triturus* embryo from which a piece from the middle of the dorsal lip was removed in the gastrula stage with a horseshoeshaped blastopore. (After LEH-MANN, 1928)

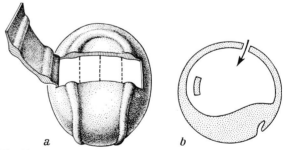

a　　　　　　　　　b

Fig. 331. Operative scheme for transplantation of medial and lateral pieces of the archenteric roof into young gastrulae. (After RAVEN and KLOOS from WOERDEMAN and RAVEN, 1946)

The dependence of the neural crest on the somites has also been demonstrated by an elegant xenoplastic transplantation experiment. A piece of presumptive somite material from an early *Bombinator* gastrula stained with Nile blue sulfate was transplanted into the corresponding region of a middle *Triturus* gastrula (Fig. 333a). During gastrulation the implant moved inside the host and later formed the trunk somites on the operated side. The somites retained their donor-specific properties, number, size, and developmental rate. The *Bombinator* somites in the transplant region were clearly shorter than the contralateral *Triturus* somites on the control side (Fig. 333b). In sections of experimental animals made seven days after the operation, the *Bombinator* somites showed further differentiation than the host somites (Fig. 333c). The *Bombinator* cells contained fewer yolk granules and already had clearly striated fibrils. On the operated side the number and position of the spinal ganglia corresponded to the foreign *Bombinator* somites, which also hastened the histological differentiation of the spinal ganglia formed from *Triturus* tissue on the operated side.

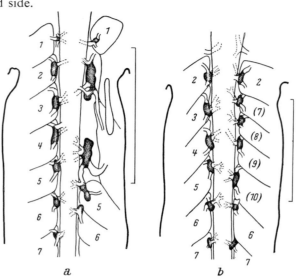

Fig. 332. Effect of the somites on the longitudinal organization of the neural crest in *Amblystoma*; spinal ganglia on the right side. a after removal of somites 2–4 in the tail bud stage; b after implantation of four smaller somites (7–10) in place of the larger ones (3–5). (After DETWILER, 1936)

These results illustrate the dependence of the determination pattern of the nervous system on local differences within the inducing chorda-somite system. But how far does this pattern of induction extend into the properties of the nervous system and of the other organs so induced? Parts of the archenteric roof, abnormal inducers, and fractions of inducing materials always evoke complex differentiation. The induced head is especially rich in its variety of organs. These emerge only in part from those parts of the brain which are always induced; others arise from the ectoderm under the influence of a part of the brain, which becomes a secondary inducer. In this way the olfactory pit is induced by the forebrain. The local and quantitative effect is clear in complex induction. Olfactory epithelium differentiates typically when the forebrain meets ectoderm. Frequently, instead of two olfactory pits, only one is formed (Fig. 334a), or else several, and at times whole fields of olfactory pits (Fig. 334b), according to whether the forebrain is symmetrically organized, or whether this part of the brain presents a small or abnormally large surface to the ectoderm. It is worth nothing that the ectoderm never responds with a giant olfactory field, but with a number of individual organ rudiments never exceeding a certain size.

The lens is induced in the epidermis by the optic cup. This secondary induction process has a number of noteworthy properties which will be discussed in the next lecture. Thus, the products of primary induction complete themselves through secondary induction.

The extensive and often astonishingly symmetrical histological differentiation of the brain evoked by abnormal inducers can only be autonomous organization on the part of the induced

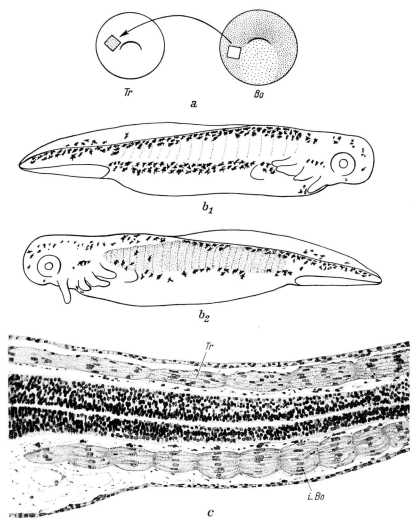

Fig. 333. Xenoplastic transplantation of presumptive somite material between *Triturus* (*Tr*) and *Bombinator* (*Bo*). a operative procedure; b_1, b_2 *Triturus* larva, seven days after the operation; b_1 control side; b_2 operated side; c frontal section of the larva in b; left *Bombinator* somites; right *Triturus* somites. (After CHEN, 1955)

tissue. So it becomes quite apparent that in the ectoderm only a small number of neural fields of self-organization are induced, and these primary neural fields then arrange themselves into a larger number of particular organ fields, which at first overlap or are separated by uncommitted zones for whose material they can compete (p. 252). This segregation of large fields into smaller ones is accompanied by a sharper delineation and the loss of regulatory ability. This restricts itself finally to regulation within an individual component field, e.g., the eye field.

The nature of such self-differentiation or self-organization of a neural field is still a mystery to us; it is certainly related to chemodifferentiation. In any event it is an autonomous process in the ectoderm, and in seeking an understanding of the events of determination we have had to look beyond the inducer, to the reacting system. An assumption in normogenesis is that the embryonic parts to which the inducers are brought by morphogenetic movements have the ability to respond, the competence to carry out the site-specific processes. This fit of the reaction system to an inducer arriving at an appointed time in development is fixed both spatially and temporally in the embryo: If a piece of presumptive ectoderm is explanted from a young gastrula, placed in Ringer's and implanted sooner or later in an inducing environment within an embryo, one finds that ectoderm of a certain age has lost its neural competence. The age in which complex neural differentiation can no longer be induced falls at the time of neurulation, when the determination of the neural fields is complete in a normal embryo. The appearance and disappearance of neural competence is thus age-dependent: it is a temporally determined process. The reactions of the ectoderm during the wearing off of neural competence also provide information on the nature of the inductive process: large pieces of presumptive ectoderm from beginning gastrulae were kept in Holtfreter's solution for varying lengths of time and then implanted into young

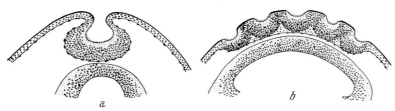

Fig. 334. Induction of olfactory sites in ectoderm by the underlying forebrain which had been induced by dorsal lip in a sandwich experiment. Response of the epithelium to enlargement of the secondary inducing contact surface. (After HOLTFRETER, 1936)

neurulae in place of head ectoderm, which included the presumptive archencephalon. Ectoderm kept as an explant until the time corresponding to the beginning of neurulation formed forebrain and associated sense organs under the influence of the prechordal plate. After this the neural competence of the ectoderm fell quantitatively, not qualitatively: the size of the brain parts and eyes formed lessened progressively. Infundibulum and epiphysis were reduced more sharply than forebrain and eyes; the latter were often perfectly developed. The reduction of the parts of the brain through the progressive loss of neural competence follows the same sequence as the reduction due to insufficiently long exposure to the inducer (p. 250f.).

As long as they retain neural competence, ectodermal explants can be induced to undergo epidermal or neural differentiation by a variety of nonspecific physical or chemical agents, such as nonphysiological salt solutions, hypotonicity, and diverse chemicals. Thus, a particular medium will cause the formation of epidermis at neutral pH, but neuralization under acid or basic conditions. In many cases it is a matter of agents which injure tissues to varying degrees so that part of the explant breaks down and then acts on the surviving part as a dead abnormal inducer. The neural structures formed are always of the forebrain type, rarely neural crest derivatives, and, remarkably, never spinal components. An entirely different influence leads similarly to the abnormal realization of ectodermal potency: centrifugation in the blastula and early gastrula stage can lead to neural and mesodermal differentiation of the ectoderm, so that parts of the spine, chorda, somites, and pronephros are produced. This response reaches its maximum in the young gastrula and drops off sharply at the time of the crescent-shaped blastopore. During this period the nature of the response changes qualitatively: after centrifugation of blastulae, neural and mesodermal structures arise together; in older stages mesodermal structures are not formed. Only neural structures and otic vesicles appear, never archencephalic components. The consequences of nonphysiological media and of centrifugation are thus quite diverse; the first resemble the effects of archencephalic inducers and the second spinal and to some extent deuterencephalic inducers (cf. Figs. 319, 321). In both cases particular competences were activated, in one case by a nonspecific trigger, and in another due to the disturbance of the cell structure and in turn of the physiological balance leading to an abnormal developmental pathway, for which the poetntial nevertheless existed.

Lecture 18

Segregation in the chordamesodermal blastema provides a lucid example of the self-differentiation of a morphogenetic field based on a gradient system. In the marginal zone of the anuran blastula explantation experiments show the "gradation" in the presumptive chorda region (Fig. 306c). In transplants of medial and lateral parts of the archenteric roof of young urodele

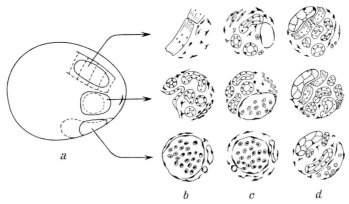

Fig. 335. Diagram of the products of various mesoderm regions of the young *Triturus pyrrhogaster* neurula in sandwich experiments: somites, pronephros, blood islands. a regions tested; b products in the embryo; c enclosed in ectoderm; d enclosed together with chorda in ectoderm. (After YAMADA, 1940)

neurulae (Fig. 331) both induction (p. 256) and self-differentiation reflect a gradient: the implants from the medial region become chorda and somites; those from the lateral region form no chorda; in contrast to the not-yet-involuted border zone (Fig. 306a), there is thus a progressive restriction of regulative ability in a dorsolateroventral arc. The displacement of the gradient by lithium is even more extreme in frog embryos than in the salamander experiments previously described: it leads (without phase and regional specificity) not only to a complete loss of the chorda, but also to a reduction of the somites, which are replaced anteriorly by pronephros. This supports the hypothesis that the qualitative differentiation series (chorda, somites, pronephros, lateral plates), which arise normally in the dorsoventral organization of the blastema (Figs. 294, 295), stems from a gradient of what Dalcq and Pasteels call "morphogenetic potential." Combined explants diagrammed in Figure 335 provide an example. From early neurulae, mesodermal fragments covered by ectoderm are isolated. In the embryo, somite muscle, pronephros, blood, and blood vessels would have come from this mesoderm (Fig. 335a, b; cf. Fig. 294). In the explants the grade of differentiation of the upper regions falls below their prospective fate: The presumptive somites produce pronephros, vessels, and blood instead of muscle; the presumptive pronephros region forms blood and vessels in addition to pronephros (Fig. 335c). But the morphogenetic potential can also be increased (Fig. 335c): When chorda is added to these explants, musculature appears not only in the somite material, but also in the presumptive pronephros

261

explant; and presumptive chorda causes the presumptive ventral mesoderm to form nephric tubules, again beyond its prospective fate. If the earlier experiments suggest a functional gradient within the blastema, the behavior of the isolated and combined explants points to the quantitative effect of a substance present only (or in the highest concentration) in the chorda, which is given off, and which influences the morphogenetic potential of other parts of the blastema. The under achievement of the isolated parts can be due to the dependence of the completion of their differ

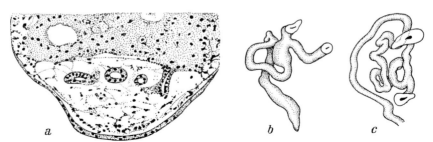

Fig. 336. Formation of pronephros from presumptive somite material of an early *Triturus* neurula after implantation in the belly of another embryo of similar age. a section through the implant region; b reconstruction of the newly formed pronepros from sections; c reconstruction of the right pronephros of the host for comparison (After YAMADA, 1938)

entiation on the maintenance of the gradient, or to the loss into the medium of a critical substance whose quantity determines the grade of differentiation. It is also possible that they lack some nonspecific, but important, influence present in the embryo.

The first or second of these possibilities is made more likely by the occurrence of a corresponding degradation in transplants. If presumptive somite material is transplanted at the end of gastrulation or at the appearance of the neural groove (Fig. 270) into the belly of another embryo

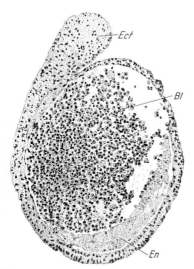

Fig. 337. Section through a ventral fragment of *Triturus alpestris* resulting from the frontal ligation of a blastula. *Bl* blood cells; *Ect* ectoderm; *En* endoderm. (After BRAUNS, 1940)

at the same stage, it forms typical pronephric tubules (Fig. 336a, b). Thus, the grade of differentiation is lowered one step, just as in the explant. And a further observation: in ventral pieces obtained by frontal ligation of a large- and small-celled blastula, a much greater proportion of the mesoderm cells become blood cells than in the normal embryo (Fig. 337). The rest of the mesoderm forms flat mesenchyme and endothelial cells. Thus, there is a degradation to the lowest step in the absence of the higher levels of the gradient in the mesoderm. Intermediate forms arise neither in the explant nor in the implant: diverse histological characters are never combined; rather the organic and histologic whole is completely unambiguous, even when it is incomplete.

We thus conclude that in the chordamesodermal blastema field the various morphogenetic reactions are induced as alternative modifications, according to a physiological or chemical gradient. For each reaction there must exist upper and lower threshold values. The reaction is always one of a set of alternatives.

While the dorsoventral organization is an autonomous process in the chorda mesoderm, the further differentiation of the lateral plate depends on its immediate environment, as is shown by a series of experiments on neurulae. In one experiment the embryo was cut in the ventral midline (Fig. 338a), and all the endoderm was removed. The ectoderm of the ventral mesoderm-free region (p. 208) is now folded in against the inner surface of the mesodermal mantle (Fig. 338b). In the flattened endoderm-free embryo, the lateral plate's inner layer (presumptive visceral layer, cf. Fig. 295), now covered by ectoderm, formed the same structures as the parietal layer, with mesenchyme and musculature; in many cases even an extra limb bud was formed inside the embryo. When, after the removal of the endoderm, the right and left lateral plates touched, the united inner layers produced only groups of degenerating epithelial cells. When the endoderm-free embryos were spread flat with their inner surface on a collodion membrane, the mesoderm attached. Ectoderm and mesoderm spread out on all sides. The inner layer of the lateral plate became a thin regular peritoneal epithelium on the collodion surface. In other experiments a part of the presumptive lateral plate material was inverted in the neurula stage, so that the original

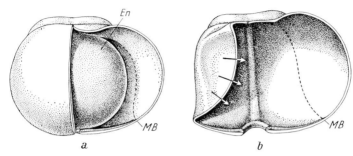

Fig. 338. Operative scheme for the testing of the determination of the lateral plates in the urodele embryo. a ectoderm cut away with the mesodermal coat and bent back on one side; b the endoderm is removed and the ectoderm is bent over the inner layer of the lateral plate. *En* endoderm; *MB* mesodermal border. (After NIEUWKOOP from WOERDEMAN and RAVEN, 1946)

inner surface lay next to the ectoderm and the original outer surface next to the endoderm. In all cases the layers developed site-specificity: the part in contact with the ectoderm became the parietal layer, and the part next to the endoderm the visceral layer. The determination of the two layers of the lateral plate thus is completed under the influence of the adjacent germ layers. And so the differentiation of mesoderm is an example of the self-differentiation of a blastema, followed by induction.

The eye is a remarkable theater of development, in which the determination of regional tendencies, field organization, induction, and competence work together in the construction of an organ, and in which a field capable of self-organization can arise anew from a part of an already differentiated organ. After the rise of the neural folds, the presumptive eye rudiments lie on both sides of the midline of the neural plate, as shown by vital staining. They are connected medially by the anlage of the optic chiasma. They occupy about a quarter to a third of the maximal width of the neural plate (Fig. 339a). In the early gastrula stage, the presumptive eye region lies on the anterior border of the presumptive neural plate and is still unpaired (Fig. 285). The presumptive retina, pigment epithelium, and optic stalk regions become delineated on the primary optic vesicle (Fig. 339b, c). The optic vesicle invaginates to form the optic cup, which is somewhat spoon-shaped at first. Ventrally it continues into the optic stalk, which becomes grooved (Fig. 339d). The edges of the optic cup close over this groove (the embryonic optic cleft) (Fig. 339e), and it becomes a tube through which the retinal artery runs. The stalk later serves as a channel for the optic nerve fibers growing out of the retina to the brain. The inner lamina of

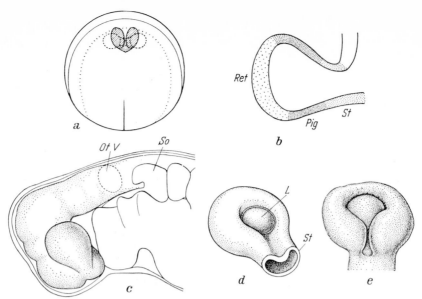

Fig. 339. Eye anlage. a presumptive eye region in the beginning and early urodele neurula; b primary optic vesicle, presumptive retina (*Ret*); pigment epithelium (*Pig*) and eye stalk (*St*) regions; c eversion of the primary optic vesicle from the midbrain anlage; d, e eye cup with lens; e beginning closure of the fetal optic cleft, diagrammatic. *Ot V* otic vesicle; *So* somite. (a, b from Mangold, 1931; c diagram after Spemann, 1903)

the optic cup is itself many-layered, but the outer lamina becomes a single layer of pigment epithelium (Fig. 341). The pigment layer and the inner layer grow by cell division over their entire extent at the beginning. Later this growth is progressively restricted to the enveloping edge of the optic cup. The inner layer soon pinches into a retinal part and an anterior thinner part from which the iris and ciliary body form (Fig. 341 d). Lens formation in the epidermis goes hand in hand with the development of the optic cup; only the basal layer of the epidermis is involved.

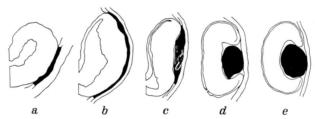

Fig. 340. Sections through eye anlagen of *Xenopus*, lens-forming cells black. (After Balinsky, 1952)

The optic vesicle touches the epidermis of the lateral wall of the head only in one small place at first, after which the area of contact spreads (Fig. 340). The number of epithelial cells in contact with the optic vesicle thus increases three or four fold before it invaginates to form the optic cup (Fig. 340 b). The cells of the lens-forming region become cylindrical and assume positions perpendicular to the contact surface (Fig. 341 a). Now, as the form of the optic cup is assumed, the morphogenetic movements of the lens-forming material begin: the cells concentrate and part from the outer layer of the epidermis. During this time there is neither immigration nor rapid mitosis to increase the cell number. Only when a lens vesicle has formed does vigorous mitosis

264

resume; the volume of the vesicle now increases. Its wall facing the retina forms lens fibers which are covered distally by the lens epithelium (Fig. 341c, d). The additional cells necessary for the continuing production of fibers during the growth of the lens arise by the mitotic division of the cells at the border of the lens epithelium.

In the neurula the eye rudiment's three major presumptive parts—retina, pigment epithelium, and eye stalk—already reveal a certain tendency for differentiation: In extirpation and transplantation experiments, retina without pigment epithelium and fragments of pigment epithelium without retina are occasionally found. But the determination is still labile. Fragments in a certain size range taken from the eye rudiment in the neural plate or from the primary optic vesicle are still capable of forming a reduced but well-proportioned eye. Thus, compensatory changes take place, in which some cells deviate from their prospective fates. Moreover, if in the open neurula

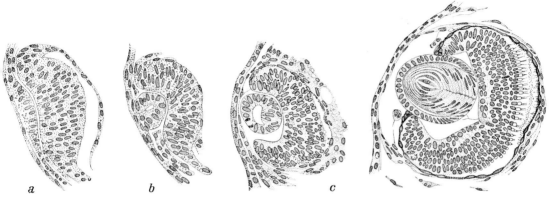

Fig. 341. Developmental stages of the eye of *Siredon pisciformis*. (After Rabl, 1898 from Spemann, 1936)

stage (Fig. 339a) a second eye rudiment is implanted behind that of the host, the two fuse completely. The vitally stained implanted material then forms the posterior half of the perfectly proportioned eye. Indeed, two primary optic vesicles can fuse similarly if their orientations are alike. The eye rudiment therefore displays the character of a morphogenetic field, rather than a mosaic, for quite a long time.

In spite of the ability to self-differentiate shown by transplants and explants, the organization of the optic vesicle can also be determined by its surroundings. At any time up to the early tail-bud stage, if the optic vesicle is turned to put the presumptive pigment epithelium on the outside and the presumptive retina inside, the layers of the optic cup differentiate according to their new positions, in contrast to their prospective fates. At a somewhat later stage, a similar inversion produces an inward-facing optic cup, but its pigment epithelium forms a secondary retina around itself where it touches the epidermis.

Lens formation is induced by the primary optic cup in most amphibians. This is shown most clearly by transplantation: if the optic vesicle is cut out and moved posteriorly under the epidermis, a lens is formed in numerous cases. If the optic cup anlage is removed in the neurula stage (Fig. 342a) the lens does not form in most amphibians, including *Triturus alpestris*. *Rana esculenta*, however, can form a lens where the optic vesicle would have reached the epidermis had it been present (Fig. 342b); the lens separates from the epidermis, sinks, and may acquire its typical structure. Many sorts of transplantation and extirpation experiments provide evidence that this apparently independent lens formation is not really an autonomous act of the presumptive lens epithelium, but rather that lens formation is mediated by other "regional factors,"

265

particularly from the anterior archenteric roof[427]. This lens formation in the absence of th
optic cup is temperature-dependent: it does not take place at high temperature (25°).

A further factor in the determination of the lens was discovered when the neural crest, fron
which the head mesenchyme is derived (p. 243), was removed together with the eye and brai
rudiments of the neurula. Free lenses now appeared in the eyeless heads, even in *Triturus a.
pestris*, presumably under the influence of the archenteric roof, and these lenses were ofte
paired and normal in structure, size, and position. The inducing effect of the archenteric roo
is therefore interfered with in the inter
vening head mesenchyme, or else th
effect decays spatially; the triggering o
lens formation is thus restricted by th
optic cup to a highly localized induction
Here also temperature is important. A
14°, 95 per cent of the animals in on
experimental group had either free lense
or at least lentoids, while at 18°, 64 pe
cent, and at 23° only 14 per cent ha
comparable structures[745].

Lens competence is not spread uni
formly over the epidermis in all form:
In some, any part of the epidermis ca
form a lens if it is placed over the opti
cup. In others only the presumptive hea
epidermis can, and epidermis from othe
parts of the body does not respond t

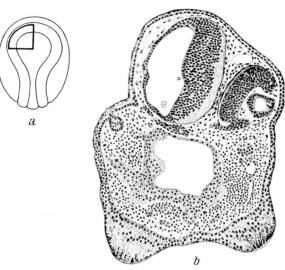

Fig. 342. Lens formation without the inductive influence of
the eye cup in *Rana esculenta*. a operative diagram; b cross
section. (After SPEMANN, 1912)

the influence of the optic cup. In sti
others lens competence is restricted to th
presumptive site of lens formation an
its immediate vicinity. Furthermore, i

the species with originally extensive lens competence, a progressive restriction takes place a
development proceeds, until lens formation can take place only at its presumptive site. This sit
thus evinces a quantitative character. The competence of the epidermis declines in a few specie:
as early as neurulation and disappears after the tailbud stage. This change in competence is a
temporally determined process in the ectoderm: it proceeds autonomously even in epidermi:
maintained as an explant for some time before implantation over an optic cup.

Lens induction depends in any event upon the release of a substance from the optic cup
Certain tissue extracts with archencephalic induction effects induce mostly independent lenses
some of which are far from the presumptive lens region, for example above the tip of the hos
liver. This suggests that "the archencephalic effect stems from a group of chemically relatec
substances, of which one functions especially as a lens inducer"[679, 48].

The size of the lens is influenced by the optic cup (Fig. 343a) and by other things as well
Even at a very early stage, when the lens has just separated from the epidermis, a fairly constant
size relationship between optic cup and lens emerges, which appears to remain fairly constant
during further development. The value of this "optic cup/lens index" varies within certain limit
(Fig. 343f). The mean value is in normal *Triturus taeniatus* eyes 17.5 ± 0.13 (two times standard
error). If the optic cup is reduced experimentally by the removal of a piece of the eye rudimen
(Fig. 343a–c) or increased by fusing with a second rudiment (Fig. 343d), the distributions and
mean values of the absolute size of the lens are displaced in the direction tending to restore this

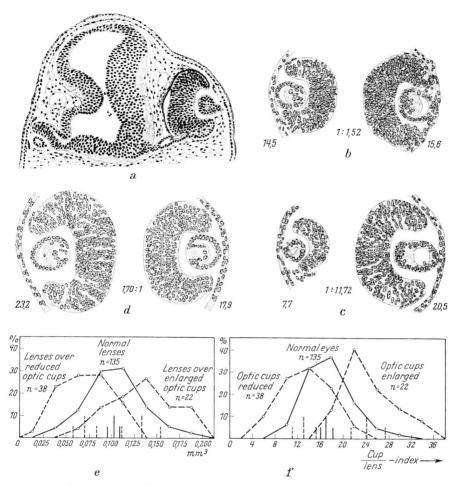

Fig. 343. Dependence of lens size on optic cup size. a *Bombinator*; b–f *Triturus taeniatus*; control eye always on right; a–c reduction of the eye anlage in the neurula (compare Fig. 339a); d enlargement of the eye anlage due to fusion with an implanted second eye anlage in the neurula. Numbers between the eyes: ratios of optic cup volume to lens volume; e variation in absolute lens size in normal eyes and over reduced and enlarged optic cups in the same stages; f distribution of optic cup-lens ratios. (a after SPEMANN, 1912; b–f after ROTMANN, 1942)

ratio (Fig. 343e). But this regulation does not generally achieve the normal size ratio between optic cup and lens: The index curve of the reduced eyes is displaced to the left (lower values) and that of increased eyes toward the right (Fig. 343f); that is, the lenses are still too large for the small eye cups and too small for the large ones. In individual cases, the ratio in a reduced eye can correspond to that in the normal eye on the other side (Fig. 343b) or it may have a much smaller value (Fig. 343c); similarly, the ratios in large eyes can run from normal to very great (Fig. 343d).

The relationship between induction and response can also be investigated in transplantations between species of different sizes, such as *Triturus cristatus* and *T. taeniatus* (Fig. 344a) or axolotl and *T. taeniatus*. The epidermis of the large *T. cristatus* is induced by the optic cup of the small *T. taeniatus* to form a lens vesicle which from the very beginning is larger than that of the host.

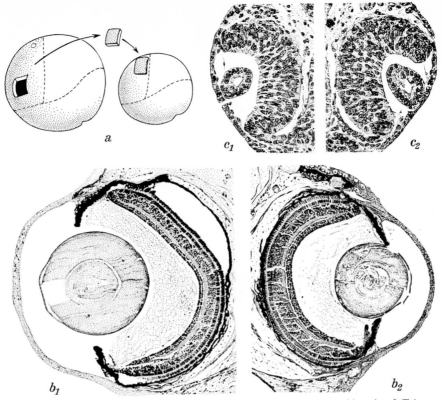

Fig. 344. Eye cup and lens size after transplantation. a operative procedure; epidermis of *Triturus cristatus* in *T. taeniatus*; b *Amblystoma* in *T. taeniatus*; c *T. taeniatus* in *cristatus*; b_1, c_1 eye with implanted lens; b_2, c_2 normal eye on the other side. (After ROTMANN, 1939, 1941)

The further growth of the implant's lens can be restricted, to be sure, so that it generally grows to a smaller final size than the lens remaining in the donor. The lens formed from the axolotl implant grows in *T. taeniatus*, in spite of its foreign environment, to its normal structure and dimensions, which far exceed those of the host lens (Fig. 344 b_1). Conversely, the *taeniatus* implant forms a lens typical of its species in spite of its induction by the large *cristatus* optic cup, for which the lens is thus far too small (Fig. 344 c_1). The epidermis responds to induction, therefore, with a species-specific norm of reaction.

The question now remains whether the epidermis responds to the inductive influence of the optic cup by mobilizing a specific number of cells or by activating a specific area. This question has been answered by a beautiful experiment. By drawing out the egg nucleus in *taeniatus* eggs before fertilization, haploid embryos were obtained, and haploid presumptive belly epidermis was used to replace presumptive lens epidermis in diploid embryos. Chromosomal counts and nuclear size measurements (Fig. 345 b) were taken to make certain that each lens was in fact made from haploid material. The haploid lenses induced by diploid optic cups (thirteen cases) were at the time of their separation from the epidermis in most cases as large as the host lenses or even somewhat larger (Fig. 345 a). The smaller cell size had therefore been compensated by a larger cell number. The haploid lenses contained on the average 69 per cent more cells than the diploid lenses at the time of separation from the epidermis. The fact that the haploid cell number

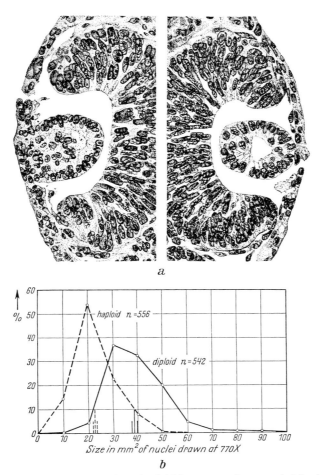

Fig. 345. Lens formation from haploid epidermis. a haploid lens somewhat retarded in development compared with the normal diploid (right); b variation in nuclear size in haploid and diploid lens cells of thirteen individuals. (After ROTMANN, 1940)

is not exactly twice the diploid cell number may be due to the somewhat slower development of the haploid tissue (Fig. 345a). Cell size therefore has no influence on the size of the lens; induction triggers the lens-forming mechanism in a particular epidermal area as a whole, and this area is species-specific.

The heteroplastic transplantation experiments also show that the induced tissue works back on the inducer. The too large or too small lens causes the optic cup to approach a better fit by accelerated or retarded growth. In Figure 344b the *T. taeniatus* eye cup enclosing an axolotl lens is much larger than the optic cup in the normal eye on the other side.

The specific morphogenetic tendencies of the optic cup and of the lens rudiment, namely, the inductive influence and the epidermal reaction, are normally in tune with one another. Moreover, the collaborative development is not a mosaic set of events synchronized at the outset, but rather continuously interacting components which actively maintain their harmony.

There is yet another opportunity for the regulation of the dimensions of the optic cup and lens: from a lateral half blastomere a normally-proportioned half-sized embryo arises. The half organizer regulates into a whole one of half size; its inductive effects and its derivatives are correspondingly reduced. Accordingly, the lens reaction of the epidermis should be reduced by half, not only at the usual site of lens formation, but also in those other places, according to the species, where a lens can be induced whose size is proportional to the optic cup. But this generalization has not yet been tested by actual measurements.

The extraordinary regulatory ability of the developing eye comes to light once more in regeneration. If the completed lens is removed, another will be formed from the dorsal margin of the iris (Fig. 346a–d). This part of the iris need not be in its normal place in order to form a lens.

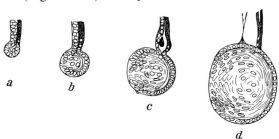

a *b*

c

d

Fig. 346. Regeneration of the lens from the dorsal border of the iris. (After WACHS, 1914)

It can form one even if it is moved into the vitreous humor after the removal of the lens. But if one places the lens and a piece of dorsal iris margin near one another in the posterior chamber of the eye, a second lens is not formed. Evidently the existing lens emits an inhibitory substance, since the inhibition does not depend on contact. The completion of lens regeneration depends also on an effect emanating from the retina: If after the removal of the lens the retina is walled off by a disc of polyethylene or cellophane (Fig. 347a), regeneration does not take place; exposure of just a small piece of retina leads to the formation of a small lens (Fig. 347b). If after a while the disc is removed (Fig. 347c_1), additional lenses form at the upper cut edge of the iris, and from the pigment epithelium. Sometimes these are of considerable size (Fig. 347c_2).

A piece of iris and retina of a young larva can also regenerate a whole eye if transplanted to a suitable place, such as the space left by the removal of a piece of the ear labyrinth of a larva of similar age. The cells of the implant multiply and, in favorable cases, form a perfect little optic cup, from whose iris margin a lens will arise (Fig. 348).

Regulation of the size of the whole eye with respect to that of the entire body takes place during growth. Extremely reduced eyes from pieces of eye rudiment grow fairly fast, so that they catch up with the normal eye on the other side in later larval stages. The corresponding size regulation also follows the fusion of two eye rudiments (p. 266): the eye which begins by being disproportionately large grows relatively slowly and thus arrives at normal size. In later stages the various developmental processes keep in step. If an optic cup and lens are transplanted in the tailbud stage from the fast-growing *Ambystoma tigrinum* into the slower-growing *A. punctatum* replacing the corresponding parts, they grow there and form an excessively large eye. This cause,

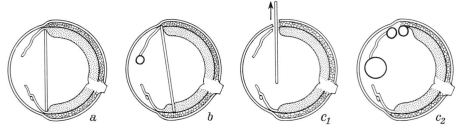

a *b* *c_1* *c_2*

Fig. 347. Test of the influence of the retina on lens regeneration in *Triturus viridescens*. a blockade of the retina; b a portion of the retina remains exposed; c_1 removal of the barrier; c_2 result of the renewed activity of the retina. (After STONE, 1958)

270

the eye muscle rudiments of the host, which can begin forming even without the optic cup, to adjust their size to that of the implant and differentiate more muscle fibers. Conversely, an unusually weak development of the eye muscles accompanies the formation of the abnormally small eye formed from a transplant from *A. punctatum* to *A. tigrinum*. Adjacent parts of the brain are also influenced, as are the orbital cartilages. Among the functionally related tissues of the head there thus exist forces which regulate their sizes. As a result, the visual apparatus finally formed is beautifully balanced[683,685].

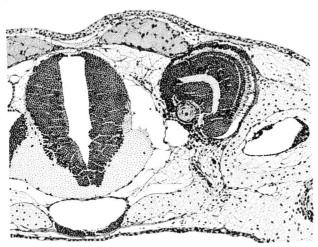

Fig. 348. Entire eye with lens formed from a piece of iris and retina transplanted into the labyrinth (*Triturus taeniatus*). (After WACHS, 1920)

So the development of the eye illustrates the collaboration of field organization, complementary induction, and quantitative effects on the growth of one part of an organ exercised by another part, and on the whole organ by the entire body—an impressive example of the morphogenetic principle of synergism, a profound and still mysterious feature of developmental physiology.

An entirely new morphogenetic principle in vertebrate development emerges from the careers of the cells of the neural crest. The morphogenetic movements during the formation of germ layers, organ rudiments, and organ components, which we consider collectively as morphochoresis (DALCQ) or topogenesis (LEHMANN), run their course in interdependent blastemas which become scattered during self-differentiation and induction. They pass through their particular morphogenetic careers in this interdependent period. The cell material of the neural crest, however, undergoes its variety of transformations in the subsequent period of dispersal. Its morphochoresis is the migration and isolation of cells, in which the cells realize their diverse competences in places far from their origin and in conjunction with cells of totally different fate and structure. The tendencies for differentiation of the migrating cells of the neural crest have been established by a series of systematic homo- and heterotransplantations. Presumptive head neural crest of the early and middle gastrula still possesses, as transplants and explants show, no single tendency, but develops according to the germ layer in which it comes to rest. During the course of gastrulation the competence for mesodermal and endodermal differentiation is restricted, and the tendency arises for the formation of neural structures, ganglion cells, ectomesenchyme (p. 243) and pigment cells. In the neurula the formation of cartilage and bone becomes the prominent tendency.

But the realization of one of these tendencies is not independent of the immediate environment. This is shown by the transplantation of neural crest material from the head into different body

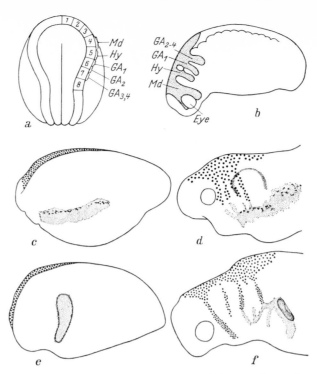

Fig. 349. Neural crest of *Amblystoma*. a prospective fate of regions of the neural crest determined by vital staining; b migration of the ectomesenchyme of the head region; c–f transplantation experiments; right neural crest stained with neutral red, left with Nile blue; c, d ventral transplant in normal orientation; e, f ventral transplant rotated 90°; GA_1–GA_2 gill arches; *Md* mandible; *Hy* hyoid arch. (After HÖRSTADIUS and SELLMANN from HÖRSTADIUS, 1950)

regions. Figure 349a and b shows the prospective fate of the regions of the neural fold and of the outward streams of neural crest cells which form the cartilage of the visceral skeleton. In the axolotl embryo, if gill bud and neural crest from the neural folds of the neurula are transplanted into the posterior trunk region, replacing the local neural crest, or into the side or belly, no cartilage will be formed; in the head, however, such a transplant always makes cartilage in place of the extirpated host neural crest. The neural crest cells in a strange place must either lack a stimulus for differentiation or else they must be inhibited. In order to distinguish between these possibilities, additional factors were introduced to the experiments, in which presumptive visceral skeletal material from the head was put in the place of trunk neural crest. If a piece of endoderm from the gill region was placed under the neural crest implant, it united with the gut endoderm and formed, due to its determination (p. 226ff.), secondary gill pouches in the trunk; and in 88 per cent of the cases (86 operations) cartilage formed in the tissue around the gill clefts. Mesoderm of the gill region also evoked cartilage formation. Evidently the competence for cartilage formation is actived by a cranial factor.

What determines the migratory path of the ectomesenchyme cells in the head? This question has been answered by a series of experiments. The left neural crest was cut out, stained with Nile blue, and implanted along the ventral surface of the head; in addition, the neural crest of the right side was stained with neutral red (Fig. 349c). When migration began, red cell material streamed over the dorsal midline and spread over the left side; and from the ventral implant the ectomesenchyme cells streamed dorsally (Fig. 349 d). If the left neural crest had been placed vertically behind the presumptive gill region (Fig. 349 e), the ectomesenchyme moved forward at first and then turned dorsal (Fig. 349 f). In both cases a typical gill skeleton was formed. Following the removal of neural crest from one side, so much cell material moved in from the

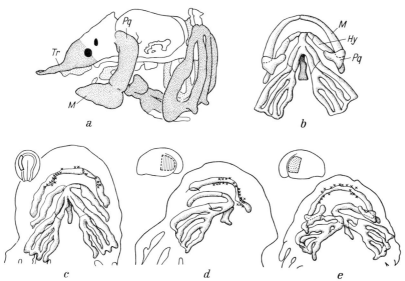

Fig. 350. Parts of the skull built from the ectomesenchyme of the neural crest in *Amblystoma*. a lateral view of the head skeleton, ectomesenchymal derivatives stippled; b dorsal view of the normal visceral skeleton; c visceral skeleton after removal of one side of the neural crest precursors of the visceral arches (compare Fig. 349a); d after unilateral removal of the pharyngeal endoderm; e after replacement of the branched mesoderm by lateral plate mesoderm on one side. XXX teeth. *Hy* hyoid arch; *M* mandibular arch; *Pq* palatoquadrate; *Tr* trabecula. (a after STONE, 1926; b–e after HÖRSTADIUS and SELLMANN, 1946 from HÖRSTADIUS, 1950)

opposite side that a complete gill skeleton could be made (Fig. 350c). Unilateral removal of the foregut wall leads to the complete absence of gill skeleton on the affected side (Fig. 350d), while replacement of the head mesoderm by lateral plate mesoderm of the trunk does not interfere with the normal migration of the neural crest cells and subsequent formation of the visceral arches (Fig. 350e). The direction of movement, the allocation of the wandering cells to the individual future gill skeletal elements, and the subsequent histological differentiation are therefore imposed on the neural crest cells by the endoderm of the pharynx, which is itself capable of autonomous differentiation (cf. Fig. 299h).

Extirpation and relocation experiments show that a regional restriction of competence already exists in the craniocaudal axis, which influences the wandering ectomesenchyme, and which corresponds in part to the regionally specific prospective fates (Fig. 349a). The presumptive material of the trabecula cannot participate in the formation of the visceral arches; mandibular ectomesenchyme cannot make trabeculae; it can make gill skeleton, but it cannot unite with the material of the basibranchial elements. For the latter, the hyoid and gill skeleton neural crest are competent; trunk neural crest cannot make cartilage.

As to the factors which govern the distribution of the wandering ectomesenchyme, we know nothing. It is surprising that half the neural crest can form an entire visceral skeleton. Not only cell movement, but also cell division evidently responds to local demands. In what manner are some of the cells held back for brain and spinal ganglion formation, while others go to ensheath

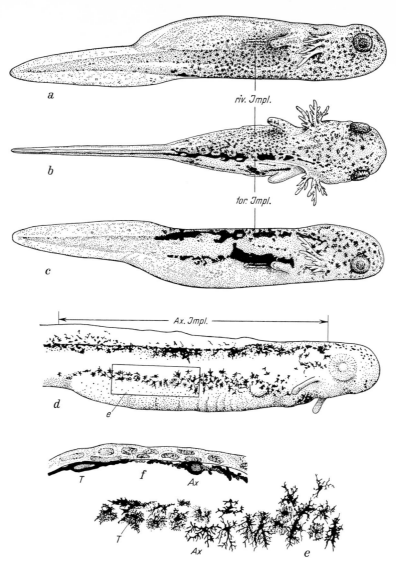

Fig. 351. Pigment pattern formation. a–c Three views of a larva of *Triturus rivularis* whose trunk neural crest (neural folds) have been removed and replaced by neural crest of *T. rivularis* on the left, and *T. torosus* on the right; d *T. palmatus* with neural crest from axolotl implanted on the right side of the trunk, *Triturus* and axolotl melanophores interspersed in the trunk; e region indicated in d greatly enlarged; f melanophores of *Triturus* (*T*) and axolotl (*Ax*) beneath the *Triturus* epidermis, section. *Ax. impl., riv. impl., tor. impl.*, implants of axolotl, *T. rivularis*, and *T. torosus*, respectively. (a–c after Twitty, 1945; d–f after Rosin, 1940)

growing nerves or travel even further to where sympathetic ganglia and adrenal medulla form, or spread far and wide as melanoblasts? Are the cells differently determined even in the neural crest or at the beginning of migration (Fig. 316) and are they then exposed to particular influences because of their specific positions? Or are they sent on in groups, still not determined, and used to fill successive local quotas? For example, when the staging area for the spinal ganglia is filled, the remainder of a group of cells might now move on and thus realize another particular competence under the influence of local conditions. We know very little about this, but the pigment cells do provide some insight.

The production of melanoblasts is not uniform throughout the neural crest; fewer come from the head than from the trunk. Accordingly the head gets some of its pigment cells from the trunk. In the skin some of the melanoblasts settle in the corium and some in the epidermis. Their route to the place where they finally become pigment is not known, nor even their main sites of proliferation. Transplants reveal the end of the trail, where the cells form pigment and thus become melanophores. The melanoblasts of some forms tend to wander more than those of others; this is shown in explants (tissue culture). The density of invading melanoblasts in a region where they

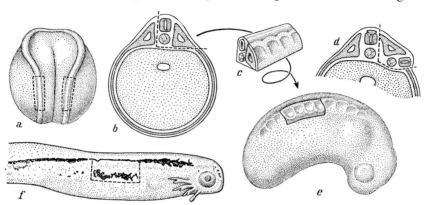

Fig. 352. Pigment pattern formation in *Triturus torosus*; after removal of the trunk neural crest material (a), and the closure of the neural tube in the absence of the neural folds (b). The mass of tissue enclosed by a dotted line in b and sketched in c was cut out, inverted dorsoventrally, and placed in another embryo where the somites had been removed (c–e); f distribution of the melanophores in the host. (After Twitty, 1945)

complete their histological differentiation depends both on the melanoblasts themselves and on the mesoderm, as is seen from a transplantation experiment previously described (Fig. 333). On the side with *Bombinator* somites the melanophores of the larva are relatively sparse. In the same way even foreign transplanted ectoderm can influence the density of the melanophores appearing within and below it.

But how is the pattern of distribution determined which is expressed in the pigmentation of the skin? In extensive transplantation experiments between anurans of various genera[5] (*Bombina variegata, Discoglossus pictus, Hyla arborea, Rana esculenta, Xenopus laevis*) the distribution of the foreign chromatophores generally follows the host pattern exactly; but the host's environment also plays a role, and melanoblasts which differentiate later can be inhibited by mature previous arrivals, and in this way the proportion of chromatophores in the pattern derived from one or the other species can vary. In transplants among various *Triturus* species the pattern is determined by the species-specific properties of the melanoblasts of the implanted neural crest. Here is one of many examples: *Triturus rivularis* embryos had their trunk neural crest cut out bilaterally; then in the ventral body wall, neural crest from *T. rivularis* was implanted on the left, and neural crest from *T. torosus* on the right. The melanoblasts migrated dorsally over the entire flank, and on the left side a *rivularis* pigment pattern appeared, while on the right the pigmentation was characteristic of *torosus* (Fig. 351 a–c). On the other hand, experiments with *Triturus* and *Amblystoma*, which are in different families, of course, show that the host species can play a decisive role in urodeles also. If axolotl neural crest is implanted into the body wall of *T. palmatus*, the diffuse axolotl pattern does not appear on this side; rather, the melanophores appear in the two zones characteristic for *T. palmatus*, where they form two more or less separate bands (Fig. 351 d). The axolotl melanophores also are organized in this pattern. They can be distinguished from the host melanophores by their size, form of branching, and color (Fig. 351 e, f). How are the zones determined in which the melanoblasts assemble and differentiate? For the longitudinal bands which are seen in many species an answer is given by an elegant experiment: after bilateral removal of the trunk neural crest (Fig. 352 a) and closure of the crestless neural tube, a block of tissue (the neural tube, chorda, and somites) was cut out, rotated 180° in the frontal plane and 90° in the transverse plane, and put in the place of the contralateral somites of another embryo (Fig. 352 d, e). Now the melanoblasts of the host neural crest invaded the implant and collected at the original dorsal margin of its somites (Fig. 352 f). Thus the level of the dorsal pigment band on the side of the *torosus* larva is determined by the somites. The parallel ventral band of the *palmatus* larva (Fig. 351 d) follows the yolk border.

In various species there are specific attractive or trapping zones in which the wandering melanoblasts collect and differentiate. When the melanoblasts alone determine the pattern (Fig. 351 a–c), their sensitivities and/or responses to local stimuli must vary among the species. But the *Triturus-Amblystoma* experiment showed that the melanoblasts of one species can respond to the localizing influences of another species. Color patterns thus reflect both the responses of invading cells and a "prepattern" of local stimuli.

Lecture 19

The neural blastema is induced to separate from the ectoderm; the self-organization of its archencephalic region leads to the segregation of the optic blastema, which acquires the character of a field again and lays out the basic plan of the optic cup. The optic cup, guided continuously by the complementary induction of the epidermis, organizes the lens. The mature nervous system and the mature eye include, in addition to various cell types of ectodermal origin, tissue from the mesoderm. This is incorporated during organogenesis; but the structural layout is determined by the ectodermal blastema alone. With endodermal organs, however, the relationship is different. The determination of the histological differentiation (histogenetic determination) of the presumptive endoderm is completed in the early gastrula (p. 230), but the normal form of the organ cannot be determined by the groups of endoderm cells alone. If presumptive endoderm cells from the vegetal floor of a blastula or young gastrula are isolated in Holtfreter's solution, the shapeless mass forms a smooth, solid ball within an hour. When such endodermal spheres contact one another, they fuse. During this fusion the endodermal cells exhibit gliding and rolling amoeboid movements. Some twenty hours after isolation the behavior of the endoderm cells changes: the entire cell mass flattens, and its margins spread in all directions over the substratum like a membrane. At the periphery cells may come loose and move a short distance further. The motility of the cells is now different. The cells are flattened and push along with more or less pointed pseudopodia. This change in behavior is autonomous and comes with the passing of time. The tendency for flattening and spreading begins at the developmental stage in which throughout the embryo endoderm spreads along the mesodermal mantle until its edges meet dorsally (Figs. 270 b, 298 c). In the embryo the spatial arrangement of the flat endodermal layer leads to the formation of the tubular gut. If pure ectodermal explants are maintained for fifteen to twenty days longer, the membrane spread upon the glass differentiates into a single layer of epithelium characteristic of the particular determined region (Fig. 301 a, b), while the central mass generally remains compact but also acquires the cell structure of the gut.

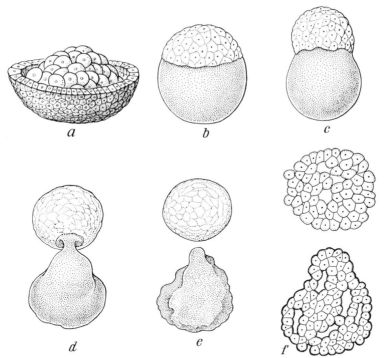

Fig. 353. Light endoderm and dark ectoderm cells from an axolotl blastula, when mixed together, separate after an abortive union. (After HOLTFRETER, 1939)

If endoderm is isolated with ectoderm, the endodermal cells at first lie in irregular heaps on the piece of ectoderm, which quickly becomes cup-shaped (Fig. 353a) and more or less encloses the endoderm (Fig. 353b). After a day and a half the form of the explant begins to change. The border between the two tissues constricts, and the endoderm is gradually squeezed out of the ectodermal sac. The connection between the two becomes tenuous, and finally the endoderm and ectoderm separate completely (Fig. 353b–e), each behaving from then on as previously separated explants (p. 228ff.). This separation is effected by an autonomous sorting out of the two cell types, beginning at a certain stage in which the as yet undifferentiated individual cells are moving amoebically in groups very much like the *Dictyostelium* cells in a pseudoplasmodium (p. 126).

In explants made up of endoderm, ectoderm, and mesoderm the germ layers behave quite differently. Such combined explants are obtained most simply by cutting pieces out of the gastrula (cf. Fig. 269) or neurula (Fig. 298c) which contain ectoderm, endoderm, and lateral mesoderm. The complex cell mass quickly rounds up once more into a ball (Fig. 354a, d), in which the mesoderm this time lies between endoderm and ectoderm. When the ectodermal mass begins to enclose the endoderm, a constriction at the endoectodermal border again forms, but in this case it does not lead to a separation. The endoderm and ectoderm do not remain solid masses; rather a fluid-filled vesicle arises into which the mesoderm moves as a loose mesenchymal tissue (Fig. 354b). The endoderm and ectoderm become the epithelial walls of the vesicle. The endoderm forms a single layer, the ectoderm mainly two layers (Fig. 354c). The two epithelia are joined at a circular border. As the mesenchyme spreads over the inner surface of the epithelium, a new reaction is induced in the endoderm and ectoderm, leading to an integrated structure. While endoderm and ectoderm do not last very long by themselves, these complex vesicles can be kept

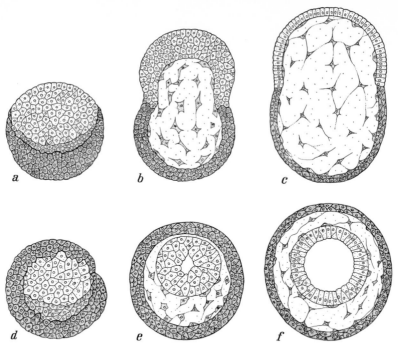

Fig. 354. Combined explant of endoderm, mesoderm, and ectoderm. a–c the ectoderm does not surround the entire cell complex completely; d–f the entire explant is surrounded by ectoderm. (After HOLTFRETER, 1939)

alive for several weeks in physiological salt solutions without added nutrients. The mesenchyme has acted here as connective tissue in the true sense by holding together two kinds of cells which originally were inclined to separate. At the same time it has a polarizing influence on the organization of the epithelial cells; for while these acquire no order when isolated, the endoderm cells now are arrayed as a single layer with the epithelial upper surface away from the connective tissue, and the ectodermal cells are arranged into a two-layered epidermis. From the mesenchyme smooth muscle cells can differentiate which cause rhythmic peristaltic contractions of the vesicle wall.

When plenty of ectoderm is present, it surrounds the endoderm and mesoderm completely (Fig. 354d). The ectoderm and endoderm now separate, and a space appears between them which expands and becomes dotted with mesenchyme. The ectoderm forms a two-layered wall, and the originally compact endoderm forms a cavity that generally expands due to secretory activity until a single layer of endothelium results whose surface faces inward (Fig. 354e, f). Now the separation of endoderm and ectoderm, their connection by mesenchyme, and the orientation of the epithelium by the mesenchyme have led to a structure covered with (histologically atypical) epidermis. It contains an endodermal vesicle inside, whose cells differentiate according to their normal fate. The vesicle is surrounded by a layer of mesenchyme and is sometimes covered by smooth muscle cells. The arrangement of the three germ layers now closely resembles a cross section of a normal gut.

The morphogenetic movements in the three types of compound explants just described depend on the relationships among the combined parts of the embryonic blastema, particularly their tendencies to unite or separate — what HOLTFRETER called "affinities." These are negative between endoderm and ectoderm, and positive between each of these and mesenchyme. In addition, the mesenchyme induces the polarization of endoderm and ectoderm and causes the uptake of

278

fluid from the external medium and its release into the connective tissue space. Conversely, the epithelia induce the orientation and differentiation of the mesenchyme cells adjacent to them. Here again we find the synergistic principle of development in this model experiment for the formation of hollow endodermal organs.

The formation of the major parts of the body (head, trunk, tail, and limbs) involves not only particular germ layer derivatives and tissue types, but a variety of organ systems as well. Their inner detail and outer form are brought to a functional unity, which itself is a part of the whole body.

Certain groups of organs show a remarkable competence for self-differentiation, as we have seen already in ectodermless exogastrulae (p. 226). If from a middle to late neurula (Fig. 355a) the neural groove is removed, together with its epidermal covering and the underlying

Fig. 355. Formation of an isolated axial system in *Triturus*. a late neurula, isolated region stippled; b product of the isolate. (After MANGOLD, 1955)

chorda and somites, this explant regularly develops into a structure with only dorsal organ regions (Fig. 355b). The brain is divided into its normal parts, and eyes are present. The trunk and tail regions develop good spinal cord, notochord, and somites. Pigment cells are plentiful in the skin. In another experiment embryos were set up with neither nervous system nor endoderm. From an early neurula the neural plate and its substratum were cut out, and the endoderm was removed; the presumptive epidermis and the lateral plate mesoderm continued to develop. The epidermis folded together and healed, giving rise to an irregular vesicle, which often was vaguely organized into anterior head, posterior head, and trunk. The epidermis has a tendency to stretch, and this can lead to folding. It is induced to form secretory cells by the mesoderm. There are no pigment cells, since they are derived from the neural crest (p. 243), which has been removed. In the anterior head region numerous balancers arise. Sections show a surprising diversity of organ structure, which is scattered chaotically, however, due to the absence of the gut and axial structures (Fig. 356). Heart fragments contract rhythmically; in the trunk are pronephric complexes with pronephric tubules, as well as muscle

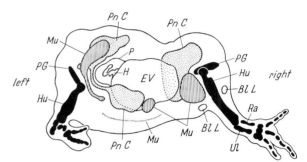

Fig. 356. Structures formed from the epidermis and mesodermal layer of *Triturus*, diagrammatic reconstruction of the more important organs from cross sections. *EV* inner epidermal vesicle; *BlL* blood lacunae; *H* heart fragment; *Hu* humerus; *Mu* muscle groups; *PnC* pronephric complex; *P* pericardium; *Ra* radius; *PG* pectoral girdle; *Ul* ulna. (After MANGOLD, 1961)

bundles, mesenchyme, blood sinuses, and frequently germinal tissue and an anus with a rectum. Forelimbs may develop as mere stumps, or they may develop so well that each of their skeletal regions can be recognized as left or right. Without the help of nerve or somite, those structures form that normally result from the activities of the mesenchyme and epidermis.

We shall now consider the experimental analysis of the development of the extremities in a little more detail.

The limb bud arises as a local thickening of the epidermis, together with mesenchyme derived from the parietal layer of the lateral plate (Fig. 357). In this flat rudiment three axes can be

279

detected which are all parallel to the corresponding main body axes: (1) the proximodistal or mediolateral axis (ML); (2) the anteroposterior axis (AP); and (3) the dorsoventral axis (DV) (Figs. 357, 358). These correspond to the later axial positions of the mature asymmetric limb: proximodistal (long axis of the limb), radioulnar (tibiofibular), and extensor side-flexor side.

The two forelimbs are mirror images, as are the two hindlimbs. Transplantation experiments tell us when the axes and the distinction between fore- and hindlimb are determined.

The determination of a limb field, and even of the properties of fore- or hindlimbs, takes place quite early, long before the appearance of the limb bud: if at the end of gastrulation a plate of ectoderm and mesoderm from the presumptive forelimb region of a urodele embryo is transplanted to another place, it can develop into a front leg, even at the site of a hind leg.

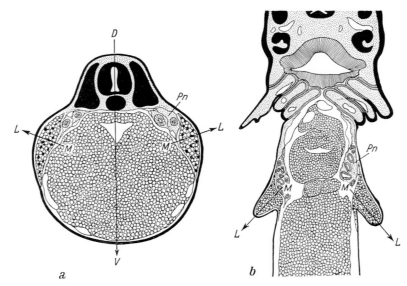

Fig. 357. *Amblystoma* embryo. a cross section at the level of the forelimb anlage; b frontal section through an older embryo whose limb buds are growing out. *D–V* dorsal ventral axis; *M–L* mediolateral axis (long axis) of the limb; *Pn* pronephros. (After HARRISON, 1925)

The axes, however, are not all fixed in the field at the same time. Ipsilateral transplantation of the limb rudiments to another embryo does not alter the axes (Fig. 358e), and the implanted limb develops like the host limb on the same side (Fig. 358c). If the flat anlage is rotated 180° and transplanted to the same side (Fig. 358f) or transplanted contralaterally without rotation (Fig. 358g), the AP axis is now backwards, and the limb forms as a mirror image of a normal limb (Fig. 358d). If the anlage is transplanted contralaterally *and* rotated 180°, the AP axis is now oriented normally (Fig. 358h). The DV axis, however, is now upsidedown (Fig. 358f, h). The AP axis is already determined in the neurula. It is apparently directed by the long axis of the body, but the inversion of the DV axis in a transplant will be corrected until the time of the beginning of limb bud growth. Thus, the transplant in Figure 358h gives rise to a normal limb. Surprisingly, the proximodistal axis remains plastic longest of all in the amphibians; the mesenchyme of the young limb bud of a salamander embryo can be turned around after the ectoderm is slit open, so that now its originally proximal side faces outward. Nevertheless, a normal limb is formed. In the hindlimb bud of a frog embryo the proximodistal axis is not completely determined even in buds 1–1.5 mm. long: an implant with this axis inverted still develops into a normal

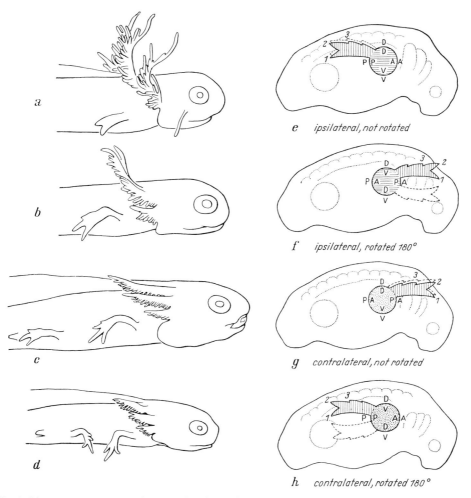

Fig. 358. *Amblystoma punctatum*. a, b normal embryos formation of the first toes of the front leg; c, d front leg transplant; c same side; d same side rotated 180°; b–h schemes of various transplants. *A–P* anterior-posterior axis; *D–V* dorsoventral axis; one to three toes. (After HARRISON, 1918, 1921)

leg. Thus, material that ordinarily would have formed foot is still capable of forming thigh instead. Somewhat later implants inverted in this way develop into stumps in which the shank forms nearer the body and the thigh at the free end.

The field nature of the young limb rudiment is also shown by the fact that normal limbs are formed from split anlagen, no matter how they are cut. Similarly, two whole rudiments experimentally united with a common AP axis fuse and can form a single well-proportioned limb.

The form-determining factors evidently lie in the mesenchymal blastema of the limb rudiment: its transplantation alone under the skin in another region can cause the formation of a typical limb. Conversely, transplantation of ectoderm from elsewhere to the presumptive limb site does not interfere with its development. Heterotransplants in early embryonic stages behave similarly. A piece of presumptive mesoderm from the border zone is removed with the region of the presumptive forelimb from a young *Triturus cristatus* gastrula (Fig. 359a) and transplanted to the corresponding place, but contralaterally, in a *taeniatus* gastrula of the same age (Fig. 359b, c).

281

During gastrulation this mesodermal implant moves under the ectodermal region of the presumptive forelimb of the host and provides the mesenchymal blastema of the limb bud. In final form the limbs of the two species differ in the length of the arm, its position, the size of the fingers, and the temporal sequence of their appearance. In the experimental animals the limb, a chimera of *cristatus* mesenchyme and *taeniatus* epidermis, has the form of the donor of the mesoderm (Fig. 359d_1–d_3). Conversely, if the presumptive forelimb epidermis of a *taeniatus* embryo is replaced by *cristatus* ectoderm, no trace of foreign influence can be seen in the limb which forms (Fig. 359c).

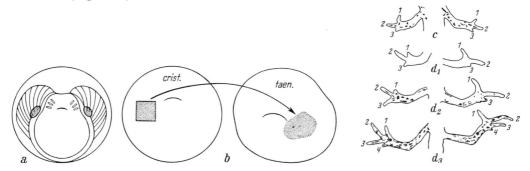

Fig. 359. Dependence of the form of the limbs on the mesenchymal blastema. a Position of the presumptive mesoderm in the front leg; b operative procedure; c *T. taeniatus* with *cristatus* ectoderm on the right leg; d_1–d_3 *T. taeniatus* with *cristatus* mesenchyme on the right leg. (After ROTMANN, 1931)

The self-organization of the anlage whose axial properties have been determined is still a mystery to us, as is the autonomous organization of every other morphogenetic field. The most we can do is follow the course of skeleton formation; Figure 360 illustrates an example, the salamander limb. The mesenchyme cells which provide the material for the cartilage come entirely or predominantly from the dividing cells of the originally cylindrical limb bud rudiment (Figs. 357a, 360a). No migration worthy of the name is seen on account of the extraordinary paucity of cells in the vicinity. In the mesenchymal blastema of the young bud, the cells are packed extremely closely (Fig. 360a). Later they gradually loosen, evidently as a result of the accumulation of liquid between the cells. In the cell mass, originally rather uniform and pervaded by capillaries, local condensations now appear, first the humerus anlage (Fig. 360b). In the earliest anlage of the smallest skeletal components, the phalanges, a single column of cells arises (Fig. 360c). Then the mesenchyme cells group into a central mass and a peripheral remainder. In cross section the cells are concentrically arranged (Fig. 360d). Only the clumps lying near the axis have an approximately circular border, and around it lie various arcs of cells. At the onset of anlage growth cell division occurs predominantly in the axial region, and the cells are forced outward on all sides. Later the peripheral cells also begin dividing, and the concentric arrangement disappears. Division now resumes in the axis and the cells are once more pushed outward. Thus, there is a series of waves of cell division which start centrally and move out. The relatively thin cartilage rudiments of the long bones of amphibian larvae contain only a few cell layers (Fig. 360e). With the onset of cartilage formation the individual elements are laid out.

In bird embryos, which we shall consider for comparison, transplantation and explantation experiments reveal a very early determination of the morphogenesis of individual skeletal elements in the mesenchymal anlage. If wing- or leg buds are transplnated into a nutritive but morphogenetically indifferent environment, such as the chorioallantoic membrane or the coelom of a host embryo, they can self-differentiate completely (Fig. 361e). Even when the wing bud has just begun to protrude, it can organize itself autonomously. The prospective fates of various meso-dermal regions have been marked with carbon. The first bulge already contains abundant material destined to form the proximal parts (scapula, coracoid, humerus) (Fig. 361 a). The rudiments of the successive distal parts arise through growth and loosening of the tightly packed mesenchyme beneath the ectodermal cap of the bud (Fig. 361 b–d). If a piece of bud is placed in the coelom of a host embryo, its differentiation corresponds to its fate map; in other words, it is mosaically

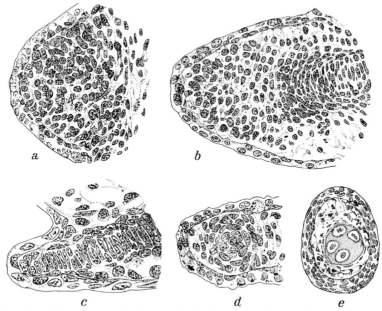

Fig. 360. Formation of cartilage anlagen in the urodele limb. a mesenchymal blastema in the young limb bud; b condensation of the humerus anlage; c–e young and older phalanx anlagen; a–c longitudinal section; d–e cross section. (After ANIKIN, 1929)

determined a good deal earlier than the amphibian limb (Fig. 361 e). The ectoderm does not influence the details of morphogenesis, but it is nevertheless important for the early development of the mesenchymal component of the limb: If the thickened epidermal cap is removed from an early wing bud (Fig. 361 a–d), the distal parts do not differentiate, and only the base achieves its prospective fate. If the epidermis is removed not from the tip but rather from either the anterior or posterior side, only those distal parts which are covered by epidermis continue to develop (Fig. 361 f, g).

If the leg bud of a four-day-old chick embryo is cut into transverse pieces, which are then placed on the allantois or in the coelom of another embryo, the proximal end forms a femur, the next part tibia and fibula, and the distal fragment a typical foot. Even the articulating surfaces are almost normal, indicating their independence of their environment.

Limb buds normally receive their musculature from outgrowths of certain somites. After transplantation they can acquire the necessary material from a different part of the body, even where there is no ventral extension of the myotome at hand. Amphibian limb buds transplanted in the tail bud stage (Fig. 358 e–h) develop in various parts of the body with more or less complete musculature. Defect experiments lead to comparable findings. If the second to fifth somites are removed completely, the dorsal and ventral body musculature will be incomplete; the forelimbs

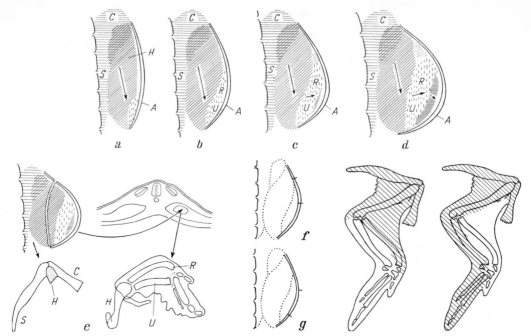

Fig. 361. Development of the wing bud of the chick. a–d presumptive regions of the mesoderm in the young wing bud; the arrows indicate the future long axis (proximodistal axis); e wing structures which form from the tip of the wing bud in the coelomic implant and the skeletal elements formed from the stump which remains; f, g effect of the removal of the epithelial cap from the wing anlage; f from the front half; g from the rear half; the parts of the wing which develop after the operation are hatched. *A* apical ectoderm; *C* coracoid; *H* humerus; *R* radius; *S* scapula; *U* ulna. (After SAUNDERS, 1948)

can develop as usual, however, which means they can draw their muscle material from the lateral plate mesoderm.

The nerves of the limbs are not necessary for the morphogenesis and histological differentiation of the skeleton and its muscles. Denervated limbs are retarded, however, ostensibly due to the lack of a trophic factor from the nerves; and the muscles atrophy at the time they normally become functional. The innervation of the limb poses an important question for the understanding of the finishing of a major part of the body: How do the motor and sensory nerves growing out from the neural tube and spinal ganglia find their way to the right places in the growing limb, even when the limb is abnormally situated? No satisfactory answer has yet been found.

Since the form of the limb is fashioned by the organization of the mesenchyme, and since the special epidermal derivatives, such as the glands in the thumb pads of male amphibians and plumage in birds, are locally controlled, the typical morphogenesis of the entire organ complex of the limb is governed by its own blastema.

The factors involved in limb determination in the amphibian embryo are not restricted to the region where the bud forms. If the growing limb is cut out and the wound permitted to close, a new rudiment appropriate to its site is formed from the material of the surrounding area. The influence of the immediate vicinity on the determination of the axes is seen in a beautiful experiment: a five-somite-sized disc of ectoderm and mesenchyme was transplanted, which contained in its center a limb bud three and a half somites in diameter. After healing, the limb anlage was rotated within the disc. A normally formed limb arose whose orientation reflected not the axis of the host, but rather that of the surrounding transplant. A morphogenetic field of a particular sort thus surrounds the limb bud as it is just becoming visible. Additional effects of this field are seen in other transplantation experiments. If an advanced tail bud with a fin is placed in the limb field of a host shortly before the onset of metamorphosis (from *Triturus* to *Triturus*, or from axolotl to *Triturus*), the implant does not develop further as a tail, but frequently is transformed into what looks like a limb rudiment with a variable number of toes. The tail organization is broken down, and in the interior an incomplete limb skeleton is put together. During this time very few or no mesenchyme cells enter the bud from the host; the induced change, therefore, results from a chemical influence of the substratum[186].

Such determinative or inductive fields are evident in the formation of other organs when pieces of presumptive ectoderm from early gastrulae (Fig. 362) are transplanted to embryos of various stages, to neurulae with well-formed ridges, to elongated embryos showing muscle contraction, or even to older larvae. In this multicompetent ectoderm, diverse and often complicated organogenesis is induced. Parts of the nervous system, sense organs, balancers, mesenchyme, and head skeleton, as well as chorda, muscle, pronephros, and tail fin can emerge from the implanted material. When the implanted ectoderm is vitally stained with Nile Blue, its derivatives can generally be characterized *in vivo*. The distribution of these organs in the host from

Fig. 362. Operative procedure, stippled ectoderm of a young gastrula transplanted to the site of presumptive epidermis in a neurula. (After von Woell-warth, 1957)

front to back or dorsoventrally either reaches a maximal frequency in particular body regions or is restricted to a single region (Fig. 363b). For the induction of these organs to be possible, determinative fields must obviously be present in late stages of development, even when the corresponding organs of the host have begun or indeed completed their development. Effective induction is limited to the region of the nervous system and the segmental plate. The definition of an individual inductive field is sharp. The induced organs correspond to the body region in which they arise. In the head region very large extra brain parts can be formed. As development of the neurula proceeds further, the implants can also join in the formation of the normal organs, as is seen from vital staining.

When the implanted epidermis covers diverse inductive regions, diverse organs can be induced, but never intermediate structures: either all one, or all another, or several different organs, each typical. Here again we see the choice among alternative package deals (cf. p. 262). We cannot hold a particular organ or tissue responsible for these field-dependent inductions. In these experiments they lead to abnormal supernumerary structures. Under natural circumstances they play a role in the events of regeneration, but their mechanism is as obscure to us as that of the self-organization of fields.

Thinking back on the last few lectures, we might ask ourselves what the basic episodes in the developmental physiology of amphibia can tell us about the causality of normal development. The ultimate goal is, of course, to understand how the content and distribution of substances in the egg cell leads step by step to the typical mature adult. In the beginning the structure of the egg is not mosaic but consists of substances distributed quantitatively according to certain axes and laid down during the growth, maturation, and fertilization of the egg. Physical fields, apparently anchored in the relatively fixed egg cortex in association with the movable endoplasm, determine during early cleavage and the blastula stage tendencies for gastrulation movements of the embryonic parts, time tables for competences, and, in varying degrees, tendencies for self-determination and induction (in the marginal zone) or unalterable histologic destinies (in the presumptive endoderm). What is being expressed here is not only the partitioning of the egg

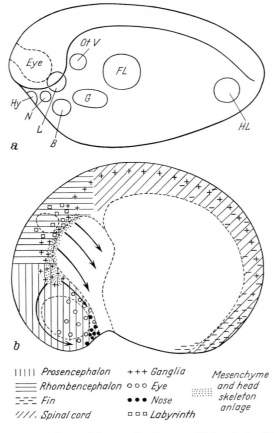

Fig. 363. Fate map in the ectoderm of the urodele neurula. *Eye* eye; *FL* forelimb; *HL* hind limb; *Ot V* otic vessel; *B* balancer; *Hy* anterior hypophysis; *G* gills; *L* lens; *N* nose. (After Huxley and de Beer, 1934.) b map of the inducing regions in the optic vesicle stage of *Triturus alpestris*. The regions are sketched which evoke the formation of certain organs in explanted gastrula ectoderm. The arrows in the head region indicate the direction of migration of neural crest cells which when induced produce extra gill arches. (After von Woellwarth, 1957)

structure to the various blastoderm regions, evoking qualitatively diverse responses to quantitatively diverse stimuli, but also an angoing chemo-differentiation. The gastrulation movements bring the blastemas of the endoderm, the chordamesoderm, and the ectoderm into physical contact, from which their tendencies and competences derive. The autonomous organization of the chordamesoderm is attributable to alternative reactions of the cell regions in a gradient system. The inductive effects of the archenteric roof, the prechordal plate, and the chordasomite complex apparently depend both on quantitative influences of a dorsoventral gradient and on a chemodifferentiation in the long axis of the gradient system, which is completed during gastrulation, and which leads to the emission of qualitatively different inducing substances. In the ectoderm epidermal and neural competences are realized, and the induced neural anlage becomes a line of fields. The organ rudiments arising from these fields act as secondary inducers. *Mutatis mutandis*, the induced and inducer work together. Positive and negative affinities between derivatives of different blastemas and blastema parts lead to physical separation and to the synthesis of complex organs.

Finally, the organism puts together its far-flung integrative system of endocrine glands and nerves, and the body is peppered with determinative fields which maintain the balance of organs and repair their losses. The onset of function influences the organ's further differentiation, and so the period of laying out the body plan, with which we have been concerned, is over.

We can no longer speak of a small restricted area from which the entire organization emanates, an "organization center" of normogenesis. The dorsal marginal zone merits in a limited sense the designation of organizer, in that it is a self-organizing field of central importance with critical inductive effects; but the organization is invariably the result of an interplay of the developmental events of the various embryonic parts. We know a few principles involved in this interplay, but as each successful experiment leads us a step closer to understanding, we are confronted with a new set of questions.

Lecture 20

The fate maps of fish and bird embryos[705], in spite of their great differences in yolk content and in morphogenetic movements, have proved to be merely variants of the amphibian plan which has been analysed so extensively (Fig. 364). Similarly, the developmental principles, field organization of the primary organizer region, primary and secondary induction, etc., appear to be essentially the same as in the amphibians on the basis of extirpation and transplantation experiments, which have already been highly informative, and which are still being extended. Even in the yolk-poor embryo of the cephalochordate *Branchiostoma* the fate map is clearly similar to that of amphibians, and a high degree of regulative ability has been demonstrated.

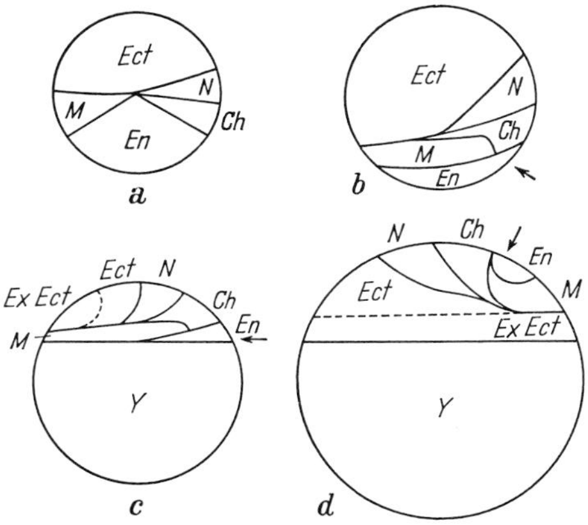

The relatively yolk-rich ascidian eggs, however, behave quite differently from those of the other chordates, although the arrangement of the germ layers agrees almost exactly with that in *Branchiostoma*, except, of course, for the lack of segmentation in the mesoderm. Ascidians develop mosaically until the larval stage. Recognized early as excellent objects of experimental investigation (particularly by CONKLIN, VON UBISCH, and REVERBERI'S group) the ascidians are among the best-known animals in terms of development.

In the full-grown oocyte, the cytoplasmic materials are arranged concentrically. In *Styela* the sur-

Fig. 364. Fate map of an ascidian (a), an amphibian, *Discoglossus* (b), a bony fish (c), and a bird (d). *Ch* chorda; *Y* yolk; *Ect* ectoderm; *Ex Ect* outer membrane ectoderm; *En* endoderm; *M* mesoderm; *N* neural tube→ site of invagination. (From PASTEELS, 1940 and WADDINGTON, 1952)

face layer contains yellow pigment granules, while the yolk mass is central (Fig. 365a). The large oocyte nucleus moves to the animal pole, and its membrane breaks down right after spawning. Clear cytoplasm appears near the animal pole; here the very small spindle develops. The clear cytoplasm spreads rhythmically toward the equator, revealing a meridional zone which in turn defines the medium plane (Fig. 365b, c). Maturation is now arrested until sperm entry, after which vigorous cytoplasmic streaming begins anew, now involving the entire egg. The yellow pigment moves toward the vegetal pole, and the yolk to the animal pole (Fig. 365d). The sperm nucleus, together with its aster, moves in an apparently predetermined meridian to the level of the equator; there it unites in a region of clear cytoplasm with the egg pronucleus, which has migrated from the animal pole (Fig. 365f). The yellow cytoplasm collects on the surface below the equator and forms the yellow crescent on one side of the egg (Fig. 365e). In this way the symmetry of the egg becomes determined. During the course of the first cleavage, further cytoplasmic movements take place: the clear cytoplasm which had collected around the pronuclei (Fig. 365f) streams into the animal half, while the yolk retreats into the vegetal half. Opposite the yellow crescent a gray

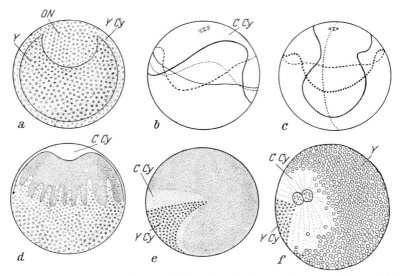

Fig. 365. Alterations in the distribution of several kinds of cytoplasm in the egg of the ascidian *Styela partita*. a–e surface views; f medial section; a newly laid egg; b, c spreading of the clear plasma after the breakdown of the oocyte nucleus and the appearance of the first spindle; b lateral view; c medial view; d streaming of yellow cytoplasm toward the vegetal pole; e, f arrangement of cytoplasm at pronuclear fusion. *Y* yolk; *YCy* yellow cytoplasm; *CCy* clear cytoplasm; *ON* site of oocyte nucleus. (a, d–f after CONKLIN, 1905; b, c after DALCQ, 1941)

crescent forms (Fig. 366a). The first spindle forms normal to the egg's plane of symmetry; accordingly, the first cleavage plane divides both yellow and gray crescents symmetrically (Fig. 366a, b). The second furrow is also meridional and normal to the first (Fig. 366c, d). The third cleavage plane is equatorial and defines the animal and vegetal halves of the embryo. The yellow crescent finds itself in the vegetal half, but the gray crescent is divided (Fig. 366e, f). Further cleavage shows strong bilateral symmetry (Fig. 366g). The diverse natural colors of the cytoplasm, vital stains, and carbon particles placed on the surface of cells (Fig. 367) permit one to follow some cytoplasmic regions from the two-cell stage right into the primitive organs (Fig. 366). In the 32-cell stage six mesoderm cells in the vegetal half separate from the endoderm (Fig. 366h); they contain the yellow cytoplasm ("mesoplasm"). In the 64-cell stage (Fig. 366i, k) further presumptive anlagen become separate: twenty-six ectodermal cells which contain yolk-poor cytoplasm ("ectoplasm"); ten neural plate cells (six from the animal and four from the vegetal half), with dense cytoplasm and the substance of the gray crescent ("neuroplasm"); six yolky chorda cells, which also have some cytoplasm from the gray crescent ("chordaplasm"); ten yolky endoderm cells; four mesenchyme-forming cells with light yellow "chymoplasm" from the yellow crescent; and eight muscle-forming cells with dark yellow "myoplasm," also from the yellow crescent. The neural plate cells form two rows, and the chorda cells an arc at one end; the muscle-forming cells form an arc at the other end of the embryo. These arcs come together at the sides

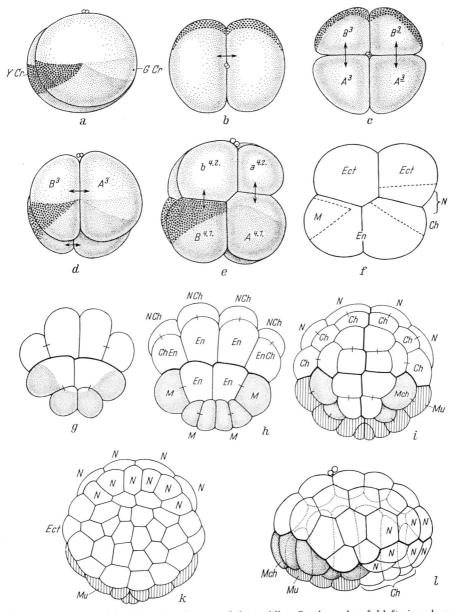

Fig. 366. Distribution of materials during the cleavage of the ascidian *Styela*. a, d, e, f, l left view; b, c, g, h, i view from the vegetal pole; k from the animal pole. a–e first-to-third cleavage; the cells in c–e have their standard designations; f the presumptive materials are diagrammed for chorda (*Ch*), ectoderm (*Ect*), endoderm (*En*), mesoderm (*M*). g sixteen-cell stage; h 32-cell stage, presumptive mesoderm region stippled; i, k 64-cell stage, l 76–112-cell stage. *Mch* mesenchyme cell; *Mu* muscle cells; *N* presumptive nervous system. (a–e, g–l after Conklin, 1905, modified according to Ortolani, 1953–1957)

of the embryo and separate the endoderm from the ectoderm (Fig. 366i, k, l). All the cell groups are symmetrical. The cells divide further until gastrulation (Fig. 366i; 368a), when the endoderm, which is ventral, invaginates forward, the tips of the arc of chorda cells meet medially, and the mesenchyme and muscle cells arrive at the posterior end of the embryo on the inside (Fig. 368 b, c). With the stretching of the ectoderm at the closure of the blastopore, the neural plate is extended in the long axis of the embryo (Fig. 368). It then closes into a neural tube. When the embryo assumes its larval form, the chorda lying under the posterior half of the neural tube grows out (Fig. 368 b), and its cells form a single row (Figs. 368 f, 372 f). The muscle cells are arranged in three rows on each side of the chorda (Figs. 368 d, e, 372 c). Most of the mesenchyme cells come to lie near the front of the chorda (Fig. 368 d–f). In the front part of the body, the neural tube

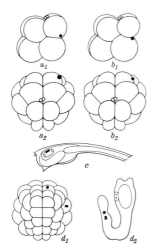

Fig. 367. Marking of cytoplasmic region of blastomeres of *Phallusia mammillata* with carbon particles, and observation of their positions in the larva. a_1, a_2, b_1, b_2 marking of presumptive brain material; c site of particles in the larva; d_1 marking of cells of the presumptive chorda (cf. Fig. 366 i); d_2 larva with marks in the chorda region. (After ORTOLANI, 1952, REVERBERI, ORTOLANI and FARINELLA-FERRUZZA, 1960)

expands into a brain containing a statolith organ and an eye spot (Fig. 368 f). A fine, solid column of endoderm runs through the tail (Fig. 372 c), having separated from the trunk endoderm. At the front end the ectoderm forms three adhesive papillae that are used in larval attachment, which just precedes metamorphosis (Fig. 368 f).

The cells of the ascidian embryo can be separated fairly easily with a glass needle, so that the degree of correspondence between potency and fate of cells or cell groups in the embryo can be tested thoroughly.

If the four-cell pairs in the eight-cell stage are isolated (Fig. 369), no cell produces any more than its prospective fate. Indeed, two of the cell pairs actually form less: no nervous system is formed, although "neuroplasm" is present in the two posterior cell pairs. This phenomenon can be studied further by combining various cell pairs from the eight-cell stage. Combinations of b4.2, a4.2, and B4.1 (see Fig. 366 e) result in defective larvae with endoderm, mesenchyme, and muscle, but neither chorda or nervous system (Fig. 370 a). The absence of B4.1 (Fig. 370 b_1) results, as expected, in the loss of the mesenchyme and musculature (Fig. 370 b_2): A brain and sense organs are formed; thus, the presence of A4.1 material is necessary for their formation; the presumptive nerve cells of a4.2 are not sufficient in themselves (Fig. 366 k, l). This suggests that A4.1 material induces brain differentiation.

If two b4.2 cell pairs (presumptive ectoderm) are combined with B4.1 and A4.1 (Fig. 370 c), no brain is formed. The ectoderm material, therefore, is not competent to form brain; A4.1 can evoke brain formation only in certain cells, the presumptive nerve cells.

The removal of certain cell groups in the 32- and 64-cell stages (see Fig. 366 h–k) carries us a step further. If the presumptive endoderm cells, derived from A4.1, are removed in the 32-cell stage (see Fig. 366 g, h), this tier of cells, which also contains nerve and chorda material, can form a brain with sense organs (Fig. 371 a). The two medial cells of this tier do not suffice (Fig. 371 b), but the two outer cells of the neuro-chordal tier do (Fig. 371 c). Surprisingly, a brain is also formed when no neural-chorda cells are left in place, but only the two outer cells which produce endoderm as well as chorda (Fig. 371 d, *cf.* Fig. 366 g, h). For the brain, now only the neural cells of the animal half remain (Fig. 366 k, l). That the chorda is not the sole inducer of brain formation is shown by extirpation experiments in the 64-cell stage (Fig. 371 c–h). The tier of nerve cells alone cannot form a brain (Fig. 371 e). The arch of chorda cells is not necessary

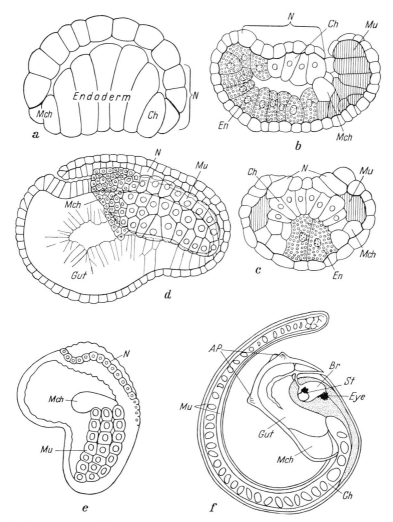

Fig. 368. Ascidian development. a medial section through a 76 to 112-cell stage of *Styela partita* (compare Fig. 366 l); b, c gastrula stage with constricted blastopore of *Ciona intestinalis*; b sections somewhat lateral to the medial plane; c cross section d, e neurula; d onset of the separation of trunk and tail in the embryo e older neurula; f larva of *Ascidiella scabra*. *Eye* eye; *Ch* chorda; *Gut* gut; *En* endoderm; *Br* brain; *AP* adhesive papillae; *Mch* mesenchyme; *Mu* muscle cells; *N* neural anlage; *St* statolith. (a–c, e after CONKLIN, 1905; d after VAN BENEDEN and JULIN, 1884; f after von UBISCH, 1939)

(Fig. 371 f–h); a brain with sense organs is formed if only two of the six endodermal cells derived from A4.1 remain. On the other hand, chorda alone can cause brain formation, as we see in embryos devoid of endoderm (Fig. 372 d–f). In order for the presumptive nerve cells to form a brain, they must come into contact either with chorda cells or with the endoderm cells, which later form the archenteric roof (Fig. 368). The resemblance to neural tube determination in the amphibia is clear, but also the difference: in the amphibia, the ectoderm of the early gastrula is multicompetent, and a neural inducer can cause the formation of a nervous system anywhere. In the ascidia this competence is restricted mosaically to the presumptive nerve cells.

It is surprising how completely the larval form is laid out, even when some major organ systems are lacking. A larva without a nervous system looks quite normal; cross sections show a well-proportioned array of organs (Fig. 372a–c).

When all the presumptive endoderm is removed from the embryo, the trunk lacking its voluminous gut is too small in relation to the tail (Fig. 372d). Nervous system, chorda, mesenchyme, and musculature can be well formed, but at times parts of the tail organs slide into the gutless

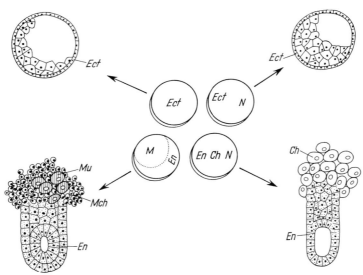

Fig. 369. Division of the four pairs of blastomeres of the eight-cell stage of the ascidian embryo after isolation (prospective fates indicated). *Ch* chorda; *Ect* ectoderm; *En* endoderm; *Mch* mesenchyme; *Mu* musculature. (After REVERBERI and MINGANTI, 1946, 1953)

trunk (Fig. 370e). The muscle cells of the tail meet beneath the chorda in the place of the missing tail endoderm (Fig. 372f). In some areas of the trunk the body wall of the endodermless larva is multilayered; this evidently stems from the excess of epidermis relative to the greatly reduced mass of tissue, which has moved inside during gastrulation, and which must be covered.

Embryos without chorda form almost normal larvae; the tail is just incompletely extended. Its epidermis is folded and thus evinces the autonomous spreading tendency of the ectoderm (Fig. 372g). Removal of the presumptive mesoderm in the 32-cell stage (Fig. 373a, b) results in larvae which have neither mesenchyme nor muscle cells, yet the larva is normally laid out, and has one, two, or three adhesive papillae (Fig. 373a). The organs can retain their normal relationships, although frequently as a result of the lack of the mesodermal masses other arrangements arise, e.g., the brain may be pushed toward the tail by the gut (Fig. 373b).

Ectodermless embryos made by the removal of the animal half in the sixteen-cell stage (and recall this includes the animal portion of the neural material) cannot gastrulate; the remaining organ-forming regions of the embryo thus do not assume their normal relationships with one another, nor are they held together by the epidermis, and so the entire form of the body is abnormal. This makes the state of determination of the individual embryonic regions all the clearer. The endoderm always forms a mass of cells in which flat cavities can arise. Neural differentiation, chorda, muscle, and mesenchyme cells can be enclosed by endoderm (Fig. 373c), or they can come to be partly or totally on the surface (Fig. 373d). The chorda cells have a strong tendency to line up single file (Fig. 373c–e) and form characteristic large vacuoles in their cell bodies. When

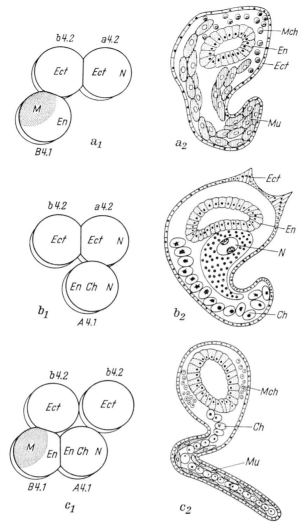

Fig. 370. Combination of the cell pairs of the eight-cell stage of the embryo (compare Fig. 366). a_1 A4.1 cell missing; b_1 B4.1 cell missing; c replacement of the a4.2 cell by the b4.2 cell; a_2, b_2, c_2 Resultant larvae. Labels as in Fig. 366. (After Reverberi and Minganti, 1952, Reverberi, 1961)

muscle cells come into contact with chorda cells, they press close, and so a long chorda can grow out with a covering of muscle cells, and often with a strand of endoderm also (Fig. 373e). We can conclude from this that in normal development the chorda is the prime mover in the extension of the tail and also serves to orient the muscle and endoderm cells. Typical mesenchyme cells can also form; they scatter and tend to move into the cavities and form loose associations. A typical differentiated nervous system never arises, due to the lack of the neural cells of the animal half, the abnormal positions of the relevant cells that remain, or both these factors.

The differentiation of the adhesive papillae is clearly dependent on induction: In embryos lacking either neural tube or endoderm, the papillae are either defective or absent. If the animal half is rotated 180° so that the animal pole is at the equator, the adhesive papillae nevertheless

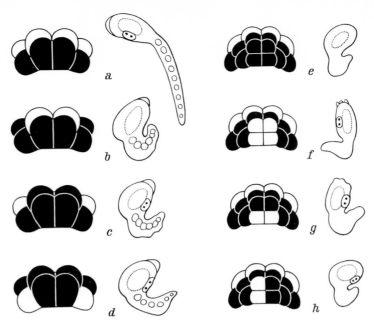

Fig. 371. Removal of certain groups of blastomeres from *Ascidiella aspersa* embryos in order to clarify brain formation. a–d removal of blastomeres in the 32-cell stage; e–h in the 64-cell stage. The blastomeres removed are black (compare the prospective fate of the cells in Fig. 366). (After Reverberi, Ortolani, and Farinella-Ferruzza, 1960)

form at the anterior end of the larva. From this it follows that all the endoderm is competent to form adhesive papillae and that an inducer of papilla formation lies at the front end of the embryo.

With the establishment of symmetry, self-differentiation is ordained for the two cells resulting from the first division. If the two blastomeres are separated, or if one cell is killed, each cell develops into a half embryo corresponding to its prospective fate (Fig. 374b). These half embryos can nevertheless transform subsequently into reduced larvae, which are outwardly normal in every respect. The ectoderm covers the originally open side of the half embryo; the half neural plate and the endodermal fragment can close into neural tube and gut respectively; but the normally paired components, such as mesenchyme and tail musculature are present only on one side of the larva. Up to three adhesive organs can be formed after the isolation of half blastomeres — again demonstrating that their site is not determined early. The cell number in the organs is not regulated, since it turns out to be half that seen in normal larvae.

The unfertilized ascidian egg is not a mosaic egg, as can be seen from the centrifugation or dissection of unfertilized eggs. In a few (not all) species, the egg can be divided by strong centrifugation into two parts, both of which can be fertilized; and each can develop into a haploid or diploid larva, depending, of course, on whether or not it contains the egg pronucleus (Fig. 375a). The egg cell can be cut in two, either parallel to the egg axis or normal to it, and when the pieces are not too small, well-formed larvae result (Fig. 375b, c). Even when small pieces of egg cortex are taken from an unfertilized egg (or with a micropipette, a portion of the egg cytoplasm), reduced larvae are formed. After fertilization, this operation is fatal. With the cytoplasmic movements after fertilization and during the first division, organ-determining substances are separated from one another and brought to specific locales. During the first division centrifuged

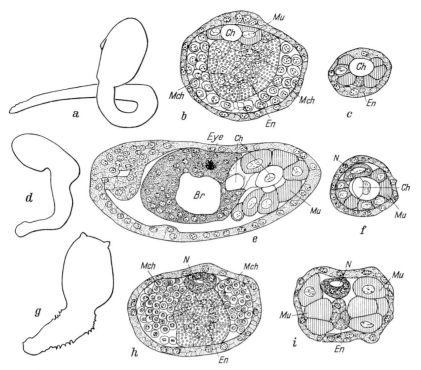

Fig. 372. Removal of organ forming regions of the *Ascidiella scabra* embryo. a–c removal of the presumptive neural material; d–f removal of the presumptive endoderm; g–i removal of the presumptive chordal material; a, d, g outlines of the body form; b, h cross sections through the trunk; e oblique longitudinal section through the trunk; c, f, i cross sections through the tail. *Eye* eye; *Ch* chorda; *En* endoderm; *Br* brain; *Mch* mesenchyme; *Mu* muscle cells; *N* neural tube. (After von UBISCH, 1940)

eggs can still yield larvae whose organ-forming cells are chaotically arranged, and thus differentiate in unusual places. Muscle cells can elaborate contractile fibrils, and ectoderm and endoderm exhibit the tendency to spread out, so that one or the other epithelium can cover the entire embryo (Fig. 376). The two-cell stage *is* a mosaic embryo.

Yet not all regulation has been excluded, as we see from the fusion of embryos (Fig. 377). This experiment is easy to carry out in ascidians: all one has to do is remove two embryos from their chorions and place them in contact. They almost always will then fuse. The fusion of two eight-cell stages of *Ascidiella malaca* produces giant larvae (Fig. 377). The organs are generally twinned, but they may be partially fused. That the larva is double is readily seen in its external structure, though sometimes a single tail with two notochords and one neural tube may form. At times even a single larva may form after fusion with three adhesive papillae, a single trunk, and a single normally proportioned tail; such a giant larva may also be untwinned within, with a completely normal arrangement of bilateral tail musculature next to the neural tube and notochord. The number of chorda and muscle cells is twice as great as in the normal larva, so this case resembles the formation of single giant larvae after the fusion of two two-cell stages in the amphibia (pp. 217ff., Fig. 282d, e). In the latter case when the two dorsal fields merge (Fig. 282, p. 218), a single gastrulation may follow; in the fused organizer material the development of bilaterality is completed regulatively. Similarly, regulation must be involved in the fusion of ascidian embryos when similarly determined regions in the two embryos are approximated during the union of the

295

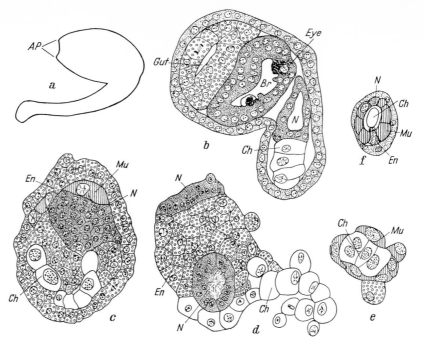

Fig. 373. *Ascidiella scabra.* a, b removal of the presumptive mesoderm. a body outline; b oblique-longitudinal section through the trunk; c–e removal of the ectoderm; f cross section through the tail of a normal larva. *Gut* gut; *Br* brain, other labels as in Fig. 372. (After von Ubisch, 1940)

two-cell stages and during cleavage. Under these circumstances, the cells reach an arrangement at the time of gastrulation which at least approaches normal. This favorable case arises in the ascidians much more rarely than in the amphibians, to be sure. For in the amphibian egg, only two gradients are laid out, the animal-vegetal yolk gradient and a dorso-ventral gradient expressed in the distinct nature of the dorsal region. In the ascidians the differentiation of the egg cytoplasm is much more extensive; therefore, the achievement of a good fit in the union of blastomeres is less probable. But in the rare fortunate cases, the chorda, muscle, and gut cells form single organs; and a neural tube is formed from the neurally determined cells. The bilateral symmetry is thus not established unalterably in the organ-forming regions of the two-cell stage; the symmetriza-

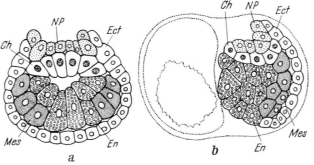

Fig. 374. Cross section through a normal *Styela* embryo before closure of the neural plate into the neural tube; b cross section through a half embryo from a blastomere of the two-cell stage. *Ch* chorda; *Ect* ectoderm; *En* endoderm; *M* mesoderm; *N* neural plate. (After Conklin, 1906)

296

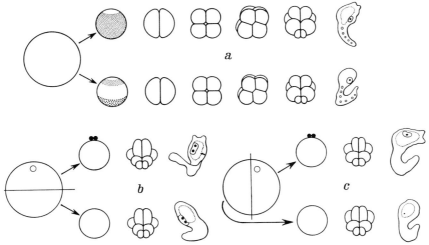

Fig. 375. Developmental potencies of pieces of unfertilized, fragmented eggs of *Ascidiella malaca*, a fragmentation caused by centrifugation; b, c by cutting; b normal to the egg axis; c near the egg axis. (a after LA SPINA, 1961; b, c after ORTOLANI, 1958)

tion remains to be completed in that material whose cells are organogenetically and histogenetically determined. And the adhesive papillae are induced as a triad by the uniform organs of the front end. The various sorts of cytoplasm in the ascidian embryo can be distinguished by their inclusions and fine structure, as well as by a number of histochemical reactions. The myoplasm is distinguished by a strong benzidine- and indophenol-oxidase reaction bound to the mitochondria. The distribution of mitochondria, which can be stained with Janus green, is one of the most striking features in the organization of the organ-determining cytoplasmic materials. In the unfertilized egg they are spread uniformly throughout. Then they migrate toward the vegetal pole and collect in the yellow crescent during the first division (Fig. 378a, b). Later, they are found in the B4.1 cells and their derivatives (Fig. 378c–e). During the division from the 16- to the 32-cell stage, they are deployed in the future mesoderm cells (Fig. 379b). In the 64-cell stage they are concentrated in the muscle-forming cells (Fig. 378f), and later they can be followed into the tail musculature (Fig. 378g, h). In embryos disarrayed by centrifugation during the first division, the mitochondria reveal the displaced myoblasts (Fig. 380a). In these cases the mitochondrial staining (Fig. 380b_1, b_2) is always found in the living but malformed larvae, where muscle contractions can be observed.

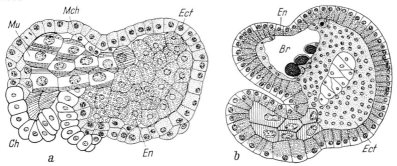

Fig. 376. Embryos from *Styela partita* eggs centrifuged during the first division, cells in disarray. *Ch* chorda; *Ect* ectoderm; *En* endoderm; *Br* brain; *Mu* muscle cells; *N* neural tissue. (After CONKLIN, 1939)

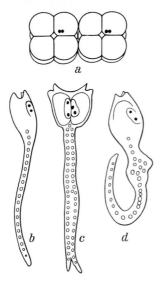

Fig. 377. Fusion of two eight-cell stages of *Ascidiella malaca*, marked with carbon particles. a experimental arrangement; b control larva; c twinning; d single giant larva with only a disturbance in the chorda region. (After GORGONE, 1961)

The histochemical reactions cannot tell us how particular substances cause differentiation. They do indicate, however, that in individual cytoplasmic regions various metabolic conditions predominate during particular developmental stages, and various cytoplasmic particles and substances are amassed. But the cytochemically demonstrable conditions are not absolutely characteristic of the differentiated cells to come. Thus, for example, the benzidine reaction declines gradually in the muscle cells, and in the larva (or in abnormally organized embryos approaching the larval age) it is the mesenchyme cells (Fig. 380c) which stain the strongest blue; these resume division, signaling the onset of a temporally determined metabolic change.

As in the echinoderms and amphibians, some insight into the events of determination comes from the use of certain chemicals. Once more, lithium ions have a particular effect. The treatment of early embryonic stages with LiCl prevents neural differentiation. The larvae then have no brain, and the adhesive papillae are also lacking. Lower dosage permits neural plate formation, but not closure (Fig. 381 b). Most of the presumptive neural cells serve as ectodermal epithelial cells (Fig. 381 c). The endoderm

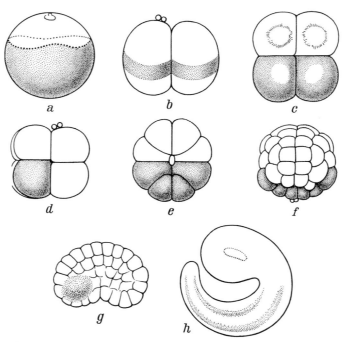

Fig. 378. Distribution of Janus-Green-stained mitochondria in the egg of *Phallusia*; and during embryonic development. Aggregation in the presumptive muscle forming cells during cleavage. (After REVERBERI, 1956, 1961)

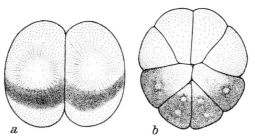

Fig. 379. Benzidine-peroxidase reaction in *Ascidiella aspersa* embryos. a end of the first division; b transition from 16- to the 32-cell stage. (After RIES, 1939)

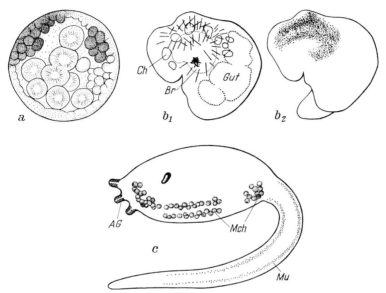

Fig. 380. Benzidine-peroxidase reaction in *Ascidiella aspersa*. a, b chaotic embryos resulting from centrifuging during the first division; a, b₂ mitochondrial staining; b₁ contractions observed *in vivo*; c larva. *Ch* chorda; *Gut* gut; *Br* brain; *AG* adhesive glands; *Mch* mitochondria; *Mu* musculature. (After RIES, 1939)

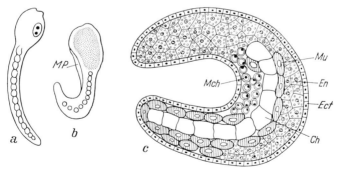

Fig. 381. Larva of *Ascidiella malaca*. a control larva; b, c after treatment of the embryos with LiCl; b treated until the eight-cell stage in LiCl; c organization of the larva after embryonic development until the two-cell stage in LiCl. *Ch* chorda; *Ect* ectoderm; *En* endoderm; *Mch* mesenchyme; *Mu* muscle cells; *MP* medullary plate before closure. (a, b after REVERBERI and FARINELLA-FERRUZZA, 1961; c after FARINELLA-FERRUZZA, 1955)

remains in a swollen trunk or spreads far into the tail, so that the whole body becomes worm-shaped.

The effect of LiCl on ascidian embryos bears a certain resemblance to that on echinoderms, where animal differentiation is inhibited; but it differs from the effects on amphibians, in which the presumptive chorda cells are mesodermalized, to be sure, but where the degree of differentiation of the nervous system is not directly affected. As to the cytoplasmic structures or determining processes acted upon by Li^+, we know nothing.

Lecture 21

One group in which an early chemodifferentiation brings about determination, as in the ascidians, comprises the invertebrates with spiral cleavage: nemertines, annelids, and molluscs.

Figure 382 shows as an example the course of cleavage to the 32-cell stage in the marine snail *Trochus*. The spindles lie neither in the transverse nor meridional planes, instead they slant. As a result, the cleavage planes are also diagonal. Spindles and cleavage furrows alternate regularly, first in one diagonal direction and then in the other. The tiers of cells separated by a division are thus rotated with respect to one another. If, when seen from the animal pole, the upper cells are displaced clockwise from their sister cells or rotated to the right (Fig. 382b, f), the division is called dexiotropic. If the rotations is in the opposite direction (Fig. 382d), the division is laeotropic. In general, the four upper (animal) cells emerging from the third division are smaller than those below (Fig. 382b, c) and so are called micromeres. The four cells below, the macromeres, are named 1A, 1B, 1C, and 1D; the upper four are 1a, 1b, 1c, and 1d. In the next division, to the 16-cell stage (Fig. 382d, e), the four macromeres pinch off a second "micro-quartet," whose cells are called 2a, 2b, 2c, and 2d. The four cells remaining at the vegetal pole are now 2A, 2B, 2C, and 2D. The daughter cells of the first quartet are now designated $1a^1$ and $1a^2$, $1b^1$ and $1b^2$,

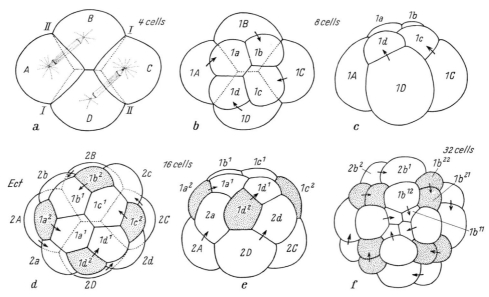

Fig. 382. Cleavage stages of the snail *Trochus* (Prosobranchia); in f only the micromeres of the b quadrant are labelled; in d and e the primary trochoblasts are stippled; in f their daughter cells are stippled. (After ROBERT, 1902)

300

1 c¹ and 1 c², and 1 d¹ and 1 d², where the cell nearer the animal pole always has the superscript 1 (Fig. 382 d, e). Because of the oblique divisions of the cells, those of the second quartet push between the daughter cells of the first quartet (Fig. 382 d–f). In the course of this cleavage determining cytoplasmic substances are parcelled out.

The mature oocyte often shows little or no polar stratification. Sometimes cytoplasmic components begin to segregate even before meiotic prophase and continue rapidly as meiosis proceeds. The egg of the sea hare *Aplysia* is a good example. Clear animal cytoplasm with little yolk separates from a zone of oil droplets and a zone of large yolk granules, which fills the vegetal half of the egg (Fig. 383 a–c). On the border between fat and yolk appears a narrow zone with highly basophilic granules (Fig. 383 b, c). The animal cytoplasm with the mitochondria is alkaline (pH 8), while the vegetal cytoplasm is acid (pH 6), and the basophilic granules are extremely

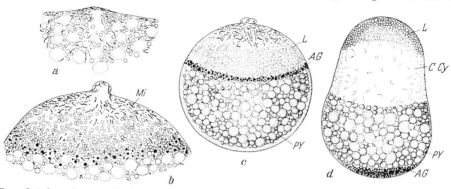

Fig. 383. Egg of *Aplysia limacina* (Opisthobranchia) stained with neutral red. a, b beginning of stratification of the egg cytoplasm at the animal pole during the first maturation division; c complete stratification in the mature egg; d mature egg after centrifugation (fifteen minutes at 3000 rpm.). *AG* ascorbic acid granules; *PY* protein yolk: *F* fat; *CCy* clear cytoplasm; *L* lipid; *Mi* mitochondria. (After RIES and GERSCH, 1936; RIES, 1939)

acid. The animal half oxidizes reduced methylene blue. In the absence of O_2, Janus green is reduced (to diethyl-safranin, which is red) beginning in the vegetal half of the egg. The red color reaches the animal half only after some time; finally the egg remains blue-green only in the vicinity of the polar bodies. Phenoloxidase-containing granules (demonstrated with the indophenol blue reaction) collect mainly in the oil droplet zone during maturation (Fig. 384e, f). Ascorbic acid (Vitamin C) can be demonstrated only in the zone of the very acid granules (Fig. 384a). It is still open to question whether these histochemical reactions are absolutely specific and whether they reveal the localization of substances precisely. It may be that granules of different sorts and diverse biological significance absorb the reagents of the reaction products preferentially and thus misrepresent the distribution of the particular substances in the living cell. But even then the reactions in the various zones of the egg cell show that each has specific chemical properties.

The further development of vitally stained embryos and the histochemical study of sequential developmental stages reveal the distribution of various substances to the respective blastomeres. The ascorbic acid granules are monopolized by certain cell lines: the very small micromeres in *Aplysia*, which are pinched off toward the animal pole in the third cleavage, contain none of these granules. Nor do the vegetal blastomeres of all four quadrants contain many (Fig. 384c). During the next division, two of the cells given off toward the animal pole by the macromeres 2a and 2b have none of the granules (Fig. 384d), which are thus reserved only for particular organs. The phenoloxidase-containing granules which lie between the ring of basophilic granules and the clear animal cytoplasm are distributed fairly evenly among the first four blastomeres.

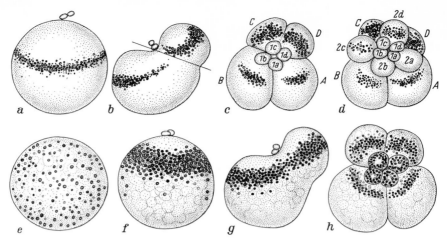

Fig. 384. Cleavage stages of *Aplysia limacina*. a–d ascorbic acid granules revealed by the silver nitrate reaction from the mature egg cell to the twelve-cell stage (where the micromeres have not yet divided); e–h indophenol blue oxidase reaction from newly laid eggs before maturation to the eight-cell stage. (After RIES, 1937)

Then the micromeres receive an especially high concentration of the granules (Fig. 384h), although previously the animal cytoplasm had been free of them (Fig. 383, c). A rearrangement of the cytoplasmic materials must therefore take place during the third division. In the larva there are a great number of the blue-staining granules in the cells of the prototroch. In *Ilyanassa* the oil droplets which start out in the animal halves of the egg cell, and of cells AB and CD also are relocated (Fig. 385). During the four-cell stage the oil droplets are found in a band which stains

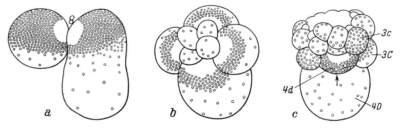

Fig. 385. Distribution of lipid droplets of *Ilyanassa* (Prosobranchia). 4D endoderm cell; 4d mesoblast (After CLEMENT and LEHMANN, 1956)

intensely with Sudan stains. At the next division very few move into the cells of the first quartet (Fig. 385b). Meanwhile, the further distribution among the daughter cells derived from the other four is fairly uniform except that 4d, the primary mesoblast (which still must give off one more endoderm cell), acquires almost all the oil droplets present in 3D (Fig. 385c).

The prospective fate of the blastomeres has been followed in several species for a great distance down the cell lines through many cell generations. In the Nemertine *Cerebratulus lacteus* vital staining can be used to mark the four first cell tiers and relate them to the various regions of the pilidium larva (Fig. 386a): 1a¹–1d¹ becomes the ectoderm of the upper half of the larva and a narrow medial piece of the prototroch which fringes the "earlaps" of the helmet-shaped larva and meets in front and in back (Fig. 386b); 1a²–1d² forms the ectoderm on the outside of the earlaps and a great portion of the prototroch; 2a–2d also contribute to the prototroch and form the oesophagus and the ectoderm of the inner side of the earlaps; 2A–2D form endoderm.

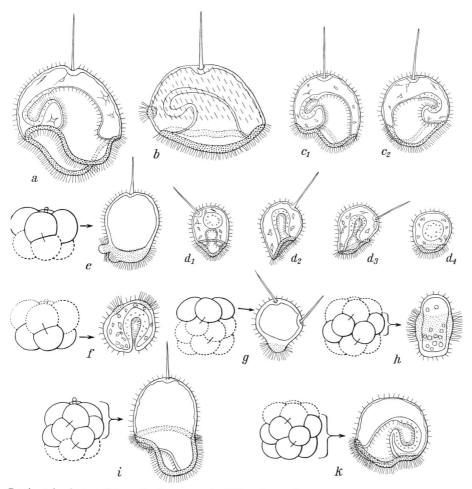

Fig. 386. *Cerebratulus lacteus* (Nemertina). a normal pilidium larva; b larva from an embryo in which the most animal cells in the sixteen-cell stage ($1a^1$–$1c^1$) were isolated, vitally stained, and reimplanted rotated 180° (as indicated by the little piece of apical tuft which apparently normally forms its most anterior part); c_1, c_2 pilidium larvae from 1/2-blastomere; d_1–d_4 larvae from 1/4-blastomeres; e–k isolation of cell tiers; e $1a$–$1d$; f $1A$–$1D$; g $1a^1$–$1d^1$; h $1a^2$–$1d^2$ and $2a$–$2d$; i embryo without $2A$–$2D$; k embryo without $1a^1$–$1d^1$. (After HÖRSTADIUS, 1937)

Isolation experiments reveal the determination of the parts of the egg and of the embryo. At the beginning of the maturation divisions in *Cerebratulus*, the animal-vegetal determination is not yet complete: If immature eggs are cut into pieces and then fertilized, miniature pilidium larvae emerge from animal, as well as vegetal fragments[753]. Separated half-blastomeres can produce two normal pilidiums (Fig. 386c_1, c_2). Even 1/4 blastomeres can form larvae which have all the typical parts, although they are generally not well put together (Fig. 386d_1–d_4). The apical organ, a long sensory filament composed of a bundle of thin fibers emerging from an epithelial thickening, is generally displaced or lacking. Moreover, the egg appears to have no bilateral organization. After the separation of the animal and vegetal quartets in the eight-cell stage, $1a$–$1d$ produce a blastula with an apical organ and a thick field or band of cilia below (Fig. 386e); $1A$–$1D$ form a gastrula, in which only the endodermal part of the gut is invaginated, however; the presumptive

303

oesophagus ectoderm bearing irregular spots of ciliated epithelium covers the larva (Fig. 386f). The most animal tier of the sixteen-cell stage ($1a^1$–$1d^1$) produces a blastula with a small piece of ciliated epithelium (Fig. 386g), corresponding to the prospective fate revealed by vital staining (Fig. 386b). The apical organ is frequently twinned; apparently the cells of the four quadrants, which contribute their most animal daughter cells to its formation, are thrown into disorder by the operation. The two interdigitating middle cell tiers ($1a^2$–$1d^2$ and $2a$–$2d$; see Fig. 382d, e) form a blastula with a thick flagellated band around the equator (Fig. 386h). If only the most vegetal cells ($2A$–$2D$) of the sixteen-cell stage are removed, only the gut is missing in the otherwise well formed larva (Fig. 386i). With the removal of the most animal tier ($2a^1$–$2d^1$) the apical organ is lost (Fig. 386k). In all experiments in which the embryonic parts are separated latitudinally, the differentiation of the parts agrees with their prospective fate, therefore. So the blastomeres are rigidly determined along the main embryonic axis.

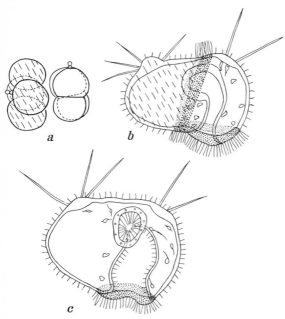

a *b*

c

Fig. 387. *Cerebratulus lacteus.* a apposition of a meridional and a vitally stained animal half, each from an eight-celled stage; b resulting larva; c same larva somewhat later; beginning of regulation. (After HÖRSTADIUS, 1937)

This is shown also by an experiment in which an animal half from a vitally stained eight-cell stage is united with a meridional half (Fig. 387a). Each half differentiates as it would have *in situ* or in isolation (Fig. 387b). During embryonic development no embryonic part influences any other embryonic part, but, surprisingly, regulation appears in the larva. A few days after the formation of the pilidium organs the ciliated band of the animal half is remodeled; only a short piece remains. This, together with the horseshoe-shaped ciliated band of the meridional part, forms a complete ring. Thus, the disharmonious larva takes a step towards unity. In the experiments of Figure 386f, h, k, the apical organ is absent because the most animal cells have been left out; it can be regenerated, however, if it is cut out of a young pilidium[753]. These observations show .that the rigid determination characteristic of mosaic development, with its cell sequences firmly fixed up to a certain point by site and cell number, does not have the final say on the fate of all the cells.

There are certain kinds of animals, including some rotifers, which exhibit cell constancy even as adults. These "mass-produced forms," prefabricated during mosaic development, are usually incapable of repairing defects; but the chemodifferentiation established in the egg cell even before cleavage often serves only to promote the fast development of a free-living larva that can feed, and that will later become capable of regulation during metamorphosis and as an adult under the influence of external factors. One need only think of the remarkable regenerative powers of the ascidians and many annelids.

We must also recognize that within the pattern of self-differentiation the formation of individual organs is not autonomous, as in the ascidian brain and adhesive papillae. In the development of mollusc larvae the shell gland is induced by the gut. LiCl and a variety of nonspecific

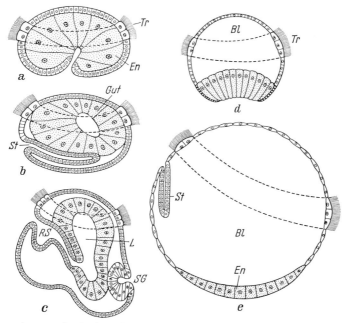

Fig. 388. *Bithynia tentaculata*. a–c beginning of normal development; d, e exogastrulation. *Bl* blastocoel; *En* endoderm; *L* liver; *SG* shell gland; *St* stomodeum; *Tr* trochus. (After HESS, 1962)

harmful factors inhibit the normal invagination of the endoderm in fresh-water pulmonates, resulting in a vesiculate exogastrula (Fig. 388). The stomodeum does invaginate autonomously. There is, however, no shell gland. After treatments gentle enough to permit the endoderm to invaginate, a shell gland is formed where the end of the gut touches the ectoderm. In abnormal embryos the tip of the gut may meet the ectoderm near the prototroch, and then the shell gland forms in an unusual place (Fig. 389). From this we see that the competence of the ectoderm is not restricted.

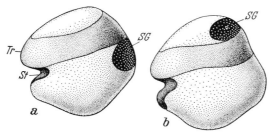

Fig. 389. Induction of the shell gland in posttrochal or pretrochal ectoderm of abnormal embryos of *Limnaea stagnalis* (Pulmonata). *SG* shell gland; *St* stomodeum; *Tr* trochus. (After RAVEN, 1952)

The extent to which the cleavage sequences and the differentiation schedules of the blastomeres can be determined is seen in isolation experiments on embryos of the marine snail *Patella*, which since the investigations of WILSON in 1904 has been a classical object of study in developmental physiology. From the macromeres of the eight-cell stage there separate, as in *Trochus* (Fig. 382), two additional micromere quartets (2a–2d, and 3a–3d), which, together with the descendants of 1a–1c, produce all the ectoderm of the trochophore larva (Fig. 390). The more equatorial

305

daughter cells of the first quartet 1a², 1b², 1d² (stippled in Fig. 382d, e) are ancestral to the prototroch, which rings the larva with flagella (Fig. 390g); they are called the primary troch-blasts. Later, other cells from the nearby embryonic regions above and below join them as secondary trochoblasts to complete the prototroch. The vegetal cells 3A–3D produce endoderm (4A–4D and 4a–4c); 3D also gives off 4d, the primary mesoderm cell or primary mesoblast.

Isolated macromeres, such as 1B, continue to cleave as if left in place (Fig. 390a–e) and pro-duce miniature larvae (Fig. 390f) in whose ectoderm large flagellated cells formed by secondary trochoblasts lie on one side. A micromere isolated from the eight-cell stage produces exactly

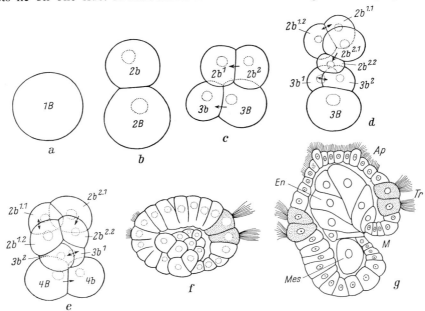

Fig. 390. *Patella* (Prosobranchia). a–e cleavage of an isolated macromere (e.g., 1 B, labeled as on p. 300); f larva derived from 1 B; g normal trochophore larva. *En* endoderm; *M* mouth; *Mes* mesoblast; *AP* apical plate; *Tr* trochus. (a–f after WILSON 1904; g after PATTON from KORSCHELT and HEIDER, 1909–10)

what one would expect from its prospective fate, a quarter of the upper half of the larva (*cf.* Fig. 390g): four primary flagellated cells of the prototroch, the usual ectoderm cells and a quarter of the apical plate (Fig. 391 a–e). An isolated primary trochoblast of the sixteen-cell stage divides only twice, and all four cells bear flagella (Fig. 391 f–k). A daughter cell of the primary trocho-blast from the 32-cell stage divides into two flagellated cells (Fig. 391 l–n), and each of the four descendants of a primary trochoblast transforms directly into a flagellated cell (Fig. 391 o, p).

By placing a cleavage stage in calcium-free sea water, one can dissociate the embryo almost completely into individual cells which can live a long time. Among these can be recognized larger primary and smaller secondary trochoblasts, individual apical cells, endoderm, mesenchyme, and muscle cells. This very model of mosaic development exhibits a rigid and far-reaching determina-tion, which extends beyond that seen in the early development of ascidians.

Extirpation and isolation experiments are very informative in embryos that form polar lobes (cf. pp. 162) which distinguish the D quadrant. When the egg of the marine snail *Dentalium* cleaves (Fig. 392c), the polar lobe can easily be cut off. Cleavage now proceeds unhindered, but no polar lobe ever forms again. With the loss of this material, the D quadrant is now no greater than the other quadrants. Gastrulation proceeds in an orderly fashion and leads to a trochophore which falls short of a normal one (Fig. 393a) in three ways (Fig. 393b): it lacks the body region behind the prototroch (the post-trochal region), which normally grows out; the apical organ is missing; and there are no bands of mesoderm, which normally come from the primary mesoblast 4d. If the first polar lobe is only partially removed, the polar lobes formed in the second and third divisions are correspondingly smaller, and the post-trochal region of the trochophore is similarly reduced. The apical organ may be present or absent. If the first two blastomeres are separated, the CD cell (Fig. 392d) produces a complete trochophore, but its post-trochal region is disproportionately large (Fig. 393c). This disproportionality is even greater in trochophores

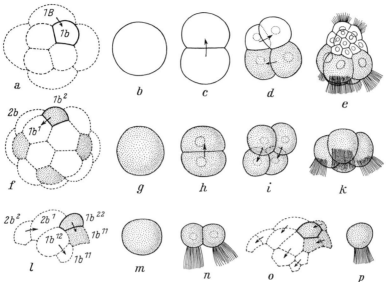

Fig. 391. Isolation of individual blastomeres from the *Patella* embryo. a–e micromere of the eight-cell stage (e.g. 1 b.); f–k primary trochoblast of the sixteen-cell stage; l–n one of the two daughter cells of the primary trochoblast from the 32-cell stage; o, p one of the four granddaughter cells of the primary trochoblast from the 64-cell stage. In the diagrams (a, f, l and o) the isolated cells are outlined heavily and the trochoblasts are stippled (compare Fig. 382). (a, f, l, o original diagrams, the rest after WILSON, 1904)

from the quarter blastomere D (Fig. 393g). Larvae from the half-blastomere AB, or from the quarter blastomeres A, B, or C resemble those from embryos whose first polar lobe was removed (Fig. 393f, h): they lack apical organs and posttrochal regions, structures at opposite ends of the larva. The factors responsible for these two body parts thus come from the D cell and are obviously present in the first polar lobe. Between the first and the second division, that is, in cell CD, the determinants of the apical organ and post-trochal region separate: removal of the second polar lobe (Fig. 392e, f) results in a trochophore with an apical organ but no posterior body region (Fig. 393c); and a miniature larva of similar appearance forms from 1d (Fig. 393i). This cell has thus inherited the apical organ factor. The isolated cells 1a, 1b, 1c (Fig. 393k) do no better at ectodermal differentiation than A, B, or C, as one would expect; in addition, they lack the endoderm, which is derived from the macromeres.

In the annelid *Chaetopterus* the first cleavage can be made equal (it is normally unequal) if the egg is compressed after the ejection of the second polar body and released when the first cleavage furrow has cut halfway through the egg. Then the first two blastomeres share the polar lobe, and the result is twinning. If two such half-blastomeres are now separated, each will form an entire larva.

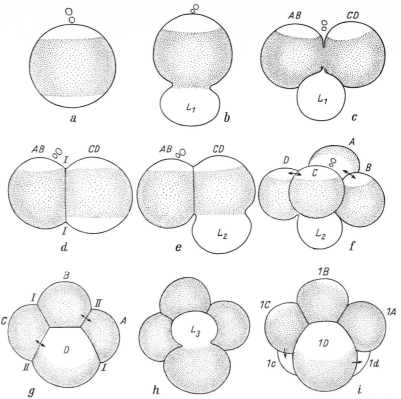

Fig. 392. Cleavage of *Dentalium* (Scaphopoda). a mature egg; b–d first cleavage; e–g second cleavage; h–i third cleavage (blastomere designations explained in text); a–f lateral views; g–i from vegetal pole. L_1–L_3 polar lobes of the three divisions; *I, II* first and second cleavage. (After WILSON, 1904)

In *Dentalium* polar differentiation is expressed even in the immature egg in the form of a white animal-pole region, a broad pigmented band, and a white vegetal-pole region. During the expulsion of the polar bodies, which is triggered by fertilization, this pigmented band becomes even more clearly defined. Now the distribution of determinants can be demonstrated already, just as in a cleaving embryo: unfertilized eggs can be cut so that the regional organization of the egg substance, and later the future of the blastomeres, can be recognized. The egg fragments can be fertilized and they develop even with only a sperm nucleus. Animal fragments cleave like eggs without polar lobes and produce larvae resembling those from AB, A, B, C, or from one of the micromeres, 1a, 1b, or 1c. Vegetal fragments always cleave like whole eggs, form polar lobes, and may become complete, normally proportioned miniature larvae (Fig. 393d). The cytoplasm of the animal pole region is thus not necessary for the differentiation of the *Dentalium* embryo; in the vegetal half of the egg the distribution of cytoplasmic substances is regulative, making possible a normal larva.

In the polar lobes and near the animal pole, the mitochondria gather as shown by Janus green staining (Fig. 394a–d). From the second polar lobe they collect in the 1 D cell (Fig. 394e). Later the cells of the apical organ and prototroch stain strongest (Fig. 394f).

The nature of determinate cleavage goes far beyond the diverse visible and chemically demonstrable localizations of cytoplasm. Moreover, the early deployment of determining substances in

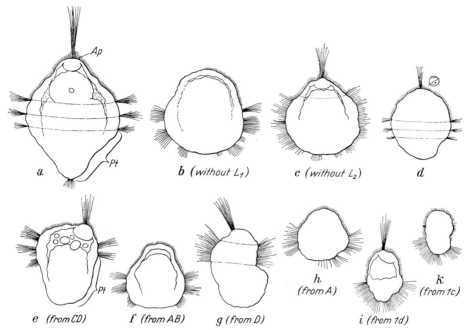

Fig. 393. *Dentalium*. a normal larva; b, c larva after removal of the polar lobe of the first or second division (compare Fig. 392); d larva from the vegetative portion of an unfertilized egg (orientation of section is sketched directly above); e–k larvae from individual blastomeres. *Ap* apical organ; *Pt* post-trochal organ. (After WILSON, 1904)

the egg cell is not decisive in itself. This is suggested by the variety of visible movements of substances during cleavage (e.g., Figs. 384, 385), and is proven by extirpation experiments: In *Dentalium* a new segregation of vegetal and animal determinants previously united in the vegetal polar lobe clearly takes place in the CD cell. We cannot decide whether further cleavage leads to the segregations of substances with progressively subtler qualitative differences, or whether the norms of reaction of the cells with alternative developmental paths are so delicately set on a stepwise quantitative scale that the larva exhibits mosaic determination.

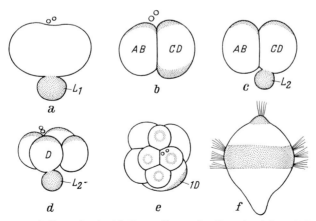

Fig. 394. Dentalium eggs vitally stained with Janus Green. L_1, L_2 polar lobes. (After REVERBERI, 1958)

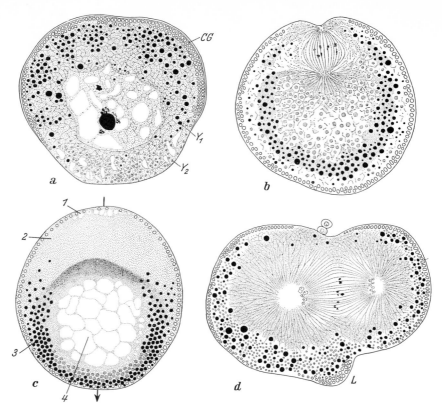

Fig. 395. *Chaetopterus pergamentaceus*. a mature oocyte in ovary; b oocyte with first maturation spindle after laying; c centrifuged oocyte (section did not cut through maturation spindle); d first cleavage, Y_1, Y_2 different kinds of yolk granules. *L* polar lobe; *CG* cortical granules; *1* oil droplets; *2* clear cytoplasm; *3* yolk granules; *4* central cytoplasmic mass (see text, p. 311). (After LILLIE, 1906, 1909)

The long reach of the determinants in the cytoplasm of a mosaic egg is revealed by the amazing occurrence of larval differentiation in an undivided *Chaetopterus* egg. The contents of *Chaetopterus* oocytes show a marked chemical polarity (Fig. 395a): a cortex containing colorless granules covers the egg, with the exception of a large region around the vegetal pole. Various types of yolk occur in polar layers. With the dissolution of the oocyte nuclear membrane, a rearrangement of the egg substance begins. As the first spindle is forming at the animal pole, the ectoplasm spreads in waves covering the vegetal part of the egg, and thus the yolk layer (Fig. 395b). At the beginning of the first cleavage division the ectoplasm parts in a latitudinal zone, and that portion in the vegetal half collects into a mass, which gathers in the rather flat polar lobe characteristic of *Chaetopterus*, and thus into the CD cell (Fig. 395d).

Parthenogenetic development can be induced in the eggs by KCl; the various cytoplasmic substances in the oocyte stream to their various positions, just as they would do before the first cleavage in a normal egg. But nuclear and cell division never take place. At the vegetal pole the modest polar lobe is formed. After some time a remarkable "unicellular gastrulation" follows: the ectoplasm gradually covers the massed endoplasm again (Fig. 396b). Cilia now form on the ectoplasm (Fig. 396c, d). The entire resulting structure (Fig. 396d) resembles the normal *Chaetopterus* trochophore larva (Fig. 396a). Its narrower part corresponds to the pretrochal region and its wider part to the post-trochal region. As in the normal larva, a ring of large vacuoles lies inside between the two regions. The position of the yolk-rich endoplasm corresponds to the position of the gut. The apical organ and prototroch of the larva are admittedly missing in the one-celled form; but its ciliary beat is strong and coordinated. Indeed, the differentiated "egg" swims around like a trochophore. The nucleus in its single cytoplasmic body becomes very large and evidently polyploid due to periodic endomitosis.

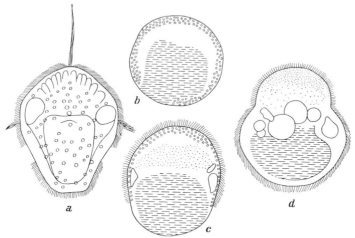

Fig. 396. *Chaetopterus*. a normal trochophore larva; b–d development of an undivided uninucleate egg; b "unicellular gastrulation"; c, d appearance of cilia in the regions of the egg covered by the ectoplasm; d unicellular larva. (After LILLIE, 1906)

This paradoxical phenomenon, whose closer investigation even with cytochemical methods would be well worth the effort, makes one thing clear. It is not necessary for cell territories first to form and then be allocated specific materials; rather, the cytoplasmic mosaic can evoke directly the differentiation which normally occurs in the blastula or later.

The mosaic at the outset of cleavage results first from the synthesis of various substances during oocyte growth, and second from a particular organization of these materials. This organization takes place, as we have seen in many eggs, during maturation or just after fertilization, by means of cytoplasmic streaming and the migration of cytoplasmic inclusions.

Some cytoplasmic components whose position depends on gravity play a role in constructing cytoplasmic mosaics in the ascidians and in the forms with spiral cleavage. This we see from the centrifugation experiments on *Styela* and *Ascidiella* embryos, which we have already discussed, and from comparable experiments on annelids, particularly on leech and Oligochaete eggs. In *Clepsine* and *Tubifex*, the pole plasms, masses of cytoplasm at the egg poles, play important roles in the determination, not only of the course of cleavage, but also of further development. Centrifugation causes them to meet and mix with other egg substances, and the result is the disturbance of development. In some eggs the arrangement of substances in the endoplasm is resistant to moderate centrifugation. In *Chaetopterus* the centrifuged egg shows four layers (Fig. 395c). Centripetally, there is a cap of fat droplets; then a zone of dense, clear cytoplasm; at the centrifugal end is a layer of yolk granules. The fourth region, in the interior, is a reticulate or foamy cytoplasmic mass corresponding to the central yolk region of an uncentrifuged egg cell (Fig. 395b). A zone of dense cytoplasm surrounds this central spongy endoplasm, within which light granules move to the centripetal border. Apparently they cannot penetrate the limiting layer. Centrifugally, the yolk granules are spun out of the spongy endoplasm. In *Chaetopterus*, a visible endoplasmatic structure is thus retained in spite of the displacement of some of the cytoplasmic inclusions by centrifugation. The cortex and its granules remain unaffected. In other eggs a comparably stable egg structure has not been demonstrated; but it has been concluded that the endoplasm is pervaded by a spongy matrix, a sort of continuation of the stable cortical structure, because of the resistance of cytoplasmic layers to centrifugal force in some cases and because of the restoration of the original distribution after centrifugation has stopped in others. Perhaps the inclusions move without destroying the matrix, or perhaps it is reestablished as a new outgrowth of the cortex. Impressive indeed are the cytoplasmic streams which gather up the inclusions displaced by centrifugation, and restore them to their normal sites. In this way the ascorbate-containing granules, after being forced to the centrifugal pole (Fig. 383 d), migrate back to their original zone (Fig. 397). When lipid droplets are forced centripetally to the opposite side, they return along the cortex to the animal half of the egg.

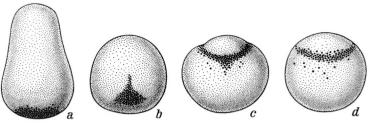

Fig. 397. *Aplysia limacina* egg. a ascorbic-acid-containing granules at the centrifugal pole; b–d return to the normal zone (cf. Fig. 383). (After PELTRERA, 1940)

Again and again we are confronted by an organizing structure in the egg cortex, and in many cases it clearly plays the decisive role. Here is a final example: In Cladoceran eggs, we have already concluded, there is a superficial cytoplasmic streaming which in centrifuged eggs regulatively restores the cleavage nuclei to a nearly normal equatorial position (pp. 164 ff.). The further development of centrifuged eggs of *Daphnia* shows that the body axes are fixed in the cortex. The long axis of the egg coincides with the long axis of the embryo. The egg's animal pole lies on the future dorsal side of the embryo, and between it and the anterior pole of the embryo is the site of the presumptive apical plate, from which the supraoesophageal ganglion is formed. Eggs transferred from the brood pouch into the water take up a position with the animal pole below, since the vegetal half contains the light oil droplet (Fig. 205 a). Eggs placed in gelatin may be fixed in any position, so that one may established any desired angle between the long axis, or the animal-vegetal axis, and the axis of centrifugal force. With sufficiently strong centrifugation a sorting and layering of the variously dense substances in the *Daphnia* egg take place. The heavy protein granules gather in one place, while the oil droplet moves in the opposite direction, and the clear cytoplasm streams into a single zone (Fig. 205 f) from the superficial regions at the animal and vegetal poles, from the layer beneath the egg cortex, and from among the yolk granules (Fig. 205). This displacement of the yolk to one side, and the clear cytoplasm to the other, brings about characteristic abnormalities in development. Figure 398 a and b shows normal embryos after formation of the limb rudiments in ventral and lateral aspect respectively. Displacement of the yolk mass toward the posterior pole and somewhat to left (Fig. 398 c_1) results in the embryo seen in Fig. 398 c_2. The embryo has not developed at the normal rate, but the malformation is already clear: the whole body is distorted in the long axis. The head is disproportionately large and contains a giant bipartite ganglionic mass more than twice as large as the supraoesophageal

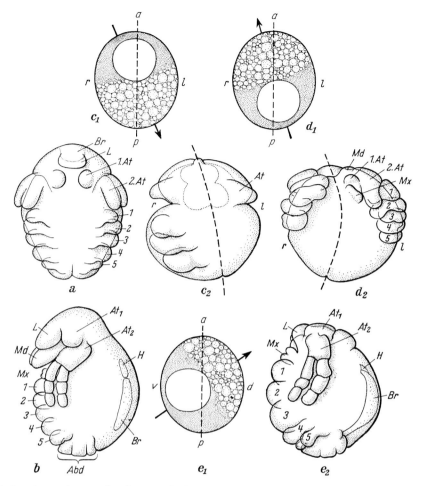

Fig. 398. *Daphnia pulex*. a, b normal embryos; a 24-hours old, ventral view; b 36-hours old; left view; c–e embryos from centrifuged eggs; c_1–e_1 diagrams of the stratification of materials; c_2 26-hours old; d_2 30 hours old; e_2 45 hours old. *Abd* abdomen; *At_1*, *At_2* first and second antennae; *Br* brood pouch anlage; *Br* brain anlage; *H* heart anlage; *Md* mandible; *Mx* maxilla; *L* labrum; one to five thoracic appendages; direction of centrifugation. (After KAUDE-WITZ, 1950)

ganglion of the normal embryo. In the thoracic region the right side has three relatively large extremities; the left side has none. The embryo in Figure 398 d_2 came from an egg exposed to centrifugal force in the opposite direction. It lacks the anterior head region with the brain; the abdominal region is too large, and the extremities on the left side are better developed than those on the right. As a result of the accumulation of yolk in the anterior-dorsal region (Fig. 398 e_1), the embryo in Figure 398 e_2 lacks that part of the head which ordinarily contains the brain and later the eye rudiments. The limbs are all laid out, those on the thorax being abnormally large.

The overloading of a presumptive egg region with yolk thus restricts the morphogenetic activities of this region. Conversely, excessive clear cytoplasm exaggerates development. Centrifugation can cause a layering of these substances and, in turn, a gradient of developmental restriction or exaggeration. The extent of the body regions and organ domains is thus not yet defined in the mature egg, although bilateral symmetry and a long axis of differentiation have already been fixed, and they cannot be affected substantially by the cell contents, which can be layered. One is led to conclude that these determinants are fixed in the topologically immutable egg cortex, which in *Daphnia* is very thin.

Of the nature of the formation and action of this cortical structure, to which a great diversity of experimental results has already called our attention (pp. 60, 136 ff., 146, 162 − 165, 200, 223), we still have just as little a notion as we have of the nature and effect of the determining components of other eggs and of their typical organization through cytoplasmic streaming.

Lecture 22

The insects, in contrast to amphibians and echinoderms, annelids and molluscs, have only recently come under the scrutiny of developmental physiologists. But today in a very broad area of development they present excellent examples for the interplay of developmental principles, from embryonic development to the relatively late pattern formation in the imaginal discs in metamorphosis and the hormonal control of the events in metamorphosis.

The eggs of insects have a lot of yolk, and their growth can serve as an example of a phenomenon seen in a great diversity of animal groups: the support of oocyte growth by auxiliary cells. In most insects, (including Hymenoptera, Coleoptera, Neuroptera, Lepidoptera, and Diptera) the eggs are formed in chains, and behind each egg in line is a group of nurse cells. At first these cells grow along with the oocyte; then they are absorbed (Fig. 399). The oocyte is in an egg compartment covered with a follicle epithelium, while the nurse cells occupy a separate compartment. From the numbers of cells it seems clear that the oocyte and the nurse cells emerge from successive cleavages of one oogonium: in the various forms examined, each egg has three, seven, or fifteen nurse cells, following the relationship $2^n − 1$. The egg cell and its nurse cells form a cell complex in which cytoplasmic strands (spindle remnants) maintain connections among the cells in the pattern in which they originally divided. Figure 400 a shows the cell complex in a butterfly as reconstructed from serial sections. The oocyte is connected in this way with three nurse cells, while the various nurse cells are connected to one another in various ways. If one arranges the cells according to their immediate history of divisions, it is clear that in three divisions four cell-generations follow one another, consisting of one, two, four, and eight cells. In one line, three determinative divisions always separate the future oocyte from a presumptive nurse cell (Fig. 400 b). As in embryonic development, during determinate cleavage there are synchronous divisions which determine the prospective fate and the arrangement of the cells.

In the same species — as a modification or as a mutation in local population — an additional division may take place, so that instead of eight, there are sixteen cells, one oocyte and fifteen nurse cells. Whether the egg-nurse-cell complexes in all insects are laid out in "polytrophic" egg tubes according to this plan has not yet been established.

The auxiliary cells provide material for the growing cytoplasm of the egg cell with its organized RNA-containing particles (ribosomes), mitochondria, and for reserve materials. Some of the metabolic activities of the nurse cells and of the follicle-epithelial cells have been established by autoradiography. If tritium-labeled cytidine, a component of RNA, is injected into a female fly, strong radioactivity appears after only a few minutes in the nuclei of nurse and follicle cells. Emanating from the chromosomes and nucleoli, it first fills the entire nucleus (Fig. 401 a) and then the cytoplasm of the nurse cells (Fig. 401 b). Now the radioactivity moves into the egg cytoplasm, where it is first seen in discrete masses coming out of the cytoplasmic bridges (Fig. 401 b). When the nutritive complex breaks down, the RNA-rich nurse cell cytoplasm streams into the

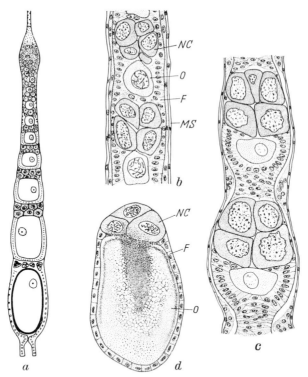

Fig. 399. Egg-nutritive cell complexes of polytrophic egg tubes of Lepidoptera. a sketch of an egg tube; b–d *Ephestia kühniella*: three stages of the egg-nutritive-cell complex. *F* follicle epithelium; *MS* mesodermal sheath of the egg tube; *NC* nurse cells; *O* oocyte. (a after WEBER, 1954; b–d after DA CUNHA, 1942)

oocyte on a broad front, and the growing egg receives the major portion of the labeled RNA (Fig. 401 c). The oocyte nucleus takes no part in RNA synthesis during this functional phase of the nurse cells; it remains free of label. And according to the autoradiographs no RNA moves from the follicle epithelium into the oocyte. The follicle epithelium, however, carries out the synthesis of yolk. When tritium-labeled amino acids are injected, the cytoplasm of both nurse and follicle cells takes up the label, but a radioactive fringe of yolk appears only at points of immediate contact between egg cytoplasm and follicle epithelium (Fig. 401 d). It is released in rather round complexes, which resemble yolk granules in size and form, and which move to the interior of the egg. The follicle epithelium, at any rate, transmits a protein synthesized in other

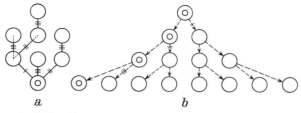

Fig. 400. Egg-nurse-cell complex in Lepidoptera. a diagram of the spatial arrangement of the cells (after KNABEN, 1934); b arrangement according to successive divisions (after HIRSCHLER, 1942). Solid lines unite cell pairs; cross lines each represent one division. Empty circles: nurse cells; circles with small inner circles: oocytes

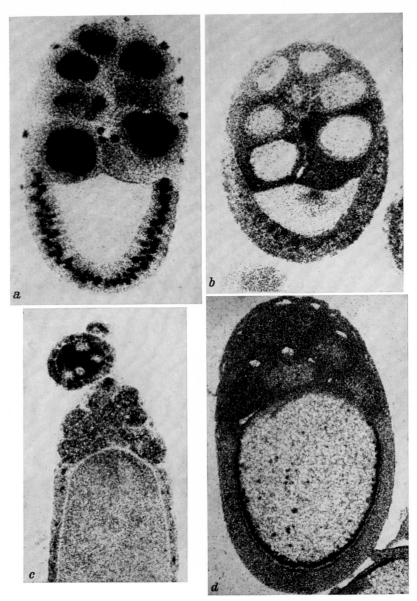

Fig. 401. Egg-nutritive cell complex in flies: a–c *Musca domestica*; d *Calliphora erythrocephala*. a–c after H³-cytidine injection: a after thirty minutes; b after five hours; c degeneration of the nurse cells with a section of a young nutritive complex above; d eighty minutes after injection of H³-histidine. (After BIER, 1962, 1963)

parts of the body from the hemolymph to the egg surface. The proteins from the hemolymph and the yolk granules are immunologically similar, but the strong RNA- and protein-metabolism of the follicle cells suggests that they also synthesize yolk protein.

For clear insight into the determination of the basic body plan, we can thank especially SEIDEL, KRAUSE, and their co-workers and students.

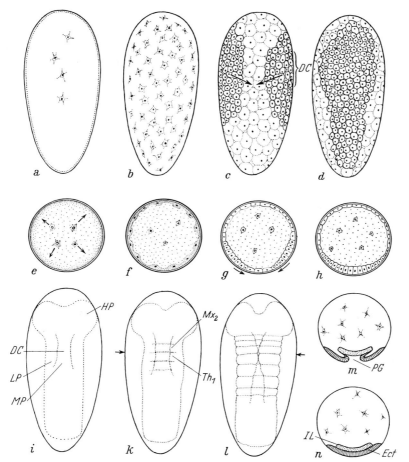

Fig. 402. Development of the insect embryo, superficial cleavage and formation of the germ band, diagrammatic. a–d ventral view: a onset of cleavage; b 128-nucleus stage, in which each nucleus is near the egg surface within a sphere of cytoplasm; c formation of the right and left halves of the germ band through frequent division and aggregation of the blastoderm cells, which takes place first in the zone of the differentiation center; d single ventral germ band after the fusion of two halves; e–h cross section through stages corresponding to a–d; i germ band with head lobes, onset of primitive gut formation; k first sign of segmental borders; l continuation of ventral groove formation and segmentation, anterior and posterior to the differentiation center; m, n cross sections at the level of the arrows in k and l. DC differentiation center; Ect ectoderm; HP Head process; MP middle plate; Mx₂ second maxillary segment; PG primitive groove; LP lateral plate; Th₁ first thoracic segment; IL inner layer. (After SEIDEL, 1936 and WEBER, 1954)

The characteristics of the course of early development can be laid out in a general way. Insect eggs cleave superficially. Nuclear division begins deep in the yolk at a place specific for each kind of egg, to which the zygote nucleus is carried by cytoplasmic streaming. At this center of cleavage the nuclei together with some cytoplasm separate and move, after a series of divisions, in certain directions toward the egg surface (Fig. 402a, e). There their surrounding cytoplasm fuses with a cytoplasmic layer at the surface. This cortical plasma, or periplasm, is generally known as the embryonic surface blastema. The nuclei divide further, and when they have reached a certain density, cell membranes arise between them. This single layer of surface epithelium is the blastoderm, in which the embryonic anlagen differentiate. Typically a pair of first embryonic

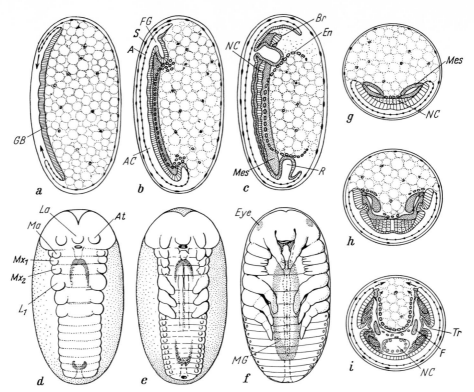

Fig. 403. Diagram of insect embryonic development from amnion formation on. a–c sagittal section; d–f ventra
view; g–i cross section. *A* amnion; *AC* amniotic cavity; *At* antenna; *Eye* eye; *L₁* first pair of legs; *NC* nerve cord;
R hindgut; *En* endoderm; *F* fat body; *Br* brain; *GB* germ band; *Ma* mandible; *MG* midgut; *Mes* mesoderm;
Mx₁, *Mx₂* first and second maxillae; *La* labrum; *S* serosa; *Tr* tracheal invagination; *FG* foregut. (After WEBER, 1954)

rudiments arises (Fig. 402c, g) in which the cells are more columnar and more rapidly dividing
than in the remaining blastoderm. These first rudiments fuse ventrally into a shield or heart-
shaped germ band (Fig. 402d, h), which spreads to the posterior part of the embryo. The remaining blastoderm is used in the formation of membranes. These cell movements are not endogenous
in the blastoderm cells; rather the cells are transported by the contractile movements of the cytoplasmic syncytium, which pervades the egg as a network among the yolk granules. The cells of
the germ band first assemble in a region corresponding to the presumptive thoracic or maxillary
segments. From this center of differentiation the aggregations of cells spread out; the definitive
morphogenetic processes have their origins here also. The anterior part of the germ band expands
into paired head lobes (Fig. 402i).

The germ band – as sketched in Figure 402 – includes a considerable part of the blastoderm,
or it may be relatively small (Fig. 404c), or even so tiny that it extends for only one-eighth to
one-twentieth of the length of the egg (Fig. 421).

Those nuclei that remain in the cytoplasmic net among the yolk granules are later bounded by
cell membranes and act as vitellophages in yolk absorption, although they can also participate in
organ formation. In some forms the yolk subsequently cleaves into giant yolk cells (Fig. 404g).

The separation of the germ layers is signalled by the separation of a medial plate from the
lateral plates. The medial plate moves inward to form a ventral groove and is overgrown by the

lateral plates, which form the ectoderm (Fig. 402i, m). This process emanates from the differentiation center and spreads fore and aft (Fig. 422a–f). Similarly, segmentation of the body spreads out from this center (Fig. 402k, l); the first segmentation furrow to appear divides the presumptive second maxillary segment from the presumptive prothoracic segment (Fig. 402k, 422f$_1$). When the ectodermal lateral plates have fused medially, forming segments, the germ band reaches its definitive form.

From that part of the blastoderm which does not form the germ band, the extraembryonic blastoderm, two embryonic membranes generally develop which cover the embryo (Fig. 403a, b, g). The outer membrane, which also covers the yolk surface, is called the serosa; the inner membrane, the amnion, emerges from the edge of the germ band and closes over it, adhering to the serosa

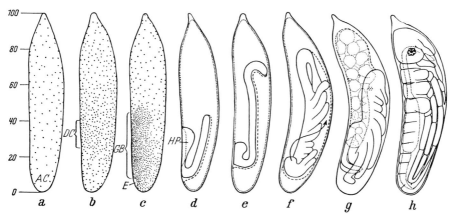

Fig. 404. Embryonic development of *Platycnemis pennipes*. a cleavage; b onset of formation of the germ band; c germ band before infolding; d, e infolding; f formation of extremities in the folded embryo; g outfolding with twisting of the embryo, yolk cleavage complete; h after posterior cleavage, shortly before hatching. *AC* activation center; *DC* site of the differentiation center; *E* origin of infolding; *GB* germ band; *HP* head process. (After SEIDEL, 1929 and KRAUSE, 1939)

as the membranes form. The amniotic or embryonic cavity is filled with a fluid on which the anlage rests and into which the extremities will be everted.

The ectoderm of the germ band invaginates near the front end and at the posterior end of the embryo to form the primitive foregut and hindgut (Fig. 403b, c). Beginning at the differentiation center, segmentally arranged groups of neuroblasts for the ganglia of the ventral nerve cord separate from the ectoderm toward the front and rear, and form the midline toward the sides (Fig. 403b, c, g, h). In the head region the paired brain rudiments differentiate from the ectoderm in the inner layer or primitive groove; a pair of lateral bands separate as mesoderm from a medial band which will generally become endoderm (Fig. 403g). The segmentation of the ectoderm likewise gives rise to mesoderm, and in each segment a coelomic cavity arises (Fig. 403g). These coelomic sacs open later in the general body cavity and various parts of their wall give rise to the rudiments of skeletal muscle, the muscular coat of the gut, the mesodermal sheaths of other organs, the fat body, and the blood cells (Fig. 403h, i).

The formation of the midgut begins with cell aggregations which arise at the front and hind ends of the medial band of the inner layer. These grow backwards and forward in crescent form and finally unite with or without the participation of the "medial cord" lying between them, to the ventral rudiment of the midgut epithelium, which later grows dorsally over the yolk mass (Fig. 403d–f, h, i). All the organs are at first laid out in the ventral embryonic band; then the

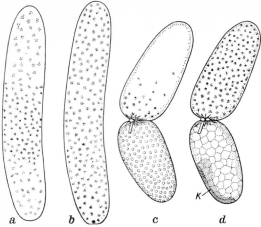

Fig. 405. Spread of anucleate cytoplasmic areas in *Gryllus domesticus* egg. c, d same in front part of an egg after ligation in the two nucleus stage; the posterior half, which retains the nuclei, shows blastoderm formation, (c) and germ band (*K*) and yolk cleavage (d). (After MAHR, 1960)

lateral parts grow dorsally together with the amnion and fuse in the dorsal midline (Fig. 403i). During development the embryo expands, especially when the germ band is very small at the beginning (in "short" embryos). During this expansion it bends and often sinks deep into the yolk mass. Later this inward curvature is reversed by an outward curvature. Figure 404 shows the major movements in the early development of the dragonfly *Platycnemis pennipes*, one of the classic objects of the experimental embryology of insects.

The experimental elucidation of the events in determination shows a surprising diversity of phenomena in the various kinds of insects.

One event, called pseudocleavage, leads to a functional structuring of the egg cytoplasm. It is found in a number of insect forms. In unfertilized eggs or after the separation of the first cleavage nuclei, there arise in the cytoplasmic net which pervades the yolk, anucleate spots of cytoplasm, whose number increases to a level roughly corresponding to that of the nucleated spots of cytoplasm in the 512-nucleus stage (Fig. 405a, b). These anucleate spots of cytoplasm do not divide, nor do they migrate. Their formation begins at the cleavage center and spreads in all directions. If, when an egg is ligated, the first nuclei are in the posterior part of the egg, a germ band develops here (Fig. 405c, d). In the anterior part cleavage spreads from the point nearest the cleavage center (Fig. 405c, d). The cleavage center is, therefore, a place in the cytoplasm from which influences gradually pass through the cytoplasmic net. Through this autonomous tendency to form spots of cytoplasm the distribution of the cleavage nuclei is achieved in normal development.

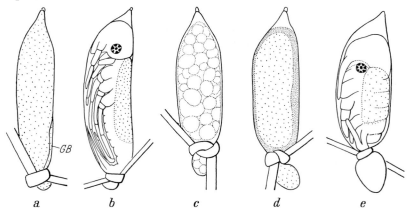

Fig. 406. Demonstration of the activation center in the *Platycnemis pennipes* egg. a, b ligation close to the posterior pole does not interfere with the formation of the germ band and a complete embryo; *GB* germ band (a); c, d ligation a little further forward prevents germ band formation; yolk cleavage (c) and development of a blastoderm without germ band (d); e ligation in the blastoderm stage anterior to the ligature in a has permitted germ band formation and the subsequent formation of a partial embryo. (After SEIDEL, 1936)

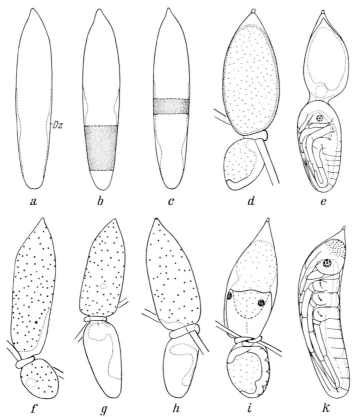

Fig. 407. Demonstration of the action of the differentiation center in *Platycnemis*. a normal site of the differentiation center (*Dz*); b, c after irradiation in the 512-nucleus stage the differentiation center is moved forward (b), or backward (c); d–i ligation experiments; d, e after tight ligation of a cleavage stage (Fig. 404a) behind the middle of the egg permits germ band formation only behind the ligature; e complete mature embryo behind the ligature; f–h ligation in blastoderm stage: f behind the differentiation center; g in front of the differentiation center; h in the region of the differentiation center; i loose ligation in a late cleavage stage; k normal embryo shortly before hatching. (After SEIDEL, 1936)

Further experiments on the *Platycnemis* egg reveal additional morphogenetic centers. During ligation before the cleavage nuclei have reached the entire egg surface one can separate only a very small part at the posterior end of the egg without preventing the development of a normal embryo (Fig. 406a, b). If the ligature is moved only slightly forward the germ band usually will not be formed. Yolk cleavage takes place, and the whole blastoderm develops into extra-embryonic tissue (Fig. 406c, d). In the posterior part of the egg, therefore, is a region, the activation center, on whose presence the formation of a germ band depends. Without it the cleavage nuclei with their surrounding cytoplasm and the cortical cytoplasm cannot form the germ band. The cytoplasmic region at the posterior end of the egg can achieve its effect only if it is supplied with cleavage nuclei; even if a ligature does not completely separate the egg into two parts, the germ band will not form if no nucleus passes through the remaining channel. But if the opening is large enough to permit cleavage nuclei to reach the activation center, embryonic development takes place. As soon as the nuclei have reached the activation center, an influence is broadcast: if an egg is ligated at the same level as in Figure 406c, but in the early blastoderm stage embryonic development proceeds in the isolated anterior part (Fig. 406e). Clearly, a chemical effector spreads forward from the activation center once it has been supplied with nuclei; as blastoderm formation proceeds, greater and greater regions of the egg may be killed with heat without preventing embryonic development. The formation of this effector substance depends upon a reaction between nuclei and a certain cytoplasmic agent in the posterior part of the egg.

A second region of particular importance in early development is the differentiation center, which manifests itself in the sequence of events in normal development. If one ligates an egg in the blastoderm stage before or behind the presumptive differentiation center, only that part of

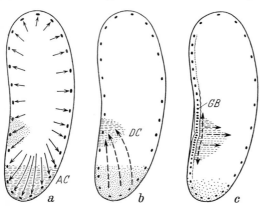

Fig. 408. Diagram of reaction sequence: activation and activity of the activation center (*AC*) and of the differentiation center (*DC*); *GB* germ band. (After WEBER, 1954)

the egg which contains the differentiation center forms a germ band (Fig. 407d–g). If the ligature is made in the area of the differentiation center, each part of the egg develops a germ band (Fig. 407h). The site of the differentiation center is not firmly fixed in a certain place in the cortical cytoplasm from the beginning; rather, it can be displaced by regulation: Ultraviolet irradiation of a ring of cortex in the region of the presumptive differentiation center during cleavage can result in the formation of another center further forward or behind the usual site (Fig. 407b, c). Also, ligation at the middle of the egg in an early cleavage stage displaces the differentiation center toward the rear, and a normally proportioned miniature embryo can arise from the posterior half of the egg (Fig. 407e).

The first alteration of the egg structure associated with the formation of a normal or displaced differentiation center is a local concentration of reticulate cytoplasm and yolk; at the site of contraction the blastoderm cells aggregate into a germ band. Accordingly, the differentiation center must be activated by the substance emanating from the activation center. A local reaction must be triggered, which, as the irradiation and ligation experiments show, is related to the disposition of the existing egg material. From the differentiation center new effects now spread toward the front and rear as well as laterally from the midline. Figure 408 is a diagram of the sequence of reactions. If a loose ligature is made during a cleavage stage, the effects of the activation center and differentiation center can spread throughout the embryo, so that the head section develops in front and the posterior part behind (Fig. 407i). Further normal development of the parts is of course prevented in these cases because invagination is hampered.

Another kind of determination is revealed by experiments on the cicada *Euscelis plebeius*: an effect on differentiation mediated by bipolar tendencies. After ligation of late cleavage stages (about 128 nuclei) until germ band formation, both parts of the egg can form embryos of varying degrees of completeness according to the site and time of ligation. If the egg is ligated in a cleavage stage more than about 60 per cent of the distance from the posterior end (Fig. 409), a complete embryo (Fig. 410a) generally develops in the posterior part, although sometimes the anterior part of the head region is incomplete (Fig. 410b). A ligature slightly further back causes the formation, in the posterior part, of an embryo whose anterior segments are missing (Fig. 410c). Also, for the formation of prothorax, mesothorax, and metathorax there are corresponding minimal distances from the posterior end of the egg for a given embryonic age (Fig. 409). The minimum distances necessary for total development and for the development of particular parts decrease as development proceeds (Fig. 409). In the pre-embryonic stage, after a ligature is placed 40 per cent of the distance forward from the posterior end, the embryos which arise are normal or have, at most, defects in the anterior part of the head (Fig. 410 d). This change in response does not suggest a progressive mosaic determination, but rather an increased ability to regulate, since the ability to form the anteriormost parts in an isolated posterior egg region increases in proportion with the length of time contact has existed with the anterior part of the egg. From this we may conclude that cytoplasmic tendencies or "anterior factors" which are necessary for the formation of the anterior structures have spread from an anterior region of the egg back into the posterior section.

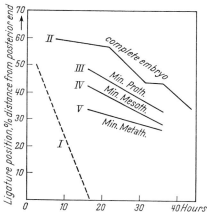

Fig. 409. Extent of differentiation of posterior portions of the *Euscelis* egg. *I* posterior border of the nuclear aggregation during cleavage; *II–IV* minimal distance of border from the posterior pole: *II* for complete embryos; *III* for prothorax; *IV* for mesothorax; *V* for metathorax. Ordinate: site of ligature; distance from the posterior end as percentage of egg length. (After SANDER, 1959)

From the anterior part of the egg, complete embryos arise independent of the age at ligation if the ligature is placed near the posterior end of the egg. If it is placed more than 40 to 50 per cent of the distance from the posterior end, no germ band develops in the front half. After ligation between about 23 per cent and 45 per cent from the rear, "complementary" partial embryos arise in the two compartments of the egg (Fig. 411). As the ligature is placed further back, the formation of the posterior parts in the forward compartment is progressively favored. The anterior isolates always begin with the anteriormost segments; the smallest ones are anterior head fragments. The posterior isolates always include the abdomen or, at the very least, just the tip of the abdomen. So anterior isolates make anterior parts, and posterior isolates make posterior parts. After ligation in cleavage stages numerous segments are missing from the total produced by both partial embryos (Fig. 411); ligation in the pre-embryo stage results in the loss of only a few segments.

The effects of the posterior factors which are found in the posterior part of the egg are elegantly demonstrated by a transplantation experiment. In numerous insects, at the posterior end of the

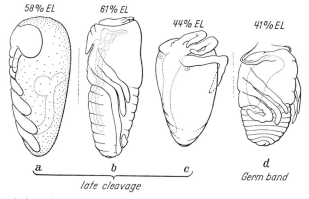

Fig. 410. Extent of differentiation of posterior portions of the *Euscelis* egg after ligation; position of ligature as in Figure 409. a late cleavage, 58 per cent egg length; b late cleavage, 61 per cent e.l.; c late cleavage, 44 per cent e.l.; d pre-germ band, 41 per cent e.l. (After SANDER, 1959)

323

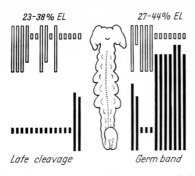

23-38% EL   27-44% EL

Late cleavage   Germ band

Fig. 411. Complementary differentiation of the posterior (black) and the anterior isolates of *Euscelis*. Above: region of ligature; below; age at time of ligation. (After SANDER, 1959)

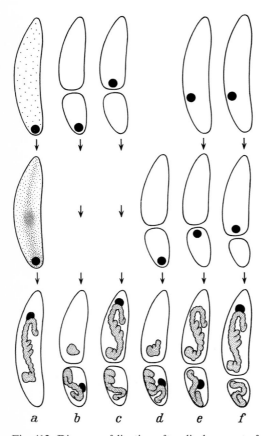

a   b   c   d   e   f

Fig. 412. Diagram of ligation after displacement of the cluster of symbionts in *Euscelis*. a normal development; the upper two sketches represent stages at ligation; b, d ligation without displacement; c early, e, f late ligation experiments. Symbionts black. (After SANDER, 1961)

egg there is a ball of symbionts transmitted to the egg by the mother. This cluster of symbionts can be moved far forward after the invagination of the posterior end of the egg by moving the invaginated part with a blunt needle. If it is placed near the periphery it remains there when, after the removal of the needle, the egg resumes its former shape due to the elasticity of its shell. Now, after the displacement of the cluster of symbionts, a ligature is applied (Fig. 412). If this is done in a cleavage stage (Fig. 412c) or in the pre-embryo stage (Fig. 412f), with the ligature behind the symbionts and about one-third of the way from the posterior end, a complete or nearly complete embryo develops in front of the ligature (Fig. 413, 414) which incorporates the ball of symbionts. In the control experiment, without the forward displacement of the symbionts Fig. 412 b, d), only scant anterior parts of an embryo are formed. If the ligature is made directly in front of the displaced ball (Fig. 412e), a complete embryo nevertheless develops in the forward compartment of the egg. The symbionts themselves are, therefore, not the critical factor. Rather the posterior polar material displaced at the same time supplies the posterior factors which complement the anterior factors already present. In the rear compartment with the displaced symbionts (Fig. 412e) the polarity of the embryonic anlage is reserved. The tip of the abdomen with the hindgut is not found at the rear end of the egg as in Figures 410 and 411; rather, it is near the ligature. This reversal of polarity due to the concentrated posterior polar material corresponds exactly to the polarizing effect exercised by echinoid micromeres, which have the maximum vegetal tendency (p. 173f.). Just as bipolar vegetal tendencies were apposed in the echinoid experiment (Fig. 221f), a vegetal mirror image embryo with a hindgut at each end and thoracic segments in the middle can be made in the *Euscelis* egg when the material remaining at the posterior pole and the nearby material which has been moved forward (Fig. 412e) exercise their influence from opposite ends of the forming embryo. Exactly the same sort of mirror image embryos can be made by centrifuging *Chironomus* eggs. When the yolk is hurled toward

324

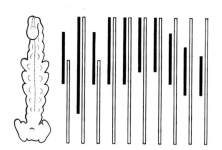

Fig. 413. Examples of the differentiation of anterior and posterior parts of the *Euscelis* egg after forward displacement of symbionts in late cleavage stages. (After SANDER, 1960)

Fig. 414. *Euscelis* embryo with defective abdomen from an anterior part including a displaced ball of symbionts (*Sy*). Ligation stage: late cleavage. Ligation site: 25 per cent from the rear. (After SANDER, 1960)

the anterior pole, embryos with two rear ends are formed (Fig. 415b). If the yolk is spun to the posterior end, the embryos have two heads. Centrifugation at right angles to the long axis does not disturb development (Fig. 415a).

These experiments on *Euscelis* lead to the rough scheme illustrated in Figure 416.

The bilaterality of the embryo is established in certain groups as early as the time of fusion of the preembryonic anlagen (half germ bands). If one of the preembryonic anlagen in the field cricket (*Gryllus domesticus*) is eliminated by UV irradiation, the remaining anlage moves ventrally and gives rise to an entire embryo (Fig. 417b₂). With the ventral and posterior movement of the half germ bands, the regulatory ability decreases: if a half germ band is destroyed shortly before fusion or just after it, a one-sided half-embryo arises (Fig. 417c₂), except that the abdomen can still be formed bilaterally. Evidently the embryonic parts are subject to a bilaterally determining process, which moves gradually back from the differentiation center.

In certain long embryos the longitudinal and lateral differentiation can already be determined in the blastoderm. In the honeybee the future site of the differentiation center can already be recognized in the distribution of cytoplasm in the egg: In a region ventral to, and behind, the front end, behind the place where the egg and sperm nuclei fuse, the cortical cytoplasm is most concentrated and the cytoplasmic net among the yolk granules is thickest (Fig. 418b). Here the first cell membranes are formed. From the 32-nucleus stage on, when the migration of the nuclei to the surface begins, this region stands out through a particular affinity for thionine. This affinity spreads from the dorsal side to the ventral side and finally encircles the egg like a belt. It suggests a chemical change in the cytoplasm of the zone which will become the differentiation center. As in dragonfly and grasshopper embryos, the center in the long embryo of the bee lies in the presumptive prothoracic area. Here the separation of the medial and lateral plates begins

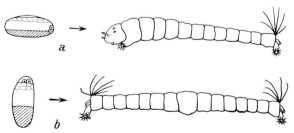

Fig. 415. Results of centrifuging *Chironomus* eggs. a at right angles to the long axis; b in the long axis, yolk displaced forward, abdomen at each end. The pole cells lie at the posterior end of the egg. (After YAJIMA, 1960)

(Fig. 419a). In the absence of contact with the differentiation center the blastoderm material is capable of no further morphogenesis except for the formation of the amnion. But in spite of the preparations in the cortical cytoplasm, the site of the differentiation center is not fixed once and for all. Until the early blastoderm stage in which a loose epithelium of flat cells is still being formed, small but normally proportioned whole embryos can follow ligation of about the front fifth of the egg (Fig. 420b). In these eggs which have been ligated early, the differentiation center relocates further back. It occupies relatively the same position in the shortened egg that it would have assumed in the normal egg, and again this position is revealed by the start of furrows between medial and lateral plates (Fig. 419b). In later blastoderm stages tight ligatures lead to

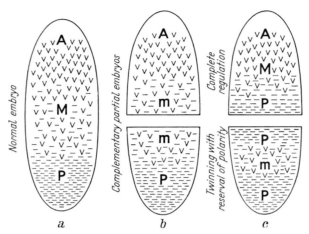

Fig. 416. Diagram of early determination in *Euscelis*. a normal development; b anterior and posterior portions cut apart; c transection after forward displacement of the cytoplasm from the posterior part of the egg. Temporal relationships are ignored. Determination of *P* (posterior), *M* (central) and *A* (anterior) parts of the embryo. *m* incomplete determination of the central portion

mosaic formation (Fig. 420c–f). By the late blastoderm stage the site of the differentiation center has been fixed, as we see from the form of the lateral plate margins, which begin to diverge posteriorly from behind the ligature (Fig. 419c). From the ligated anterior fifth of the egg the brain and anterior part of the head now develop. The influence of the differentiation center has thus already determined the anterior end. The determined, relatively long anlage of the brain can also be ligated so that parts of the brain develop in front of the ligature and behind it (Fig. 420c). The more posterior the ligature, the more organs are formed in front of it (Fig. 420d–f). Thus, in the bee a mosaic of determination spreads over the blastoderm and permits no more regulation in the embryonic parts. So it is evident that a phase runs its course, in which one part of the blastoderm region is allotted the tendency for the formation of certain sections of the heterogeneously segmented body, but the differentiation of individual segments is not yet determined: embryos are successively produced with the mouthpart region entirely intact (Fig. 420d), with a complete thorax and with a complete abdomen. Later on, even these groups of segments become mosaically distinct (Fig. 420f), or else individual segments at the site of the ligature do not form (Fig. 420e).

In the Lepidoptera[415, 428] and Diptera[502] the cortical cytoplasm of the egg is clearly already organized into a pattern of rigidly determining factors: destruction of even a small part of the egg surface before the arrival of the cleavage nuclei leads to mosaic development of the remaining parts of the embryo.

326

The embryo of the greenhouse cricket (*Tachycines asynomorus*) provides the extreme example of this early determination. Here the germ band possesses great regulative ability, even while it is differentiating. In spite of its small size the young embryo of this species can be manipulated with fine glass needles. If the egg is pricked, small parts of the embryo come out as an "exovate": actual cutting can separate embryonic parts from one another with little or no loss of substance, and these can be observed in their further autonomous development. The heart-shaped germ band (Fig. 421) moves toward the surface during the formation of the embryonic cavity and comes to rest near the micropyle; it can be observed clearly *in vivo* until it sinks into the yolk. Figure 422 a–f shows in surface view and, in diagrammatic cross-sections, the development of the embryo up to the separation of the lower layer and the appearance of the first segmental margins. The setting out of the basic body plan and the organ anlagen takes place in the turned-in embryonic cavity stage (Fig. 421 c). Figure 422 g illustrates the anlage-plan in the embryonic streak.

Pieces of the germ band, which in the first days after the onset of the heart-form stage are separated in front of, or behind, a middle zone and are thus cut off from the effect of the differentiation center, are capable only of forming empty amniotic cavities or disorganized tissue. The major portion of the germ band which remains after an anterior or posterior cut, or after being trimmed laterally, can produce smaller but complete embryos if enough of the cells at the center of the band are present. The center of the band thus still behaves as a morphogenetic field, capable of controlling the proportional self-differentiation of a reduced band.

With the formation of the head lobe depressions there begins a distribution of morphogenetic tendencies along the egg axis: a piece removed from the

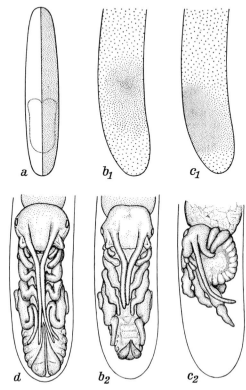

Fig. 417. Elimination of pre-germ-band material and half germ band in *Gryllus domesticus* by UV irradiation. a diagram of irradiation; the margin of the germ band is dotted; the dotted half is irradiated; b_1, c_1 irradiated stages; b_2, c_2 the resulting embryos; d normal embryo. (After Sauer, 1961)

head lobe can now build an incomplete ectodermal anterior part of the head. The rest of the embryo can still form a complete set of embryonic structures (Fig. 423 a). Thus, a considerable amount of material can still be lost, without preventing the formation of a complete embryo. But here a certain minimum of the head lobe region and the segmentation zone must be intact. The head lobe region is distinguished by the tendency to form a bilaterally symmetrical anterior part of the head, without itself differentiating as a region. The segmentation zone has the tendency to develop a heterogeneously segmented abdomen, but individual segments are not yet determined in it. Even at the onset of the invagination of the material of the lower layer (Fig. 422 c) the segmentation zone is still not divided into areas capable of self-differentiation into a metameric series. It develops into an entirely harmonious partial embryo from the mandibular segment to the anal segment, even when some material is missing from the front or rear end.

With the further invagination and the closure of the ventral groove, determination continues; the segmentation zone becomes able to self-differentiate. But at first, regulation can still take place within groups of segments: the loss of material from the front cut surface leads to a reduced but complete mouthpart region (Fig. 423 b). After the appearance of the prothoracic border, the pattern of segmentation is longitudinally determined. If now a large part of the germ band is removed from the region of the differentiation center (Fig. 423 c₁), certain segments will not be formed. For example, Figure 423 c₂ illustrates the absence of the maxillae and the first thoracic segment while the remainder of the embryo continues to develop normally.

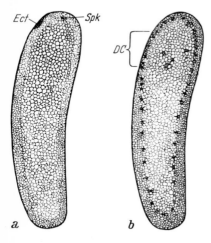

Fig. 418. *Apis mellifica*, sagittal section through egg. a newly laid egg; b 512-nucleus stage. *DC* differentiation center; *Ect* site of the egg nucleus; *Spk* site of the sperm nucleus. (After SCHNETTER, from SEIDEL, 1936)

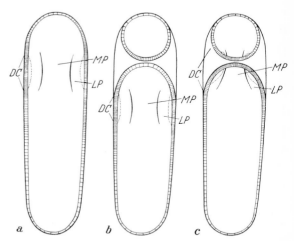

Fig. 419. Diagram of the results of ligation of bee eggs. a normal embryos; b after ligation in early blastula stage; c after ligation in late blastula stage; all three embryos at time of onset of early formation of the inner layer. *DC* differentiation center; *MP* middle plate; *LP* lateral plate. (After SCHNETTER, 1936)

The bilaterally symmetric determination pattern is still not present when the inner layer has invaginated and the ventral groove closed. From the heart-shaped embryo to the age of eighteen hours and beyond, complete twins or double anterior or posterior parts can still be obtained by sagittal sectioning (Fig. 424). The ability of a head lobe to regulate into a whole head can be demonstrated by dividing the ingrowing material of the inner layer, so that for each head lobe the ventral groove can form a front end. Longitudinal division of the ventral groove suffices to cause twinning of the posterior end (Fig. 424 b).

The course of twinning varies according to the age of the embryo at the time of the operation and according to whether both the germ band and amnion or just the germ band are cut. Figure 425 is a schematic illustration of the various possibilities. After complete cutting (Fig. 425 a–c) the cut border of the germ band and the cut border of the amnion can unite, so that two completely separate embryonic cavities are formed. When the cut is made early, two embryos usually develop side by side and completely separate twins result (Figs. 424 a, 425 a). After cutting in later stages, most pairs of embryos reunite, either as they bend in, or as they bend out again and move backwards. They become similar or dissimilar Siamese twins (Fig. 425 b₄), or one embryo overgrows the other in the course of its dorsal closure (Fig. 425 c₄). The way in which bilateral regulation takes place varies with the age of the isolated embryonic halves. In the youngest stage (Fig. 425 a₁)

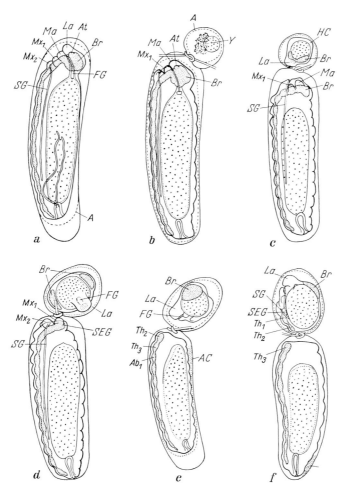

Fig. 420. Results of ligation of bee eggs in the blastoderm stage. a normal embryo after differentiation of the organ anlagen; b miniature embryo after ligation in the early blastoderm stage (cf. Fig. 419b); c–f ligation in late blastoderm stage: c ligature 16 per cent from anterior end, forehead missing behind the ligature; d ligature at 23 per cent, forehead and mandibles missing behind the ligature, head capsule and brain missing in front of ligature; e ligature at 30 per cent, head, jaws and first thoracic segment missing behind ligature; f ligature at 36 per cent. *A* amnion; *Abd$_1$* first abdominal ganglion; *At* antenna; *AC* amnion cells; *Y* yolk; *HC* head capsule; *Ma* mandible; *Mx$_1$*, *Mx$_2$* first and second maxillae; *La* labrum; *Br* brain; *SG* silk gland; *Th$_1$*, *Th$_2$*, *Th$_3$* first, second, and third thoracic ganglia; *SEG* sub-esophageal ganglion; *FG* foregut. (After SCHNETTER, 1934, from SEIDEL, 1936)

the middle plate of the germ band is not yet differentiated. As development proceeds in the halves, the differentiation of the middle plate in each of the half germ bands proceeds posteriorly from the site of the presumptive middle plate toward the center of the rest of the embryo (Fig. 426a, b). The germ band behaves as a morphogenetic field and is remodelled after the cut is made. The bilateral differentiation of the ectoderm of the embryo is accomplished entirely with the material of the half germ band (Fig. 426c). In a later sagittal cut the inner layer has already invaginated and is sliced in two (Fig. 426d). The cut surface of each half now remains medial and differentiation proceeds undisturbed toward the intact side. The cut surface of the germ band unites with

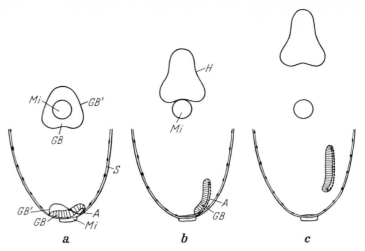

Fig. 421. Early stages of embryonic development in *Tachycines*, with position in the egg; above, the outline of the germ band and position with respect to the micropyle; below, schematic sagittal section through the micropyle pole of the egg. a three days, b nine days, c sixteen days after heart-shape stage (at 26°). b beginning invagination; c invaginated germ band. *A* amnion; *GB* germ band; *GB'* border of germ band on the surface; *H* head lobe depression; *Mi* micropyle; *S* serosa. (After KRAUSE, 1939)

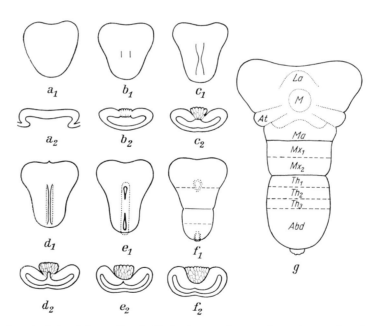

Fig. 422. Stages of ventral streak formation in *Tachycines*. a_1–f_1, g surface view; a_2–f_2 schematic cross section through the region of the differentiation center. Age after reaching the heart-shape stage: a3, b6, c9, d12, e15, f18, g27 days. a amniotic fold; b middle plate; c infolding; d ventral groove; e groove closure; f ectodermal prothoracic border; g fate map of invaginated ventral groove. *Abd* abdomen; *At* antenna; *Ma* mandibular segment; *M* mouth; Mx_1, Mx_2 first and second maxillary segments; *La* labrum; Th_1–Th_3 thoracic segments. (After KRAUSE, 1953)

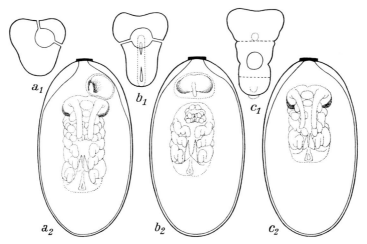

Fig. 423. Development of *Tachycines* embryos after small medial exovates and horizontal tearing. a_1, b_1, c_1 operation diagrams; a_2, b_2, c_2 resulting endpoints. a_2 anterior tip and complete embryo; b_2 anterior tip and miniature jaws with thorax-abdomen; c_2 absence of maxillae and prothorax. (After KRAUSE, 1953)

the cut edge of the amnion, and through complementary induction the medial margin of the amnion is converted into half of the remodelled germ band (Fig. 426e, f). Now the remaining embryonic half behaves during bilateral regulation as an induction field which recruits an unusual tissue in order to complete its work. This tissue is normally not involved here, but as the result shows, it is still competent to respond. When only the germ band is transected longitudinally

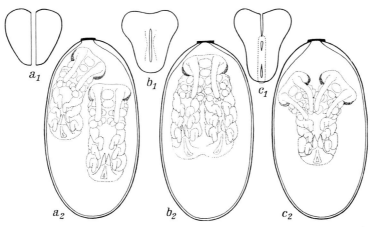

Fig. 424. Twinning after longitudinal sectioning as a function of age when cutting was made in *Tachycines* (cf. Fig. 422). a_1–c_1 operated stages; a_2–c_2 endpoints. (After KRAUSE, 1953)

(Fig. 425d–f), the embryonic cavity remains intact, and the embryonic halves are held together by the amnion. Their cut surfaces reunite sooner or later. Cutting before the formation of a middle plate results in the formation of two laterally displaced ventral grooves. Two germ bands arise, and after they speed out, the yolk is contained in a common midgut. "Janus forms" appear with two ventral sides (Fig. 425d_4). If, after the transection of the ventral groove, a piece

331

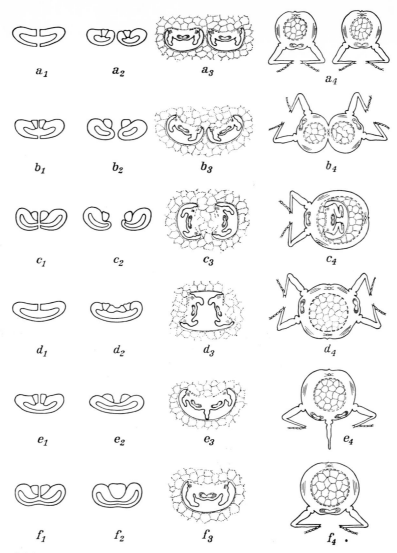

Fig. 425. Degrees of deformity in separated *Tachycines* embryos. a_1–f_1 before healing; a_2–f_2 after healing; a_3–f_3 types of parallel twinning in invaginated condition; a_4–f_4 after evagination; diagrammatic cross section through prothorax. a–c median cutting of the germ band and the amnion; d–f median cutting of the germ band alone; a_1, d_1 during formation of the amniotic cavity (cf. Fig. 422a); b_1, e_1 during formation of ventral groove (cf. Fig. 422b, c); c_1, f_1 during closure of ventral groove (cf. Fig. 422e). (After KRAUSE, 1953)

of ventral ectoderm forms between the halves by regulation, it is never sufficient to push the induction fields far enough apart, but rather remains under the influence of both, so that a mirror-image sector of a segment is invaginated, with an unpaired ventral organ (Fig. 425e_3, e_4). If the cut edges grow together directly, a normal single embryo is formed (Fig. 425f).

A lot has been learned from recent explantation of insect embryonic stages into specific culture media (KRAUSE[354], [354a]). Young single-layered half germ bands of the silkworm *Bombyx mori* can form a germe band. Membrane formation and mesodermal cleavage take place without the usual yolk syncytium for a substratum. Young germe bands can elongate and form limb rudiments. Additional experiments are needed to show whether the absence of further development is due to suboptimal nonspecific conditions or to the lack of specific inducers.

After the outer layer of the insect embryo is underlaid by mesoderm (Fig. 402 m, n), reactions begin between the ectoderm and mesoderm. These can be demonstrated elegantly in the superficial long embryo of the lace-wing *Chrysopa perla*. Parts of the germe band are killed with a hot needle in the ventral groove stage. Such a defect can be restricted to a single segment and does not extend from one side of the embryo to the other. Figure 427 shows in a_1–a_3 normal development up to organ differentiation; in b–d the consequences of mesodermal defects, and in e–g the

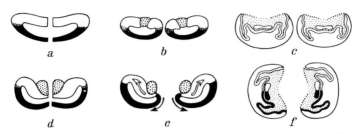

Fig. 426. *Tachycines:* rearrangement of the embryonic material during early (a–c) and late (d–f) longitudinal cutting (cf. Fig. 425a, c). Amnion and, in d–f, amnion-derived embryonic parts in black. (After KRAUSE, 1952)

results of eliminating ectodermal regions. If the mesoderm is fully eliminated (Fig. 427 b_1), the ectodermal organs differentiate to a histologically complete end point: the ectoderm can self-differentiate. The body form and the form of the individual organs is abnormal, however. Without the underlying mesoderm the normal lateral extension on the surface of the embryo does not take place. The mesoderm is necessary for the morphogenetic extension and differentiation of the extremities, for the stretching and bending of the abdomen and for the posterior movement as the yolk is surrounded.

If half of the mesoderm is removed (Fig. 427 c_1), the defective side will not resemble the normal side; the ectoderm of the mesoderm-less side becomes histologically differentiated, but remains condensed (Fig. 427 c_2). If only a small part of the mesoderm, medial or lateral, is eliminated (Fig. 427 d_1), cells from the remaining mesoderm of the operated side move as they would have normally. The movement of the ectoderm is a function of the mass of the underlying mesoderm (Fig. 427 d_2). Whether the mesoderm cells come from the medial or lateral region of the medial plate makes no difference for the differentiation of mesodermal organs: the mesodermal cells which move to the edge of the ectoderm (Fig. 427 d_2) form cardioblasts and midgut muscle; if they remain close to the midline, only muscle, and, according to the concentration, fat body will be made. The mesoderm cells are therefore isopotent; that is, they are competent for all mesodermal differentiation.

When the ectoderm of a half segment is completely missing (Fig. 427 e_2), the mesoderm spreads laterally as usual, but it cannot differentiate, and finally degenerates (Fig. 527 e_1). The movement of the mesoderm cells is therefore independent of the ectoderm, but the differentiation of the mesoderm, including the formation of coelomic sacs, does depend on the ectoderm. If enough medial (Fig. 427 f_1) or lateral (Fig. 427 g_1) ectoderm is removed, so that the wound is not closed by the ectoderm nearby, the missing part is not replaced regulatively. The mesoderm differentiates only where it underlies ectoderm, and in accord with the particular region: mesoderm under the

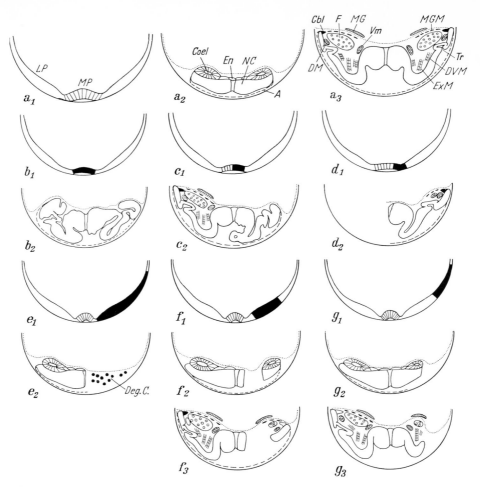

Fig. 427. *Chrysopa perla* (Neuroptera): diagrammatic section through the ventral side of a thoracic segment (cf. Fig. 403). a_1–a_3 stages of normal development; b–g treatment during ventral groove formation; eliminated areas in black in b_1, c_1, d_1, f_1, g_1; b–d elimination of mesoderm; e–g elimination of ectoderm. *A* amnion; *NC* nerve cord; *Cbl* cardioblasts; *Coel* coelom; *Deg.C* degenerating cells; *DM* dorsal muscles; *DVM* dorso-ventral muscles; *En* endoderm; *ExM* muscles of extremities; *F* fat body; *MG* midgut rudiment; *MGM* midgut msuculature; *MP* middle plate; *LP* lateral plate; *Tr* tracheal invagination; *Vm* ventral muscles. (After SEIDEL, BACH, and KRAUSE, 1940)

medial ectoderm develops only into muscle and fat body (Fig. 427 g_3); under lateral ectoderm it develops into cardioblasts and gut muscle (Fig. 427 f_3).

The mesoderm is therefore directed regionally by the ectoderm. The ectoderm of the ventral groove is an induction field which induces according to its determination template, the organ formation of the underlying mesoderm.

Ligation and extirpation experiments show, according to the insect group, various grades of regulation or mosaicism in embryonic development, leading to the organization of the larva. But to what extent is the organization of the adult already determined in the embryo? Normal development shows that among the insect orders there is great diversity in the degree of

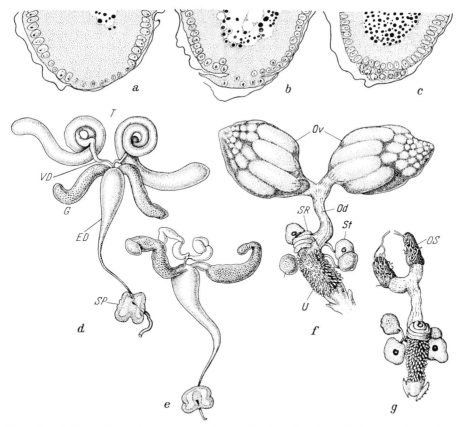

Fig. 428. Genitalia of *Drosophila melanogaster*. a–c longitudinal section through the posterior pole of the egg; a, b UV-irradiated in blastoderm stage and fixed one and a half to two hours later; c normal earliest germ cells in corresponding stage; d, e male genitalia; d normal; e castrated by irradiation of the posterior pole in the blastoderm stage (as in a, b); f, g female genitalia; f normal; g castrated as in a, b. *ED* ejaculatory duct; *G* glands; *T* testes; *Od* oviduct; *OS* ovarian sheath with tracheal ends; *Ov* ovary; *SR* seminal receptacle; *SP* sperm pump; *St* spermathecae; *U* uterus; *VD* vas deferens. (After GEIGY, 1931)

differentiation during embryonic development of the cell groups, which produce during metamorphosis the imaginal organs which the larvae lack.

In many insects, at the time of the movement of the cleavage nuclei into the periplasm, primitive embryonic cells are already determined by the specific action of the periplasm at the posterior pole. The nuclei which find themselves in this cytoplasmic region become pinched off to the outside from the blastoderm, into "pole cells" (Fig. 428 c). Later they are reincorporated into the interior of the embryo and arrive in the mesoderm of the genital gland rudiments. If the "pole plasm" is irradiated with UV, the nuclei in that region degenerate (Fig. 428 a, b); they are extruded with the damaged cytoplasm, and the individuals are castrated. Gonads with no germ cells arise with a thin capsule of connective tissue (Fig. 428 e, g) in the sterile ovarian anlagen; rudimentary egg tube anlagen are formed, together with indications of empty egg chambers. The ducts and accessory glands are formed normally in both sexes (Fig. 428 e, g; cf. d, f).

In holometabolous insects, according to the kind of larva, imaginal discs for the legs, wings, compound eyes, and the ectodermally derived parts of the genitalia, are formed, which generally originate as epidermal invaginations and sink more or less deeply into the larva (Fig. 429). In Cyclorrhaphan flies the entire larval epidermis, which grows solely or mainly through cell enlargement supported by extreme chromosomal polyteny (Fig. 103, 104), is replaced: from epidermal imaginal discs imaginal cells emerge and overgrow the larval epidermis, which is broken down from below by phagocytes. In other holometabolous insects, e.g., Lepidoptera, the epidermis grows by periodic cell divisions, which are in step with the moulting periods, and the same tissue forms larval, pupal, and adult structures successively. The larval legs of Lepidopterans are made into adult legs in the pupa. From certain parts of the larval legs, predetermined adult leg material arises by epithelial growth[374].

Indeed, well before the formation of the imaginal discs, clearly observable processes in the blastoderm or the ectoderm of the young germ band prepare the determination of the imaginal parts. In the Diptera, UV irradiation during the germ band stage can result in damage to the adult thorax, wings, legs, and abdominal segments, as well as the duplication of whole legs or of leg segments. According to the time, the degree and the kind of imaginal changes, a change in sensitivity emanates from the thorax and proceeds posteriorly: early stages develop only wing and thoracic defects, and when one pair of legs is defective all three pairs are. Radiation in intermediate stages reveals a posterior movement of defect production in the embryo axis: the wings are normal, the first pair of legs is generally normal or shows small anomalies, such as defective claws; the development of the second pair of legs is greatly distorted, and one or both of the metathoracic legs is completely absent. The irradiation of still older eggs affects only the metathoracic legs. Accordingly, a change in ectodermal conditions runs its course in successively more posterior regions, just as the determination of larval differentiation in *Platycnemis* moves backward from the differentiation center (p. 322). Temperature shock (four hours at 35°C) and ether treatment produce changes in the metathorax and in the halteres which grow from it. The halteres become more winglike. In extreme cases they resemble wings closely in form, venation, and bristle pattern (Fig. 581).

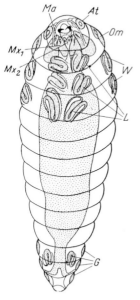

Fig. 429. Imaginal discs in last larval instar of the wasp *Encyrthus fuscicollis. At* antennae-, *L* leg-, *Om* compound eye-, *W* wing-, *G* genital-, *Ma* mandible-discs; *Mx₁*, *Mx₂* anlagen of the first and second maxillae. (After BUGNION, from EIDMANN, 1941)

Some of the results of irradiation can be interpreted as stemming from the elimination of certain cells of the blastoderm, or the ectoderm of the germ band in a sensitive stage, through death or the loss of competence. Twinning of parts may be due to the separation of cell groups by regions of cell death. The observations that the effects of irradiation move posteriorly with embryonic age, and distally in the legs, and that the metathorax is modified to resemble the mesothorax, make it highly likely that in certain cells a process taking place during a sensitive period is either inhibited or diverted into another path, and that this process is directly concerned with determination.

Further commentary on the mechanisms of determination of adult structures during embryonic development can only be made when we understand how the onset of the events of metamorphosis is triggered.

Lecture 23

The *postembryonic*, or *juvenile*, *development* of insects begins with hatching and extends through the periods of growth, which are separated by molts. Juvenile development is *indirect development*, or *metamorphosis:* in the series of juvenile stages, the characters of the sexually mature adult stage stem from changes in external morphology and from the transformation of the internal organs. In insects with incomplete metamorphosis (Hemimetabola) the adult organs appear without a profound transformation of the larval organization. Certain larval structures are dismantled, the rudiments of adult organs, wings, genital appendages, etc., grow quite gradually (e.g., Odonata, Heteroptera), or nymphal stages with sheathed external wing rudiments differentiate from completely wingless larvae. In the Holometabola early development ends in a nonfeeding pupal stage. In the molt leading to the pupa (*pupal molt*) the larva's internal rudiments of wings, other extremities, and genital appendages are everted; and in the pupa, the last major process of metamorphosis is carried out. The demolition of larval organs proceeds simultaneously with the growth and differentiation of adult parts.

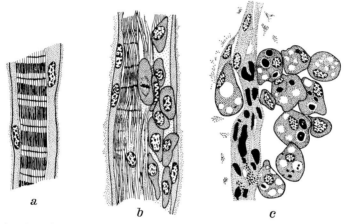

Fig. 430. Histolysis of a thoracic muscle of *Ephestia kühniella* in the pupa. a intact muscle, b last vestiges of cross striation; invading lymphocytes; c advanced breakdown and phagocytosis

The behavior of individual tissues varies from case to case. Parts of the larval musculature are often retained throughout the pupal stage (e.g., abdominal muscles of Lepidoptera). Others are discarded entirely and replaced by new ones tailored to the specific adult needs with respect to movement and feeding. The onset of tissue breakdown is always a chemical *autolysis;* then begins generally the activity of phagocytic hemolymph cells (Fig. 430). Apparently the phagocytes also secrete a histolytic enzyme. Even during the rapid growth of the larva, organs have grown and new structures have been built. Cyclic dedifferentiation and redifferentiation during growth has been seen in muscle fibers of a bug during the molting cycle (Fig. 431), and this probably takes place in other insects also.

The life of insects is determined by the tempo of metamorphosis. In contrast to vertebrates, juvenile development is almost always much longer than the reproductive period. The larval stages and the resting period of the pupa can last for weeks, months, and even years, the exact time depending upon environmental conditions; on the other hand, insects frequently spend only a few days as adults, rarely weeks; and years only in unusual cases like that of the queen bee. In many insects metamorphosis to reproductive maturity is very closely tied to a *natural death*, just as the blooming of annual and biennial plants. In some orders there is great variation. Among the Lepidoptera the spectrum ranges from forms whose females emerge from their pupal cases ready to lay mature eggs (e.g., Saturniids) to forms which at the time of eclosion generally have no mature eggs at all in their ovaries (most Rhopalocera). In many cases the adult does not have to eat. The nutritive materials stored in the fat body suffice for the maturation of the gonads and sometimes even for the continuing development of successive waves of eggs.

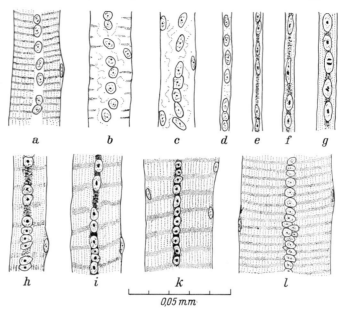

Fig. 431. *Rhodnius prolixus*. Changes in an abdominal muscle of the first larval stage after the molt to the second stage. a immediately after molt; b after two days; c after 4 days; d after 14 days, immediately after a blood meal; e 2f, 4g, 5h, 6i, 7k, 9l, eleven days after the blood meal (right after the molt to the second stage). (After WIGGLES-WORTH, 1956)

In many insects the course of development is temporarily halted by a resting stage called diapause. Diapause can, according to ecological circumstances, take place during the embryonic stage (overwintering eggs), a larval stage, or, in the Holometabola, most frequently in the pupa.

In the events of metamorphosis even in the larval molts the changes taking place in all parts of the body are precisely ordered. The onset of each phase of metamorphosis and the coordination of the individual processes is controlled through the ordered activity of the hormones of a system of endocrine glands. Although many of the questions raised by the consideration of metamorphosis are still unanswered, the experiments of the last decade have, nevertheless, revealed a most informative picture of the hormonal control of development.

Since 1934, when WIGGLESWORTH began his investigation of metamorphosis with a series of experiments on the hemimetabolous insect, *Rhodnius prolixus*, many groups of insects have come under study. The Lepidoptera, which are such favorable experimental animals, have been the most informative; they will, therefore, be used as the main examples.

In all insects three endocrine organs play a clear-cut role in the control of metamorphosis (Fig. 432): neurosecretory cells in the brain, the prothoracic glands, and the corpora allata.

The neurosecretory cells in the Hemimetabola (*Rhodnius*) and in the Lepidoptera (Fig. 433) are few in number and lie in the dorsal and lateral parts of the brain. They have neural connections like ordinary ganglion cells but are already much larger than these at the time of the first larval instar. In the individual larval instars the neurosecretory cells pass through periods of secretory activity. During and immediately after a molt the cytoplasm stains weakly and contains many vacuoles (Fig. 434a, d, h). During the period of larval growth the cytoplasm stains strongly and is dense (Fig. 434 b, e). Gradually the vacuoles appear, reducing the stainability of the cytoplasm, and the secretion becomes liquid (Fig. 434 c, f, g). The period of secretory activity in the last larval instar concludes with secretion in the prepupal stage; so that we can consider as fully grown a larva which has stopped eating, started spinning, and which shows the internal beginnings of metamorphosis by the movement of the larval ocelli away from the cuticular lens (Fig. 437). In the meal moth, *Ephestia kühniella*, the young pupa enters into still another period of secretory activity (Figs. 434 i–m), which comes to a conclusion before adult emergence.

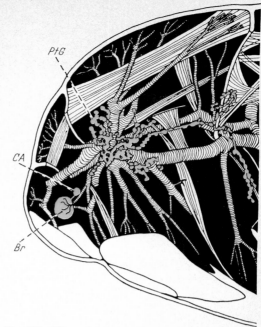

Fig. 432. The endocrine glands of metamorphosis in the head and thorax of a pupa of *Platysamia cecropia*. Fat body and other tissues have been removed. *CA* Corpora allata; *Br* brain; *PtG* prothoracic gland. (After WILLIAMS, 1948)

The prothoracic glands (Fig. 432) are branched, chainlike organs, surrounded and penetrated by tracheae, and innervated from the subesophageal ganglion. The gland cells are already undergoing periodic changes during the larval instars; their nuclei swell and change in structure; the cytoplasm grows and becomes vacuolated.

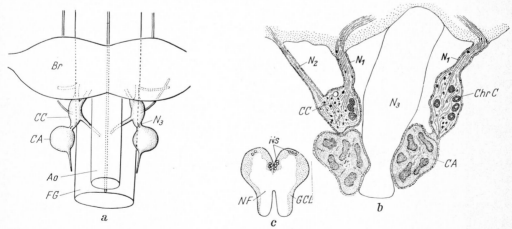

Fig. 433. Lepidopteran endocrine glands. a *Hyloicus (Sphinx) ligustri*, pupa, diagram; b *Pieris brassicae*, adult, frontal section; c *Ephestia kühniella*, fourth larval instar, frontal section through the brain. *Ao* aorta; *CA* Corpus allatum, *CC* Corpus cardiacum; *ChrC* chromophil-ganglion cells; *Br* brain; *GCL* ganglion layer; N_1, N_2 nerves from brain to corpus cardiacum; N_3 nerves from corpus cardiacum to corpus allatum; *NF* nerve tracts; *Ns* neurosecretory cells; *FG* foregut. (a, b after CAZAL, 1948; c after REHM, 1952)

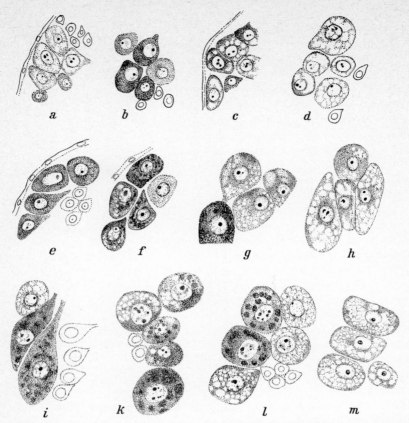

Fig. 434. Secretory cycles of neurosecretory cells in the brain of *Ephestia kühniella*. a larva at the end of the third larval instar; b–d fourth (penultimate) larval instar; b just after molt; d during larval growth; d during the laying down of the new larval cuticle; e–h fifth (last) larval instar; e during pigmentation of the head of a newly molted larva; f in the growing larva; g in the almost fully grown larva; h in the young prepupa (second stage, cf. Fig. 437); i–m in pupa; i newly molted pupa; k 24-hour old pupa (20°); m before the outgrowth of the first scales on the head. 750×. (After Rehm, 1952, modified and slightly schematized)

After the molt to the last larval instar, the cytoplasm once more becomes homogeneous and the nucleus shrinks and becomes highly structured. During feeding and the larval growth period of the last instar, the nuclei and cytoplasm in the prothoracic glands grow again, and in the prepupa vacuoles form and then empty. In the young pupa still another period of secretory activity runs its course. When the scales appear, the prothoracic gland degenerates.

The corpora allata in the *Lepidoptera* are paired packets of gland cells which are ensheathed in connective tissue (Fig. 433). They lie behind the ganglion-like corpora cardiaca and receive nerves from them. Sensory activity of the corpora allata, in rhythm with the larval molts, is made evident by structural changes of the nucleus and of the cytoplasm, which becomes vacuolated before every molt (Fig. 435a–d). In the last larval stage vacuoles appear in only a few cells, whereas most cells retain dense cytoplasm, and the nuclei do not show the structural changes characteristic of secretory activity; accordingly, only a small amount of secretion takes place. The volume of the corpora allata increases greatly, however, in the last larval stage, and in males much more than in females. This sex difference becomes even more pronounced in the pupa, although concomitant secretory activity is not evident. At any rate, the corpora allata have still another role to play in adult life. In *Ephestia*, the cell number remains constant in all glands from the first larval instar on; their growth is achieved only through great enlargement, and this, in turn, is made possible by nuclear polyploidy.

If one compares the secretory circumstances in a variety of endocrine cells during the larval stages, one finds a clear sequence of secretory periods: in a new larval instar, first the neurosecretory cells become active, then the cells of the prothoracic glands, and finally those of the corpora allata. Extirpation, ligation, and transplantation experiments demonstrate the nature of the interplay of the hormones from these endocrine glands.

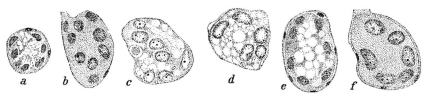

Fig. 435. Secretory periods of larval corpora allata in *Ephestia kühniella*. a newly molted pupa of the penultimate (fourth) larval instar; b growing larva of the penultimate instar; c during the retraction of the epidermal cells from the cuticle; d at the deposition of the new larval cuticle; e newly molted larva of the last instar; f growing last instar larva. (After REHM, 1952)

The activity of the corpora allata can be determined by removing them from various larval instars; when these glands are removed, the next normal larval instar does not ensue; instead, pupation takes place prematurely (Fig. 436). Naturally, the resulting pupae are very small, and they produce correspondingly reduced moths. If after extirpation new corpora allata are implanted into the abdomen, they function normally, so that premature pupation does not occur. Rather, further larval molts take place. Now, if one implants into larvae at the *beginning of the last larval instar* corpora allata from younger larvae, their secretion has the effect of causing the host larva, not to pupate normally, but to molt once more into a perfectly typical larva. After this extra larval molt, yet another is possible, and finally the resulting giant larva can still undergo a pupal molt. These giant pupae turn into very large moths. In the larval body, then, the stage is set for a molt independent of the corpora allata, but the corpora allata specify that it be a larval molt. Their hormone acts as a *juvenile* hormone in preventing the onset of metamorphosis to pupa and adult, an event for which the larva is actually prepared at the time. The factors that have brought about this readiness to molt will be demonstrated in further experiments.

If one ligates prepupae in the various intervals defined by the degree of retraction of larval eyes from the cuticle (Fig. 437), one finds a *critical period* in which pupation is determined. When larvae are ligated prior to tanning, the abdomen never pupates (Fig. 437a), but in subsequent age classes the capacity of the ligated abdomen to pupate increases. Even as late as the third interval, some of the experimental animals

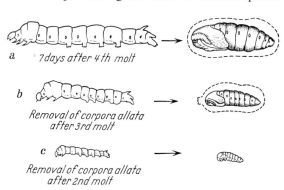

a   *7 days after 4th molt*

b   Removal of corpora allata after 3rd molt

c   Removal of corpora allata after 2nd molt

Fig. 436. Effect on extirpation of the corpora allata in the silkworm. a Normal pupation; b, c premature pupation after the operation. (After BOUNHIOL, 1938)

have unpupated abdomens (Fig. 437a, b). A few hours later, almost all of the ligated animals pupate completely (Fig. 437 a, c); the critical period is over. The front half always metamorphoses to pupa and moth; in it, therefore, are the factors which determine pupation and adult development.

During the critical period the capacity to pupate gradually spreads from front to back: in the third age class the further back one ligates the larva, the greater the proportion of unpupated posteriors parts (ligation behind the second thoracic segment 14 per cent, behind the third thoracic segment 44 per cent, in the middle body region about 50 per cent, behind the ninth segment 86 per cent.

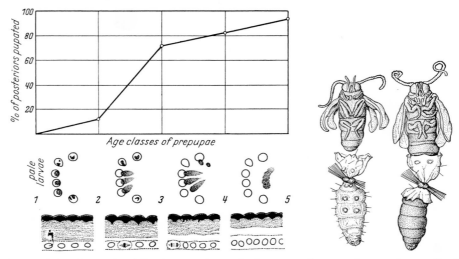

Fig. 437. *Ephestia kühniella*, results of ligation experiments in the prepupal stage. a increase in pupation frequency of ligated abdomens during the critical period; age classes according to appearance of larval ocelli. Status of the epidermis diagrammed for each age class; b, c emergent moths after ligation; b in the third age class; c in the fourth age class. (After Kühn and Piepho, 1936, with additions)

The *epidermal changes of the pupal molt* begin even before the end of the critical period: while the larval eye pigment begins to sink into the deeper tissues and the number of isolated abdomens capable of pupating increases, the epidermis retracts from the larval cuticle, and the epidermal cells begin to divide (Figs. 437a, 438b). Cell division in *Ephestia* does not begin simultaneously throughout the epidermis, but it starts first in the thorax in the second interval and spreads posteriorly (Fig. 439). In the third interval the mitotic count reaches its maximum, and mitoses are seen throughout the epidermis. During the fourth interval mitotic figures are once more rare. After the completion of the mitotic period there is liberated into the exuvial space between the epidermis and the larval cuticle a secretion which breaks down the chitin of the larval cuticle (Fig. 438c–e). Only the thin epicuticular layer remains intact and will later be torn off. This molting fluid is secreted by the epidermal cells and special epidermal glands (Verson's glands) whose secretory activity is triggered by the hormones of metamorphosis[370]. The pupal epidermis, which by now has increased in size, elaborates in turn the pupal cuticle. The epithelial cells undergo a change in form (Fig. 438d, e). An "epithelium on stilts" composed of fibrillar bundles of elongating cells appears. It apparently serves to protect the epithelium by damping mechanically the movements of the pupa in the molt to come. Before the molt a new epicuticle and exocuticle are laid down in the exuvial space (Fig. 438d, e). This follows the removal of the rest of the now weakened and colorless larval endocuticle. In the newly molted, greenish pupa, the epithelial cells resume their usual form. The exocuticle becomes stiff and pigmented, and the thicker, lamellar, unpigmented endocuticle (lamellosa) is laid down (Fig. 438g). The mitotic period leads into the changes of the epidermal molt. By means of ligation its occurrence in the abdomen may be prevented completely. If the block is made at the beginning of the critical period, mitoses can take place, but further epidermal changes do not. These changes must require, therefore, a *continuing supply of "molting hormone."* The determination of development of the imaginal extremities, the antennae, the mouth parts, legs, wings, and adult eyes is also achieved in steps: if one implants into *older larvae of the last instar* corpora allata of younger larvae, mosaics result in

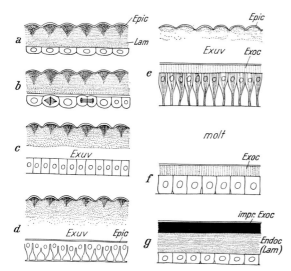

Fig. 438. Diagram of pupal changes in the epidermis and cuticle in *Ephestia* and *Galleria*. a mature larval integument; b mitotic period; c secretory period; release into the exuvial space of a section which breaks down chitin; d, e transformation of the epidermis into "epithelium on stilts": d deposition of new epicuticle; e deposition of exocuticle; detachment of larval cuticle from epicuticle by molting fluid; f retraction of the "epithelium on stilts" in the young (green) pupa; g impregnation of exocuticle with pigment and encrusting material and differentiation of endocuticle (lamellosa). (After KÜHN, 1939)

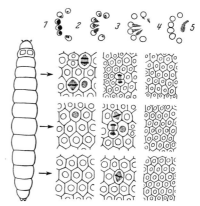

Fig. 439. Spread of mitotic activity over the epidermis of *Ephestia kühniella* in the prepupa (cf. Fig. 337a). (After KÜHN, 1939)

which both larval and pupal characters are developed. The various organs and their parts are not all in the same intermediate stage, but the individual features of legs, wings, and eyes fall somewhere between those characteristic for a larva and for an adult. The experimental animals that have molted into mosaics most like larvae after the implantation of younger corpora allata can undergo further molts and reach the pupal condition after passing successively through further mosaic stages.

Similar results follow the implantation of corpora allata into Hemimetabola. When in *Rhodnius* these glands from third or fourth instar larvae are implanted into larvae of the fifth and last instar; an extra molt to a sixth larval instar results in a giant animal (Fig. 440c). At times this is an intermediate form between larva and adult (Fig. 440d). The giant larva of the sixth instar becomes, during the next molt, either a seventh instar larva or a giant adult.

343

This reaction is, therefore, not all-or-none, but proceeds along a continuum reminiscent of the transitional steps between worker and queen honeybees that can be caused by a change in food.

Where do the pupation hormones come from?

One can prevent pupation not only by ligation, but by removing the brain from a mature larva before the critical period (Fig. 437, interval 1). The brainless permanent larva can continue to live more than two months, during which time the control animals have long since pupated, emerged as adults, and died. If one now inplants another brain into the abdomen of a permanent larva which had been debrained at the end of the feeding period of the last larval stage, it can now pupate. The brain consequently produces a hormone, and the time at which the neurosecretory cells are seen to release their product (p. 339) agrees with the time at which the brain hormone introduces the period of pupation. The brain hormone is, however, not itself the metamorphosis

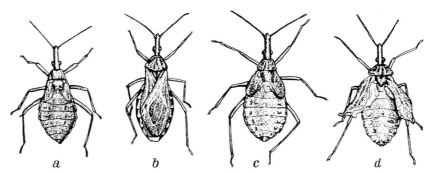

Fig. 440. *Rhodnius prolixus*. a normal fifth instar larva; b normal adult; c giant "sixth instar" larva obtained by implanting a fourth instar larva's corpus allatum into a fifth instar larva; d a similarly obtained "sixth instar" with larval cuticle on the abdomen, wings on the thorax, and other characters intermediate between larva and adult. (After WIGGLESWORTH, 1954)

hormone that causes pupation; it *activates the prothoracic glands*, which in turn release another hormone. If, in the beginning of the critical period, silkworms are ligated between the first and second thoracic segments (Fig. 441 a), only the part behind the ligature pupates. After ligation of a similar larva between the second and third thoracic segments, only the anterior part pupates (Fig. 441 b). And a further experiment: from a mature larva the prothoracic glands are removed and implanted into the abdomen of a younger larva (Fig. 441 c). When the host has entered the critical period, and therefore the neurosecretory cells have released their hormone, the last segment, with the implant, is ligated: now it pupates (Fig. 441 d). In controls of similar age the ligated posterior part retains its larval character (Fig. 441 e).

That the *adult development of the pupa* is also under hormonal control becomes clear from brilliant experiments on the giant silk moth *Platysamia* (*Hyalophora*) *cecropia*, which normally undergoes a long diapause. The larvae pupate sometime between June and September, and the pupae remain inactive for at least eight months. The heart beats very slowly; metabolism is reduced to a minimum, and its character (enzyme activities) changes; development stops completely. At room temperature the animals remain dormant. Further development must be triggered by a period of chilling: if the diapausing animals are held at 5° for awhile and then returned to room temperature, the onset of adult development follows in a few days, and after about three weeks the moths emerge. If now a *chilled brain* is implanted into an unchilled, brainless, and therefore permanently diapausing pupa, the host develops as if it had been chilled. Even the part of the chilled brain which has the neurosecretory cells suffices by itself to initiate this response. Experiments with parts of pupae are also extremely informative. Diapausing pupae are cut into two parts, an anterior part terminating just posterior to the wings and a posterior part which includes the last six abdominal segments. The cut surface is covered with a plastic cover slip and sealed with paraffin. A hole is made in the center of the cover slip (Fig. 442a, b) for the introduction of the implants. If chilled brains are placed in each half, the front half will develop into the front half of a moth, but the posterior half does not develop. The brain alone does not suffice to initiate adult development. If a posterior part is now fastened to the rear end of an entire diapausing pupa (Fig. 442c) and this pupa is caused to develop by the implantation of a chilled brain, the added isolated abdomen will accompany it on its course of adult development. If, in addition to a chilled brain prothoracic glands are also placed in an isolated abdomen (as in Fig. 442b), a mature moth abdomen results which can even lay eggs. Now if a brainless anterior half of a diapausing pupa is connected by a glass tube to a diapausing abdomen (Fig. 442d), development will not take place. But if a chilled brain is then placed in the abdomen, diapause will be broken and both parts of the preparation are transformed into the corresponding parts of a moth: from the implanted brain an activating hormone moves to the front half and triggers the secretion of the prothoracic glands; the prothoracic gland hormone spreads throughout the body and evokes, in the various tissues, the changes of adult development (Fig. 442e).

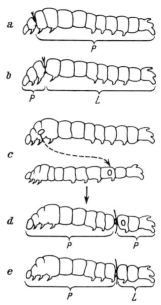

Fig. 441. Activity of the prothoracic gland in the silkworm. a, b ligation at the onset of the critical period; a between first and second thoracic segments, b between second and third thoracic segments; c, d transplantation of prothoracic gland; c transplantation from a mature larva into the sixth abdominal segment of a larva at the beginning of its last instar; d ligation of host in the critical period; e control larva ligated but without transplant. *P* pupation; *L* permanent larva. (Diagrams according to results of experiments; a, b from BOUNHIOL and c–e from FUKUDA. After BOUNHIOL, 1948)

These experiments reveal a three-part metamorphosis-hormone system. The hormone of the neurosecretory cells activates the secretion of the prothoracic glands, and the prothoracic gland hormone initiates the tissue alterations of metamorphosis; the juvenile hormone of the corpora allata has an effect which depends on its concentration. In high concentration it makes the tissues' response to the prothoracic gland hormone another larval molt, rather than pupation. In the last larval stage the activity of the corpus allatum is reduced, and in the pupal stage it stops entirely. If larval corpora allata are implanted into young pupae, metamorphosis to the adult is prevented; instead, a second pupal molt occurs. If an extract of corpora allata is placed topically on a wound in the pupal integument, pupal cuticle arises in the area of the wound instead of adult integument[604]. *The normal sequence of stages results from a normal sequence of particular concentrations of this hormone.*

The responding tissues are not intrinsically programmed. The sequence of events is not unalterable. This is demonstrated by direct experiments on the ability of the tissue to respond: the implantation of pieces of integument into hosts of various stages. If a small piece of larval epidermis, with its covering cuticle, is implanted into another larva, numerous cell divisions take place at the edge of the implanted epidermis, and a flat blastema is produced (Fig. 443a)[512]. Meanwhile, the growing epithelium bends in such a way that the outer side of the epithelium is on the inside, and the implant finally closes, forming a hollow sphere (Fig. 443b). The newly formed

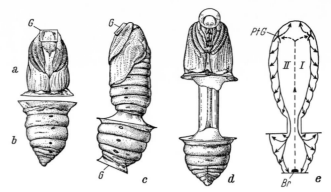

Fig. 442. Experiments with *Platysamia cecropia*. a isolated front half; b rear half of diapausing pupa; c abdomen of diapausing pupa connected to the end of cut brainless diapausing pupa, with windows (*G*) on both sides for observation of internal structures; d same experiment as c, but two pupal halves which are connected by a glass tube; e diagram of the experimental results after implantation of a chilled brain in the rear piece. *Br* implanted brain; *PtG* prothoracic glands; → I path of brain hormone; → II path of thoracic hormone. (After WILLIAMS, 1947)

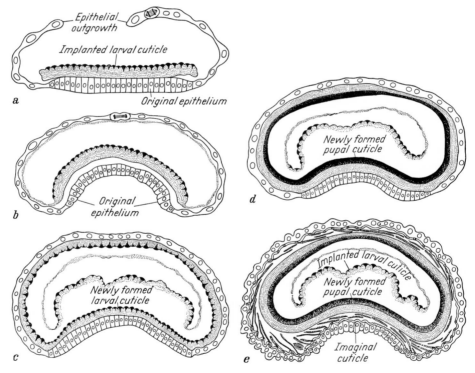

Fig. 443. Development of integument transplant in Lepidopterans; *Ephestia, Galleria,* diagrammatic section. a–c larval integument in a larva of the penultimate instar. Formation of a hollow epithelial vesicle (a, b). Lifting off of the cuticle of the implanted original epithelium and formation of a new larval cuticle by the original and derivative epithelium of the implant (c); d a piece of larval integument (implanted into last larval stage) after pupation of the host; e similar implant, completely metamorphosed after completion of metamorphosis by the host. (After KÜHN, 1939)

346

epidermis now deposits a thin cuticle on its inner (formerly outer) surface. Although the epidermal implant is topographically inside out compared to the integument of a normal animal, physiologically it is correctly oriented: its lower layers retain contact with the hemolymph. This little sphere, which consists of the original implanted epidermis and a greater proportion of derivative epidermis, can now pass through further metamorphic changes and thus demonstrate its ability to respond to the hormonal influences emanating from the host. In a larva which has another larval molt to go the implant molts when the host does. The larval cuticle and the thin cuticular layer deposited over the derivative epidermis are pushed into the center of the sphere. Meanwhile, the original epidermis releases molting fluid, which removes the lower layers of the larval cuticle as in the normal integument, and then a new typical larval cuticle is laid down in its place (Fig. 443c). If the implant is put into a last instar larva, it undergoes a pupal molt when the host pupates. After the larval cuticle lifts off, a typical pupal cuticle is seen, with exocuticle and lamellar endocuticle (Fig. 443d; cf. Fig. 438g). While adult development next proceeds in the host, the epidermal implant is transformed into adult epidermis. Such a fully metamorphosed implant (Fig. 443e), which has developed from a piece of larval skin implanted into a last instar larva, includes four layers. On the outside, an adult cuticle with scales; next, the adult epidermis; then the retracted intact sphere of pupal cuticle; and within this the larval cuticle, which has lifted off at the time of pupation. In this case the implant has followed the normal program of metamorphosis. However, this need not always be the case. If a piece of integument from a mature larva is placed in a larva of a earlier instar, it will molt several additional times along with the host. When the host reaches the last instar, the implant may be removed and placed in another young larva, and this can be done a number of times (Fig. 444a), so that the implant can apparently undergo a limitless number of molts. One can also bypass all the larval molts: the skin of a newly hatched first instar larva will, upon implantation into a mature larva, pupate immediately and then metamorphose into adult integument complete with scales (Fig. 444b). The entire course of larval development with its four or five molts is eliminated. On the other hand, the pupal molt can be repeated. If a piece of integument from a young pupa is put back in a mature larva, the implant will make another pupal cuticle during the pupation of the host. Moreover, even the sequence can be reversed: the course of larval, pupal, and adult epithelial differentiation is not a one-way path. If the integument of a young pupa is implanted in a larva which still has another larval molt to go, the pupal cuticle will be pushed aside in the implant and another larval cuticle will be built. Especially striking is the result of the following experiment: implanted spheres which have made a pupal molt with their hosts and have begun adult development and even laid down scales are placed in larvae of the last instar. In every case, the imaginal cuticle with its scales is molted, and the implant accompanies its host through metamorphosis for a second time. It makes yet another typical pupal cuticle when the host does and then makes a second adult cuticle with scales (Fig. 444c). Until shortly before the emergence of the moth, even the cells of the imaginal epidermis have, therefore, not lost the ability to carry out qualitatively distinct differentiation processes repeatedly and out of their natural sequence. The epidermal cells of the implant are a ping-pong ball for the hormones of the host, and they are permanently competent to perform a series of qualitatively different reactions. The competence is not lost when a reaction is completed.

The transition from larva to pupa is explained on the basis of a reduction in the secretion of the corpus allatum. This alteration in the function of the gland is clearly not autonomous, but depends in turn upon the activity of the neurosecretory cells of the brain. This is shown most strikingly by an experiment on the cockroach, *Leucophaea maderae*, which also sheds light on the role of the corpora cardiaca. Specific staining shows that the secretion of the neurosecretory cells passes into the corpora cardiaca and from there to the corpora allata, as well as being stored

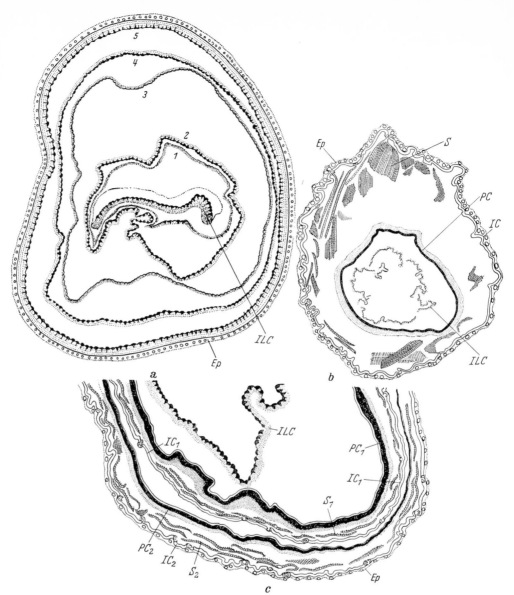

Fig. 444. Sections through an implanted vesicle in the wax moth (*Galleria mellonella*). a integument of a mature larva previously implanted into a young larva and after two molts transplanted into another young larva and allowed to remain until the last larval instar; b skin of a first instar larva implanted into a mature larva, the host then having pupated and developed into an adult; c integument of a mature larva implanted into a last instar larva, the host having then pupated and then developed scales, the implant now returned to a mature larva and pupating and developing adult cuticle a second time, together with the new host. *1–5* the cuticle of extra larval molts; *Ep* epithelium; *IC* imaginal cuticle; *Ic₂* second imaginal cuticle; *ILC* implanted larval cuticle; *Pc* pupal cuticle; *Pc₂* second pupal cuticle; *S* scales; *Sch₂* second layer of scales. (After Piepho, 1938 and Piepho and Meyer, 1951)

in the corpora cardiaca (Fig. 445). When the nerve from the brain is transected, the secretory material is dammed up in the cut nerve and disappears from the isolated corpus cardiacum

(Fig. 445). While connected to the brain, the corpus allatum displays a normal resting condition; cut off from the brain, it swells up, and the cell nuclei enlarge. The gland consequently evinces increased activity. The products of the neurosecretory cells of the brain inhibit the function of the corpora allata.

The question still remains as to the mechanism of regulation of the activity of the neurosecretory cells which initiates each step of metamorphosis. In any case, a certain nutritional condition of the larva is necessary for the neurosecretory cells to be active. In *Rhodnius* the synthesis of brain hormone follows by a few days a blood meal sufficient to distend the animal, which must occur once in each larval instar. Regeneration experiments indicate that even from the developmental condition of certain growing organs of the larva, signals are given off which influence the onset of activity of the neurosecretory cells of the brain. If hindwing imaginal discs are removed from

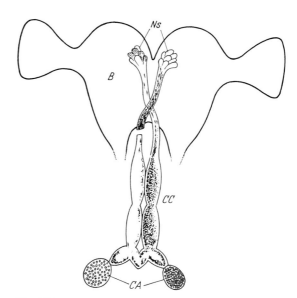

Fig. 445. *Leucophaea maderae*, diagram of the connection of the brain (*B*) with the corpus cardiacum (*CC*) and further with the corpus allatum (*CA*); *Ns* neurosecretory cells. On the left side, the nerve to the corpus cardiacum has been cut. (After BERTA SCHARRER, 1952)

larvae before the beginning of the prepupal stage, the wings are usually regenerated. In regenerating animals development takes longer, regeneration of both wings requiring somewhat longer than regeneration of one wing (Fig. 446). This is clearly not merely a matter of repair, which delays development as do several other forms of interference; for the simple injury effect without removal of the wing disks was slight (Fig. 446). On the other hand, however, when after the removal of the disks, regeneration did not take place, the moths emerged at the same time as the controls.

Several experiments suggest that nerve impulses also influence the rhythmicity of secretion. The prothoracic glands receive nerves from the subesophageal ganglion. In every case, the corpora cardiaca play a role in the events of metamorphosis (Fig. 433a, b). We do not understand

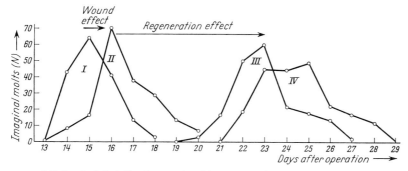

Fig. 446. Emergence of *Ephestia kühniella*: *I* controls without operation (*n* = 166); *II* after cutting the body wall with loss of blood (*n* = 215); *III, IV* adults with regeneration; *III* after one wing disk (*n* = 189); *IV* after removal of both wing disks (*n* = 213). (After POHLEY, 1960)

this role in detail, to be sure, but the corpora cardiaca themselves contain neurosecretory cells as well as serving in the transport and storage of brain hormone. The pericardial cells, also, which produce secretory material and empty into the hemolymph in register with molting[619], appear on the basis of extirpation experiments to be involved in the control of the rhythm of metamorphosis[180]. In addition, we still do not know whether each gland releases only one hormone or whether each secretes a number of active substances. In spite of all the insights which have been gained, the metamorphosis of insects is still not an open book.

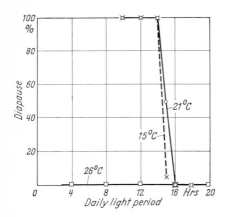

Fig. 447. *Pieris brassicae*, influence of daily light periods on the onset of diapause. Substantial temperature independence; only with a serious rise in temperature is the photoperiodic control broken

The corpus allatum hormone is not species-specific, not is it limited to a single insect order: in mature wax moth larvae (*Galleria mellonella*) an extra molt may be induced by implanted corpora allata from other Lepidopterans (*Ephestia kühniella, Achroea grisella, Bombyx mori*) and also from beetles (*Tenebrio molitor*) and hemimetabolous insects (*Carausius morosus*) as well[516]. The fact that these various juvenile hormones can work in distantly related forms suggests that they are chemically identical or at least very similar. The first hormone about whose chemical nature we know something is the prothoracic gland hormone called ecdysone. In 1953 a 25-mg. concentrate obtained from one-half ton of silkworm pupae was shown, after painstaking work, to be a steroid with the empirical formula $C_{27}H_{44}O_6$[330]. It works in extremely low concentrations. Apparently, insects make it from cholesterol. An extract with juvenile hormone effect can, interestingly enough, be obtained from the abdomens of male *Platysamia* moths. In addition, other insect orders, as well as mammalian organs, particularly adrenal and thymus glands, have yielded potent extracts. This further extension leads to the suspicion that to some extent, we may be dealing with a master key which happens to have the same effect on insects as the natural juvenile hormone.

A particularly ecological and developmental question is raised by diapause. What causes its onset? There are forms with obligate diapause and those in which it is facultative. In these latter forms, development — whether immediate further development or the onset of diapause — is nearly always controlled by light periods. Many Lepidopterans enter diapause in seasons with short days, but omit diapause if they are developing when days are longer. Thus the measure of day length by means of the "*physiological clock*" (BÜNNING) is surprisingly sharp and independent of temperature over a broad range within which development takes place (Fig. 447). In this way a sequence of spring and summer generations is brought about in some insects. Accompanying this change there can also be a distinct change in morphology, as in the patterns of the summer and winter generations of the Lepidopteran *Arachnia levana* (-*prorsa*). This difference has long been known, but only recently has it been demonstrated that it stems from a photoperiodically-controlled diapause effect (Fig. 448). Larvae raised in a greenhouse under constant conditions (temperature 18–22°) produce nondiapausing pupae and moths of the *prorsa* form exclusively after long days (greater than sixteen hours), and diapausing pupae and *levana* moths exclusively after short days (eight hours). It does not matter which type of parental generation gave rise to the larvae. The condition in which metabolism and development are blocked is cured by chilling and subsequent warming, a process which is clearly reminiscent of vernalization in plants. After warming to normal temperature further development begins, accompanied by a sharp rise in respiration and a reversal of protein metabolism. These measurable metabolic changes are certainly not the first reactions but rather consequences of the activities of ecdysone on cells. Experiments on Dipteran larvae (Chironomids) show that some doses of ecdysone evoke puffs at particular chromomeres of giant chromosomes, which would arise normally only in the prepupa. This phenomenon will be discussed at length later.

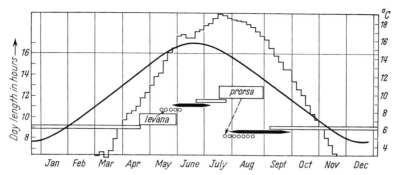

Fig. 448. The seasonal generation cycle of *Arachnia levana*; *levana-* (spring-) and *prorsa-* (summer-) generations. circles-eggs, black spindles-larvae, light rectangles-pupae, continuous curve-day length, stepped curve-temperature. (After Müller, 1955)

The histological, cytological, and biochemical experiments on insect metamorphosis are moving rapidly. Soon the chemical nature of additional hormones will be known, and this will open new possibilities for experiments on animals. Schneiderman and Gilbert are correct: "The amplitude of the developmental changes induced in insects by endocrine manipulation is without parallel. The changes produced in mammals are modest when compared with the hormonally controlled transformation of a caterpillar into a moth or a maggot into a fly" (1959 [604, 157]).

Lecture 24

The formation of adult organs is gradual in the Hemimetabola; in the Holometabola it takes place in sharply separate steps. A tissue complex can either produce a single array of organs only once during development, as is true for the imaginal discs; or different organs can be produced successively from the same tissues, as in the epidermis of butterflies. Implanted vesicles show this impressively. According to the hormone content of the host's blood, the cuticular structures of the larval molt, the pupal molt, or the adult integument are formed. The dimensions of the structure are not determined. This is shown by the premature metamorphosis of the implanted integument taken from a first larval instar (Fig. 444b), by the excessive growth of vesicles during extra molts (Fig. 444a), and by the relatively small pupae and moths formed after the extirpation of the corpora allata and the large ones formed after the implantation of these glands. The number and the qualitative nature of the steps in metamorphosis are not determined in the epithelium, and three modes of response are always present. What is determined, however, is the regional character of the integument. In the posterior sections of the pupa and the adult, the intersegmental membranes differ from those found elsewhere in the body. The segments of the soft larval skin are scarcely demarcated in the Pyralids. Flat bands, which serve as muscle attachments, are their only indication. The cuticular structure is only slightly altered here. In the pupa the intersegmental connections of the abdomen are structured according to whether the segments are rigid, as those covered by the wings (1 to 4, and the terminal segments), or whether they move with respect

351

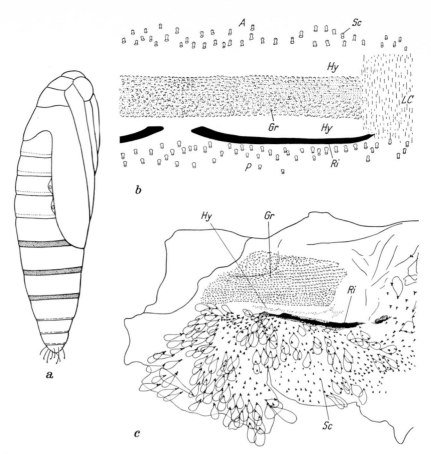

Fig. 449. *Ephestia kühniella.* a pupa seen from the right side, 8× ; b intersegmental structure of adult moth, 80× ; c adult integument of an implant from the anterior border of a segment including some intersegmental membrane, 27× ; *Gr* granulated cuticle; *Hy* hyaline cuticle; *Ri* chitinous thickening; *LC* lateral, longitudinally folded cuticle; *Sc* scales. (After Yosɪɪ, 1944)

to one another 4/5 to 6/7). The intersegmental membranes between the anterior segments from the third thoracic to the fourth abdominal are folded inward, hyaline and rigid; from the fourth to the seventh abdominal segment the membrane is thin and folded deeply inward with a well-developed cuticular pattern, reflecting the borders of the epidermal cells. From the posterior edge of the seventh abdominal segment the segments are thickened and closely bound to one another. In the adult moths the intersegmental membranes of the abdomen all fold in, and the epidermis forms various cuticular structures (Fig. 449b). At the anterior edge of each segment is a pair of chitinous ribs. The cuticle of the folded epidermis is dotted in transverse stripes. In front and behind is a thin hyaline zone which bears only tiny spurs of exocuticle. Toward the sides the cuticle is thin and irregularly rippled. If a piece of larval skin from the middle of a segment is implanted, the implant becomes covered with scales after metamorphosis. But if the implants include the region of a segmental border, when the pupal cuticle is formed, it will bear the structure associated with the edge of the segment. After the host metamorphoses the adult cuticle of the implant contains parts which are densely covered with scales as well as chitinous ribs and dotted and plain cuticle (Fig. 449c). If very small pieces from the segment edge are implanted, the implanted vesicles may form only the structures associated with the intersegmental membrane. The dimensions of each region can be far greater than normal, as we have already seen from the luxuriance of the overgrowing epidermis, as well as from the further growth of implanted epidermis removed from later stages and implanted in younger larval stages. The implants can also be cut up the resulting pieces implanted in younger larvae, and this multiplication of epidermis can be repeated again and again. Thus, one can—theoretically—culture a square centimeter of cuticle with scales or a similar amount of intersegmental membrane. The *regional determination* of the cells lasts through all the processes of cell division, growth, and molting.

The imaginal discs of the Holometabola do not differentiate until the pupal stage. During the larval stages they grow by cell division. At this time their growth rate, until they are competent to differentiate, is surely not rigidly determined in many cases. This is shown by the fact that butterflies of various size, due to premature or delayed pupation, have normally proportioned legs, wings, etc. Pieces of the male genital imaginal discs of *Drosophila* grow, when implanted into hosts which are not quite ready to pupate, to the typical size of whole imaginal discs, so that fragmenting them must trigger increased cell division. Even when an imaginal disc is determined for the formation of an organ or an organ complex, it still does not have to be a self-differentiating mosaic in order for the organ parts or whole organs to be formed. For example, X-irradiation in the third larval instar of *Drosophila* can still evoke duplication of the palps and wings. Evidently the X-rays result in local cell death and thus lead to the partitioning of the imaginal anlagen. The imaginal discs that respond in this way are, therefore, still self-differentiating fields whose parts can still restore themselves to the original whole field condition.

An excellent analysis of the state of determination has been achieved in the male genital imaginal discs of *Drosophila*. This disc is found ventrally under the larval gut just in front of the anal opening and is in the form of a slightly bent dumbbell (Fig. 450a). It consists of a great number of closely packed cells arranged around a lumen. The imaginal discs produce the seminal ducts and their accessory glands (see Fig. 428d), as well as the external genitalia and the last abdominal segment with a portion of the gut. The inner genitalia consist of paired *vasa deferentia* and accessory glands (paragonia), a single ejaculatory duct, and the complex, bilaterally symmetrical seminal pump, which nestles in the duct. The external genitalia include the penis, together with its chitinous plate and two pairs of symmetrically arranged genital plates: the medial plates, which have powerful thornlike appendages, and the anal plates, which are covered with fine hairs (Fig. 450). When whole imaginal discs are implanted in young (e.g., second instar) larvae, they form essentially all of the structural components of the normal genital apparatus (Fig. 450b). If the implants have been cut medially (Fig. 451a), in the majority of cases a complete and typical genital apparatus is formed by each part (Fig. 451b). The division of the little imaginal discs by two cuts on either side of the midline, resulting in a middle piece and two outer pieces (Fig. 452a$_1$), is very difficult to accomplish with precision. If it is done exactly, a complete and typical genital apparatus can arise from the middle piece (Fig. 452a$_2$: α), while the lateral pieces form one or two medial and anal plates (Fig. 452a$_2$: β, γ). At times there are also formed, in each case due to diagonal cutting, seminal pumps in two pieces; or paired organ elements may be found in several places. After a transverse cut which divides the tissue into an anterior and posterior part (Fig. 452$_1$) the products of the cut parts are not made as well as after cutting in the other direction; apparently the longer cut damages the blastema more than when only a short cut surface is produced. Only the anterior piece can produce a fairly complete apparatus (Fig. 452b$_2$: α); in the best case two medial plates and two anal plates can be formed both by the anterior and by the posterior parts. These experimental results lead to the conclusion that the larval genital disc consists of a number of individual fields (Fig. 453a). When such a field is cut in two, the original field condition is restored in each of the parts. A field, or a group of fields, for the unpaired structures lies medially in the disc. Longitudinal cutting permits regulation and typical self-differentiation of the structures. The medial plate field and the anal plate field must lie transversely like a dumbbell, with lateral organization centers. If the field is cut medially (Fig. 453b), a new, bilaterally symmetrical field organization can arise on each side, resulting in the formation of two mirror-image plates in each half (Fig. 451). If the longitudinal cut is made laterally, a new

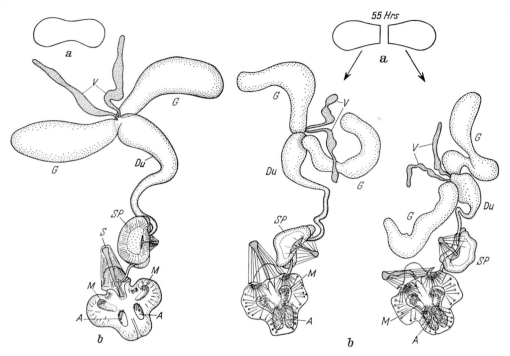

Fig. 450. Transplantation of male genital discs of *Drosophila melanogaster*. a genital disc; b typical sex organs derived from an implanted disc. *A* anal plate; *G* accessory gland; *Du* ejaculatory duct; *M* medial plate; *SP* sperm pump; *S* penis support plate; *V* vas deferens. (After HADORN, BERTANI, and GALLERA, 1949)

Fig. 451. *Drosophila melanogaster*. a divided male genital disc from a 55-hour larva; b typical sex organs resulting from the implantation of each half into a larva of similar age. (After URSPRUNG, 1959)

bilaterally symmetrical field can also arise (Fig. 453c) and, instead of the normal two, a total of six plates form in symmetrical pairs (anal plates in Fig. 452a). When the imaginal disc is divided into an anterior and a posterior section, a field running across the disc may be divided in two, as in Figure 453d. Clearly, the more evenly the field is divided, the easier the restoration of a complete field in both parts will be. But even pieces far from the center of the field can produce all the structures generally formed by an entire field. In the case of the asymmetric division of the male genital disc of *Drosophila montium* into two lateral parts (Fig. 454), pairs of plates are formed by both parts. The large plate on the right, the only one which is normal in size and detail, has obviously been formed from the center of the intact right field. The three other plates come from the divided field material, whose center has formed a pair of plates regulatively and whose peripheral portion has formed one plate. The influence of the quantity of blastema material on the structural completeness of the plate is clear. These experiments show in addition that

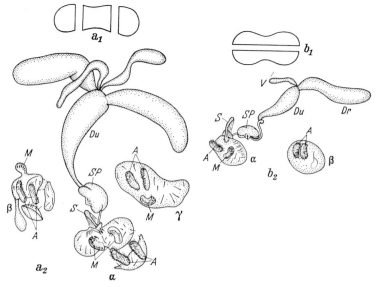

Fig. 452. State of determination of a male genital disc of *Drosophila melanogaster*. a₁, a₂ result of paramedial trisection of the disc; b₁, b₂ result of transverse section of disc. Labels are as in Fig. 450. (After HADORN, BERTANI, and GALLERA, 1949)

regulative power depends on the age of the implanted half-disc and on the age of the host. Half-discs from 72-hour larvae transplanted into larvae of the same age (Fig. 455) form an incomplete set of genitalia: *vasa deferentia* and accessory glands do not develop in pairs; the plates are

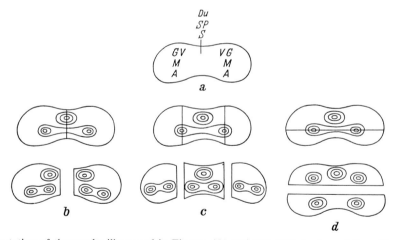

Fig. 453. Interpretation of the results illustrated in Figures 451 and 452. Distribution of the self-organizing fields in the male disc, diagrammatic. (After HADORN, BERTANI, and GALLERA, 1949, with additions)

generally paired, but the one on the cut side is smaller and has fewer bristles than its counterpart (Figs. 455, 456). A masterful experiment in both logic and technique has demonstrated that neither the age and experience of the disc nor influences directly attributable to the host's age are responsible for this lessening of regulative power. Rather, it is the short period of time in which the implant must do its best to replace the missing half blastema through cell division and reorganize its field structure before the onset of pupation. Sagittal halves of imaginal discs were removed from larvae ready to pupate and transplanted to 81-hour hosts (Fig. 457b). In metamorphosis they produced only highly defective sex organs in which not even the genital plates were paired (Fig. 457b). Now several transplants were carried out (Fig. 457c): after various intervals the implants were removed and placed in new 81-hour hosts. This procedure was repeated several times until finally the implant was left in an 81-hour-old larva for good. The result was a complete set of genitalia (Fig. 457c) comparable to that produced after the implantation of

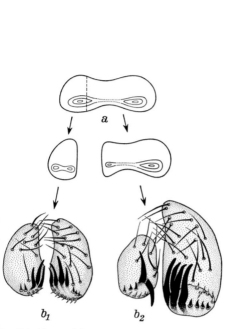

Fig. 454. *Drosophila montium*. Anal plates of the two partial sex complexes resulting from a laterally divided disc. (After HADORN, 1953)

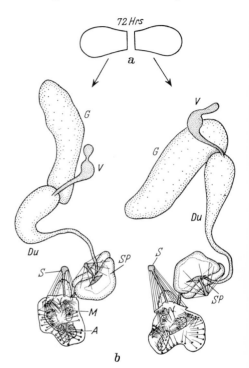

Fig. 455. *Drosophila melanogaster*. a divided male disc of a 72-hour-old larva; b incomplete apparatus results from implantation into larva of similar age. (After URSPRUNG, 1959)

a half-disc into a second instar larva (Fig. 457a, cf. Fig. 451). Indeed, after a four-day sojourn within an adult fly (followed by implantation in an 88-hour larva), a half disc was able to produce a complete set of genitalia. If the fragments of the imaginal discs are left in an adult for eight to fourteen days, they grow tremendously without differentiating. A cell mass up to fifty times as large as the original implant is often found. This can now be cut up and the fragments implanted into third instar larvae to metamorphose with the host. Other fragments have been "cultured *in vivo*" through transplantation to other adults. In this way, pieces of male genital discs containing $1/2$, $1/4$, $1/5$, and up to $1/512$ of the original cell material have been tested for their regenerative and regulative abilities. Repeatedly divided fragments can often produce symmetrical and complete genitalia. With increasing fragmentation and duration in the adult hosts, a tendency for the differentiation of "monocultures" appears however, in which individual structural elements, for example, anal plates or medial plate chaetae, appear predominantly or exclusively and in great numbers, of course. And what is really surprising is the formation of structures outside of the prospective fate of these derivative cultures, and these unexpected products may include imaginal head, leg, or wing structures. It is certain that such structures arise from the cells of the

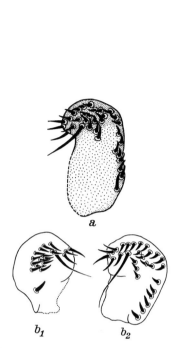

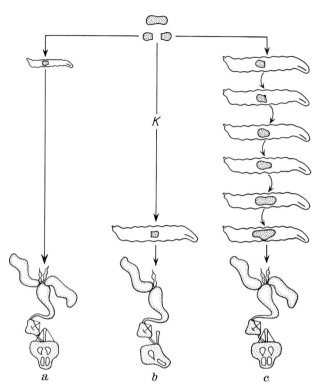

Fig. 456. Male medial plates of *Drosophila melanogaster*. a normal plate; b_1, b_2 pair of plates from sex apparatus derived from a half-disc, left (on the cut side) plate of reduced size, right, an almost normal plate. (a after HADORN, BERTANI, and GALLERA, 1949, b after URSPRUNG, 1959)

Fig. 457. *Drosophila melanogaster*. Serial transplantation experiments with larval, intermediate, and final hosts. Implantation of half-discs from a mature larva. a into a second instar larva; b as a control (K) into an 81-hour larva; c six transfers into 81-hour hosts, the last being permitted to develop into an adult. (After URSPRUNG, 1962)

implanted male genital disc whose determination has been upset rather than from incorporated host material: when legs arise unexpectedly, they bear male sex combs, while the fragments have always been cultured in female flies[235a, 235b].

Each of the paired imaginal discs in the head of *Drosophila* produces a compound eye, a lateral ocellus, half of the median ocellus, an antenna (Fig. 458b), a palp, and some head epidermis. If such eye-antenna imaginal discs from mature larvae are fragmented (Fig. 458c), and the pieces implanted into larvae of similar age, almost all the pieces show mosaic development. Regions can be distinguished which produce either compound eye facets and head capsule, or two ocelli, or one antenna (or just the arista), or a palp. On rare occasions, two palps are formed by the lowest fragment (in Fig. 458c$_2$). In contrast to the other organ fields, therefore, the sectioned palp field exhibits some regulative ability. But all the pieces from imaginal discs that are two days younger, taken from the beginning of the final larval instar (Fig. 458d), behave quite differently. The formation of ocelli is restricted to the dorsal half of the longitudinally split disc (Fig. 458d$_1$) just as in the mature larva, but the appearance in a few cases of three ocelli instead of two shows that the determination is not yet rigid. If the cut is moved dorsally from the midline (Fig. 458d$_2$),

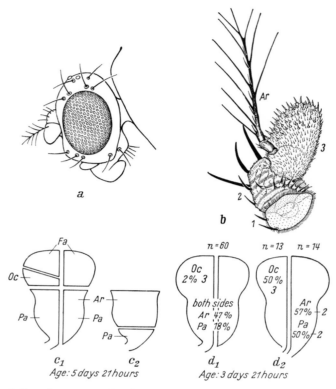

Fig. 458. *Drosophila.* a, b *D. melanogaster*: a head somewhat diagrammatic; b left antenna, medial aspect; c, d diagrammatic representation of the potencies of eye-antenna discs cut up at various ages and implanted. *1, 2, 3* antennal segments; *Ar* arista; *Fa* facets; *Oc* ocelli; *Pa* palps. (b after LE CALVEZ, 1948; c, d after the experiments of VOGT, 1946)

three ocelli arise very often: when the field is cut, a reorganization clearly takes place. Similarly, if the arista and palp field of the young third instar larva is split evenly by a longitudinal cut, an arista and palp often are formed by each half (Fig. 458d$_1$). If the cut is moved dorsally, the smaller (dorsal) piece makes neither arista nor palp, but in the large (ventral) piece, each of these organs may be represented twice (Fig. 458d$_2$). These field reorganizations result in more organs being formed by a part of the disc than normally arise from an eye-antenna disc. But split compound eye fields achieve true regulation: from one donor larva, one imaginal disc is implanted whole and the other in two longitudinal halves into the same host larva. The sum of the facets formed by the two implanted halves is up to 46 per cent greater than the number formed by the undivided implant. This enlargement of the eye in the direction of normality is obviously achieved at the expense of the presumptive head epidermis. Thus, the organ fields of the imaginal discs are still labile well into the last larval instar and can renew their field structure after fragmentation.

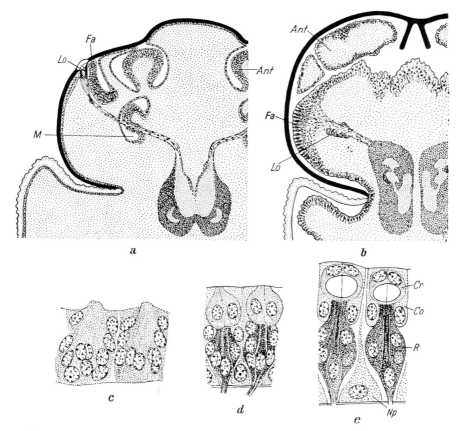

Fig. 459. Imaginal anlagen in the head of *Ephestia kühniella*. Horizontal sections through the head. a last instar larva; b prepupa just before pupation; c–e longitudinal sections through ommatidial anlagen; c from ventral border of the eye; d from dorsal border of the eye of a seven-day-old pupa; e of an eight-day-old pupa. *At* antennal anlage; *Fa* ommatidial anlage; *M* mandible anlage; *LO* larval ocellus. (After Umbach, 1934)

By means of regulation after the isolation of parts of the eye-antenna imaginal disc, the number of individual organ anlagen in the disc can therefore be modified. But even the determination of the competing fields for particular organs is not fixed in the early larval stages: exposure to mustard gas after seventy to eighty hours of larval life transforms a prospective arista into a prospective leg. This transformation is not all-or-none, but rather there are a number of intermediate steps between a normal arista and an almost perfect leg[54].

When the eye field is completely determined, the cell groups become the individual ommatidia. This process was first described in *Ephestia*. The eye disc, from which only the compound eye comes in the Lepidoptera, is an epidermal invagination just in front of the group of larval ocelli (Fig. 459a). Behind this the epidermal layer is thickened. During pupation after the retraction of the epidermis from the larval cuticle the eye disc spreads out on the surface to be occupied by the compound eye (Fig. 459b). In the pupa, after the retraction of the pupal cuticle, the eye anlage stretches, reducing the thickness of the epidermal layer. The grouping of the cells of the individual ommatidium anlagen begins in the prepupal eye disc (Fig. 459b). This does not take place synchronously throughout the whole disc, but begins in a dorso-caudal region and spreads

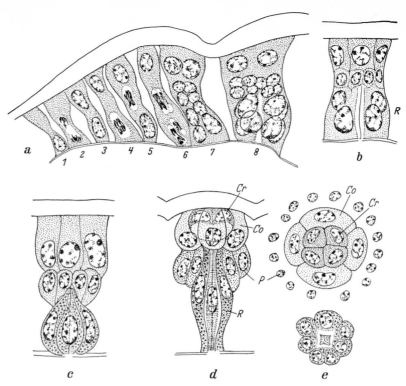

Fig. 460. Development of the ommatidia of the pupa stage of the worker of the ant *Formicina flava*. a *1–8* sequential stages at the edge of the eye anlage; b–d further stages and longitudinal stages; e cross section, same stage as d, above cross section through the upper part of the eye, below through the retinula. *Co* corneagen cells; *Cr* crystal cone cells; *P* pigment cells; *R* retinula. (After BERNARD, 1937)

from there. The later steps in the differentiation of the ommatidia (Fig. 459c–3) also proceed in this spatiotemporal sequence. Afterwards, when the definition of the individual cell elements is clear, all the cells grow vigorously.

An important question is that of the origin of the cell groups of the ommatidial anlagen. This can be answered only by the elegant use of a favorable experimental object — a small subterranean ant, *Formicina flava*, whose workers have only ninety to a hundred ommatidia. The epidermis of the eye anlage consists of a single layer of large cells in the young pupa. Ommatidial development begins at the edge of the anlage and proceeds toward the center (Fig. 460a). From a single ommatidial stem cell, after a set sequence of differential divisions, the whole ommatidium is produced. After the first division a stem cell for the distal cell group and one for the proximal group are formed (Fig. 460a, *1–3*); then the proximal stem cell divides into a pigment stem cell and a retinula stem cell (Fig. 460a, *4, 5*). The distal stem cell divides twice, forming four cells (Fig. 460a, *7, 8*, b, c), each giving rise to cornea cells and a quarter of the crystalline cone (Fig. 460c–e). Further division of the proximal stem cells (Fig. 460a, *6–8*, b) forms the sixteen pigment cells which surround each ommatidium, as well as seven retinula cells (Fig. 460a, *7–8*, b–e). Presumably, either one of the divisions of a retinula mother cell does not take place, or one of the final division products does not persist. Figure 461 summarizes the sequence of divisions. These divisions occur in specific places within the cell group. The development of the com-

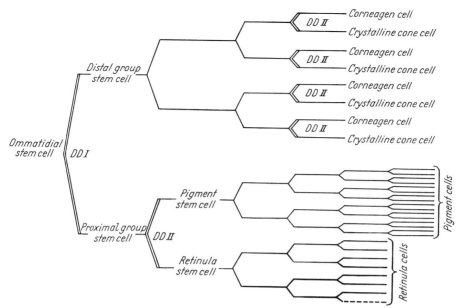

Fig. 461. Lineage of the ommatidial cells (cf. Fig. 460); the differential division steps are always marked by a double line

pound eye is a good example of the autonomous organization of a morphogenetic field whose individual cells embody the principle of differential cell division, so that each produces a specific number and arrangement of cells in four functionally disparate categories.

Lecture 25

The development of the insect wing disc is a veritable kaleidoscope of differentiation. The mature wing is a light but sturdy laminated plate consisting of two chitinous lamellae and their underlying epidermal layers which merge at the edge of the wing. The wing veins, which pervade the structure, form originally as blood lacunae in the epithelial wall containing tracheae and nerves. In the epidermis cells divide into sensory receptors, bristles, hairs, and scales. In the Lepidoptera the scales make a rich pattern. The total organization of the wing results from the interaction of a great variety of morphogenetic processes.

The lacunar system of the wing is laid down in the last larval instar when the imaginal disc is still a pouch opening to the inside of the body. The epithelial layers become appressed to one another, leaving only the primary lacunae (Fig. 462a). These lacunae merge proximally (Fig. 462b) to some extent and thus set up the branching system of the Lepidopteran wing veins. The strips of fusion of the wing lamellae which extend in from the edges have a particular relationship with one another, reflecting a transverse periodicity in the wing, whose nature is not understood. The rhythm is particularly striking when the number of lacunae is altered in certain mutant strains. In *Ptychopoda seriata* the mutation *Va* upsets the formation of the primary lacunae, so that there may be too many or too few longitudinal veins. Nevertheless, the intervals between these longitudinal veins are kept uniform over the entire wing. If extra veins are inserted, all the veins crowd closer together; if veins are missing, all the spaces between the veins are larger (Fig. 463).

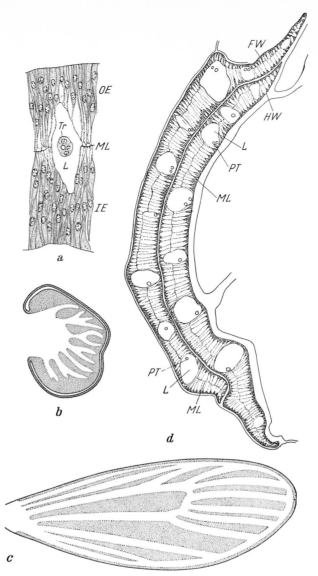

Fig. 462. Lacunar system in the wings of *Ephestia kühniella*. a part of a cross section through an imaginal disc through the forewing in the sixth larval instar; b diagram of the lacunar system in the imaginal disc in the sixth larval instar; c diagram of the lacunar system in the larval wing; d cross section through the wings of a six-hour-old pupa (75×). *HW* hind wing lacuna; *L* lacuna; *ML* middle lamella; *OE* epithelium of upper surface; *PT* primary tracheae; *Tr* trachea; *IE* epithelium of lower surface; *FW* forewing. (a, b after Köhler, 1932; c, d after Kühn and von Engelhardt, 1933)

We do not find a corresponding broad region without a vein in it. Evidently, the sequence of fusions beginning in the outer edge follows a particular lateral periodicity; and it appears that the length of this period can be modified.

In the prepupa the wing discs grow quickly as the cells divide vigorously. Mitosis begins to increase immediately after the eversion of the wing disc (Fig. 464). The cells divide predominantly in the long axis of the wing, and it is in this direction that the wing mainly grows (Fig. 464). The young imaginal disc corresponds to the basal region of the future wing. Evidently growth brings the descendants of the young imaginal disc's basal cells to the outer end of the wing, as can be seen from the positions of genetically determined somatic chromosomal aberrations (loss of whole chromosomes or of chromosomal regions). These occur in individual cells at different developmental stages, e.g., in the cells of young wing discs (Fig. 465).

After mitosis has slowed down again, the wing anlage stretches considerably. Contractions of the thoracic muscles force blood into the growing wing so that the epithelium becomes thinner and more extensive (Fig. 466). Secondary changes in the lacunar system take place in the prepupa. Cross connections are formed between certain lacunae, and a lacuna is also formed around the edge of the wing. Moreover, some lacunae are actually obliterated. The lacunar system has this new, altered form at the onset of the pupal stage (Fig. 462c, d).

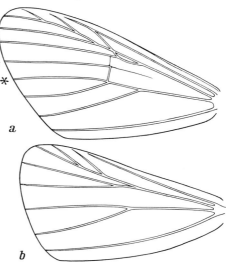

Fig. 463. *Ptychopoda seriata*. Right wing seen from below. a veins of normal wing; b wing veins in the mutant. The veins in *a* marked with *Va*, in which the veins marked "*" are not present. (After Kühn, 1948)

Within the primary lacunae, during their formation in the last larval stage, the first pupal tracheal system forms from the tracheal arches of the longitudinal tracheae of the larval body. In the pupa, while the primary system is still operant, buds develop at the base of a part of the wing tracheal system, from which a new system of secondary or imaginal tracheae grow out into those lacunae which are not resorbed (Fig. 467a). At branch points in the pupal lacunar system the secondary tracheae ramify. As the secondary tracheae grow out (Fig. 467b), lateral outgrowths appear, the first signs of branching in the secondary tracheae. At the same time, primary tracheae begin to degenerate. When the adult emerges, the mature secondary tracheal system (Fig. 467c) fills with air for the first time.

The nerve trunks of the wing run the same course as the tracheae in the lacunae. They emerge in groups from the sensory receptors and grow in a basal direction. At the branch points of the lacunae they unite. In contrast to the branching of the tracheae, the nerve pattern is one of repeated joining, the confluence of originally separate elements. And so the morphogenesis of the three systems is characterized: that of the blood lacunae by even spacing; that of the respiratory system by branching; and that of the sensory nerves by confluence. Thus, all three pervasive systems are laid out within the wing.

If a cut is made in the wing disc of a mature larva, reconstruction of the lacunar system follows in many cases. The number of lacunae is as a rule reduced in such cases. The wing, however, is provided with displaced or newly formed lacunae just as in the case of genetically determined absence of lacunae (Fig. 463). Thus, the overall pattern remains harmonious. Experimental cases of this sort show that the tracheae and nerves grow in patterns determined by the lacunae. Empty lacunae apparently exercise an attraction on the secondary tracheae, so that tracheal branches enter these lacunae in a way not seen in the normal wing. An oversupply of tracheae for a particular lacuna is apparently prevented by the mutual inhibition of growing tracheae. At the base of a lacunar branch several tracheae may be present which have branched off the main tracheal trunk, one after another. But only one tracheal branch actually penetrates the empty lacuna, while

363

the others stop short behind its point of entry (Fig. 468a). The nerves also follow the modified pattern of the lacunae, which apparently attract the neurons growing from the sense cells down toward the base of the wing. If a lacuna is broken by injury, the oncoming nerves either coil up or seek out other lacunae. In such cases nerve tracts sometimes find themselves heading back out toward the apical part of the wing in another lacuna. And so, a ring forms around the entire intercostal field (Fig. 468b).

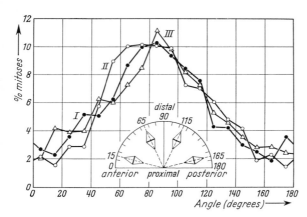

Fig. 464. Mitotic orientation on the upper surface of the hind wing of *Ephestia* in age classes I, II, and III (see Fig. 466). Abscissa angle in degrees as illustrated in insert. (After POHLEY, 1959)

The onset of the pupal stage is followed by a short period of cell division. Then begins the visible development of the wing epithelium with scales. The scales are homologous to single cell hairs. The mature scale is the empty, chitinous skeleton of a great flat extension of a vastly hypertrophied epidermal cell. The scales are little plates, 0.5 to about 5 microns thick, thirty to over a hundred microns wide and of quite diverse length. They often exceed 20,000 square microns in surface area; and they are built to maintain their structural rigidity. The chitinous layer of the upper surface is thrown into longitudinal thickenings, which continue around the edges of the scale to the undersurface. Between the upper and lower layers are reinforcing trabeculae. These are formed in the cytoplasm and joined to the cuticular lamellae.

The larger the scales, the more trabeculae are formed. These stiffening structures were first seen clearly in the electron microscope[361, 365]. Figures 469a and 476c illustrate a simple form of trabecular structure. Thin cross struts connect the 0.1- to 0.2-micron-thick ribs, which run the length of the slightly curved upper lamella. The chitinous lamella between these struts has holes in it here and there. The trabeculae sometimes meet the upper surface at the ribs, and sometimes between the ribs. They often branch as they approach the upper layer. The large scales, which

Fig. 465. Right forewing or *Ephestia kühniella*, genotype *dz Mo* (dark central field, mosaic). Bands of unpigmented scales from the base to the tip of the wing, absence of a chromosome (or chromosomal fragment) in the young imaginal disc. (After KÜHN, 1960)

need even stronger stiffening, have reinforced ribs, and the trabeculae are sturdier and joined into a network (Fig. 469b). The discontinuities in the upper lamella become quite large, so that it takes on the form of a grating.

The differentiation of the scale-forming cells in the pupal epithelium takes place in two differential cell divisions. In the epidermis of the wing surface of the pupa (at forty hours at 18°, at eighteen to twenty hours at 25°) scattered cells appear, which are distinguished from the ordinary

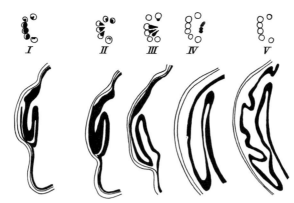

Fig. 466. Prepupal age classes I–V defined by the retraction of the larval ocellus; below, medial longitudinal sections through the wing anlage in the same stages. (After QUERNER, 1948)

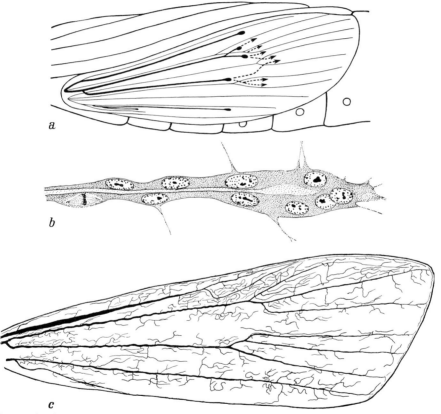

Fig. 467. Wing tracheal system of a pupal wing, with the principal pupal wing tracheal system (thin lines) and the growing secondary adult tracheal system (thick lines) (18×, diagrammatic). b growing tip of a secondary trachea at the same stage, 550×; c secondary tracheal system in the adult wing, 12×. (After BEHRENDS, 1935)

epidermal cells by the size of their nuclei and by their pear shape (Figs. 470, 471 a). These are the first-order scale stem cells. A period of mitotic activity now begins in the wing epithelium.

The spindles in the ordinary epidermal cells are tangential (Fig. 470c, d). In the first-order scale cells the spindles are perpendicular to the wing surface, and a small cell, which later degenerates, divides off in the direction of the middle membrane (Fig. 471b). Comparative studies show that in insects which produce hairs under similar circumstances, sensory stem cells form in this position. The daughter cells near the upper surface are now the second-order scale stem cells. They divide again, and now the spindle axis is inclined at an angle of about 30° to the upper surface. The upper end of this angle always faces the distal end of the wing (Fig. 471c). This division separates a larger scale-forming cell from a similar socket-forming cell. Thus, a scale complex is set up by two differential cell divisions. Now the scale process begins to grow out of the scale-

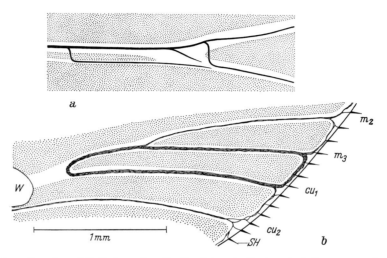

Fig. 468. Sections from the wing of *Galleria mellonella* after injury to the imaginal disc. a secondary tracheae with branching where a lacuna forks; b abnormal nerve pattern after lacunar interruption. *SH* sensory hair; *W* wound scar. (After CLEVER, 1959)

forming cell (Fig. 471d, e). It grows in the space between the pupal cuticle and the retracted epidermis. At first, it is club-shaped, but then it flattens into a thin plate and grows in length and width. Meanwhile, the cell body and nucleus of the scale-forming cell are growing vigorously (Fig. 471d–g). When the scale has reached its final form, the chitinous structures are laid down. Then the cytoplasm disintegrates, and air enters the interior of the scale through the holes in the chitinous lamella. The socket-forming cell grows around the base of the scale extension and forms a chitinous socket for the scale (Fig. 471e–g).

Although many butterflies and moths have only one kind of scale, the Pyralididae have several types, which, due to their diverse sizes, form several scale layers: the top layer, the middle layer, and the deep layer (Fig. 472a). The deep scales are the most numerous. They make up about fifty per cent of the scales of the upper surface of the front wing. The difference among the size classes of scales becomes apparent at the beginning of scale development. During the differential divisions of the formation of the scale outgrowth, the first- and second-order stem cells and then the scale-forming cells of the top layer precede those of the middle and deep layers. In *Ephestia* the three scale types are not found throughout the wings. The top scales are not present on the undersurface; and in great areas of the upper surface of the hind wing only the deep type scales are found. The scale size classes correspond to the size classes of scale-forming cells. Moreover, the nuclei, which are especially active just when the scale extensions are growing out, as can be

their vigorous growth and by the deposition of metabolic products in the cytoplasm, contain chromocenters, whose number is an indication of the number of genomes present. The counts suggest that the scale-forming cells are polyploid. While the ordinary epithelial cells are 2n, the deep scales, middle scales, and top scales form from 8n, 16n, and 32n cells, respectively. The socket-forming cell nuclei are either 2n or 4n, the polyploid cells being associated generally with the largest scale-forming cell classes. Figure 473 summarizes these relationships schematically.

Where all three types are present, scales of the three size classes are arranged in recurrent identical groups. Zigzag rows run across the wing from front to back (Fig. 472b). Here the deep scales are distal to the larger types, and the standard grouping consists of one top scale, one middle scale, and two deep scales. The developmental mechanism underlying this grouping has not been elucidated. The hypothesis[276] that a series of determinative cell divisions leads to a group of scales and ancillary epithelial cells has not been confirmed. The demonstrated periods of cell division in the wing epithelium contradict the hypothesis, and the distribution of UV-induced somatic mutations in nests of mutant scales excludes the postulated cell lineage[182, 519].

Besides the scale complexes there are a variety of other "little organs," or *organules*, on insect wings. These are formed by determinate divisions and include epidermal glands with ducts and sensory bristles with one or more sensory receptors. The sequence of divisions leading to the formation of a sensory complex in the wing margin of *Galleria* (Fig. 474) illustrates the prospective fate of individual cells, although it is not clear when and how determination of the individual cells occurs. The

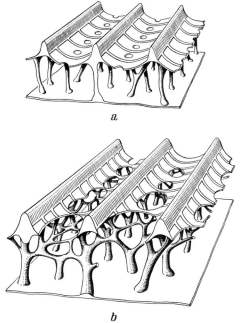

a

b

Fig. 469. Chitin structure of moth scales deduced from electron micrographs. a basic type; b larger scales in Pyralids. (After KÜHN, 1946)

injection of crystalline methylene blue into young pupal wings results in the strong selective staining and death of sensory cells. Even their immediate neighbors remain unstained and undamaged. If the sensory cells are destroyed in the young pupa by this method, a portion of the sensory bristles are missing in the adult wing. The stain may affect all or only some of the sense cells. Indeed, if all the sensory cells are eliminated, no further bristles will form (Fig. 474c). In this case, the bristle-forming cells, which normally would have become polyploid, remain diploid just like the ordinary epithelial cells. The survival of only one sensory cell is sufficient to permit a normal bristle to form. Thus, the developing sense cells appear to emit a morphogenetic influence during the development of the little organ. Differential division sequences are thus integrated with the dependent differentiation of particular cell types. On the upper surface of the forewing in *Ephestia*, the top scales form a pattern (Figs. 474, 475a). At the level of the middle scales the pattern fades and the deep scales are uniformly pale grayish brown (Fig. 472a). Differences in color are caused by varying pigment concentrations and distribution in the chitin of the scale. Eight color types can be distinguished in the scales (Fig. 475d), ranging from the modest pale grey or tan type I to types with more striking pigment. This reaches its highest expression in the dark scales of type IV which are prominent in the wing pattern.

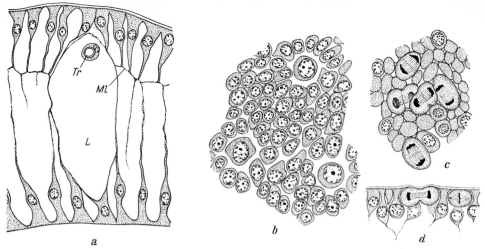

Fig. 470. Epithelium of the pupal wing of *Ephestia*. a two hours after pupation, cross section through the wing; b, c horizontal sections; d longitudinal section through the upper surface epithelium; b 56–58-hour pupa; c, d 63-hour pupa, 400×. (a after BEHRENDS, 1935; b–d after KÖHLER, 1932)

Type VI has a narrow white tip. The brightest scales have a broad white tip (VII, VIII). These scale types differ not only in their pigmentation, but also in their form and structure. The dark and bright pattern scales differ the most: Type IV is relatively long and narrow, while type VIII is shorter, broader and more triangular. The structural differences are clearly seen in electron micrographs: they reflect variations on the pattern diagrammed in Figure 469 b. The dark scales (Fig. 476 b) are covered with numerous highly pigmented ribs, which run closely packed on the

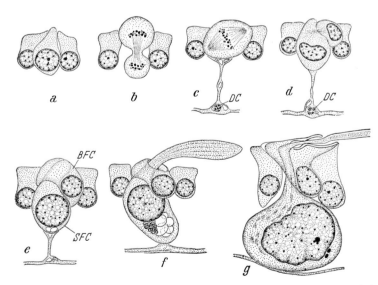

Fig. 471. Scale formation in *Ephestia kühniella*. a primary scale stem cell; b first differential mitosis; c second differential mitosis; d–g scale forming cell (*SFC*) and socket forming cell (*BFC*); *DC* degenerating cell. (After STOSSBERG, 1938)

368

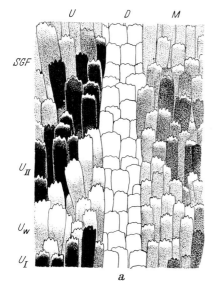

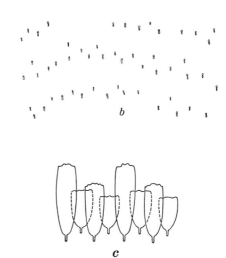

Fig. 472. Wing scales of *Ephestia kühniella*. a part of the forewing upper surface; in the central part, upper (*U*) and middle (*M*) scales have been removed. On the right only the upper scales have been removed. *D* deep scales; b arrangement of the scale sockets in the upper surface of the forewing; c diagram of scale groupings. *SGF* surrounding grey field; U_I, U_{II} proximal and distal transverse band; *Uw* bright cross band of the wing pattern (cf. Fig. 475a). (a after KÜHN and HENKE, 1932; b after HENKE, 1946; c after POHLEY, 1953)

upper surface of the scale to its tip; between the two lamellae is a highly branched network of trabeculae. In the light scales (Fig. 476a) the ribs are fewer and farther apart; they are not pigmented; and relatively few, weakly branched, pigmentless trabeculae end in swellings at the upper lamella. The knobby upper surface resembles white powder in its diffuse reflection of light. Thus, there is more to the determination of the wing pattern than the mere fixing of a variety of pigment aggregates. These diverse arrangements of pigment stem in turn from variations in the morphogenesis of the scales. The elaboration of a particular structure-color type is a response of the scale-forming cell to local influences at a particular time during development.

In *Ephestia*, black and white elements in the pattern stand out from the brownish-gray background composed of the modest types I, II, III, and V. The components are arranged in patterns, of which the most important are sketched in Figure 475c. In the symmetry system there is a central field placed between two symmetrical bands, the distal band, and the proximal band. Each band consists of a white middle zone, and inner and outer border zones. A central spot system lies in the interior of the central field at the crossvein; it consists of two black spots with a white spot in between. A row of marginal spots lies between the veins at the outer margin of the wing. The symmetry system is also influenced by the wing veins, in that over each vein the zigzag bands point away from the central field.

Operations on the wings of various Lepidopteran species, as well as mutations affecting the layout of the wing veins, show that the localization of the central and marginal spots is

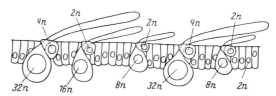

Fig. 473. Diagram of the epidermis with polyploid classes of nuclei. Top scales (*32n*); middle scales (*16n*); deep scales (*8n*); scale sockets (*2n* and *4n*) and epidermal cells (*2n*)

369

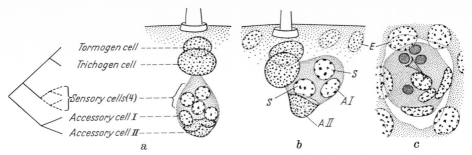

Fig. 474. Marginal sensory bristle apparatus in the wing of *Galleria mellonella*. a diagram of development; b, c elimination of the sensory cells with methylene blue; b two sense cells eliminated; c all 4 sense cells killed; *E* epithelial cells, *A I*, *A II*, accessory cells; *S* sense cells. (After CLEVER, 1960)

entirely dependent on the lacunar system and thus is determined in the larva or prepupa. When the number of veins reaching the wing margin is increased or reduced, there is always a spot between each two adjacent veins; and if the crossvein is missing, the central spot is missing, too. The symmetry system is set up in the pupa at a particular time when a wave of determination is spreading over a part of the wing surface. The spread of this determination can be monitored by microcautery, in which tiny burns are made in the pupal wing at various times and places. If a tiny burn is produced on the first pupal day, the nearby white symmetry band is displaced from it toward the central field (Fig. 477 b, c). The burn has stopped the spread of the determinative wave; the white band just reaches the barrier, and its black border extends beyond the burn. If the burn is made at the posterior border, an anterior field forms, edged in white (Fig. 477 d). In melanic (dark) mutant strains, the barrier effect is particularly clear (Fig. 477 e). From such forms it may be concluded that a stream of determination emanates from the front edge and hind edge of the wing, and spreads distally and proximally (Fig. 477 a). This wave of determination lays out the field of symmetry (Fig. 475 c). When a burn is made anywhere in the wing of

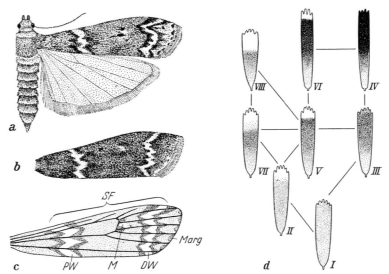

Fig. 475. *Ephestia kühniella*. a body and right fore- and hindwings of the wild type; b forewing of the black mutant strain; c diagram of the pattern system. *DW*, *PW* distal and proximal white symmetrical bands; *M* central spot; *Marg* marginal spots; *SF* field of symmetry; d scheme of the color types of the top scales of the forewing outer surface. (After KÜHN and HENKE, 1932)

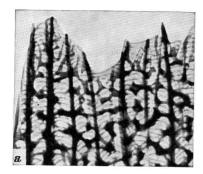
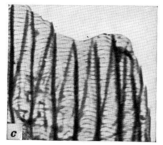

Fig. 476. Electron micrographs of *Ephestia kühniella* scales. a, b pattern scales; a light; b dark; c deep scale about 5,000×. (Original preparations of Kühn and von Engelhardt)

a 48 to 60 hour pupa (at 18°), the wave of determination, which is in progress, stops dead (Fig. 475 k–o).

The resulting patterns in the mature wings also reveal the field organization of the symmetrical bands and within the central field: When the spread of the determination is stopped while it includes only a small field in the middle of the wing, this region is covered completely by white scales. When the determination has advanced further to the anterior and posterior borders of the wing, black regions appear within the white (Fig. 477 k, l). After further spread, there appears within the black region an area with brownish-gray scales, first at the anterior and posterior margin (Fig. 477 l) and later also in the center of the wing (Fig. 475 n). Then the white region narrows into a band which moves distally and proximally, the black following to border it, and the central field, with its browhish-gray scales, gradually reaches its final extent (Fig. 477 o and p). A larger burn stops the spread of determination completely, so that the only pattern seen is made by black scales above the longitudinal wing veins (Fig. 477 i), as is normally found in some species. With the spread of the field of symmetry, the black vein pattern breaks up and is moved farther and farther out until it finally forms the outer border of the white band. The field organization pattern of the symmetry system thus comes to exist through its own organization, which takes place during the spread of this determination field toward the field border, and through the organizing effects on the field border from the surrounding field, whose elements are initially reflected in an entirely different way, namely via the veins. The determination of the symmetry system spreads over the wing epithelium immediately before the cell division period in which the determination division of the first-order scale-forming cells takes place. It influences not only the form and color of the scales in the pattern, but also the frequency of epithelial cell division. These divide more frequently in the "light-determined" zones than in the "dark-determined" zones; the narrow, dark scales consequently are packed closer together

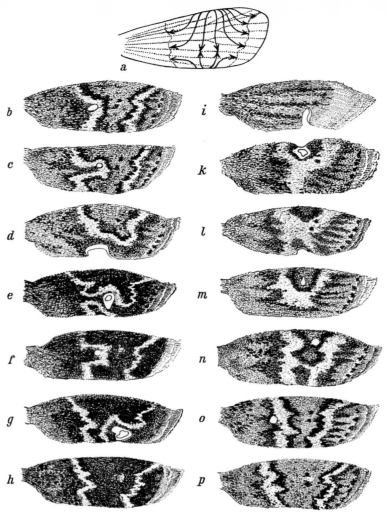

Fig. 477. Determination of the system of symmetry in the forewing upper surface of *Ephestia kühniella*. a scheme of the spread of determination; b–p colored wings of a moth, ready to emerge, which has been liberated from its pupal cuticle; b–g results of microcautery on the first pupal day; k–o after 48 to 64 hours of pupal life; h, p normal control wings; e–h black mutant strain with no other mutant characters. (After KÜHN and VON ENGELHARDT, 1933)

than the broad, light ones. The determination of the central spot, which has been completed much earlier, stands out clearly when the spread of the symmetry determination is arrested at an early stage, so that the white determination area is right at the crossvein region: the black scales of the central spot are formed without regard to their surroundings.

In the *Saturniidae* the central spots develop as several large zones surrounding "eye spots." In the giant silk moth *Philosamia cynthia*, the center of the eye spot has almost no scales and is clear as glass; around it is a white zone and around that a yellow zone. This whole system of scale differentiation is demarcated by a branch of the black marginal band (Fig. 478). The pupal wing epithelium in the region of the presumptive eye spot can be cut into two or more pieces. It then becomes clear that the formation of the eye spot at a particular place with respect to the

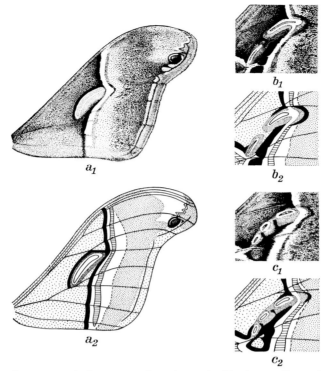

Fig. 478. *Philosamia cynthia.* a normal wing pattern, forewing underside; b, c ocellus region after the young pupal wing has been cut in the presumptive ocellus region. (After Henke, 1933)

lacunar pattern has already been determined. From each part of the blastema a partial eye spot arises. Each part does not correspond to a piece of a mature eye spot; rather, each part forms the entire system of zones (Fig. 478 b, c). The blastema in which the eye spot anlage has been determined thus behaves – just like a piece removed from a *Dictyostelium* pseudoplasmodium – as a harmonious equipotential system, whose parts always turn into normal miniatures.

The pattern of the wing as a whole, its outline, the pattern of its supply system, the repetition of identical groups of different scale types, and the system of wing markings are, therefore, not determined together by a single event, but rather by the orderly sequence of a number of more or less independent processes involving a great diversity of morphogenetic principles. So the determinative events in the formation of the insect wing – clearly a relatively simple organ – represent an impressive model for the determination of organ patterns in the development of animals in general. But let us not forget this: We can define a series of determinative events and conditions, but we know nothing of their chemical nature.

Lecture 26

In the plants we encounter a series of morphogenetic principles which in animals either do not occur or remain obscure throughout development. They stem in large part from the fact that plant cells have cellulose walls, which permit little or no movement of the cells with respect to one another. This fact eliminates in one fell swoop the migration of cells and cell masses, the

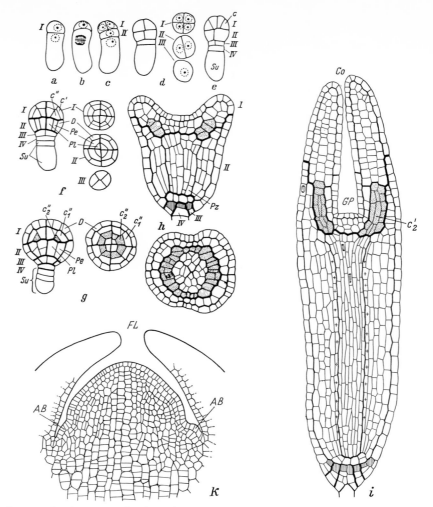

Fig. 479. Angiosperm development. a–i embryonic development of *Biophytum dendroides*; a two-cell stage; b transition to three-cell stage; c four-cell; d eight-cell, left longitudinal section, right cross section at levels I, II, and III; e sixteen-cell stage; f embryo after separation of the cells into Dermatogen (*D*), Periblem (*Pe*); Plerom (*Pl*) and Suspensor (*Su*), longitudinal section and cross sections; g older spherical embryo, longitudinal and cross sections; h beginning of the outgrowth of the cotyledons, longitudinal and cross sections (*Pz* = pericycle). Longitudinal section through an embryo about 530× long, endodermis cells marked +. Root cap initials in h and i are hatched. *GP* growing point; *c, c', c'', c₁'', c₂''* cell sequence from which the mass of cotyledon tissue emerges; a–h about 280×; i about 200×; k growing point of a bean embryo; *AB* axial bud of the first leaf (*FL*). (a–i after NOLL, 1935; k after SACHS)

inward or outward folding of epithelium and other morphogenetic movements characteristic of embryonic development in animals. So the overall organization of plants and the orderly formation of organs and tissues are achieved only through regular cell growth, cell division, and cell differentiation within coherent cell masses. This situation is not unknown in certain stages of animal development, e.g., in the development of skeletal cartilages, but in plants it represents *the* developmental principle. For communication among the organs, it is essential, since the plants lack a general circulatory system like that of the higher Metazoa, that all the cells be bathed in the same medium. In spite of this, there exist transported chemical influences, most striking during development. The phytohormones, like animal hormones, evoke morphogenetic effects in minute concentrations in particular target tissues.

In contrast to growth in animals, plant growth is long lasting but localized. Plants do not grow as a whole in the manner of animals in embryonic and post-embryonic stages, where the organs develop harmoniously until an appropriate size is reached. In the animals, the order of the tissues and organs, the detailed and overall organization of the body plan is established during embryonic development; but in the plants new organs are continuously formed in certain growth zones. As these organs grow, the overall form spreads to a certain developmental conclusion in annual and biennial plants, or it reaches a size limited only by secondary factors, as in the trees.

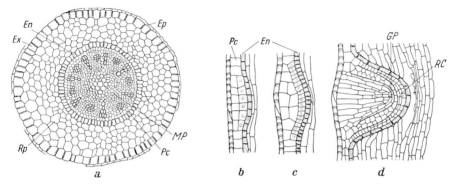

Fig. 480. Angiosperm root. a cross section (aerial root of *Hartwegia comosa*) (Liliaceae); b–d formation of a lateral root, diagrammatic. *En* endodermis; *Ep* remainder of the epidermis; *Ex* exodermis (outermost cortical layer with corky cell walls); *MP* medullary parenchyma; *Pc* pericycle; *Rp* cortical parenchyma; *GP* growing point of the lateral root; *RC* root cap. (a after Wiesner from Rothert and Jost, 1934; b–d after Holman and Robbins, 1944)

Annual and biennial flowering plants are particularly good examples of the influence of internal and external factors on the course of development, which we have only touched upon in the insects and have otherwise fairly well neglected in animal development, in favor of considering the laying out of the basic body plan in the embryo.

The embryonic development of angiosperms proceeds with particular cell division sequences in which the embryo proper becomes separated from the suspensor and divides into the three layers called dermatogen, periblem, and plerom (Fig. 479). The fertilized egg cell is polarized: It is usually longer in the apical-basal direction, and cylindrical or pear-shaped. The nucleus moves to the apical or shoot pole, and in the first division a transverse wall forms, cutting off a small apical end cell, which is rich in cytoplasm (Fig. 479a, *I*). The "lower," longer cell of the "pro-embryo" participates to varying degrees in embryogenesis according to the species. We shall take as an example the development of *Biophytum dendroides*. The next wall formed divides the end cell lengthwise, while the other longer cell is divided transversely again (Fig. 479b, c). Now a vertical wall forms in the next highest cell also (Fig. 479d), and after two further divisions of the original basal cell, a four-tiered complex forms (Fig. 479e, *I–IV*), from which the embryo is made. The lower cord of cells becomes the suspensor. It pushes the embryo into the developing endosperm in the embryo sac and can also bring it nutrients. The embryo now becomes spherical (Fig. 479f, g). It divides further with *periclinic* cell walls running parallel to the surface and *anticlinic* cell walls perpendicular to these. The spherical embryo is radially symmetrical (see cross sections in f and g). Then the embryo elongates and acquires bilateral symmetry: the two

375

cotyledons arch forward. These arise from the lateral parts of the first tier and from the upper and outer parts of the second tier (Fig. 479h, i). They are formed from dermatogen and periblem, as the true leaves will be also (Fig. 479k). The major portion of the cotyledon tissue comes from cells which are labeled (in Fig. 479e–g) c, c', c'', c_1'', and c_2''. The shoot growing point arises from the middle of the top tier (Fig. 479i). From the second, third, and fourth tiers come the part of the stalk below the cotyledons, the hypocotyl, and the root (Fig. 481a). The central cylinder is

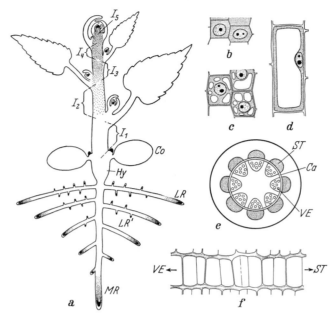

Fig. 481. a diagram of a dicot; b–d transformation of the meristem cells with increasing distance from the growing tip; e diagram of the second thickening of a dicot stalk of radial column of cambial cells. *VE* vessel elements; *MR* main root; *Hy* hypocotyl; *I* internodes; *Ca* cambium; *CO* cotyledons; *ST* sieve tubes; *LR, LR'* primary and secondary lateral roots. (a–d after Sachs, modified; e, f, Rothert and Jost, 1934)

derived from the plerom. Its outer layer of cells soon separates from the inner plerom cells as the pericycle (Fig. 479h), from which the endogenous formation of lateral roots later proceeds (Fig. 480). In the periblem the future cortex, the innermost cell layer, later differentiates as endodermis (Figs. 479i, 480). In contrast to the two upper tiers, the third and fourth tier cells in the embryo divide very little. They form the middle part of the root cap (Fig. 479h, i) on the root tip. The borders of the tiers of the embryo can be followed over a fairly long period of time. In this way it has become clear that the basic organization of the embryo into cotyledons, hypocotyl, and roots is not associated with their borders. The position of the first longitudinal wall in the upper half of the embryo (Fig. 479d) also has no relationship to the later bilaterality: the cotyledons can occupy any position with respect to the first longitudinal walls in I and II. Only the root-shoot polarity is seen to be determined in the fertilized egg cell. Since the embryo assumes a particular position in the bipolar-differentiated embryo sac, it is apparent that this polarity has already affected the egg before fertilization.

Persistent embryonic undifferentiated cells capable of further division remain in the *meristems*. In the growing points or growing cones of the shoot and root tips and in the axial bud rudiments of the leaves lie *primary* meristems, which are immediately responsible for the further morphogenesis of the plant (Fig. 481). In the shoot's growing tip, the border between the first and second tiers (Fig. 479 h–k) disappears at the end of embryonic development, and the tip becomes a block of cells from which the epidermis, the cortex, and the pith differentiate basally. At the very tip lie several layers of cells in which the cytoplasm is of considerable size; they represent the *apical meristem* (Figs. 479 k, 481 b). The cells of the uppermost layer divide almost solely anticlinically during further development and thus serve as the so-called *tunica* for superficial growth. In the body of the growth cone, the so-called *corpus*, the cells divide in various orientations, but predominantly periclinically and anticlinically; divisions oriented obliquely to the surface are rare.

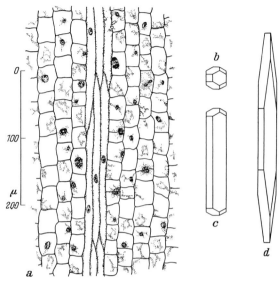

Fig. 482. Growth in length of prosenchyme cells. a origin of a fiber bundle in the supporting tissue of *Sanseviera*; b basic form of a meristem cell; c, d elongated forms. (After Meeuse *et al.* from Frey-Wyssling, 1945)

As the ratio between anticlinic and periclinic divisions of the meristem cells changes, the growth cone broadens basally (Figs. 479 k, 481 a). This primary thickening growth soon stops below. Basally, the anticlinic divisions (transverse walls) predominate and so vertical columns of cells arise (Fig. 479 k). In spite of growth and the continuing division of its cells, the primary meristem as a whole retains a constant volume and a fairly constant cell number, for in the very mass in which growth and cell multiplication proceed, the cells farthest from the tip lose the character of primary meristem. Those primary meristem cells which replenish the cells in their layer by dividing are called *initials*.

The cells that remain behind as the growth cone goes forward become larger, mainly through the uptake of water, which collects in vacuoles (Fig. 481 c, d), and their divisions become less frequent. Now tissue differentiation begins. The cells which ordinarily form the major substance of the shoot divide after elongation, forming mainly transverse walls; they therefore remain relatively short and broad *parenchymal cells*; they form the basic tissue (Figs. 480 a, 482 a). In other columns of cells scattered throughout the parenchyma or gathered in groups, longitudinal divisions begin so that bundles of elongated *prosenchymal* cells (Figs. 482 a, 483 a, b) are formed. The prosenchymal fibers remain in contact with the parenchymal cells (Fig. 482 a) and therefore do not stretch uniformly over their entire length. Their elongation growth is much greater where neighboring cells are dividing than where cell division has paused locally. The wall of a long stretching cell, therefore, does not grow uniformly in its tissue, but rather has local mosaic regions of extension. This produces an alteration of cell form. The basic form of a meristimatic cell is in general fourteen-sided (a combination of an octahedron and a hexahedron) with curved edges (Fig. 482b). When such a cell stretches, not all the surfaces expand, due to the relationships with neighboring cells. Either the sides stretch more than the ends (Figs. 482c, 483a), or the ends grow especially fast (Fig. 482d). The local growth of a cell membrane is thus influenced by the

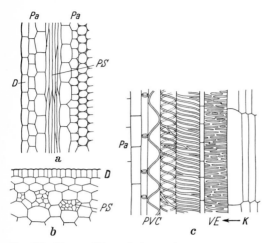

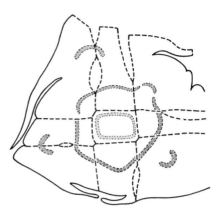

Fig. 483. Tissue differentiation. a longitudinal section; b cross section through a very long leaf of *Pandanus*; c radial longitudinal section through the stem of *Oenothera*. D dermatogen; *VE* vessel elements; *PVC* primary vessel cells; *K* cambium; *Pa* parenchyma; *PS* strands of prosenchyme. (After HABERLANDT from ROTHERT and JOST, 1934)

Fig. 484. Cross section through a *Primula polyantha* shoot in which the center of the growing cone has been isolated by a vertical cut. (After WARDLAW, 1952)

neighboring cells. This *symplastic* growth illustrates the first developmental relationship among the cells of a differentiating meristem. In all vascular plants, conducting bundles differentiate in the ground tissue. Frequently, supporting tissues also form with the formation of specials walls (sclerenchymal fibers, *et al.*). In the embryo the formation of conducting tissue is first signaled in the basal parts of the plerom of the cotyledons by the longitudinal division of the cells (Fig. 479i, c_2'). The differentiation of conducting bundles follows gradually from the growth cone. First, individual columns of cells in narrowly limited places in the prosenchymal bundles are transformed into relatively short, small-bore xylem initials and phloem initials. These initials can still elongate; during further lengthening of this part of the shoot their walls acquire thickened rings or spirals (Fig. 483c). Later, after elongation is over, the final vessel elements of the xylem and sieve tubes of the phloem differentiate.

The sequence of cell divisions and layer formation that can be established in the embryo has no final determining influence on the zone of conducting bundle differentiation. If four longitudinal cuts are made from the growing tip to separate an inner tissue mass from the lateral tissue in which the conducting bundles will arise, a conducting bundle system differentiates from the top downward in the isolated central block, and its position is determined in relation to the newly formed surfaces (Fig. 484). In addition to determining influences emanating from the growing point, therefore, factors operating radially are also critical in the formation of conducting bundles.

378

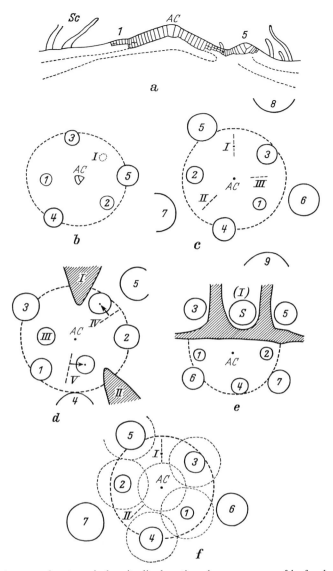

Fig. 485. *Dryopteris aristata.* a shoot cap in longitudinal section; b, c sequences of leaf anlagen; d displacement of anlagen *IV* and *V* after destruction of *I* and *II*; e formation of a shoot bud (*S*) after isolation of the site of leaf anlage *I* by a cut; f diagram of the inhibitory fields. *AC* apical cells; *Sc* scales; *1–7* and *I–V* leaf anlagen. (After WARDLAW, 1952)

The elements of the conducting bundles of the supporting tissue and of the parenchyma are permanent forms in that after their differentiation, they are either totally incapable of division, or capable only in extraordinary circumstances. In the open vascular bundles of the dicots, the outer phloem is separated from the inner xylem by a layer of primary meristem, the cambium (Fig. 481 e). This tisssue consists of radially flattened, elongated cells rich in cytoplasm (Figs. 481 f, 483 c). The cambium undergoes secondary thickening growth, giving off phloem on the outside and xylem on the inside.

At the growing point the leaf rudiments are formed right next to the shoot bud rudiments (Figs. 479k, 481a). When the leaf rudiments form close together in a flat, apical meristem, as in the fern *Dryopteris*, operations can lead to information relating to the factors which influence leaf position. In the center of the apical meristem lies a triangular apical cell, whose daughter cells increase the cell number on the apical surface (Fig. 485a). At particular distances from the apical cell, leaf rudiments arise which lead to a helical arrangement of leaves. This helix may be right-handed or left-handed (Fig. 485b, c).

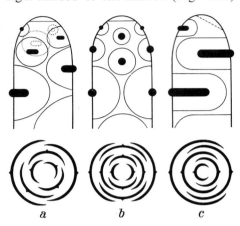

Fig. 486. Diagrammatic representation of the three kinds of leaf arrangements and their inhibitory fields. a spiral arrangement (as in *Dryopteris*); b opposite paired leaves; c alternate leaves. (After von DENFFER, 1962)

Newly formed leaf rudiments arise at specific distances from those which have formed previously and are now developing. The apical cell, with its contiguous daughters, and each young leaf rudiment form growth centers which are always surrounded by an inhibitory field. New growth centers can arise outside of these inhibitory fields. Bud rudiments can be formed between the leaf rudiments, but while the leaf rudiments are developing the shoot buds are restrained. The influence of the rudiments on others is shown by isolation and extirpation experiments. If, for example, the site of a rudiment which we will call *I* is isolated from *3* and *5* by a cut, *I* grows faster than normal. If leaf rudiments *I* and *II* (Fig. 485d), for example, are destroyed, the subsequent rudiments will be displaced in the direction of the missing inhibitory field and will develop faster than normal. If the apical cell is destroyed, the leaf rudiments are displaced toward the center of the apical meristem. If the presumptive site of a rudiment is separated from the center of the apical meristem and from the neighboring rudiments by a cut (Fig. 485e), a shoot bud forms instead of a leaf rudiment, and this bud rapidly develops into a shoot. The effect of the growth centers on newly forming structures thus influences their very nature. In Fig. 485f the inhibitory fields suggested by these experiments have been diagrammed as circles. Theoretically, other leaf arrangements can also be understood in terms of inhibitory fields (Fig. 486).

As the shoot stretches, the nodes, where the leaves branch off, become increasingly far apart and the internodes reach their final length (Fig. 481a). In the cells of the zone of elongation a cell wall may increase its surface area tremendously in the course of a few days; the increase may be 20-fold or even greater. The elongation of the epidermal cells of the *Avena* coleoptile (the stalk sheath, Fig. 487a, b) may even increase the surface area 150-fold. Since BOYSEN-JENSEN's discovery (1910) we have known that chemical effectors are produced in the meristem which induce and regulate growth in the zone of elongation. These are called auxins or growth hormones. When the movement of auxin is prevented, for example, by cutting off the coleoptile tip of one of the grasses (Fig. 487a, c) in which the auxin is made, cell elongation ceases. If an agar block is placed on the cut edge of the shoot, elongation resumes. If the agar block is placed to one side, the coleoptile bends, since the cells on the side receiving the auxin grow faster (Fig. 487e, f). The angle of curvature reached after a certain time serves as a quantitative criterion of the growth-promoting effects of solutions or extracts of particular substances. The natural growth substance of plants is not species-specific. The most important auxin is β-indoleacetic acid (indole-3-acetic acid), which apparently occurs in all higher plants. In addition, other indole derivatives with growth-promoting effects have been found.

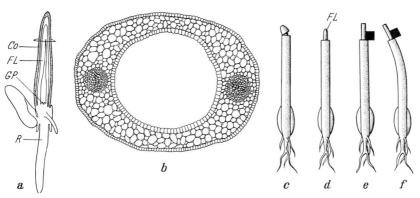

Fig. 487. Effects of auxin on the oat coleoptile. a oat seedling coleoptile. Coleoptile (*Co*), first leaf (*FL*) and growing point (*GP*) cut longitudinally; *R* root; b cross section through the coleoptile at the level indicated in a; c–f curvature test; c removal of the coleoptile tip; d after a further shortening of the coleoptile 3 hours later; e lateral placement of a growth substance, containing agar block next to the first leaf which had grown further and was then decapitated; f curvature of the coleoptile. (From J. BONNER and GALSTON, 1952)

The β-indoleacetic acid works as a phytohormone in extraordinarily small concentrations — from 10^{-10} to 10^{-8} M. The mechanism of auxin action in evoking elongation is not known. It is certainly not a direct effect on the cell walls, which would not even be covered by a monomolecular layer at the effective concentrations, but rather must be mediated by metabolic events in the cytoplasm of the responding cells. The essential consequences of these events are increased expansibility of the wall, strong cytoplasmic growth, high-water uptake by the swelling vacuole (Fig. 481d), and a corresponding increase in osmotically active material. This increase must be egual to the 20-fold increase in cell volume, since the turgor amounts to 3–4 atmospheres both at the beginning of elongation and at the end (during the growth of wheat embryo roots in liquid culture). The cell wall is not merely stretched plastically by the turgor; rather, it increases in substance, and therefore undergoes true growth. During elongation oxygen is required, and high-energy molecules are consumed in rapid metabolism[191]. Thus, even the elementary developmental process of cell elongation turns out to be a complicated response.

In growing green plants, auxin is made mainly in the growing point and moves from there into the lower parts of the plant. In addition, growth substances are also made in the root tip and in the leaves. In the young broad bean the relative concentrations in various parts are as follows: terminal bud, 12; the youngest leaf, 2.2; the next older leaf, 1.5; older leaves, 0.3–0.4. Auxin movement is polarized in the apical-basal direction. If a cylindrical piece of coleoptile below the tip is cut out and provided with an auxin-containing agar block on its apical cut surface, auxin will accumulate in a second agar block at the other end. Gravity is irrelevant; the same effect is observed when the whole assembly is turned upside down. The mechanism of this polar auxin movement is unknown. The auxin movement is temperature-dependent, and its rate is on the order of ten to twelve mm. per hour. It thus moves against a concentration gradient when the basal block contains more auxin than the apical block. Auxin transport in the presence of oxygen is clearly no simple diffusion process.

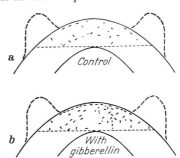

Fig. 488. Position and orientation of mitoses in the meristem of the shoot tip of *Solmolus parviflorus*. (After SACHS, BRETZ, and LANG, 1959)

The auxins are not the only phytohormones with growth-promoting effects. More recently, the *gibberellins* have been studied extensively. A pathogenic increase in elongation was found to be produced in Japan by a plant pathogen *Gibberella fujikuroi*, a fungus which attacks flowering plants. Extracts from *Gibberella* culture can produce the abnormal growth. At present, we know of five different gibberellins that promote shoot growth. They act in tiny amounts: in corn 0.001 γ suffices; in peas and oats 0.01 γ produces the minimal measurable effect. Many plants which under certain conditions produce only rosettes of leaves can be induced to form shoots by gibberellins; thus, lettuce plants six feet tall have been produced. Citrus saplings treated with gibberellins can reach six times their normal height. The promotion of elongation seems to begin with

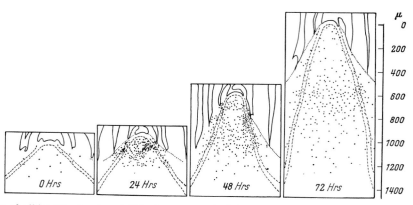

Fig. 489. Effect of gibberellin treatment (25 micrograms of gibberellic acid administered at 0, 24, and 48 hours) on cell division activity in the subapical region of the shoot tip meristem of *Solmolus parviflorus*. Each point stands for a mitosis in a central tissue slice 64 μ thick. Dotted lines are the borders of the conducting-bundle-forming tissue. (After SACHS, BRETZ, and LANG, 1959)

an increase in cell division in all the layers of the growing point and particularly in the subapical region. In the tunica, more anticlinic divisions are seen (Fig. 488); in the corpus, numerous divisions are seen in various orientations (Figs. 488, 489). The determination of the tissue layers is not altered, but in each tissue there is an increase in the number of cells which subsequently elongate and differentiate. We are not sure of the number of intermediate steps in the effect of gibberellin on cell elongation.

Just as certain animal hormones trigger different tissue-specific reactions in different parts of the body (e.g., male sex hormones evoke comb growth, development of spurs, and cock-feathering in capons), the presently known plant hormones also have diverse effects. Thus, in dicot stalks, auxin causes secondary thickening of the differentiated, elongated parts (Fig. 481e). The primary meristem cells within the conducting bundles, between the xylem and the phloem, are caused to divide rapidly; and between these bands of conducting bundles, bordered by *fascicular cambium*, a secondary *inter-fascicular cambium* forms by the tangential divisions of parenchyma cells (Fig. 490b₂). Thus, a closed cylinder of cambium (Fig. 481e) arises, which makes sieve elements on the outside and vessel elements on the inside. Within this cambium cylinder is a layer of initial cells which gives off transforming daughter cells toward the inner and outer surfaces. Since the output of xylem cells to the tissue mass enclosed by the cambium tube increases, the diameter of the cylinder of initials must increase by means of some intercalary divisions in the radial plane. Figure 481f shows a radial row of cambium cells from a cross section of a shoot. The young xylem and phloem elements are generally not fully differentiated but divide once or

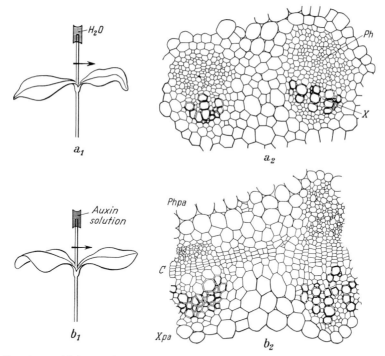

Fig. 490. Auxin effect in cambial growth. a_1, b_1 diagram of the experimental design with sunflower seedlings; a_2, b_2 two conducting bundles in each case from cross sections through the stem at levels indicated by the arrows in a_1 and b_1. (After SNOW from J. BONNER and GALSTON, 1952)

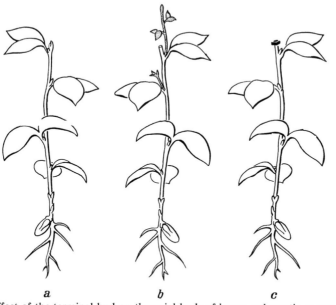

Fig. 491. Inhibitory effect of the terminal bud on the axial buds of leaves, schematic. a growing young pea plant; b growth of axial bud after removal of the terminal bud; c inhibition by an agar block containing auxin. (After J. BONNER and GALSTON, 1952)

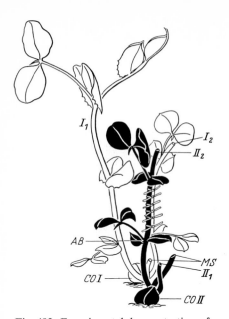

Fig. 492. Experimental demonstration of the transport of a lateral-bud-inhibiting substance. Explanation in text. (After SNOW, 1940)

more as they differentiate. Figure 490 diagrams an experiment through which the cambium-activating effect of auxin is demonstrated.

The gibberellins, too, influence cambium activity, particularly in collaboration with β-indoleacetic acid, as experiments on various trees have shown. Here, the gibberellins appear to work primarily on cell division in the cambium, while auxin seems to cause the differentiation of the xylem.

In some plants, β-indoleacetic acid also causes the development of roots at cut shoot surfaces. Either lateral root anlagen, already present in the shoot segments, are activated, or organ formation is induced in the pericycle (Fig. 480 b–d). The concentration of auxin required for root formation, 10^{-7} to 10^{-5} M, is considerably higher than that for cell elongation.

The development of the shoot system is also controlled by a phytohormone. In most herbaceous plants, the axial leaf buds (Fig. 481 a) do not sprout at once, but remain inactive as long as the main growing point is highly active (Fig. 491 a). If the terminal bud is removed, the axial buds now grow out (Fig. 491 b). But if one places an auxin-containing agar block on the cut surface, the swelling of the lateral buds is inhibited (Fig. 491 c).

However, auxin's effect here is not an immediate one; rather, another phytohormone is activated by the auxin. This is shown by a very logical experiment (Fig. 492): two pea embryos were planted next to one another, and the main shoot (H) was removed from each. The lateral shoots now grew out from each cotyledon ($CO\ I$ and $CO\ II$). In one plant one of the two lateral shoots (II_1) was removed and the tip of the other lateral shoot (II_2) was cut away. In the other plant the tip of one lateral shoot (I_2) was cut away and the other lateral shoot (I_1) was left intact. The decapitated shoots (I_2 and II_2) were decorticated over the length of an internode and held in close connection with thread. The inhibitory influence of the intact growing shoot of one plant (I_1) on the axial buds (AB) below the united internode was measured. The growth of the buds averaged 3.8 mm. in several experimental preparations, vs. 8.4 mm. in the controls, which were treated as II, but not bound to another plant. The inhibitory agent was thus transported down from the growing shoot (I_1) and back up through I_2 to the axial buds of II_2, where it influenced terminal growth at the tip. Since auxin transport is known to be highly polar, auxin itself cannot have exercised this inhibition directly.

Lecture 27

In most herbaceous plants the development of the individual comes to an end at the conclusion of the flowering phase. The growing point of a main or lateral shoot is consumed in the formation of flowers. The modified leaf structures, calyx, corolla, stamens, and pistils arise in series on the extended stalk axis, the pistils being at the end. The transition to flowering is regulated by a variety of external and internal influences, depending upon the species.

In some plants, peas, for example, flowering is almost independent of external factors and is determined early in the seedling. As a rule, the first flower anlagen are formed in the axis of the ninth or tenth leaf rudiment (Fig. 493), irrespective of soil conditions, nutrients, day length, temperature, humidity, and the treatment of the mother plant. If the cotyledons of young seedlings are excised up to the age of five days, two or three additional leaf rudiments are formed before flowering, which is understandable, since the normal food supply has been lost by the plant. In five-day-old seedlings (Fig. 493b) the anlage of the seventh leaf protrudes clearly. In this stage the development of the subsequent rudiments from the growth cone is determined and can no longer be altered by the removal of the cotyledon.

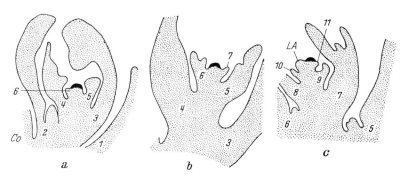

Fig. 493. Longitudinal sections through terminal buds of pea plants. a embryo in an ungerminated seed after six hours in water; b five-day-old seeding; c nine-day-old seedling. *1–11* leaf anlagen; *LA* flower anlage; *Co* cotyledon, growing point black. (After HAUPT, 1952)

The exact opposite is seen in other herbaceous plants whose development is dependent on certain temperature and light effects. In biennials a cold period must precede flowering. In nature these forms diapause during the winter and bloom for the first time during the year after sowing. For example, while in an annual strain of henbane (*Hyoscyamus niger*) an extended stalk and flower form during the first year of growth (Fig. 494a), a biennial strain of the same species produces only a rosette of leaves on a compressed stalk (Fig. 494b) it grows tall and blooms only in the second year. Thus, the growth of the stalk is also altered at the onset of the flowering phase. An appropriate cold treatment can cause these biennials to bloom in the first year. This cold-induced alteration of development is called vernalization. In almost all plants it is the meristematic cells of the growing tip which must be cooled; cooling of leaves does not lead to premature flowering in the biennials, nor does cooling of tuberous roots. Cold treatment may be replaced by the transplantation of a scion of an annual or an already vernalized biennial near the nonvernalized growing point of a biennial (Fig. 494c). The scion does not have to be of the same species or even the same genus (Fig. 494e). Five days of union between scion and stalk suffice to determine the growing point for sprouting and flowering. The transition to flowering is thus triggered by the transmission of a non-species-specific but function-specific substance, a flowering hormone whose own production is triggered by vernalization. This flowering hormone has been called *vernalin*.

In certain species the flowering hormone can also be made in the leaves as in unifoliate species of *Streptocarpus*. Here the entire exposed vegetative plant consists of one of the two cotyledons, which becomes huge (Fig. 495a). When the plant is ready to bloom after four-to-eight weeks of cold, a regular arrangement of flowers arises at the base of the cotyledon (Fig. 495a), which is finally consumed after massive seed production. If the leaves of plants ready to flower are cut into pieces, and planted as cuttings, the results depend upon the place in the leaf from which

Fig. 494. *Hyoscyamus niger*. a annual; b biennial race in first year of growth; c–e transplants; c 1st year scion of the annual race (Hy ⊙) on a first year non-vernalized plant of the biennial race Hy ⊙; d control graft Hy ⊙ on Hy ⊙; e scion of annual *Nicotiana tabacum Ni. ta* ⊙ on a first year Hy ⊙ plant. (After MELCHERS and LANG, 1948)

the cutting is taken. From the wound callus, flower stalks appear either at once (Fig. 495b) or from the leaf axes of leaves which form first (Fig. 495c); sometimes only vegetative leaves develop from this callus (Fig. 495d), with further vernalization being required for blooming. Figure 495e illustrates the distribution of the flowering hormone throughout the leaf, deduced from numerous experiments of this type. If the cold period has been relatively short (e.g., six weeks), the distribution of flowering hormone is different from that illustrated in Figure 495e: flower stalks arise infrequently, and then only from pieces taken out of the middle of the leaf. With progressively longer vernalization, the amount of flowering hormone increases and moves toward the base of cotyledons. The development of flower stalks after new leaves have formed (Fig. 495c) depends on the previous movement of flowering hormone from the piece of cotyledon into the leaf stalk; for it is required that the leaves remain joined to the piece of cotyledon.

In many plants flowering depends upon day length, the photoperiod. In the "long-day" plants there is a critical shortest-day length beyond which flowering sets in (Fig. 496L); they bloom

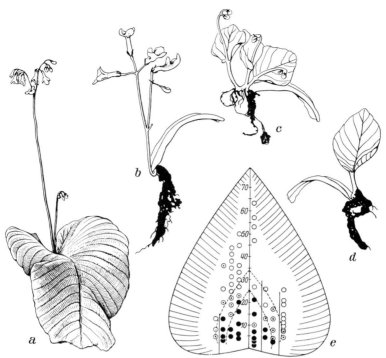

Fig. 495. *Streptocarpus wendlandii*. a normal plant at onset of flowering; b–d plants from cuttings of a leaf; e diagram of the distribution of products of cuttings taken from a leaf just before flowering. The scale along the central rib gives the size of the leaf in cm. The lateral ribs are shown only at the edge. Black-filled circles as in b, circles with dots as in c; hollow circles as in d. Each symbol stands for an experiment. (After OEHLKERS, 1955, 1956)

optimally in continuous light. For the "short-day" plants there are two critical values for daily light; a very short light period exists as a minimum for flowering, and a greater day length for the maximum beyond which flowering is inhibited (Fig. 496 S). The critical day lengths vary among species and strains. Thus, among the long-day plants, e.g., *Hyoscyamus niger*, the critical value which must be exceeded for flowering is between 10 and 11 hours; for spinach, it is between 13 and 14 hours. Short-day plants always require a certain dark period; if the light intensity is high enough, the daily light requirement for the initiation of flowering in some extremely sensitive plants (e.g., *Kalanchoë*) may be as small as a second. Moreover, the dark period must be continuous to be effective; even a one-second flash of light during this period can *inhibit* flowering. Both long-day and short-day plants need only a certain number of 24-hour cycles with the appropriate day-lengths to initiate the flowering phase. They form flowers after this even if they are given unfavorable day lengths. Thus, *Hyoscamus* will sprout and bloom after 72 hours of constant light even if exposed to days afterwards. *Chrysanthymum* blooms in long days after 30 short-day cycles and sometimes far fewer than that. The Composite *Xanthium pennsylvanicum* blooms 6 to 14 days after a single short-day, long-night cycle. These quantitative light effects thus bring about a physiological condition in the meristem cells of the growing point of the shoot, so that further divisions lead to a particular program of morphogenesis.

Photoperiodicity is an adaptive characteristic. Our native long-day plants generally do not bloom in the short tropical day, and similarly, the tropical short-day plants do not flower in our long summer days, although they will during the short days of the winter months in green-houses. The elaboration of a large vegetative shoot system before flowering, followed by an abundance of blooms, is closely tied to seasonal changes and photoperiods in the plants' native habitats.

The onset of the flowering phase is not an all-or-none reaction, but depends on a quantitative gradual change in the physical condition of the plant. The following may serve as an example:

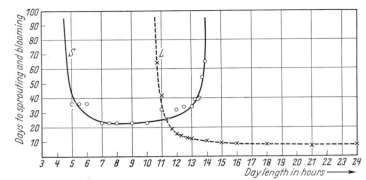

Fig. 496. Photoperiodic blooming reactions: *S* a *Crysanthemum*-species (short-day plant); *L Hyoscyamus niger* (long day plant). (From MELCHERS and LANG, 1948)

Kalanchoë blossfeldiana, a Crassulacean from Madagascar, needs at least 12 hours of darkness every day in order to bloom. Under a continuous regime of long days, it forms a highly branched shrub more than a meter high in the course of a year. Plants kept on a short-day regime become dwarfs only a few centimeters high, with no branching and just a few pairs of leaves on a compressed shoot with a few flowers on a stalk. Figure 497 shows two plants of identical age, the left one having developed in short days, the right one in long days. The greater the proportion of long days during vegetative growth, the more stalks of flowers can be produced; and the greater the number of short days, after the long-day period, the stronger is the tendency to bloom and the more flowers formed. The intensity of the impulse to bloom is reflected in the time of onset of the first visible flower rudiment, as well as in the degree of development of flowers and their stalks. On a normal stalk (Fig. 498a), one flower (*1*) is formed at the end of the main stalk; below it are two tiny opposed leaves whose axial buds grow out as lateral shoots, which again terminate in flowers (*2*) and form two second-order lateral shoots in a plane perpendicular to those of the first order. This diaxial branching generally is repeated several times; then the bifurcation is replaced by a helical arrangement of flowers. When conditions are unfavorable for the onset of

Fig. 497. *Kalanchoë blossfeldiana*. a short day; b a long-day individual, both planted March 21, photographed August 2. (After HARDER and VON WITSCH, 1904, from a photograph)

Fig. 498. *Kalanchoë blossfeldiana*. a normal flower stalk; b flower stalk with leaves (reduced flowers); c unilateral flower stalk on a plant growing in long day after keeping one leaf (*K*) in short day conditions. (After HARDER, VON WITSCH and BODE, 1942; HARDER, 1953; drawn from original photographs of HARDER)

flowering, e.g., when the photoperiodic effect is weak, due to a gradual shortening of the day-length, the stalk is simpler. There is no helical bunch of flowers, the number of branches is reduced, and the little scalelike leaves become more similar to ordinary leaves in size and shape. The number of flowers is reduced, and the individual flowers themselves are poorly formed. The formation of corolla, stamens, and pistils fails; finally, only an empty calyx remains (Fig. 498b). In other members of the *Crassulaceae* the petals themselves may even become leaflike; thus, the competition between the vegetative and flowering tendencies extends even into the formation of individual leaf rudiments at the growing point. The site of action of the light which evokes the photoperiodical flowering response is the leaves. With a short-day plant growing in long days, a single leaf can be put on a favorable short-day regime by covering it with a black sack at about 3:00 pm and removing it at about 8:00 am. The plant now blooms on one side: at its next bifurcation, the branch exposed to short days develops a flower stalk (Fig. 498c).

We can thus conclude that a critical agent is formed in the leaves under appropriate light conditions which moves *up* the shoot and triggers flowering.

Transplantation experiments lead clearly to the same conclusion: the short-day plant *Nicotiana tabacum* (Maryland Mammoth strain) can be induced to bloom in long days if a leaf of the long-day plant *Nicotiana silvestris* is grafted on to it (Fig. 409). And similarly, the long-day plant *Hyoscyamus niger* will bloom in short days if a shoot of the short-day plant *Nicotiana tabacum* (Maryland Mammoth) is grafted to it. Thus, the flowering hormone evoked by the appropriate light stimulus is the same in both long- and short-day plants; it has been called florigen.

Subthreshold photoperiodic stimuli can be summated: the number of long-day cycles required to induce flowering is not substantially increased by the interspersion of some short days. From this we may conclude that in each adequate cycle something is made that does not break down again during ineffective cycles. When the amount of the substances exceeds a threshold, flowering is induced.

The unique dependence of the onset of flowering on day length has been explained as the interplay of an endogenous daily rhythm in the plant with the light-dark rhythm in the environment (BÜNNING). The critical daily rhythm takes place in the leaves; removal of the leaves leads to the complete absence of day-length dependence in the plant.

The endogenous rhythm of higher plants is manifest in numerous cyclical phenomena which have a daily (circadian) rhythm even under constant conditions[88, 89, 94]. There may, however, be some external influences, as yet unknown, which have an impact even under conditions of constant temperature. The 24-hour rhythm does not remain precise under extended constant conditions. Thus, the endogenous rhythm is actually regulated by external cues, particularly daily light rhythms, and kept to an exact 24-hour period. Of the two phases of the internal rhythm, one is associated with the light period of the normal day and the other with the dark period. A plant's biological clock thus generates a rhythm with a photophilic phase and a scotophilic phase.

The photophilic phase is marked by vigorous biosynthesis and uptake of substances, low respiration, and often by resulting higher pH. In the scotophilic phase hydrolytic activity increases, starch and sugar are broken down more, respiration increases, and consequently pH goes down. Properties at higher levels of organization, such as permeability and growth, are closely related to these metabolic oscillations. The autonomous alteration between the two phases seems to result from one metabolic pattern's gradual creation of cytoplasmic conditions which bring about its alternative.

Progressive and periodic autonomous changes in cytoplasmic conditions have become familiar to us in various development stages of animal embryos. Moreover, autonomous rhythms persisting through long periods of adult life are also known in animals (daily, tidal, and lunar cycles) (BÜNNING[94]); but there is nothing so striking as the rhythms in higher plants. In many plants, leaf movements provide a very clear expression of the daily rhythm. In all the plants investigated to date, the photophilic phase produces a lifting of the leaves and the scotophilic phase a lowering of the leaves.

The photophilic phase favors flowering, and the scotophilic phase either inhibits it or has no effect either way. In the short-day plants the photophilic phase begins immediately, or very soon after the "entraining" light stimulus, which may be a very short one. After a certain time, the photophilic phase is succeeded by the scotophilic phase. Figure 500a shows the rise and fall of the photophilic and scotophilic phases and the rise and fall of the leaves, as indications of the current condition of the inner rhythm. The leaves are raised as soon as the plant is in the photophilic phase and lowered at the onset of the scotophilic phase. A very short light stimulus switches on the photophilic phase quickly in short-day plants. The promotion of flowering is greater when the light stimulus is longer, and when it includes the entire photophilic phase or at least its peak, some six hours after the onset of the phase. Up to a certain level, then, the time required for sprouting and blooming becomes shorter as the light period increases (Fig. 496 *S*), and more, better-formed flowers appear (p. 388). But if the light period extends past the photophilic phase by even a little bit, it has an inhibitory effect, and this is strongest in the maximum of the scotophilic phase (Fig. 496). The degree of inhibition caused by a flash of light during the dark period reflects quantitatively the rise and fall of the scotophilic phase.

In long-day plants, the photophilic phase begins only several hours, often 10 to 12 hours, after the onset of the entraining illumination (Fig. 500b), and it is during this period that a daily

Fig. 499. Induction of flowering in a short-day plant (*Nicotiana tabacum* Maryland Mammoth) under long-day conditions by the grafting of a leaf of a long-day plant (*Nicotiana silvestris*); b control graft graph, leaf of Maryland Mammoth on a similar individual in long days. (After MELCHERS and LANG, 1948)

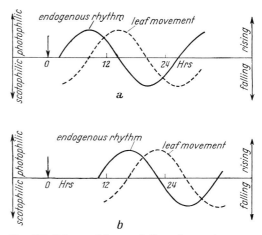

Fig. 500. Scheme of the regulation of an endogenous rhythm and the motion of the leaves in response to light conditions. a in a short-day plant; b in a long-day plant. (After BÜNNING, 1946)

light stimulus must work in order to promote flowering (Fig. 496 *L*). Then the autonomous phase change continues to oscillate in a 24-hour rhythm. The same promotion of flowering, however, is also reached when the entraining illumination is permitted to work for only a short time, provided that a second light stimulus is given after the onset of the photophilic phase, which is, of course, revealed when the leaf moves (Fig. 500b). The effect of the second illumination, which need not be a long one, increases as it approaches the maximum of the photophilic phase. For example, *Kalanchoë blossfeldiana* can be induced to flower by a 9-hour light period, which in itself is too short. If a second illumination lasting as little as a minute is provided anywhere in the 7th to 9th hour of the dark period, i.e., some 17 hours after the onset of the previously administered 9-hour light period, *Hyoscyamus niger*, which requires at least an 11-hour

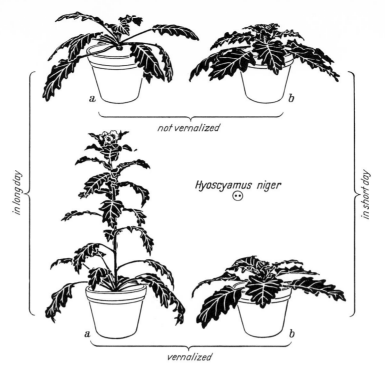

Fig. 501. Relationship between vernalization and photoperiodism in the biennial race of *Hyoscyamus niger*. (After MELCHERS and LANG, 1948)

day-length in order to bloom, can be made to flower with an aggregate illumination of 6 hours divided into an entraining light period of 5 hours and an additional hour of light in the middle of the photophilic phase. The autonomous oscillation continues, so that 24 hours after this maximum sensitivity to an additional light stimulus, a second maximum can be demonstrated. In long-day plants, there is no light-sensitive flower-inhibiting phase; otherwise, constant light would not produce maximal flowering (Fig. 496).

The effects of low temperature and of a particular day-length are both required for flowering in many plants: biennial long-day plants bloom only after a cold period; and even when they are vernalized, they bloom only under long-day conditions (Fig. 501).

Apparently, two different chemical effects work in a series; the flowering hormone "vernalin" evoked by cold is a prerequisite for the light-induced "florigen." This is suggested by the fact that the short-day plant *Nicotiana tabacum* (Maryland Mammoth) can be induced to flower in long days by biennial *Hyoscyamus* grafts only if the grafts have been vernalized. The annual short-day plants contain the transmissible substance which the unvernalized biennals lack: Grafted tobacco shoots or leaves induce flowering in unvernalized biennial *Hyoscyamus* plants, even when the grafts have been exposed to long days and thus cannot flower themselves. We still do not know whether vernalin is a precursor of florigen or whether vernalin acts to make cells competent to respond to florigen.

The gibberellins also couple into this mechanism. In long-day plants, cold and long-day effects can be replaced by gibberellin treatment: year-old rosettes sprout and bloom. Gibberellins can be extracted from various higher plants in amounts which rise sharply with the transition to the flowering phase. Various chemicals inhibit the growth processes which are stimulated by the gibberellins and which serve as indicators of an increase in gibberellin concentration. Gibberellin

392

reverses this inhibition. But although the gibberellins have a role in flowering, we know that no gibberellin is identical with florigen. This is clear from experiments with *Bryophyllum daigremontianum*, a long-short-day plant, in which the onset of blooming requires a certain period in long days followed by short days. The plants do not bloom in an extended long-day rhythm nor in an extended short-day rhythm. A graft from a plant ready to flower can cause its host to bloom either in long or in short days; but the addition of gibberellins promotes flowering only in short days but not in long days. Thus, gibberellins replace the long-day component of the flowering stimulus (florigen) but not the short-day component. The response to florigen thus differs from that to gibberellin.

Environmental factors influence more than the transition to the flowering phase in many plants: Somatic structures may also be modified extensively in the course of development. Many local modifications come to mind. Sun and shade leaves of the same plant are frequently quite different. The water crowfoot *Ranunculus* (*Batrachium*) *aquatile* makes two radically different kinds of leaves. The submerged leaves are divided and fringed, while the floating leaves are kidney-shaped. If the water level changes as

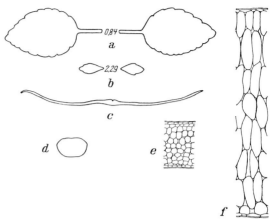

Fig. 502. Leaves of *Kalanchoë blossfeldiana*. a, b leaf outlines (numbers = succulence values); c, d leaf cross sections, actual size; e, f leaf cross sections, 14×; a, c, e, long-day leaves; b, d, f short-day leaves. (After HARDER and VON WITSCH, 1940).

the leaves are being formed, transitional types may arise. The daily light rhythm also affects the form of leaves. *Kalanchoë blossfeldiana* will not bloom in a long-day cycle; moreover, its entire appearance is different from that of a plant growing in short days. In long days (e.g., 16 hours) the plant forms large flat leaves with scalloped margins and petioles (Fig. 502a, c, e), separated by long internodes (Fig. 497b). In short days (e.g., 9 hours) the leaves are much smaller, with entire margins, without petioles, and so succulent that they have a cross section like a loaf of bread (Fig. 502b, d, f). They point almost straight up and have almost no stalk, and therefore very small internodes, between them (Fig. 497a). If an immature leaf on a plant that has been growing in long days is now exposed to short days, its growth lags, and it emits a substance which rises on its side of the plant: the leaf directly above always becomes relatively succulent (Fig. 503b, c). Toward the tip of the plant, the effect also spreads laterally; leaves that form at right angles to the experimental leaf grow slower and become more succulent on the side nearer the experimental leaf, so that these affected leaves become curved (Fig. 503b, c). The converse experiment on *Sedum kamtschaticum* is illustrated in Figure 504. The shoot of a plant previously exposed to short days is cut just above 4 succulent leaves and one of these leaves (*L* in Fig. 504) is exposed to long days. Now in a short-day regime, the bud in the leaf axis above the long-day leaf assumes the long-day character as it sprouts; the two neighboring sprouts are intermediate in form, and the sprout on the opposite side develops all the characteristics associated with short days.

We can conclude that a morphogenetic substance is made in the leaves and moves upward in the stem; it has been called *metaplasin* by HARDER. The substance is transported in the vascular system, as we learn from model experiments with berberine sulfate solutions, which fluoresce in UV light and can thus be seen in very low concentrations. If a leaf is cut off and its petiole (still attached to the plant) placed in berberine sulfate solution, the latter rises in the tracheal bundles with the transpiration stream; it is first seen directly above the petiole, where it enters and later appears in the lateral leaves. This distribution is mediated at first by the conducting elements, and secondarily, it would appear, by the parenchyma. The quantitative distribution of berberine sulfate coincides completely with that of the metaplasin effect (Fig. 503 c).

Florigen seems to move in the same way as this short-day substance, and in *Kalanchoë* the responses follow the same time course. The triggering of flowering, however, appears to be a different process from the induction of a new kind of leaf structure; two different substances,

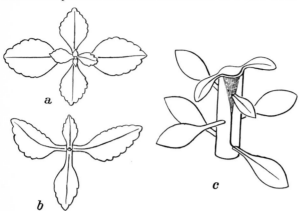

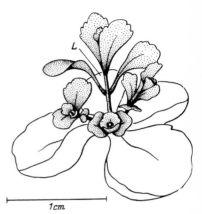

Fig. 503. *Kalanchoë blossfeldiana*. a view of the tip of a normal long-day plant with successive opposed leaf pairs displaced 90°. b view of nodes *7* and *8* of a plant in which a lower leaf in node 4 has been kept under short day conditions; c diagram of the upward movement of the active substance from a short-day leaf. (After HARDER, 1953)

Fig. 504. *Sedum kamtschaticum*. growing in short days, the upper part of the last two succulent leaf pairs removed, after which one leaf (*L*) was placed in long day conditions; outgrowth of the axial shoot. (After MEYER, 1947)

both formed in the leaves, are involved. This is shown in other Crassulaceans (*Sedum* spp.), in which flowering occurs in long days, although succulent leaves are formed in short days (Fig. 504). Even in *Kalenchoë* the flowering response can be separated from the formation of succulent leaves by general narcosis. Using plants with 8 to 10 pairs of leaves, the tips were enclosed in airtight cellophane bags into which either normal air or a chloroform-containing gas mixture could be introduced. The lower pairs of leaves remained outside of the chamber and thus could take in CO_2 and supply the enclosed tip of the plant with carbohydrates, even when photosynthesis was inhibited in the upper levels. The plants were held in a short day (8 or 9 hours) for a few days and narcotized during this time, either during the light period or during the dark period, but not both. After the course of several cycles of narcosis and normal air the plants were now exposed to long days in a normal atmosphere. In all cases the developing leaves became more succulent (a short-day character), while flowering was either prevented or strongly inhibited, according to the chloroform concentration[251].

Lecture 28

In the determined parts of the plant, particular tissues form in a particular order from the primary meristem of the shoot and root tip and from the secondary meristem in the shoot, in the root, and in the leaf rudiments; and individual cells differentiate in particular ways.

Tissue organization and cell differentiation are typical for certain phyletic groups and even for species.

As a means of testing the determination and competence of meristems and differentiated tissues, explantation has been used long and extensively with animal embryos, but it is relatively new to botany. Many of the early experiments were in vain; but the culture of plant tissues does succeed when they are placed on a gel substratum containing the necessary salts and carbon source rather than in nutrient solution. For growth of most tissues, the evidence is clear that substances regularly associated with the promotion of growth and cell division (Vitamin B_1, cysteine, and β-indolacetic acid) are required. But the mass of information acquired from tissue culture experiments includes little that bears on tissue differentiation.

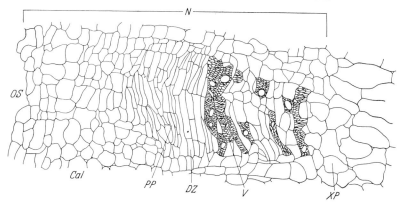

Fig. 505. Section through an explant of xylem parenchyma cultured in a medium containing 10^{-7} molar indole acetic acid. *DZ* differentiation zone; *V* vessels; *XP* xylem parenchyma; *Cal* callus; *N* new growth; *OS* surface of the callus; *PP* phloem parenchyma. (After GAUTHERET, 1950)

A piece of undifferentiated parenchyma, e.g., from carrot or Jerusalem artichoke (*Helianthus tuberosus*), undergoes vigorous superficial cell division, which results in a uniform parenchyma, a callus, similar to that produced by an entire plant in wound closure. After some time the callus begins to differentiate. From this we can see that the parenchyma has various tissue-forming properties, according to the region from which the parenchyma comes. In a callus formed by xylem parenchyma from the vessel element region (cf. Fig. 490b$_2$, *Xpa*), at a particular moment, the cells at the border between old and newly formed tissue begin to divide rapidly with tangential walls. They replace a meristematic zone, a new cambium which in rapid sequence gives off xylem parenchyma with more or less aberrant vessel elements inwardly, i.e., toward the old tissue; and phloem parenchyma toward the outside (Fig. 505). In this way, islands arise within the callus which build conducting bundles with fascicular cambium (Fig. 506a). In phloem parenchyma callus (cf. Fig. 490b$_2$, *Phpa*), islands of phloem parenchyma form in the vicinity of the old tissue, and they are surrounded by an outer cambial zone. This produces more sieve elements toward the inside and vessel elements outside (Fig. 506b). Later, the newly formed cambium spreads out around the vessel elements and forms sieve tubes throughout the area surrounding the island of differentiation. In this way, although admittedly in an abnormal form, a round conducting bundle is produced.

These experiments contain two important results: One is that the nature of the explanted tissue governs the first differentiation in the newly formed parenchyma. The second is that the newly differentiated tissue has the tendency for further tissue organization just as in a normal organ. A "differentiation gradient" of a substance diffusing from the explanted cells into the callus, or a tendency brought into the callus by the descendants of a determined parenchyma, causes the formation in the neighboring newly formed parenchyma of either vessel elements or sieve tubes and, to the component made via newly formed cambium, adds the other component of the conducting bundle.

Organs also form in the explants: shoot buds and roots. A bud emerges from the intense division of a small number of cells in the upper region of a parenchyma explant. In the cells of

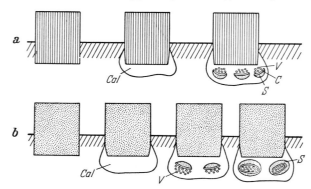

Fig. 506. Diagramatical representation of the histological differentiation of explants. a of xylem parenchyma; b of phloem parenchyma. *V* vessels; *C* cambium; *Cal* callus; *S* sieve tube. (After GAUTHERET, 1950)

the bud rudiment, the nuclei are larger than those in differentiated tissue and the large vacuoles in the cytoplasm fragment; the cells take on the character of primary meristems at a growing point (Fig. 507, cf. Figs. 479k, 481). The shoot buds grow into shoots with leaves (Fig. 508).

The differentiation of shoot and root in most explants reflects the original polarity of the tissue. Often the callus formation itself is polarized. In a carrot parenchyma explant the newly formed parenchyma grows out only on the side toward the root, no matter what orientation the explant has with respect to the medium. If pieces of endive root are placed horizontally on the medium so that the shoot and roots poles are placed under identical conditions, callus forms earlier and more extensively at the root end (Fig. 508a, b), after which, at about the tenth day, sprouts appear at both poles; shoots with leaves appear at the shoot end; and at the opposite end the sprouts remain small at first, but then grow into roots when the leaved shoots have developed to a certain stage (Fig. 508d, e). If the buds are cut off, root growth lags until new buds have grown out. The buds influence root formation from afar.

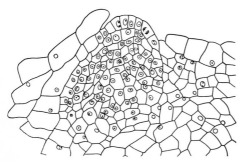

Fig. 507. Section through an explant of an elm terminal bud which has produced a growing plant in the parenchyma. (After GAUTHERET, 1945)

This morphogenetic effect of buds is shown very nicely in the following experiment (Fig. 509): a piece of endive root is placed with its shoot pole in the nutrient medium. In the course of about 15 days, buds sprout laterally. If these are cut off, new buds appear at different places in the explant, occasionally even on the root side which faces upward. Those on the shoot side grow more vigorously than the others, even though the shoot pole is submerged in the medium. The buds at the other sites develop very little. At the same time, roots are formed at the root pole (Fig. 509e, f). If all the buds and shoots are removed except those at the root pole, these now resume their growth, but stop

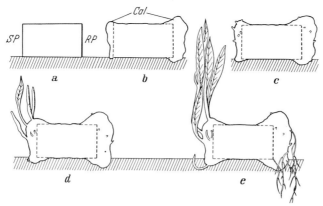

Fig. 508. Differentiation of new structures from a piece of endive root. a start; b after 8 hours; c after 11 hours; d after 14 hours; e after 28 days. *Cal* callus; *SP* shoot pole; *RP* root pole. (After Camus, 1949)

as soon as new buds arise at the shoot pole. If these latest buds are removed once more, then the buds at the root pole quickly resume further development; after this, no more buds are formed anywhere else. It now appears that the polarity has been reversed. But this is not really the case: no roots appear at the shoot pole, even though it is in contact with the medium. From this experiment we conclude that a substance is produced by leaved shoots which promotes root formation and which moves only in the shoot-root direction.

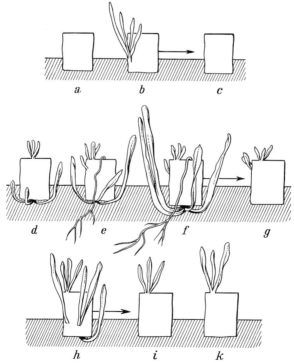

Fig. 509. Effect of buds in an explant of endive root a start; b after 15 days b→c ablation of the shoot buds; d after 45 days; e after 57 days; f after 75 days f→g ablation of all new structures except the upper shoots; h after 89 days h→i ablation of all new structures except the upper shoots; k after 109 days. (After Gautheret, 1945)

397

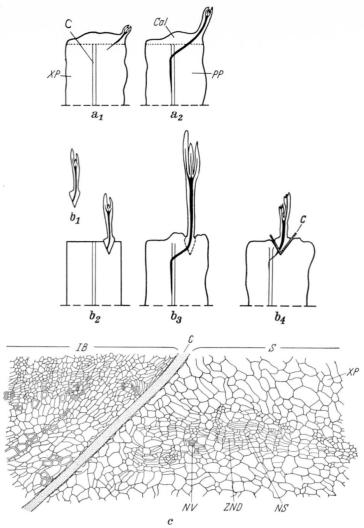

Fig. 510. Histogenetic effect of buds on an explant of endive root. a, b schematic longitudinal sections, new conducting bundles black; a_1, a_2 buds separating from the callus; b_1–b_3 implanted buds growing together with the substratum; b_4 bud placed on a substratum, but separated from it by a cellophane membrane; c part of a cross-section through an explant like b_4. C cellophane membrane; IB implanted buds; C cambium; Cal callus; ZND zone of new differentiation; NV new vessels; NS new sieve tubes; PP phloem parenchyma; S substratum; XP xylem parenchyma. (a, b after GAUTHERET, 1950; c after CAMUS, 1949)

These experiments provide insight into the process of polarization of an explant: at the time of its isolation the root fragment contains a certain quantity of material uniformly distributed. This material has two properties; it evokes root formation, and it inhibits bud development. The substance moves only toward the root pole, collects there, and leaves the opposite pole free to form buds. The developing leaved shoots form the phytohormone anew; it continually moves rootward, inhibiting new buds on the way. These results agree entirely with those obtained using pieces of entire plants, although the explants are more revealing because they provide us with a simpler model.

398

The buds also have a histogenetic effect: an endive root explant is used which contains xylem parenchyma, phloem parenchyma, and a cambium zone in between. After a while, buds develop from the callus above the phloem parenchyma, and reveal regions which differentiate conducting pathways among them in the parenchyma, vessels, and sieve tubes leading diagonally inward. When they reach the cambium zone of the root explant, the cambium cells begin to divide vigorously and form ever more basally numerous vessels, which finally anastomose with the vessels of the root (Fig. 510a$_2$). If a bud from an older explant is grafted onto a young one, the graft evokes the same sort of differentiation in the parenchyma of the stalk as a bud arising at that site would have done (Fig. 510b$_1$–b$_3$). This effect occurs even when a thin cellophane membrane is placed between the implanted bud and the host explant (Fig. 510b$_4$, c).

Explants are also useful assays for the effects of phytohormones. Tissue cultures from tobacco shoot pith have thus revealed the surprising effects of kinetin, a plant hormone isolated from crude yeast nucleic acid preparations and from various animal organs, and now synthesized and characterized as 6-furfuryl adenine. In the presence of auxin, kinetin induces cell division and, depending on its concentration, differentiation. A tissue kept on a medium containing 2 mg/l indole acetic acid and 0.2 mg/l kinetin remains a callus forever; with 0.02 mg/l kinetin, roots form from the callus, and with 0.5 mg/l, a great number of shoot buds.

These experiments are extreme examples of the diverse effects of various quantities and combinations of chemical agents and suggest that in the natural course of events growth and tissue differentiation are often not caused by individual functionally-specific phytohormones but by the collaboration of various substances, in particular in quantitative and temporal combinations. We do not understand the effect of any hormone at the cell level. ANTON LANG is right in saying "that we are well into the pharmacological phase of the work and have yet to enter the physiological phase."[382]

Lecture 29

In the formation of complex plant tissue, such as conducting bundles, and in the development of little organs (organules) in the epidermis — rather analogous to similar little organs in the epidermis of the insect wing — the differentiating cells in many cases emerge from their mother cells by differential divisions; and frequently a polar division of the cytoplasm, or of certain of the contents of the mother cell, signals the disparate determinations of the daughter cells. We shall consider only a few examples. The first relates directly to the inductive differentiation of the conducting bundle anlage: After the sieve elements have been formed from the cambium, certain cells are detailed to form a sieve tube and an accessory cell (Fig. 511). Cell polarity is shown

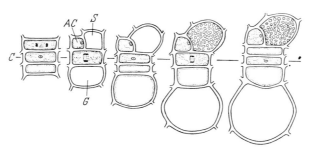

Fig. 511. Diagram of the differentiation of tissue elements from the cambium. *G* vessels; *AC* accessory cell; *C* cambium cell; *S* sieve tube. (After HOLMAN and ROBBINS, 1944)

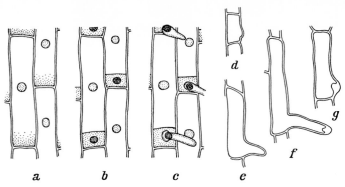

Fig. 512. Differentiation of the root hair. a–c diagram of the monocot type; d–g root hair of timothy (*Phleum*); d, e normal formation; f, g inhibited by water treatment. (a–c after BÜNNING, 1952; d–g after BOYSEN-JENSEN, 1950)

clearly in the formation of the root hairs of monocots. The density of the cytoplasm increases toward the root pole. The nucleus moves to the region of denser cytoplasm, and a small "tricho-blast cell" divides off. From this a root hair grows out (Fig. 512a–c). A dense mass of cytoplasm moves to the tip of the growing hair (Fig. 513). The expansion of the cell membrane is clearly limited to this region; so the formation and placement of newly formed cellulose in the expanding membrane takes place only here. This is shown by the following experiment: If growth is inhibited by the treatment of young root tips with water, congo red, or other means, cellulose is

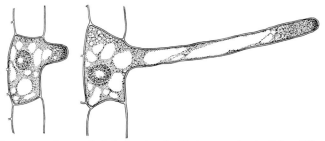

Fig. 513. Two stages in root hair formation in *Elodea canadensis*. (After BÜNNING, 1953)

laid down at the tips of the root hairs which have stopped growing, or at the site of the trichoblast where the hair would have formed (Fig. 512f, g). Thus the accumulation of cellulose-forming enzymes is closely tied to the polar determination.

The pattern of differentiation of the leaves of the peat moss *Sphagnum* bears the mark of polar determination and differential division very clearly. Immature leaves consist of a single-layered network of highly stretched, living chlorophyll-containing cells, in whose mesh lie large dead cells with distinctive thickenings in their walls. The youngest leaves show a uniform construction from rectangular cells with large round nuclei. Each contains 4–6 oval chloroplasts distributed evenly among the corners of the cells. At the apex of the leaf is a two-edged cap cell. In further divisions it divides off cells, which themselves divide further (Fig. 514a). After a certain time the leaf apex begins to differentiate. From each of the fairly similarly sized cells, in the course of two differential, unequal divisions, arise two small chlorophyll cells with much cytoplasm, as well as a larger "hyaline cell" with very little cytoplasm (Fig. 514a). At the beginning of the first differential division, when the nucleus enters prophase, the cytoplasm becomes denser at the apical cell

400

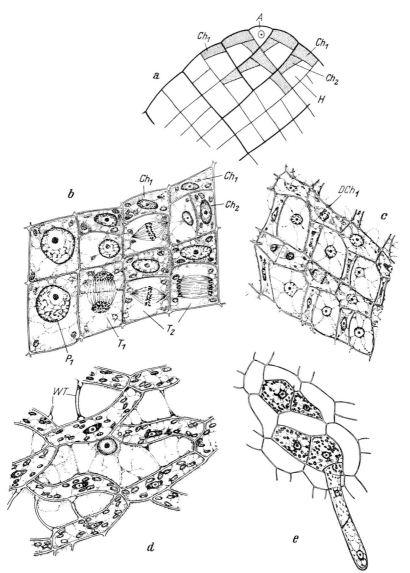

Fig. 514. *Sphagnum cymbifolium.* a young leaf in which differentiation has begun in the leaf tip; b–d sections of young leaves; b zone with first and second differential divisions; c older leaf with some chlorophyll cells in equal division; d still older leaf, chlorophyll cells and hyaline cells fully formed. Beginning of the formation of wall thickenings in the latter; e formation of a protonema from a regenerating hyaline cell. *A* apical cell; Ch_1, Ch_2 first and second chlorophyll cells; *H* hyaline cells; *P* prophase of a first division (T_1); T_2 second division; DCh_1 division of a primary chlorophyll cell; *WT* wall thickenings. (After ZEPF, 1952)

pole to which the nucleus moves (Fig. 515b). The spindle is eccentric, and the difference in cytoplasmic density between the two cell poles becomes strong. The apical cytoplasm acquires a distinctive basophilia. The cell wall, which arises as the smaller apical cell separates from the larger basal cell, is somewhat slanted (Fig. 514a, b). After a short time the second differential division begins. The spindle lies perpendicular to the axis of the previous spindle (Fig. 514b). Again a cytoplasm-rich cell divides off apically. Now vigorous cell growth begins. The chlorophyll

401

cells (Fig. 514c) divide rapidly, the hyaline cells rarely. While the chloroplasts multiply regularly in the chlorophyll cells, they dwindle in the hyaline cells, in which the cytoplasm becomes arranged in strips at the cell wall to lay down its cellulose stiffening (Fig. 514d).

The determination of the cell pattern of the sphagnum leaf thus has an apical-basal polarity which pervades the entire leaf. In the derivatives of the cap cell, each cell at a certain time in the

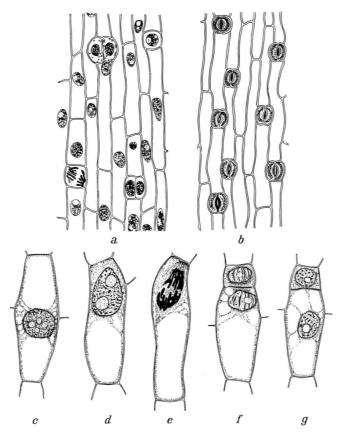

Fig. 515. Formation of stomata in members of the Liliales. a, b sections of leaves of *Leucojum vernum*; a younger; b older stage; c–g *Allium cepa*; c epidermal cell before differential division; d displacement of cytoplasm during prophase; e–f differential division; g stoma initial and ordinary epidermal cell. (a, b after BÜNNING, 1953; c–g after BÜNNING and BIEGERT, 1953)

growth of the leaf reaches a position where it acquires a polarity that confirms its differentiation. The determination of cells which normally would become permanently differentiated is not irrevocable, however: the chlorophyll cells of excised, old, fully grown sphagnum leaves can resume division in the appropriate nutrient solution and grow into protonemata. It is surprising that even hyaline cells, indeed even those which have begun to lay down reinforcing strips in their walls, can dedifferentiate. The chloroplasts, which have become tiny, grow again and multiply; new cytoplasm is formed; and cell division begins. The thickenings in the wall are more or less dismantled, and a protonema, threadlike at first, grows out (Fig. 514e). Under favorable conditions the protonema gives rise to *Sphagnum's* flat prothallium-like early gametophyte – an impressive example of the labile determination found even in higher differentiated plant cells.

A variety of morphogenetic principles can be seen in the distribution pattern of the stomata in the leaves of flowering plants.

The individual stoma is laid out by a set sequence of cell divisions. The epidermis of young leaves consists of small polygonal cells which contain dense cytoplasm and divide vigorously. After a region of the leaf has grown a certain amount by cell division, the cells begin to differentiate. In monocots the epidermal cells are stretched in the long axis of the leaf (Fig. 515). Their only cytoplasm is a thin peripheral layer, together with cytoplasmic strands leading to the nucleus in the interior of the cell (Fig. 515a, c). Differential cell divisions of the stomatal stem cells now take place, evenly spaced, among the epidermal cells (Fig. 515d–g). The nucleus moves to the apical pole of the cell, and here the cytoplasm becomes denser. The nucleus divides with the spindle in the long axis of the cell. The stem cells divide unequally; one daughter cell is the stomatal initial of the sphincter mother cell, and the other resumes the character of an ordinary epidermal cell capable of elongation. The initial is rich in cytoplasm, and after some time it divides equally and transversely into two cells, which will form the sphincter cells (Fig. 515a). Finally, the stomata are found at quite uniform intervals among the elongated epidermal cells in the mature leaf (Fig. 515b).

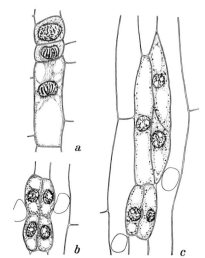

Fig. 516. Abnormal division of two initials next to each other from a stem cell in *Allium cepa*. (After Bünning and Biegert, 1953)

Once in a while a stem cell can divide off a second initial and very rarely even a third. Then two (Fig. 516a) or even three initials, and subsequently pairs of stoma-forming cells (Fig. 516b), lie in a row. But one never sees stomata in this arrangement in a mature leaf. As soon as the development of a more basal sphincter cell pair of such a group has reached a certain point, the more apical pair enlarges and resumes the character of normal young epidermal cells. After a certain time, these two enlarging sphincter cells grow independently, i.e., they no longer reflect the unity of a single developing stoma. Thus, for example, one of the pair may divide while the other does not (Fig. 516c). The basal cell pair generally forms a normal stoma, but in some cases it may also enlarge. These observations show two things: First, the determination of the initials and their two daughter cells is not final; and second, the developing sphincter

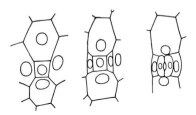

Fig. 517. Formation of the accessory cells in the leaf on *Tradescantia*. (After Bünning, 1953)

cells release an agent which can lead to the dedifferentiation of nearby stoma-forming cells.

When nearby cells join themselves to the stoma, still further unequal divisions are carried out beyond the aforementioned production of a larger cell from the sphincter mother cell. In epidermal cells bordering the initial, the nuclei move toward the initial and divide, so that a very small cell is walled off (Fig. 517). Evidently the initial releases a substance which causes the movement of the nucleus and the subsequent unequal division.

In dicot leaves the stomata are scattered among the polygonal epidermal cells, whose edges are often wavy (Fig. 518b). The distribution of stomata on the leaf surface is not as regular as in the monocots, but it is not random either: surrounding each stoma is a stoma-free area whose size does not vary greatly.

What determines the sites of the initials of the epidermis? The unequal divisions that produce the initials at angles to the epidermal cells (Fig. 518a) take place when the general mitotic activity of the epidermal cells has ceased. The sequence of divisions is clearly an expression of differentiation: if a cut is made releasing "wound hormone," cell division resumes and initial formation is simultaneously prevented (Fig. 519). Topical application of young bean-pod homogenate has the same effect.

The stomata are frequently not formed synchronously in the growing leaf: When the epidermis has grown in area to a certain extent, the first initials thus being sufficiently separated, new initials may arise among them (Fig. 518b). Clearly each initial is surrounded by an inhibitory

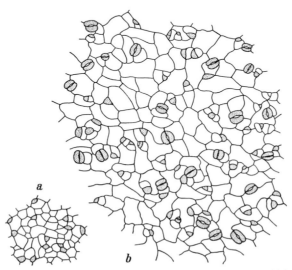

Fig. 518. Pattern of stoma initials in *Alliaria officinalis*. a in a young leaf (0.6 cm wide); b in an older leaf (1.5 cm wide), formation of new initials after growth in area. (After BÜNNING and SAGROMSKY, 1948)

field in which no further stomata can arise. The initials induce divisions leading to accessory cells and attract the nuclei of these nearby epidermal cells towards the sphincter cells. The nearer the epidermal cell is to the sphincter cell or to the initial, the stronger the effect (Fig. 520a). The gradient which underlies this positive chemotactic effect on the nuclei thus begins, not with the metabolic activity of the sphincter cells, but even in the undivided stomatal initials.

Zones of nuclear displacement and inhibitory fields are seen around the initials of other epidermal structures as well; and the interaction between inhibitory fields results in the arrangement of hairs, glands, and other structures in the total epidermal pattern. The inhibitory field around a stomatal initial also prevents the formation of a hair; similarly, a growing hair prevents the nearby formation of stomata, as well as of other hairs. The inhibitory fields surrounding large hairs, gland scales, and similar structures are generally larger than those around the stomata (Fig. 520b).

This formation of epidermal patterns is apparently initiated after the end of mitotic activity (which is itself perhaps due to the exhaustion of a cell-division-promoting substance). The cells enter a labile state, in which a gentle stimulation can lead to one of several alternative revolutionary modifications. The switch mechanism in the formation of an initial operates stochastically,

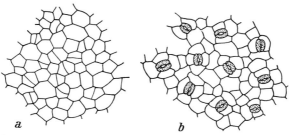

Fig. 519. Prevention of the formation of stoma initials in the leaf of *Theobroma cacao* by topical application by a tissue homogenate from young *Phaseolus* shells. a ten days after the onset of treatment; b untreated portion of the same leaf. (After BÜNNING and SAGROMSKY, 1948)

leading to the appearance of a few initials; the subsequent formation of initials is now restricted by the inhibitory fields of those which have already begun further development.

Still other factors participate in the formation of the epidermal pattern. In many species hairs or glands develop in place of stomata, particularly in the region of conducting bundles. They obviously come from originally similar initials whose development, however, is switched into a different pathway.

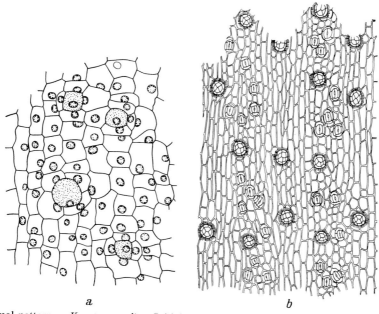

Fig. 520. Epidermal pattern. a *Karatas carolina*. Initials of one secretory hair and three stomata in a young leaf; b *Vriesea hieroglyphica*, stomata with four accessory cells each and scale hairs. (After BÜNNING and SAGROMSKY 1948)

The long axis of the opening on the leaf surface is arranged randomly in most dicots (Figs. 519 b, 520 b), in contrast to the monocots (Figs. 516 a, b, 521 b). One group of dicot species, however, does show some order in the orientation of the stomata. In most of the Proteaceae with parallel-veined leaves the stomata are oriented parallel to the long axis of the leaf (Fig. 521 a). At times even net-veined leaves also display a relationship between stoma orientation and the nearest vein.

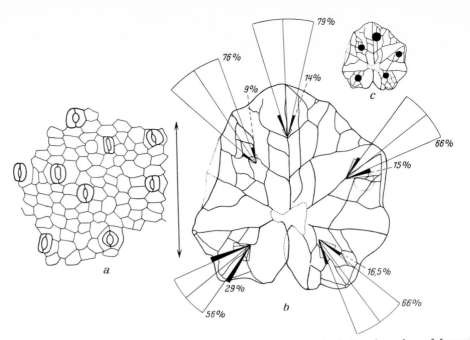

Fig. 521. Relationship in orientation between stomata and veins. a piece of a leaf undersurface of *Leucadendron*; b, c young leaves of *Ficaria verna*; b results of measurements: the length of each sector is proportional to the number of stomata on the surface studied whose orientation lies within the given angle. The central line indicates the direction from which the angles were measured; c areas studied. (After G. E. Smith, 1935)

In the lobed leaves of *Ficaria verna* the stomata are oriented exactly in the direction of the main veins of the lobes. This orientation varies only within a small angle (Fig. 521 b, c). In all cases the position of the opening depends upon the orientation of the initial, due to stresses arising between the epidermal cells and the underlying mesophyll as a result of the young leaf's growth. And here, clearly, the differentiation of the conducting bundles, which are capable of stretching the least, plays a role. It is not certain whether the plane of unequal division of the initial is determined by the stretching of the leaf tissue.

The stomatal openings show an even clearer relationship to the underlying mesophyll. In *Iris japonica* the stomata are only on the underside of the leaf. Here the neighboring mesophyll cells have irregular wavy borders. They form air spaces among them and a large respiratory cavity under each stoma (Fig. 522a). The epidermis on the upper surface lies over mesophyll cells, which are rounded and among which there are only small intercellular spaces. Clearly, the cell organization of the mesophyll does not influence stoma formation: often developing stomata are found over mesophyll cells without corresponding air spaces. In such cases the mesophyll cells generally divide once more and form a respiratory cavity along the new transverse wall. Here is a conclusive experiment: the action of kinetin, the chemical touchstone that has so many effects, causes the formation of stomata even on the upper side of the leaf, and an air cavity always forms underneath it (Fig. 522c), without an alteration in the form of the mesophyll cells. The cavity is thus induced in the mesophyll by the stoma initial. In other cases the mesophyll also acts back on the stomatal pattern: in some plants (e.g., *Begonia*), several stomata generally occur in close groups separated by stoma-free zones, just as individual stomata are separated in most plants. The initials leading to the formation of such a group do not develop simultaneously, but clearly one after another.

406

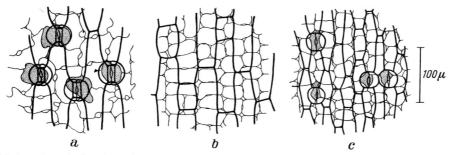

Fig. 522. *Iris japonica*. a leaf undersurface. The sphincter cells are sunk below the bordering epidermal cells; b leaf upper surface; c leaf upper surface of a stage younger than a and b treated with kinetin. Epidermal cells and sphincter cells heavily outlined. Air spaces stippled. (After JANTSCH, 1959)

Below the first initial, an intercellular space arises early and separates the initial from the mesophyll. The space enlarges quickly, and the region in which the further initials develop is coextensive with this intercellular space. A new initial never forms before its place in the epidermis is separated from the mesophyll by the intercellular space. It would seem that the intercellular space nullifies locally the inhibiting effect emanating from the initial.

In particular cases, other epidermal patterns influence the stomatal pattern, and a very nice example of this is seen in *Sedum*. As the young leaf rudiment grows out, light cells (Fig. 523a, b) appear among the previously uniformly dark cells, especially in the region of the leaf apex. The differentiation of the epidermal cells spreads from the apex to the base of the leaf. Even when the whole epidermis is already sprinkled with light cells, additional cells in dark regions can become

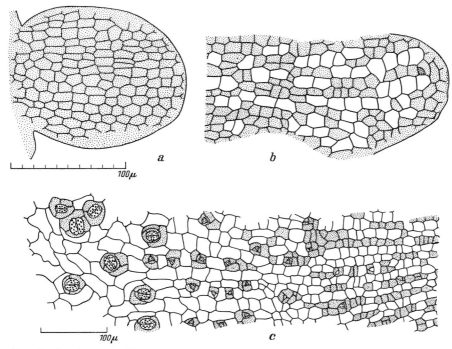

Fig. 523. Epidermis of *Sedum rotundifolium* leaves. Upper scale for a and b, lower for c. (After SAGROMSKY, 1949)

407

light. An unknown agent evokes this change. The derivatives of light cells are also light. Finally, only relatively small islands of dark green cells remain. Stomatal initials can be formed only by dark cells, and so their pattern derives from the pattern of differentiation in yet another way (Fig. 523c).

These few examples from the developmental physiology of plants show a number of diverse morphogenetic principles, some of which are reminiscent of those seen in animals, such as the distant action of hormones, the induction by determined tissue of nearby competent tissue, and differential cell division. Some unilateral or multilateral effects among plant tissues stem from the specific body plan of plants, in which there is no general body fluid bathing the cells, and where cell movement is usually prevented by the stiff frame of cellulose walls. Moreover, we are repeatedly struck by the ability of plant cells to dedifferentiate and redifferentiate. The development of patterns in leaves shows particularly clearly the collaboration of diverse developmental principles in the expression of tissue patterns; and in this way leaf patterns represent a case complementary to the patterned differentiation of the Lepidopteran wing.

Lecture 30

One of the most puzzling aspects of developmental physiology is regeneration: What adaptations make it possible for an organism to replace parts lost by mechanical removal or by internal reorganization, so that a harmonious whole once more results?

As to the major regenerative abilities of plants[92a], we have become familiar with some impressive examples (p. 395ff, 402). Among the animals, with their much richer internal differentiation, and their fixity of outer structure, the ability to regenerate an entire organism from a small part is all the more surprising. It is always impressive to see, for example, how a single arm ripped off a starfish regenerates the central disc and the remaining arms (Fig. 524). Compared to that, the regeneration of a salamander limb or a lizard tail appears quite modest.

We can consider in the space of these lectures only a few examples of regeneration if we wish to analyze them thoroughly. So we shall restrict ourselves to investigations on hydroids and planarians.

The hydroids provide a unique example of the continuous repair of missing organization. A general organizing principle of their body plan is the use of cells scattered throughout the ectoderm and the endoderm and which have remained "embryonic," i.e., undifferentiated and undetermined (Fig. 525). These interstitial cells (I-cells for short) have relatively large nuclei for their cytoplasmic mass. Their cytoplasm is strongly basophilic as a result of their high RNA content; thus, they can be stained selectively. The I-cells are mainly in groups in the ectoderm (Figs. 525, 526). They serve as replacement cells or neoblasts for somatic cells. The latter can also still divide, to be sure (Fig. 525), but their life span is restricted. The I-cells give rise to germ cells (Fig. 527a), and in asexual reproduction through budding they provide the material for the new individuals and for growing stolons.

The I-cells multiply in the ectoderm and go where they are required. They meet the massive need for nematocysts, especially in the tentacles of the polyps and medusae. They collect where buds form in the ectoderm (Fig. 527b). There they multiply and move into the endoderm (Fig. 527c), their derivatives giving rise to the parts of the differentiated medusa buds (Fig. 527d). The I-cell mechanism finds unique application in the solitary, continuously budding, and intermittently sexual fresh-water *Hydra*. *Hydra* has a continuous center of growth in the upper body region just below its wreath of tentacles. Here mitoses in the ectoderm and endoderm are most numerous, and the interstitial cells are multiplying rapidly and closely packed (Fig. 528). From

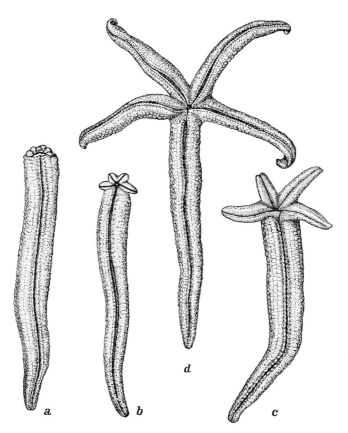

Fig. 524. Arms removed from the starfish *Linckia multifora* undergoing regeneration. Stages of the redifferentiation of the central disc and the four missing arms. (After RICHTERS from KORSCHELT, 1927)

this growth zone outward the tentacles are renewed, additions to their bases compensating for losses at their tips. Additions to the body wall are continuously consumed by budding, and the "old" tissue moves basally and forms the stalk (Fig. 529). The cells in the mixed tissues of the ectoderm and endoderm, respectively, are constantly replaced. Thus, morphological and physiological individuality is retained by the use of interstitial cells as neoblasts, and so a potential immortality is conferred on the individual *Hydra*; and this case, more profound than the immortality of generation after generation of Protists, is unique in the animal kingdom as far as I know. Naturally, *Hydra* can generally express this property only in aquaria, lasting year round essentially without limit; in nature, they form thick-shelled resistant eggs under unfavorable conditions and the mother *Hydra* generally dies.

The distribution of the I-cells (Fig. 528) of the basipetal growth movement (Fig. 529) reflects a polarity in *Hydra*. The mouth region is the highest point in an apical-basal gradient system. This polarity also dominates the process of regeneration.

The regenerative competence of *Hydra* is a classical subject of developmental investigation and may well be the oldest: TREMBLEY carried out a systematic series of surgical experiments as early as 1740, and many investigators have followed in his footsteps, seeking the answers to the new questions which arose as fast as the *Hydra* regenerated.

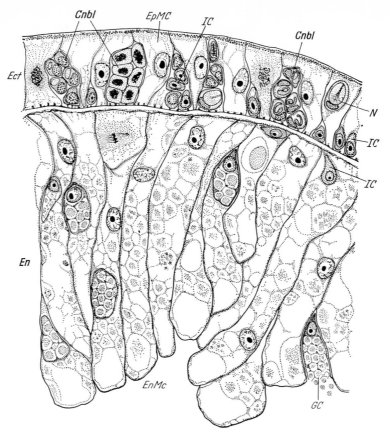

Fig. 525. Portion of a cross section of the body wall of *Hydra attenuata*. *Cnbl* cnidoplasts; *GC* gland cells; *Ect* ecto-derm; *En* endoderm; *EpMC* Epithelio-muscle cells; *IC* interstitial cells; *EnMc* endoderm muscle cells; *N* nemato-cysts. (After BRIEN, 1953)

Even very small pieces of a *Hydra* are fully capable of regenerating. If a ring is sliced out of the body (Fig. 530), both the upper and lower openings are quickly closed by the contractions of the circular muscle cells. In this way, within a few hours, a sphere or a closed cylinder is formed. After a day or two the structure stretches in conformance with its original polar axis and soon transforms into a tiny polyp. At one end tentacles grow out, and a functional holdfast forms at the other. The stretching is, of course, accompanied by thinning until the normal proportions have been reached in miniature (Fig. 530). Even a piece of the oral field with a tentacle can regenerate into a small polyp (Fig. 531). In this case, the tentacle remains intact and its tissues take no part in the morphogenesis.

The I-cells play a central role in regeneration. The movements and local transformations of the I-cells have been followed step by step, by means of serial experiments employing a counting procedure developed together with the use of an unrolled, flat piece of ectoderm of standard size. Figure 532 illustrates the course of regeneration after a series of experiments on eleven hydras whose apical portions had been cut away below the tentacles. In the first few hours, the average I-cell count rose in the region just below the cut surface (Fig. 532*A*), due to immigration and division. The count remained higher than that in a non-regenerating hydra for some time. When,

20–30 hours after the operation, the body had stretched out and the tentacles sprouted, the I-cell count in this upper region fell sharply below normal: the I-cells had now been incorporated into the regenerating tissue, losing their characteristic form and staining properties as they differentiated.

I-cells are attracted to wounds in the hydra, just as they are to normal morphogenetic sites (buds, gonads). Such foci of morphogenesis compete with one another for I-cells, and this amounts to indirect mutual inhibition. Thus, the development of male and female gonads ceases when the polyp is forced to regenerate a new head. The growth of a regenerating structure is strongly inhibited if a bud is growing on the same stalk. And, indeed, it may stop entirely until the bud completes its growth and drops off. The greater the distance between the regenerating site and the bud regeneration, the weaker is the inhibition. The inhibition is completely removed if the hydra is ligated between regenerate and bud.

Experiments on the reconstitution of hydras from aggregates of small fragments have revealed the organizing effect of the hypostome. In an aggregate of 30–60 hydra fragments, clumped together by centrifugation (Fig. 533), randomly oriented fragments join primarily at their endodermal surfaces and fuse. Ectodermal cells move over the mesogleal surface of the endoderm and cover it. In the interior of the aggregate, ectodermal clumps and endodermal cells without mesoglea are broken down. In the course of 7–10 hours the gastral cavity forms, swells up (Fig. 533b), and flattens again, expelling all the tissue remnants, especially cnidoblasts that have not been incorporated (Fig. 533c). The regenerant becomes smaller,

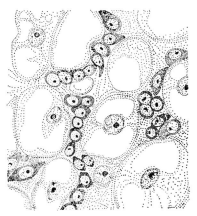

Fig. 526. Section of an ectodermal layer preparation of *Hydra viridissima* from the apical trunk section; groups of strongly basophilic I-cells among the large epithelial cells. (After Tardent, 1954)

and after two or three days new tentacles sprout at places where hypostome material, often with pieces of tentacle, comes to lie. At such places a mouth is formed (Fig. 533d, e). In young aggregates, irregular muscle movements course through the fragments. During the formation of the gastral cavity they stop. and the contractile muscle regions of the epitheliomuscular cells now become oriented along the developing polar axis, apparently through dedifferentiation and redifferentiation. Then the aggregates begin a quick division into as many individuals as there were mouth-tentacle complexes (Fig. 533f), and finally the little polyps separate and attach to the substratum with their holdfasts (Fig. 533g). According to the histological evidence, the induction of an apical-basal gradient appears to emanate from places where tissues containing the specific secretory epithelium of the hypostome have been incorporated into the reorganizing body wall. We do not know, however, whether the incorporated basal regions have a comparable inducing effect, or whether the holdfast arises at the aboral end of each apical-basal gradient.

In any event, polarity cannot be reversed by transplantation: the basal region of a hydra is amputated and replaced by the apex of another hydra; its own apical region is now removed (Fig. 534). If the polarity had been reversed by this operation, we would expect a holdfast to regenerate in the place of the excised mouth region and tentacles, but, in fact, what forms is a new mouth and set of tentacles. Two hydras now separate from one another, the implanted head commandeering material for its base from the host. Another experiment illustrates the integrity of the individual polar system once more. Five hydras were grafted together in series to form a

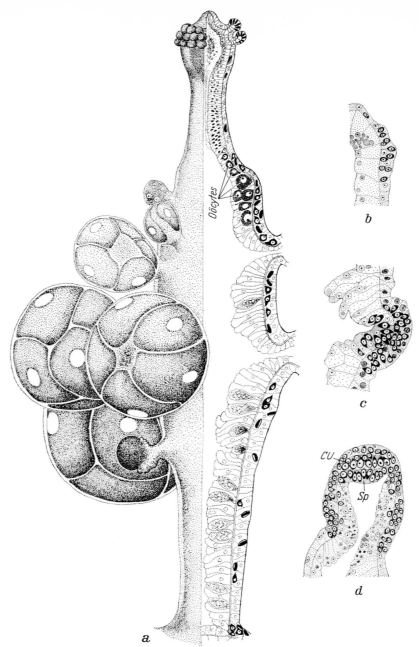

Fig. 527. Gonophore-bearing polyp (blastostyle with styloid gonophores) of *Hydractinia echinata*, left surface view, right longitudinal section; movement of the germ cells into the endoderm and development into egg cells in the gonophore zone. b–d development of medusa buds of *Cladonema radiatum*; c I-cells dark. *CU* center of umbrella; *Sp* spadix (= anlage of the endoderm of the mouth stalk). (a after MÜLLER, 1964; b–d after WEILER-STOLT, 1960)

giant polyp (Fig. 535). The first polyp retained its head, the last its holdfast, and the others were allowed to keep neither terminal structure. Sooner or later the original pieces separated from one

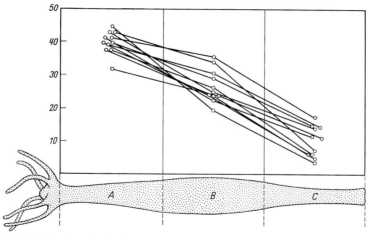

Fig. 528. Distribution of I-cells along the body axis in ectodermal layer preparations of *Hydra viridissima* in eleven normal individuals. Ordinate: average number of I-cells per $6,400\,\mu^2$ in each third of the body. (After TARDENT, 1954)

another, regenerating apical and basal organs in the regions of experimental junction. The secession of individual components begins with the one farthest from the apical end of the super-polyp, suggesting that perhaps the apical organs have an inhibiting effect on regeneration which decreases with distance. If the head of the first hydra is cut off, there is no evidence of such in-hibition: regeneration starts simultaneously in all of the individual components. The polarity of the hydra is also expressed in histological differentiation. Here the nematocysts are formed in

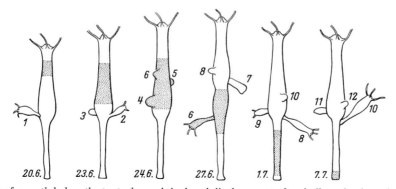

Fig. 529. Stages of growth below the tentacles and the basal displacement of a vitally stained portion of the body of *Hydra attenuata*. Numbers stand for day and month as well as for a series of buds. (After BRIEN, 1953)

nests in the ectoderm from I-cells (Fig. 525). Synchronous divisions of stem cells produce cnido-blasts, which are seen as groups of cells in similar stages of differentiation. *Hydra vulgaris* and *Pelmatohydra oligactis* have relatively small volvents and glutinants, as well as two different sizes of penetrants. The stem cells of the cnidoblasts containing large penetrants pass through one less division than those containing small penetrants. The size difference in the functionally mature nematocysts can ordinarily be seen during their development. Large penetrants form in the upper part of the polyp and smaller ones in the lower part (Fig. 536a). If a hydra is cut in two trans-versely, two complete polyps will regenerate; and in the upper one only large penetrants will be

413

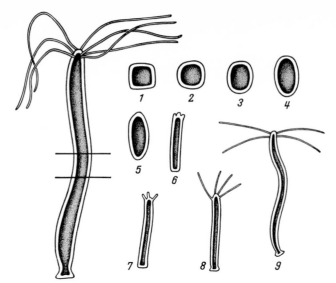

Fig. 530. *Hydra viridissima*. Normal individual and regeneration stages from a cross section. (After MORGAN, 1907, somewhat modified from KORSCHELT, 1927)

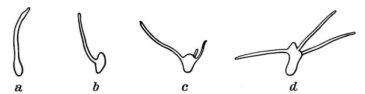

Fig. 531. Regeneration of *Hydra* from a piece of the mouth region with an attached tentacle. c formation of two new tentacles. (After KOCHLITZ, 1910)

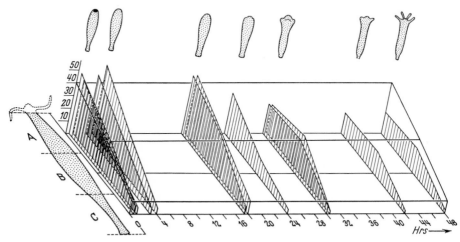

Fig. 532. Axial distribution of the I-cells in regenerating stages of *Hydra viridissima*. The block outlined represents the axial I-cell distribution of a non-regenerating individual. Ordinate: average number of I-cells/6,400 μ^2. Abscissa: duration of regeneration. (After TARDENT, 1954)

414

seen, while in the lower one, only small ones. The penetrants which form in the next couple of days retain this pattern; evidently the stem cells have been determined before the polyp was cut in two. After 3–5 days, cell groups with small developing cnidoblasts appear in the newly formed region of the upper regenerant; conversely, groups of large cnidoblasts appear just below the tentacles in the polyp regenerating from the original basal half. This corresponds to a new determination of the stem cells, reflecting the new polar gradient (Fig. 536 b, c).

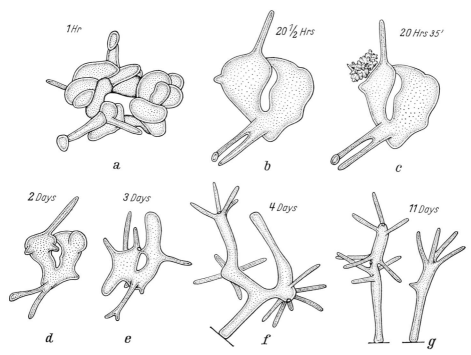

Fig. 533. Course of reconstruction of a new individual from an aggregate of *Pelmatohydra oligactis* fragments. a union of a number of fragments has proceeded a good way; b seams have disappeared, and two parts of the aggregate have now puffed up; c one half has burst, expelling a mass of cells; d, e formation of tentacles and mouths; f attachment with a pedal disc, three mouth regions with tentacles; g separation into three individuals corresponding to the oral fields; one hydra has already become completely separate. (After LEHN, 1953)

Investigations on *Tubularia* add further detail to this picture. In *Tubularia larynx*, the hydranths are actually larger than a whole *Hydra*. Each hydranth is situated at the tip of a stem up to 4 cm long, and it bears two sets of tentacles, a distal or oral set and a proximal or aboral set.

The distribution of the I-cells in the stem suggests an axial gradient (Fig. 537), just as in *Hydra*: the I-cells are most numerous in the hypostome and become progressively scarcer basally. Just below the hydranth their numbers drop off sharply. The rate of regeneration of various stem sections corresponds to this gradient in I-cell distribution.

During regeneration, a hydranth differentiates within the tube of perisarc from a region of the stem just below the cut (Fig. 538 a). No blastema grows out; rather, the transformation of the tissue and the laying out of the hydranth occur *in situ* as the I-cells are recruited and put to work (Fig. 538 a). The tentacles arise as ridges, which later pinch off at their inner surfaces

415

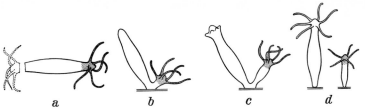

Fig. 534. Transplantation experiments on *Hydra vulgaris*. Head and foot regions have been removed from one hydra, and the head has been implanted at the basal cut; regeneration of a head at the site of its removal and separation into two hydras. (After TARDENT, 1954)

(Fig. 538c). The regenerated hydranth is now pushed out of the perisarc tube by the extension of the stem.

The formation of a hydranth on a cut stem is inhibited if a mature or regenerating hydranth is grafted to the opposite end of the stem. Aqueous and alcoholic extracts from homogenized *Tubularia* hydranths inhibit regeneration according to their concentration. A similar effect is shown by water in which hydranths have remained for some time. This suggests that the hydranth

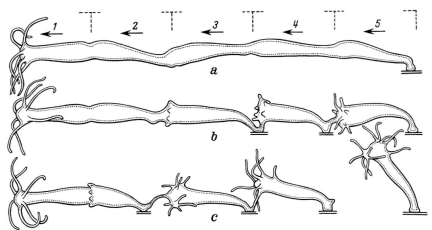

Fig. 535. *Hydra vulgaris*. "Giant polyps" achieved by the grafting together of five individuals (arrows point anteriorly). The oral end of #1 has not been removed. The pedal end of #5 has not been removed. All other oral and pedal ends have been removed. a 20 minutes after grafting; b 2 days after grafting, showing successively greater regeneration posteriorly; c 5 days after grafting. (After TARDENT, 1954)

makes some chemical inhibitor of further hydranth formation. If there is an endogenous inhibitory substance, this serves as an antagonist to the great regenerative ability of *Tubularia*, which permits it to make a new hydranth at the site of even a very minor wound in the stem. Such a substance might also limit the density of the polyps, which often grow in fields like grass, so that a detrimental competition for food is avoided.

Normally a piece of stem regenerates a hydranth at its apical cut surface and a stolon at the basal surface. This occurs invariably when the stem is placed erect in the sand. But if the stem lies free in water, and particularly if it is a very short piece, hydranths can form at both ends (Fig. 539a). Perhaps in the very short pieces the axial gradient does not result in a large enough difference to distinguish the two cut ends from one another. Isolated pieces of stem 3 cm long have formed hydranths up to 15 times (in 74 days) after repeated cutting, and they have

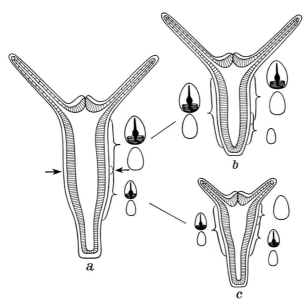

Fig. 536. Diagram of the distribution of the cells which form "penetrant" nematocysts in *Hydra vulgaris*. a normal hydra, arrows indicate level of cut; b regenerated upper part; c regenerated lower part; cnidoblasts and mature nematocysts. (After LEHN, 1951)

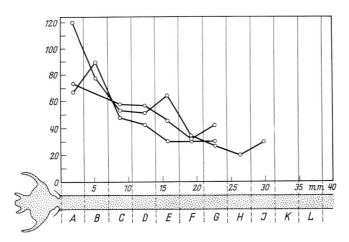

Fig. 537. Distribution of I-cells in an axial gradient in the ectoderm of the stem in a flat ectodermal preparation of *Tubularia larynx*. Ordinate: average number of I-cells per $6,400 \mu^2$. (After TARDENT, 1954)

formed apically as well as basally. In the course of this experiment the mature hydranths become progressively smaller, and the speed of regeneration is reduced. Hydranths are formed more often apically than basally, particularly in longer pieces. In a ligated stem, the dominance of the apical end disappears: hydranths regenerate as often basally as apically. These repeated regenerations lead gradually to the consumption of the stem tissue, especially of the endoderm; the I-cell supply is depleted, and the gradient becomes less steep. The *Tubularia* stem in this respect differs from *Hydra*, whose continuous ability to bud comes from continuous I-cell production[667].

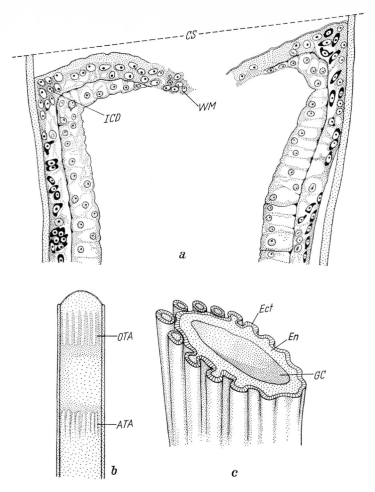

Fig. 538. *Tubularia larynx*. a longitudinal section through the wound region of the stem; b sketch of a regenerate; c oblique section through the regenerated aboral tentacle anlage, diagrammatic. *ATA* aboral tentacle anlage; *GC* gastral cavity; *Ect* ectoderm; *En* endoderm; *ICD* differentiated I-cells; *OTA* oral tentacle anlage; *CS* cut surface; *WM* margin of wound. (After TARDENT, 1954)

If a very short piece of stem (about 2 mm) is isolated, it transforms within its tube of perisarc into a "polar heteromorph," a double polyp with two sets of proximal tentacles (Fig. 539b) or with only one such ring of tentacles (Fig. 539c). Distal to each set of tentacles, medusa buds can form. In more extreme cases, the entire lower part of the hydranth is missing, and only a double mouth region is formed (Figs. 539d, e, 540). The wound effect clearly works equally strongly at both ends and meets the same developmental tendencies at both ends of this short piece; the material at hand is evenly divided between the two regenerants. We unfortunately still do not know whether only interstitial cells are active in this tiny morphogenetic theater, or whether differentiated cells of the stem dedifferentiate and participate in tentacle formation. The nature of the developmental tendencies along the gradient remains a total mystery.

418

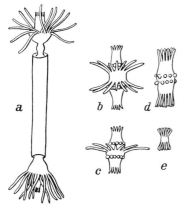

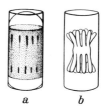

Fig. 539. Antipolar regeneration (polar hetero-morphosis) in *Tubularia*. a hydranth formation at both ends of a cut stem; b–c heteromorphic regeneration of a short piece of hydranth. (a after KING and MORGAN; b–c after MORGAN and CHILD from KORSCHELT, 1927)

Fig. 540. Heteromorphosis in *Tubularia mesem-bryanthemum*. a anlage of two rings of oral tentacles; b mature heteromorphosis (as in Fig. 539e) within a tube of perisarc. (After MORGAN, 1907)

Lecture 31

Planarians, like hydroids, are classical objects in regeneration experiments: in 1814, DALYELL found them "immortal under the edge of the knife." Still earlier, PALLAS (1778) had worked with them; and to this very day, new regeneration problems as they arise are taken to the planarians for investigation.

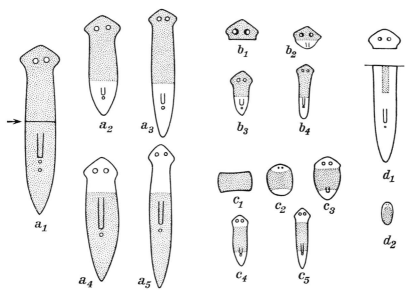

Fig. 541. a–c *Planaria maculata*; d *Curtisia simplicissima*; a regeneration after transverse cutting more or less in the middle of the body; b regeneration of an excised head; c regeneration of a transverse section; d regeneration of a piece removed from the center of the body. Regenerated portions unstippled. (After MORGAN, 1907)

419

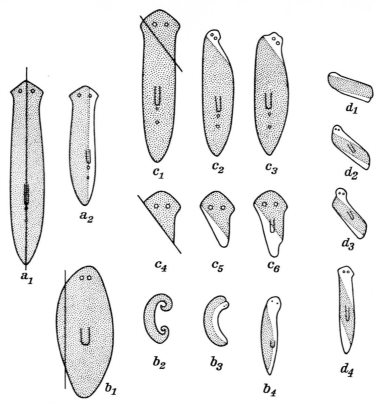

Fig. 542. Regeneration experiments. a in *Curtisia simplicissima*; b, c, d in *Planaria maculata*. a_1 medial; a_2 lateral cuts; b regeneration of a narrow lateral piece; c oblique cut, regeneration of both pieces; d regeneration of oblique sections from the body. (After MORGAN, 1907)

Large and small transverse sections of planarians regenerate posterior parts at the posterior cut surface and anterior parts at the anterior cut surface (Fig. 541 a–c). Even very small pieces taken from the body surface can regenerate in this way (Fig. 541 d). If a planarian is cut in two sagittally, the missing half is rebuilt (Fig. 542 a), and even a narrow lateral piece reorganizes into a tiny whole planarian (Fig. 542 b).

Diagonal sections from the center of the body also regenerate at the front and back cut surfaces (Fig. 542 d), and the polarity of the fragment is imposed on the regenerated parts.

Regenerative ability is not distributed uniformly over the length of the body. In most species there is a more or less steep axial gradient falling off posteriorly (Fig. 543 a, c). In a few species regenerative ability dips in the center but rises again toward the tail (Fig. 543 b), often in those which reproduce by transverse fission just behind the pharynx (Fig. 544). The speed of regeneration varies in proportion with the ability to regenerate (Fig. 543 d).

The stimulus to regenerate emanates from the wound; it stems from the destruction of tissue at the open site. If pieces of a planarian are kept in isotonic saline, they survive for weeks but evince no regular signs of regeneration[655].

The course of regeneration begins with the assembly of regeneration cells into a regeneration blastema in the region of the wound. This blastema is composed of neoblasts normally scattered through the connective tissue and attracted to the wound. The neoblasts resemble the interstitial

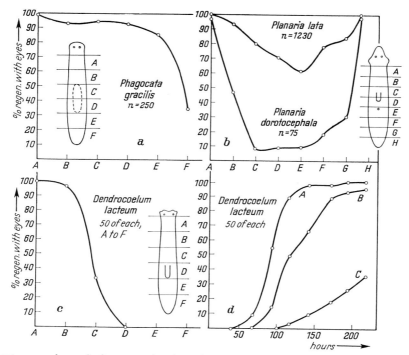

Fig. 543. Competence and speed of regeneration in various body transects in planarians. (After ŠIVICKIS, 1931)

cells of hydroids in their form and intense staining. The first act of regeneration is wound closure; the wound is drawn together by muscle contraction. The epidermis at the edge of the wound extends over it, and the neoblasts come together to form a new epidermis on the outer surface.

The invasion of the neoblasts from distant body regions has been established by irradiation. X-irradiation of the total body in doses higher than 3,500 r prevents regeneration; after an amputation, the animals die in 3–7 weeks. If the anterior half of the body is irradiated (Fig. 545a) and then the head is cut off, a regeneration blastema forms in front of the irradiated region (Fig. 545b), and the anterior part regenerates, eyes and all. The regeneration blastema takes longer to form than in comparable unirradiated animals, and the extra time corresponds to the period needed for the neoblasts to pass through the irradiated region, since, as we shall see, they come only from unirradiated portions of the flatworm. Lateral movement also takes place when the planarians are irradiated unilaterally, with a lateral piece being cut away from the irradiated half (Fig. 545b). The movement of the neoblasts has been demonstrated elegantly by the combined use of irradiation and transplantation, which works very well in planarians. In order to distinguish host– and transplant tissue, different colored planarians are used: e.g., a dark host and a light donor. The host is totally irradiated, the head is cut off, and an implant is lodged in the center of the body (Fig. 545c). After some time, a regeneration blastema is formed which produces a light-colored regenerate; clearly the regeneration cells have come from the implant. In the blastema, and in the zone just behind it, many neoblasts are dividing. In unirradiated, regenerating animals, the mitotic activity of the neoblasts is maximal 3–5 days after amputation. But when the neoblasts have to pass through an irradiated stretch up to a third the length of the animal, a sharp increase in mitotic activity is seen in the regenerating region only after 4–6 weeks.

Fig. 544. *Planaria dorotocephala* undergoing normal binary fission. (After CHILD from STEINMANN, 1916)

In a few experimental series the amputations were repeated 15 times on *Dugesia*, and the time required for regeneration was always the same as that in a control group operated on for the first time. We can thus conclude that the rate of regeneration is not slowed by the repeated decapitation. The supply of neoblasts seems practically limitless[56].

During the first days of regeneration, a sharp rise in DNA synthesis throughout the body has been demonstrated biochemically; this is consistent with the vigorous cell division taking place. In the subsequent period of growth and tissue differentiation, the content of RNA and free nucleotides also rises substantially[411].

The regeneration blastema grows out from the edges of the wound as a cone or ridge, according to the form and size of the wound, and it gradually replaces in outer form and inner organization the lost body parts (Figs. 541, 542). These events have been called *epimorphosis* by MORGAN. In addition to the new construction around the edge of the wound, however, other important changes take place in the intact remainder of the body, and these changes are called, collectively, *morphallaxis*. If the form of a little piece is so changed that the normal proportions must be recreated on a smaller scale (Figs. 541 b, c, 542 b–d), some organs that are normally far apart must be crowded together; organ parts or whole organ complexes, such as the reproductive apparatus, are dismantled if they are now disproportionately large. They are then formed anew. The quantitative relationship between epimorphosis and morphallaxis depends of, course, on the proportions of the body which must be regenerated; it also depends on the species. Epimorphosis can predominate (Fig. 541a), or morphallaxis can restore the typical organization by itself, practically, without the formation of a regeneration blastema (Fig. 546). Generally, though, epimorphosis and morphallaxis work hand in hand.

This is seen in the regeneration of the gut. Neogenesis and transformation are employed to make the gut functional as quickly as possible so that food can once more be utilized to support further regeneration, which otherwise must depend upon reserve material and the breakdown of

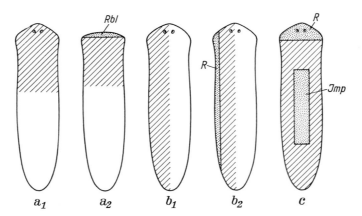

Fig. 545. *Euplanaria lugubris*. Regeneration after irradiation. a, b hatched region irradiated followed by (a) removal of the head or (b) removal of a portion of the left side; c total irradiation, decapitation, and subsequent implantation of a piece of a normal animal in the center of the body. *Imp* implant; *R* regenerant *Rbl* regeneration blastema. (After DuBOIS, 1949)

existing tissues. A severed gut is provided with missing parts by neoblasts, which first form cords of cells (Fig. 547e). These structures now become hollow and unite with the existing gut. The arrangement of these cords does not simply follow the direction of the older parts of the gut; rather, it corresponds to the plan of the regenerating portion (Fig. 547c, d). The cords fork or unite in order to replace the lateral, posteriorly directed branches of the gut as the anterior half regenerates; as the posterior part regenerates forward, anteriorly directed branches of the gut are formed. Sometimes a few anteriorly directed branches appear in post-pharyngeal posterior ends; these are then partially resorbed in favor of a single foregut of typical form. A pharynx arises very early in anterior and in posterior regenerates. At first entirely independent of the rest of the gut, it arises from a ball of neoblasts, which then differentiate into pharyngeal epithelium, pharyngeal sheath epithelium, and pharyngeal muscles and glands. At this point, the gut and pharynx meet, and anastomose. In anterior pieces, the ends of the branches of the foregut are resorbed (Fig. 548) to conform with the new proportions of the regenerant (Fig. 541 b). At the same time, the two posteriorly directed major branches and the pharynx begin to develop in the regenerate, which is still very small.

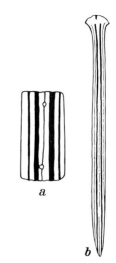

Fig. 546. Reconstitution by means of morphallaxis of a transverse piece removed from the center of the body of *Bipalium kewense*. (After MORGAN, 1906)

Manifestations of the cephalocaudal gradient are seen after transverse cuts are made in the body. The regenerate in the anterior part is always a head (Fig. 549a), and at the posterior end a tail is frequently formed (Fig. 549b). If the wound is kept open, regenerates can form on both sides. In that case, a head grows forward, and a tail

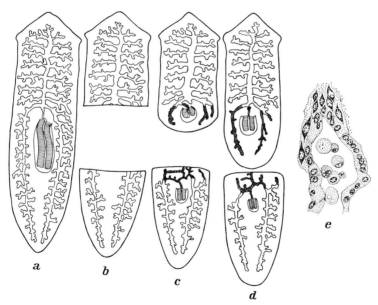

Fig. 547. *Planaria gonocephala*. b–d course of gut regeneration in pieces cut anterior to the pharynx and posterior to it; regenerating gut regions dark; e regenerating gut with proliferating neoblasts. (After STEINMANN, 1926)

grows backward: both can arise from the same wound (Fig. 549c). A head regenerating from such a wound imposes its gradients on the rest of the body as it develops: the gut reorganizes to fit the new head, and a second pharynx often is formed (Fig. 549c, d).

In addition to the cephalocaudal gradient, a mediolateral gradient can also be demonstrated. In various species it is associated with a broad field behind the head region. Any part of the head region which is isolated by cutting can regenerate a new head (Fig. 550a, b) with a complete brain (Fig. 550c). The anterior branch of the gut is split in two (Fig. 550d). But if the head is cut off transversely, only a single median head will regenerate (Fig. 541). The greatest tendency for head formation lies in the midline. Neoblasts are organized for head regeneration here, and it is inhibited elsewhere, where the tendency is weaker. If this medial portion is transplanted laterally (Fig. 551a), the head regenerates at the site of the transplant (Fig. 551b). The following experiment shows that inhibition does play a role in this process: if a *Bdellocephala punctata* is beheaded and an additional median piece is removed (Fig. 552), each anterior prong regenerates a head with eyes. We now vary the time after beheading at which the additional piece is removed (Fig. 553). The longer the interval, the slower is eye formation in the lateral positions (Fig. 554a). An inhibitory factor thus spreads out from the self-organizing field of regeneration. When the eyes are fully formed in the head anlage, the inhibitory effect disappears (Fig. 554b).

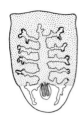

Fig. 548. Regenerating head portion of *Planaria alpina*; formation of the extreme ends of the gut branches. (After STEINMANN, 1926)

Like *Tubularia* polyps, planarians can regenerate polar heteromorphoses, heads at each end (Fig. 555a, b), from excised heads or from relatively thin slices taken from the anterior of the body. Again, as in the case of *Tubularia* (pp. 415ff) we can only suggest an explanation: the extent of the cephalocaudal gradient is so short that the two cut surfaces do not differ sufficiently in their developmental tendencies to proceed down gradient pathways. At first, it was believed that such two-headed individuals must starve. But further observations showed that the remarkable regenerative ability of planarians can save the day. The older head dominates the newer one,

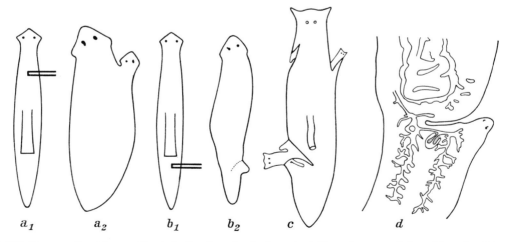

a_1 a_2 b_1 b_2 c   d

Fig. 549. Regeneration after small cuts in the planarian body; a, b, d *Planaria gonocephala*; c *Planaria alpina*; a, b wounds in front of and behind the pharynx; c oblique cuts; d head regeneration showing the influence of the gut in the main portion of the body. (a, b, d after STEINMANN, 1916; c after VOIGT, 1899)

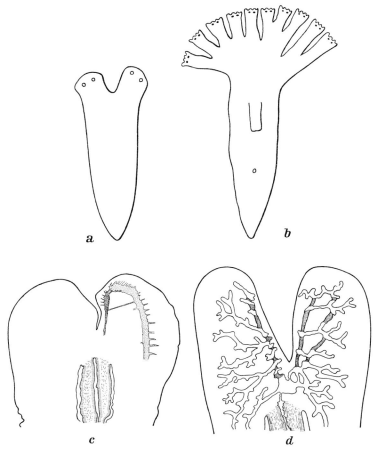

Fig. 550. Multiple head regeneration. a twin heads after decapitation and a medial incision in *Curtisia simplicissima;* b numerous heads formed after decapitation and numerous incisions in the anterior part of the body in *Dendrocoelum lacteum;* c, d *Planaria gonocephala* decapitated and then cut medially. After the formation of the central part of the gut in the head region, the branches have reunited by means of regenerating connections. (a after MORGAN, 1907; b after LUS, 1924; c, d after STEINMANN, 1927)

and induces the formation of a lateral trunk anlage. The regenerated head is first pushed to the side and then resorbed (Fig. 555c). The tiny bit of foregut present in the regenerant is transformed and equipped with a pharynx (Fig. 555d).

Excised tails or short pieces from the posterior body region can give rise to heteromorphic trunks in species with a short rapid posterior decline in the head-regenerating gradient. At first, pointed tails are formed at each end, and sometimes a pharynx forms in each tail, always directed toward the tip[459].

Abnormal proportions may also be produced between regenerating parts experimentally. In *Euplanaria tigrina*, a sagittal cut is made from just in front of the pharynx to the tip of the tail and then the head is cut off (Fig. 556a, *1, 2*), so that the two halves of the body are connected only tenuously. The end of the cephalocaudal gradient is thus split, the pharynx is removed, and the two halves of the body are prevented from rejoining. The anterior part of the cut draws together, and thus the area of junction between the two body heads splits (Fig. 556b). At the third

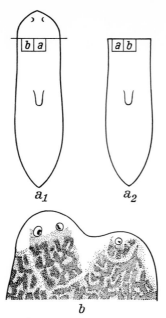

Fig. 551. *Euplanaria lugubris*. a decapitation and exchange of medial and lateral pieces; b result after regeneration. (After Brondsted, 1956)

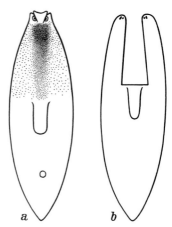

Fig. 552. *Bdellocephala punctata*. a "time graded" mediolateral regeneration field; stippling density corresponds to the gradient; b operative scheme: the high grade portion of the regeneration field is cut out; the lateral portions regenerate eyes. (After Brondsted, 1954)

or fourth day, light-colored, regeneration blastemas appear in the cut surfaces. Where the two body halves are joined, one head regenerates forward and another backward (Fig. 556c). As healing continues, the body halves stretch, and a symmetrical, two-headed animal arises (Fig. 556d), which is reminiscent of the so-called "*duplicitas cruciata*" seen in vertebrates. The symmetry is seen also internally: both heads contain brains and lateral nerve trunks (Fig. 556f),

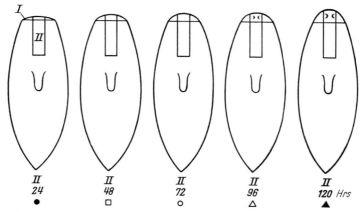

Fig. 553. Operative scheme on *Bdellocephala punctata*; *I* decapitating cut; *II* excised part of the body. Times given below represent intervals between decapitation and operation II, that is, the interval during which head regeneration has been allowed to proceed. Symbols appear in Fig. 554. (After Brondsted, 1954)

426

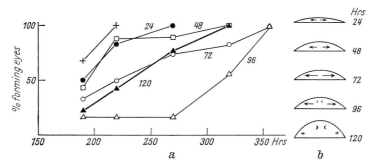

Fig. 554. Times of eye formation in the isolated lateral parts of *Bdellocephala* after different periods of head regeneration. Symbols: +-+ operation *II* (see Fig. 553) immediately after decapitation, other symbols refer to Fig. 553; b diagram of the changes in degree of inhibitory effect in the time-graded mediolateral regeneration field. (After BRONDSTED, 1954)

and the gut branches undergo corresponding regulation (Fig. 556g). This peculiar inner organization of the two-headed animal, and in particular the mutual independence of the brains, leads to the persistent efforts of each head to go its own way. This results in twistings and turnings which extend to the very ends of the body. In *Polycelis nigra* the normally situated regenerant head is generally dominant. The head formed at the wound points backward or is curled anteriorly with the rest of the lower surface (Fig. 556e). The symmetrical *duplicitas cruciata* is usual. But once

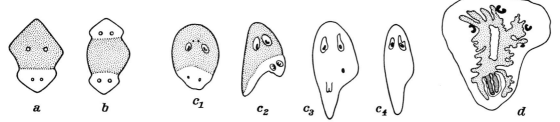

Fig. 555. Polar head heteromorphoses. a, b *Euplanaria tigrina;* c *Planaria polychroa;* d *Curtisia simplicissima.* a head regenerate on the cut surface of the excised head; b bilateral head regeneration on a cross section taken from the anterior head region; c transformation of a head heteromorphosis into a normal body; d section through a transforming regenerate, (a, b after MORGAN, 1907; c, d after LANG, 1913)

in a while additional abnormal heads are formed at the wound: one head always forms in the crotch, but this can broaden and acquire more than two eyes (Fig. 556h). In addition, extra heads can form generally close to the symmetrical head and rarely at some distance from it (Fig. 556i–l).

The symmetrical two-headed regenerates can be understood in terms of the gradient system whose existence we have previously concluded, but the extra heads and eyes are hard to understand. Perhaps the regenerating nervous system plays a role here[630].

In any event, the reorganizing region as a whole is responsible for constructing and maintaining the stable structure out of an abnormal situation. This is seen in the resorption of supernumerary body parts when the normal balance of abnormal structures is destroyed. If a lateral part of the *duplicitas cruciata* is cut off (Fig. 557a), the wound closes, and the two heads draw together and ultimately fuse (Fig. 557). The two eyes which are now in a median position are also resorbed, and thus morphallaxis gradually results in a single head.

427

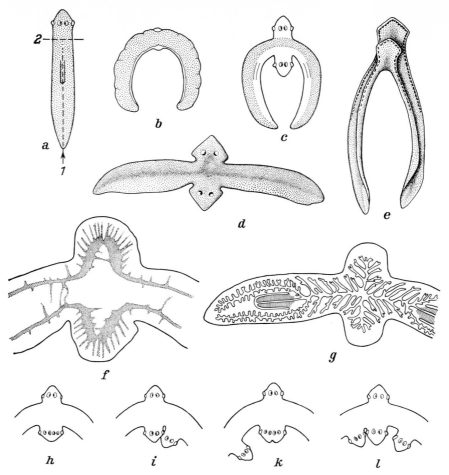

Fig. 556. Experimental formation of "*duplicitas cruciata*" in planarians. a–c, h–l *Euplanaria tigrina;* d–g *Planaria gonocephala;* e *Polycelis nigra;* a–e operative procedures and results; f, g frontal sections through fully formed head regenerates; f brain; g extent of branching of the gut; h–l extra heads. (a–c, h–l after SILBER and HAMBURGER, 1939; d–g after BEISSENHIRTZ, 1928)

A series of experiments reveals inductive effects during regeneration. A piece taken the from head region of *Dugesia lugubris* is transplanted to the postpharyngeal region of another individual (Fig. 558a, b). The head of the host is removed in order to reduce body movement during healing of the transplant. The implant induces an extra pharynx in the host near the tail (Fig. 558c). Then the head implant is excised, and the induced pharynx is excised also. At the site of the secondary pharynx another regenerates. This experiment shows that the implant induces a new pharynx *field* rather than a definitive pharynx. It is the pharynx field that possesses the property of carrying the induction further. The pharynx is regenerated, just as it is when it is removed from its normal place in the intact flatworm after the removal of the head and the region directly in front of the pharynx[581].

A direct chemical effect emanates from the brain, as the following experiments show: in *Polycelis nigra*, the brain and a number of eyes are removed (Fig. 559a). The eye regenerates only when a fully differentiated brain has formed from a mass of neoblasts. If, after removal of the

Fig. 557. Regulative reduction of the heads of a double planarian (*Planaria gonocephala*). (After Beissenhirtz, 1928)

brain, the brain region is X-irradiated every two days (this kills the neoblasts, the most sensitive cells in the tissue, Fig. 559b), the eyes are not regenerated until irradiation is discontinued. The neoblasts which now find their way to the region initiate regeneration of the brain, and eye regeneration follows seven days later. After the removal of a row of eyes, the interior part of the flatworm can be irradiated (Fig. 559c). The eyes regenerate; thus, their replacement does not depend on the arrival of neoblasts from the brain region, but rather upon an agent coming from the brain, which is not destroyed by the irradiation. We can learn even more from further transplantation experiments. A lateral piece of *Polycelis* is implanted into the posterior brain region (Fig. 559d); then the eyes are cut away. The implant grows into an appendage whose border is dotted with eyes. If a similar implant is placed more centrally in the body, eye regeneration also takes place; but not if the piece is implanted near the posterior end. If a brain is implanted in addition (Fig. 559e), eye formation does occur at the posterior end, but only when the implant comes from a region that normally bears eyes. The brain does not produce a similar effect on other body regions: only the eye region is competent to respond. This is not species-specific: if a piece of the

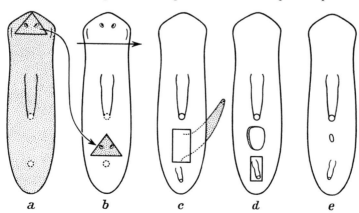

Fig. 558. Experiments on *Dugesia lugubris*. Transplantation of a portion of the head, → decapitation of the host at the arrow; excision of the implant and of the induced pharynx in the induced pharyngeal field. (After experiments of Sangel, 1951, from Wolff, 1953)

429

eye region of *Polycelis* is placed near the brain of the two-eyed *Dugesia lugubris*, a previously removed *Polycelis* eye will be regenerated in the implant.

During regeneration, there are no neural connections between the brain and the induced eyes. Only when the eyes are complete do the photoreceptors send nerves to the brain. The regeneration of the eyes is accomplished by neoblasts at the site of eye formation. In a xenoplastic transplant of *Dugesia* to *Polycelis*, essentially no neoblasts move from the host into the implant. The brain thus acts via an inducing substance which diffuses through the tissue.

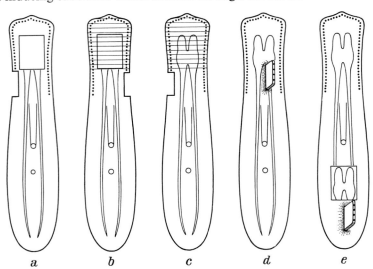

Fig. 559. Diagrammatic representation of experiments on *Polycelis nigra*. a the brain and a portion of the row of eyes have been cut out; b same as a, followed by irradiation of the anterior part of the body (hatched); c a portion of the row of eyes has been cut out and the anterior portion of the body irradiated; d transplantation of a part of a row of eyes into the region of the brain; e transplant of a brain into the posterior part of the body, followed by the transplantation nearby of a portion of the row of eyes. (After LENDER, 1952)

If the brain and the eyes are removed from *Polycelis nigra* and *Dugesia lugubris* and the animals are kept during regeneration in a medium containing a head region homogenate, or the particulate material of such a homogenate, the eyes will regenerate normally, but not the brain, whose site is loaded with regeneration cells which do not differentiate into nerve cells[400], [401]. Thus, a substance is given off from the brain which on one hand induces eye formation and on the other inhibits brain regeneration, thus imposing organization in a regenerating area.

The influence of the brain is also seen in the elaboration of a row of eyes in *Polycelis nigra*. Both the sizes of the eyes and the intervals between them increase along the eye-bearing border from front to back (Fig. 560; not represented faithfully in Fig. 559). The eyes near the brain are close together and small; they have only one or two photoreceptors. The eyes further back have three photoreceptor cells. The distance between the eyes and the size of the eyes are both proportional to the distance from the brain. If a piece of the posterior eye-containing border is implanted near the brain after two or three of its large eyes have been removed, 5–9 eyes, each containing only one photoreceptor cell, are regenerated in the implant (Fig. 560a, b). Conversely, if a piece of anterior border is deprived of 5–7 eyes and transplanted posteriorly, it regenerates two or three eyes with three photoreceptor cells each (Fig. 560c, d). As the inductive effect tapers off posteriorly, the eyes become larger and farther apart. If a piece of posterior border is transplanted forward without the removal of its eyes, new ones will appear between

430

those already present. Starvation, which leads, of course, to a general reduction in body size, also causes a reduction in the number of eyes (Fig. 561a). In the spaces between the eyes that remain, new eyes are induced when the animal begins growing again after feeding; interestingly enough, new eyes are induced also in response to the general regeneration stimulus resulting from the excision of the tip of the tail. Either way, the intercalated eyes arise at typical distances and reach the appropriate size (Fig. 561b). Perhaps each complete eye emits an inhibitory influence which opposes the formation of a new eye in its immediate vicinity and also regulates the size of an eye forming at some particular distance.

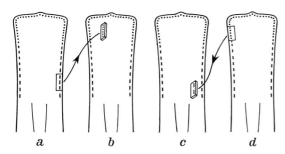

Fig. 560. Transplantation experiments in *Polycelis nigra*.
(After PENTZ, 1963)

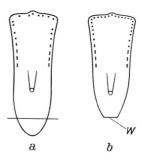

Fig. 561. *Polycelis nigra*. Resorption of the eyes during starvation and formation of new eyes after regeneration has been stimulated.
(After PENTZ, 1963)

We have now seen signs of gradient systems and particular inductive and inhibitory effects at work during regeneration; but we still do not have a clear idea of the nature of the overall direction which produces an entire miniature, normally proportioned individual through the epimorphosis and morphallaxis of a fragment.

Lecture 32

All morphogenetic events are based on the inhomogeneous distribution of material. The origin, replication, aggregation, and separation of components of the organism at all levels result from all sorts of spatial orientations, in diverse ways, and over a wide range of intensities. This is true for organic molecules in the transient and permanent structures of the cells, for inorganic molecules in the skeleton, for cells, and for tissues.

A general uniform growth would lead to a sphere. So at once the tubular hyphae of the fungi pose the problem of the orientation of substances. A certain basic orientation is already established in the egg cell or imposed at the beginning of embryonic development. Isolated blastomeres of annelids and molluscs have particular spindle orientations in each of a set of sequential divisions and thus produce stereotyped, spatially ordered cell classes (p. 305f). Explanted plant tissues retain firmly the polarity they have displayed from embryonic times (p. 396f). The earlier the directions of morphogenesis are determined, the harder it is to discover their nature. But even the distribution of materials that occurs later in development and is influenced by the embryonic environment is mysterious enough. In the young limb bud, unordered individual mesenchyme cells come together and form the structural precursors of the limb skeleton in an orderly fashion; and they shape these structures also when they have been transplanted into neutral places in embryos or explanted into appropriate nutrient solutions (p. 281ff.).

The influences on the organization of materials can be seen a bit more clearly in the simple component processes of histogenesis, which occur in many varieties under ordinary circumstances, and which are also accessible to experimentation.

The connective and supporting tissues of the body are constructed so that their architecture seems engineered to meet local mechanical stresses. The trabecular arrangement of bones is an old example of a structure which has an optimal strength-to-weight ratio, according to the rules of theoretical mechanics, and, of course, considering the specific distribution of stress which the particular bone must bear. The proof that the structure of the bones is really a developmental response to stretching and differentiation is seen in the consequences of the abnormal healing of damaged bones. When a broken bone is set crookedly, it gradually builds a new trabecular structure which is as finely tuned to its demands as the normal structure was to normal conditions. The dismantling of struts in the spongiosa and the formation of new ones take place simultaneously in the connective tissue. But how do the mechanical demands influence the orientation of the material? The crystal axis of the calcium salts is determined by the organization of the collagen fibers which are first laid down in the hyaline ground substance of the connective tissue. What determines the spatial organization of the fibers and the direction of growth of the connective tissue cells which are precursors of the trabeculae? The trabecular structure arises well before the skeleton experiences the stresses it is defined to meet. The "functional" structure is set up for its function in the embryo by a genetically determined mechanism. Does it seem possible that the consistency between the actual function in postembryonic life and the embryonic provision for it can be related to a single morphogenetic principle? RHUMBLER has arrived at a close approximation of the trabecular structure in a model based on stretching: "If one extrapolates from the structure of the acetabular proces of an immature femur by 'stretching' it into its fully grown form, one finds the trabeculae in their normal arrangement"[545,114]. But what sort of process in the living tissue corresponds to the mechanical orientation of the model?

A biological model makes a real contribution when it shows how to generate specific form from unorganized living cells by the application of specific influences.

In the culture of connective tissue cells ("fibroblasts"), the tissue formed is amorphous and chaotic. Mitoses are seen in the tissue and in the cells which have wandered out into the medium from its edge. Growth declines gradually and comes to a standstill when the cell mass has reached a certain size, ostensibly as a result of the accumulation of metabolic products. If the cell mass is cut up, the fragments resume growth after being placed in new culture medium. The connective tissue cells reveal a tremendous growth activity beyond comparison with that seen in the intact organism. A tissue culture which doubled its mass in 24 hours and whose fragments were transferred regularly would reach astronomical proportions, theoretically, in the course of a year. But within the organism the amount of growth is limited in amount and directed in space.

The chaotic growth of tissue cultures can be oriented if directional stress is applied. The culture medium is stretched in a rectangular or triangular frame made from a 1-mm-thick glass rod (Fig. 562a, b). First, the frame is dipped into plasma; upon removal, it encloses a thin film. Next, a little bit of embryo extract is added to the remaining plasma and mixed with a fine glass needle. Some of the medium is now drawn off so that only a thin layer remains. Now the cell mass is placed in the medium, and when this medium has coagulated to a certain degree, it is placed in the center of the frame. If it sticks, a drop of embryo extract is placed on the culture to keep it going. The culture frame, with its stem fused to a thicker glass rod, is inserted into a small glass flask whose stopper is sealed with paraffin (Fig. 562a). The air inside the flask is kept saturated by a drop of water which does not touch the culture. An uneven stress arises within the membrane. The static form of the frame determines the distribution of stress. The stretch of the membrane can be divided into surface tension and elastic stretch; the distribution of the two components is

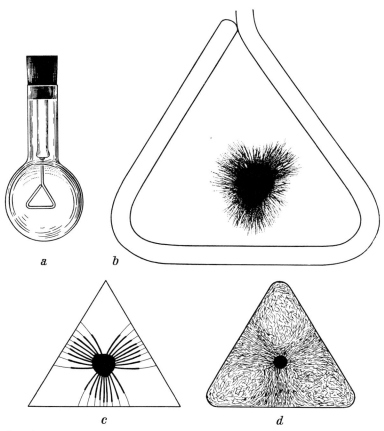

Fig. 562. Tissue culture in a triangular frame. a culture vessel; b fibroblast culture from an embryonic chick heart after 72 hours; c diagram of the lines of stress in the stretched membrane loaded in the middle; the intensity relations are represented by the length of the heavy lines; d diagram of the arrangement of colloid particles in the stressed membrane loaded in the middle. (After Weiss, 1928, 1929)

similar. The surface tension must naturally have its major effect before and during coagulation. Between the culture in the center and the frame itself, the membrane forms a biconcave meniscus as long as it remains liquid. The curvature of this meniscus is greatest between the culture and the nearest parts of the frame. Since the stress on the surface is proportional to the curvature, the maximal stress is found along perpendiculars to the frame. As the coagulum stiffens, the elastic tension arises, the medium is now a stable thin layer adhering to the entire frame and loaded with the culture in its center. When the membrane is place horizontally in the flask, which has one flat wall, the stress is perpendicular to the membrane surface. The tension is once more not uniform, but is distributed as a function of the form of the frame; and, of course, as theoretical calculations and model experiments with sheets of rubber show, the maximal stresses are again in the directions perpendicular to the size of the triangle. In the zones in between, the lines of tension are displaced from the radii toward the main stress axes (Fig. 562c). The distribution of the surface tension and of the elastic stress arising upon coagulation, an ordering of macromolecules or micelles primarily composed of fibrin, is set up within the membrane. Optically isotropic colloids show polarization, whose optic axis lies in the direction of stress, as soon as

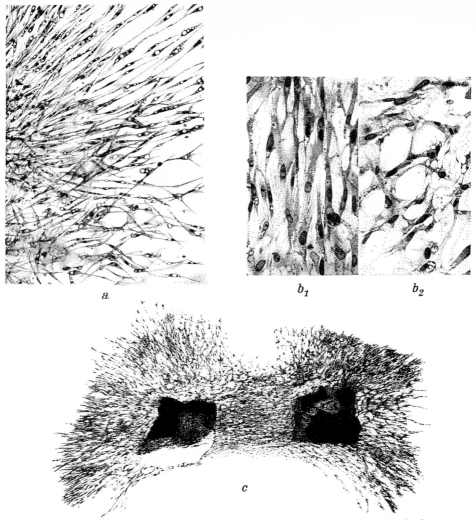

a.   b₁   b₂

c

Fig. 563. Fibroblast cultures in triangular frames. a ordered and random growth at the border between a taut and a relatively unstressed region, 120× ; b tissue structure; b₁ in a stressed region; b₂ in an unstressed region, 285× . c two cultures in one triangular frame, 16× . (After WEISS, 1928, 1929)

they are stretched or compressed. This effect results from a parallel orientation of polar components. In our experimental membrane, the fibrin micelles must orient in the axes of stress; and the extent and degree of this polarization is in proportion to the orienting stress; thus, its maximum is in the major stress axes which extend from the culture as perpendiculars to the sides of the triangle. In the regions of minimal stress the colloidal elements show no order. Figure 562d illustrates the organization of micelles in a membrane stretched on a triangular frame. The fibrin micelles, as in a congealing blood clot, unite only into very small fibrillar aggregates, which are arranged in parallel in the zones of great stress and which show much less order in other zones. The tissue culture in its triangular frame now grows, not as a regular expanding disc, but rather primarily in the direction of stress: growth directly toward the sides of the triangle is stronger, and the radial and orientation of the cells is generally displaced toward the main stress axes (Fig. 562 b). The movement and multiplication of cells contribute to the growth at the edge of the culture. In the oriented rows of cells, the cells themselves are spindle shaped, and their equatorial plates are perpendicular to the cell axis. Thus, the daughter cells form in a row along the original cell axis, so that chains of cells are formed. In intensively growing cultures, the cells make contact with their neighbors by means of fine cytoplasmic processes, so that a network is formed (Fig. 563 a). In the zones of stretch this tissue is composed of parallel strands of cells (Fig. 563 b$_1$). Where there is little stress, there are few cells; most cells become multipolar, pseudopodia extending in various directions, and numerous cross-connections form (Fig. 563 b$_2$). While these parallel oriented components of the culture suggest the first step toward tendon tissue, the random zones look more like the kinds of connective tissue that contain a great deal of ground substance. The transition between cell order in zones of stress and lack of it in relatively unstressed regions can be very sharp (Fig. 563 a), and it reflects precisely the distribution both of tension and of the micellar organization of the medium in its frame (Fig. 562 c, d).

The growing culture itself is not under mechanical stress. The fibrin network of the membrane has acquired a specific structure, determined by the stress distribution, before the culture begins to grow. The cells grow on the surface of the membrane or in its liquid-containing crevices, and in this way they reflect the structure of the medium. If a membrane coagulated under stress is removed from its frame, the cells of a culture now placed upon it grow in such a manner in the stress-free medium to reflect the previously established fibrous structure. Their submicroscopic fibrils serve as tracks for the growing cells to follow.

The stressed and relatively unstressed regions differ in cell concentration as well as organization (Fig. 563 a, b). Apparently the parallel orientation of cells permits more rapid growth because a movement of fluid along parallel fibers can take place more easily than in a network of tangled fibers. The following experiment shows that the culture satisfies its water requirements more effectively in the parallel structured regions of the medium; a membrane coagulated under stress and bearing a culture is removed from its frame and placed on a coverslip with a drop of Ringer's solution between the membrane and the glass. Now the membrane shrivels up into a three-cornered structure due to the loss of water from the zones of stress. The culture itself causes the removal of water by its growth activity. If additional embryo extract is not placed on the membrane after coagulation, cells do not grow out from the edge of the culture; and if the culture does not grow, the membrane does not get deformed after removal from the frame.

The growing culture acquires structure autonomously also. If two cultures are placed on the membrane instead of one, they draw out the fluid in the region between them, with the result that the micelles become ordered in parallel here and make a track for the growing cells (Fig. 563 c). A piece of dead tissue, or a piece of paraffin or agar, produces no such effect. The intense outgrowth of cells between the two cultures suggests that a mutual production of growth-promoting substances may also be playing a role here.

Morphogenesis in tissue culture can proceed still further. In fibroblasts, argyrophilic fibers can be formed which stand out after silver impregnation [55, 430]. In cell-free culture media they do not appear; but within 24 hours after the establishment of a heart fibroblast culture, they appear in the growing zone, oriented radially for the most part. Digestion experiments with trypsin show that the fibrin threads are not involved. After a period of digestion in which the cells and the fibrin of the coagulum have been broken down, the argyrophilic fiber structure remains intact. Whether the cells release enzymes into the medium which produces the fibers or whether the fibers are extruded from the cells remains undecided at present.

It is quite apparent that these experiments with tissue cultures are valid models for the development of functional connective tissue structures *in vivo*, where the cells are also ordered in a colloidal substratum. The ground substance of the connective tissue contains a great many collagen fibers arranged in bundles, primarily along axes of stress. Surgeons have found that after a tendon has been cut, the blood thrombus that forms between the cut edges acquires a fibrillar structure matching that of the tendon. This fibrillar structure is passed on via the movement of cells to the reconstruction of the tissue.

Can the same morphogenetic principle—formation of growth tracks by structures formed as a mechanical response to stress in a colloidal substratum—also explain the functional morphology of the skeleton? Detailed histogenetic investigations, especially on the early phases of the remodeling and new construction of the trabecular structure in bone healing after improper setting, have not been performed, as far as I know. Rhumbler's model, whose original is in the Zoological Institute at Göttingen, purports to show that the colloidal connective tissue of the superficial periosteum has already reached a state during fetal development where it determines the trabecular trajectories that will subsequently appear[545, 114]. This model is thus concerned with precisely the factors to be decisive for the development of connective tissue structures in tissue cultures on a stressed substratum. It is quite reasonable that those factors should play a role in the morphogenesis of the bone spongiosa, perhaps the decisive role. The role of the scaffolding, corresponding to the triangular

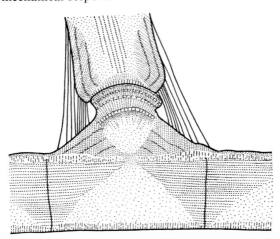

Fig. 564. Pattern of orientation of the main trabeculae in the calcareous framework of a plate and the spine it supports in a sea urchin (*Echinus esculentus*). Meridional section through the plate of a young animal, 16×. (After E. BECHER, 1924)

frame of the tissue experiment, would then be to determine the surface contours of the bone during prefunctional morphogenesis. Here we approach the mysterious determination of a species- and site-specific growth pattern, primarily expressed as an orienting influence, and making use of the developmental mechanism of an elementary histogenetic process.

Another excellent example of the organization of materials which has been analysed from a number of different angles is provided by the echinoderm skeleton, which is made of ordered crystals of calcite. The skeletal elements laid down by the cytoplasm are of course never bounded by crystal surfaces; their species- and site-specific form is imposed by the cells which form them; but techniques employing polarization, etching, and the study of cleavage planes show that they behave as crystals just as if they had been cut out of a piece of Iceland spar. A regular relationship appears, moreover, between the position of the crystal axes and the morphology. We call such structures (with W. J. SCHMIDT) biocrystals, and they present us with a unique developmental problem: how can the organism influence the molecular order of a growing calcite crystal so that a specific biological structure emerges?

The large skeletal elements which continue to grow as the body grows, such as skeletal plates, spines, and dental mill ("Aristotle's lantern"), have a reticulate structure almost throughout. The cavities are pervaded by mesenchyme, which appears to be a syncytium and which contains numerous nuclei and a variety of connective tissue fibers. In the sutures between skeletal elements the mesenchyme is especially plentiful, and it is oriented approximately perpendicular to the adjoining skeletal surfaces. This mesenchymatous syncytium, which also encloses each element,

436

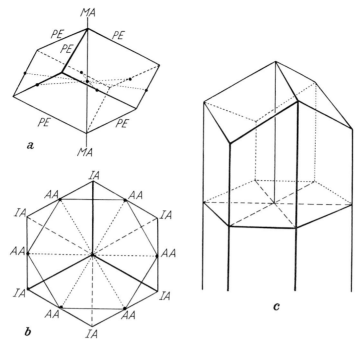

Fig. 565. Crystallographic relationships of calcite a rhomboidal cleavage planes; b projection of a vertex of the rhomboid on its median cross-section. The inner hexagon corresponds roughly, in its relation to the edges of the cleavage rhomboid, to a protoprism; the outer hexagon, to a deuteroprism; c combination of the rhomboid with the deuteroprism; on the cross-section, the crystallographic intermediate axes are shown. *MA* crystallographic (optical) main axis; *AA* accessory axes; *PE* polar edges; *IA* intermediate axes. (After W. J. SCHMIDT, 1930)

is responsible for its growth, regeneration, and the resorption associated with remodeling. Like the bone spongiosa, the skeletal meshwork in the echinoderms is often structured to meet mechanical stress. A good example of this is provided by an element of the capsule of a sea urchin, together with a spine (Fig. 564). A thin, hard, compact pigmented layer covers the surface of the plate. At the sutures, where further growth produces pressure, the girders run in perpendicular bands. The first anlage of a skeletal plate does not have these bands; rather, it is arranged in an irregular grating. On one side of the plate, a knob develops with the convex surface that articulates with the spine. Stretching and compression are both experienced by the spine apparatus. Pressure is brought to bear by muscles which run between the edge of the knob and the base of the spine, surrounding both structures as a muscular capsule. These muscles keep the articular surfaces closely appressed. At the sites of muscle attachment, at the joint, and in the region of the concave articular surface of the spine, the components of the skeletal mesh run mainly in the axes of stress. Within the plate the knob and the spine, the spongiosa is arranged randomly. This spongy character and the formation of structural trajectories along lines of stress in no way interfere with the crystalline character of the skeletal elements, for the structural components themselves always contain the typical calcite crystal structure. Polished sections of echinoid spines behave just like calcite plates polished parallel to the base, in spite of the extraordinary complexity of the girders and grating that make up the skeletal elements. Calcite rhomboids (Fig. 565a) and prisms (Fig. 565c) can be obtained by smashing spines, and their main axes conform to the main axis of the spine.

437

The morphogenetic principle of these biocrystals can be recognized in the early anlagen of skeletal elements. A hexagonal net with a branching angle of 120° underlies the reticular skeletal structure. It begins with a rod or 3-rayed element with terminal dichotomous branching (Fig. 566). As the branches grow and themselves branch, the first spaces of the network become circumscribed. If the network is to express the regularity of the scheme, it is necessary for the branching angle always to be exactly 120°, and the length of the branches must be constant. These conditions are in fact rarely met after the further growth of the skeletal network, but in early stages, they are (e.g., Figs. 567a, 570a–d). Since the outgrowth of some branches interferes with the growth of others, the further deposition of calcite leads to a curved skeletal element (Fig. 567b).

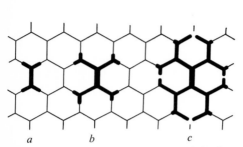

Fig. 566. Diagram of the formation of the hexagonal net from the primary rod by means of 120° forking. (After W. J. SCHMIDT, 1930)

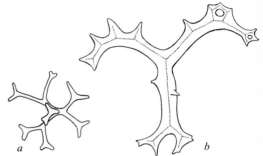

Fig. 567. Experiments on bifurcation in skeletal elements. a anlage of an interambulacral plate in *Echinus miliaris* 460×; b calcareous structure from the integument of *Holothuria tremula*, 160×. (a after GORDON from SCHMIDT; b after W. J. SCHMIDT, 1930)

The relationship of the primary branches to the axis of the calcite cleavage planes can be recognized in certain skeletal elements. In the Holothurian *Pseudocucumis africana*, reticulate plates occur which bear on both sides short three-surfaced spines (Fig. 568a, b), all of whose corresponding surfaces are parallel on each side of the plate, but in such a way that the spines on one side seem to be rotated 180° with respect to those on the other side. Since the crystal axis is normal to the skeletal plate surface, these spines correspond to the vertices of rhomboids (Fig. 568, cf. 565a). The spines are arranged around holes in the plate, so that three spines are placed equidistantly on the upper surface and three are similarly placed, but with their positions rotated 60° on the lower surface. Their positions thus correspond to the interstices of a hexagonal net, and the branches are directed toward the sides of the rhomboid spines (Fig. 568c). And since one may regard these as split rhomboids, the axes of branching fall into crystallographic intermediate axes (Fig. 568c, cf. Fig. 565a, b). Above the corners of the hexagons in many-layered skeletal elements, rods arise in the main crystal axis and branch at the next level (Figs. 567a, 569). In various echinoderms, during spine formation, 6-rayed elements sometimes appear in addition to the usually 3-layered structures (Fig. 569). Here also the branching follows the crystal axes. The cytoplasm of the mesenchymal syncytium begins with a tiny initial which crystallizes out and lets it grow according to its general and specific rules of crystal growth, regulating its form at a higher level of organization.

The formation of the skeletal element is species- and site-specific. The formative cytoplasm permits the deposition of further materials in certain vectors in the crystal axis system, imposing its genetically determined bias. The organism can thus disregard the crystal axes at any time, so that the girders are laid down in patterns other than a hexagonal network and along paths other than the main crystal axis. The calcite itself, however, retains its crystal structure.

438

The arrangement of the skeletal elements into a complete skeleton is another problem, and it has less to do with biocrystals. The first skeletal rudiment, which is formed according to the rules of crystal vectors, acquires a certain orientation, and this influences the further growth of the element, in collaboration with regulatory factors from other parts of the organism and mechanical effects.

Quite a different problem is posed by the nonfunctional correlations among skeletal elements of diverse forms which are laid down within the same syncytium. In the integument of the Synaptid Holothurians little perforated plates are arranged normal to the body axis beneath the epidermis in the loose connective tissue; these plates articulate with anchor-shaped skeletal elements (Figs. 570f., 571c). The end opposite the curved part of the anchor is formed into a so-called handle, which is roughened where it faces the epidermis. When the body wall is stretched, and the anchors are pressed

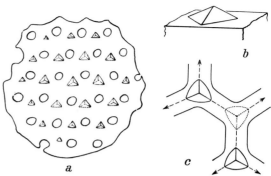

Fig. 568. Perforated plate of *Pseudocucumis africana*. a surface view, 220×; b, c diagrams; b vertex of the rhomboid on the upper flat surface; c bifurcation. (a after W. J. SCHMIDT, 1930)

against the plates, the handles produce a gooseflesh effect. Before the shaft terminates in the handle, it bulges toward the inner connective tissue layer of the body wall. The curve of this ridge is the articulation between anchor and anchor plate. It rests on the tip of the plate and is further supported by a little arch which grows out of the plate (Fig. 570e, f). Both anchor and plate arise in a multinucleate syncytium, which is distinct from the surrounding connective tissue (Fig. 571). The anchor precedes the plate in development. The plate is quite symmetrical in most Synaptids. The free end is perfo-

rated by 7 main holes in *Leptosynapta bergensis*. Of these, 3 lie in a median row with 2 on each side, so that in all 6 holes are arranged around a central hole. The plate is a section of a hexagonal network. A primary rod which lies normal to the shaft of the anchor rudiment forks at both ends at a 120° angle and a series of further branches circumscribe the 7 main holes. At the other end, 2 or 4 symmetrical holes are added; the arch forms here, as well as smaller irregular perforations (Figs. 570a–e, 571).

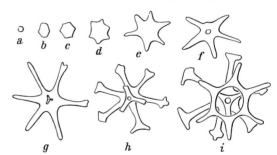

Fig. 569. Early development of the spiny skeleton of *Echinus miliaris*. (After GORDON, 1926 from W. J. SCHMIDT, 1930)

It is striking that anchor-plate complexes can be formed in one region in the integument in a variety of sizes, but always with the same proportions. A big anchor has a big plate, and a little anchor has a little plate. There are always the 7 main holes; the hexagonal network may arise from long or short branches. In addition, variant forms of the plate arise which further show its dependence on the anchor rudiment. It sometimes happens that the primary rudiment is not laid down normal to the anchor shaft, but rather parallel to it (Fig. 572a$_1$). When this happens, the subsequent branching does not follow the symmetry set up by the primary rod, but rather follows that of the anchor. Thus, a typical plate forms, but as a result of the regulative changes

439

in its branching pattern (Fig. 572 b_1–c_1), the 6 holes surrounding a central hole are rotated 60° from the normal arrangement (Fig. 572 d_1). Clear-cut anomalies in anchor structure lead to even more striking indications of the anchor's influence on the development of the plate. Occasionally, the anchor shaft forks, and two curved ends (Fig. 573 a) or two handles (Fig. 573 b) are formed.

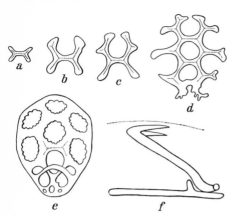

Fig. 570. a–e Development of the anchor plate in *Leptosynapta*, somewhat diagrammatic, with bifurcation axes dotted in; f diagram of the apposition of anchor and plate, side view

In the first case, the free end of the plate broadens, following the anchor hooks; in the second case, the other end of the plate is doubled. Moreover, bipolar anchors are sometimes laid down in the syncytium (Fig. 573 d, e). Again, plate formation follows the anchor anomaly. Figure 573 d illustrates a double plate conforming to a bipolar anchor. The two free ends of the plate are well developed; the fixed end, which normally bears the arch, has not formed. In the same syncytium, however, a second anchor also has formed, whose broken arch end has no counterpart. The handle of the shaft lies over a part of a double plate, and this part of the plate forms a small but typical arch end right under the handle. Figure 573 c illustrates an abnormal anchor with only one hook; its plate appears underdeveloped on the side where the hook is missing. All of these malformations, these "spontaneous experiments" which provide us with fortunate alterations in the standard form of the anchor, reveal in their positive and negative aspects the dependence of the form of the plate on the form of the anchor. This dependence now being clear, it is surprising that self-differentiation of the plates can also take place in the syncytium: occasionally, plates form in the absence of anchors. Their development is exactly like the normal plate development carried out in the presence of a developing anchor (Fig. 574 a–d). Indeed, primary rods that form transverse to the long axis of the body, rather than parallel to it, can give rise to complete plates (Fig. 574 a_1–d_1), just as those do which arise in the exceptional orientation parallel to the anchor shaft (Fig. 572). In these cases, the symmetry of the plate conforms to the orientation of the parallel transverse strands which can be seen in the connective tissue. These strands may

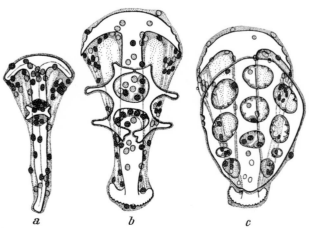

Fig. 571. Development of anchor and plate in the syncytium of *Leptosynapta inhaerens*. (After WOODLAND, 1907)

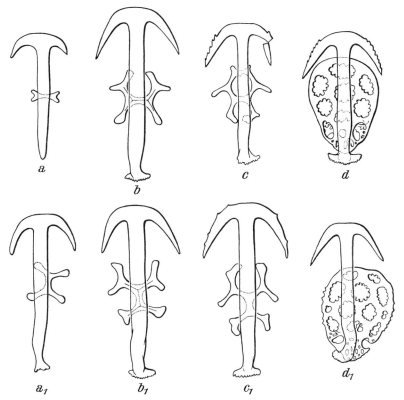

Fig. 572. Development of the anchor plate in *Leptosynapta bergensis*. a–d normal relations, a_1–d_1 primary rod rotated 90° with respect to the anchor. (After S. BECHER, 1911)

reasonably be recognized as providing orienting stresses; the common occurrence of malformations shows, however, that the influence of the anchor itself is substantially stronger. In the double structure shown in Figure 573a the single shaft section makes a 45° angle approximately to the longitudinal and transverse lines in the body wall. The arch end of the plate has formed in response, undisturbed by this unusual orientation; and, similarly, the almost diametrically opposed curved ends of the anchor have brought about the formation of two free ends of the plate. In the case of Figure 573b, where two handles have arisen almost at right angles, two corresponding portions of the plate have also formed.

When we say that the form of the anchor exercises an inductive effect on that of the plate, via the syncytium, we are only describing what happens; we know nothing of the nature of this effect. Both the experiments with tissue culture in media under stress and the analysis of the formation of the calcareous skeleton in echinoderms have led us to elementary mechanisms of the ordering of substances. Each suggests that stress in colloidal growth substances can lead to an ordering of micelles, which in turn triggers the growth of mesenchyme cells and finally results in ordered tissue formation. What remains unclear is the origin of stresses during development. Calcareous skeletal elements show that the vector properties of a crystalline material dominate its structure and can be channeled by the organism. Again, the manner in which crystal growth is steered along a species- and site-specific path remains a mystery.

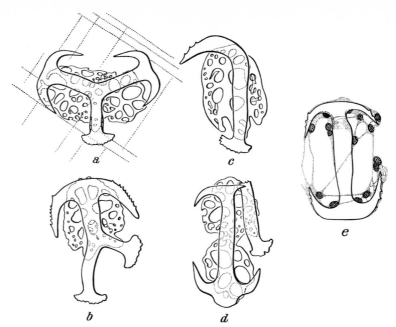

Fig. 573. Malformed anchor and plates. a from *Leptosynapta bergensis*; b, d from *Labidoplax thomsonii;* c from *Labidoplax digitata;* e from *Labidoplax buskii.* (a, b, e after S. BECHER, 1911, 1912; b, c after WILHELMI, 1920)

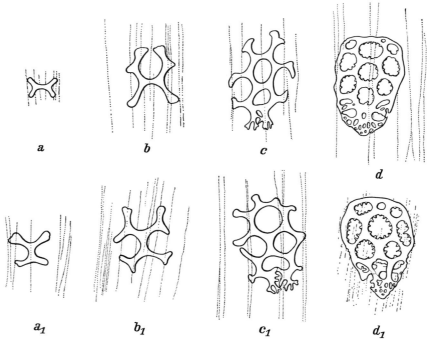

Fig. 574. Autonomous development of anchor plates without anchors. a–d normal; a_1–d_1 rotated (90°) position of primary rod with respect to the transverse folds of the connective tissue. (After BECHER, 1911)

Let us take a moment now to recall the morphogenetic principles which we have learned from multicellular organisms. *Polarity* has been expressed in the points of departure of animal and plant ontogeny, eggs and spores, and later in individual cells capable of differentiating. Polarity is basic to differential division, which takes place at the beginning of embryonic development or later in the laying out of this pattern. *Gradients*, progressive, quantitative variation in the concentration of substances or in the rate of processes, result in qualitatively diverse developmental reactions. Opposing gradients govern the body plan in many cases. *Induction* triggers the determination of fields and organ formation where it meets *competence*, which may be present everywhere, or just at certain places or times. *The principle of the self-organizing fields* is encountered in the organization of simple multicellular systems and in certain developmental phases in highly differentiated multicellular organisms. *Inhibitory fields*, seen in the determination of leaf position and of the arrangement of special components in the leaf epidermis, certainly play important roles in many theaters of animal and plant development. An ordering of cells and of extracellular structures takes place as a *response to mechanical stress* in colonies of algae in a gelatinous matrix, as well as in the skeleton of echinoderms and vertebrates. Now we can only establish the operation of these principles and their collaboration. To describe a chain of causality in development in physicochemical terms from beginning to end, however, is a very distant goal! The task of causal analysis appears possible to us as long as we are engaged in the experimental study of the interplay of cells. But when we think of differentiation with cells, e.g., merely the formation of a complicated skeleton, in a single cytoplasmic mass, as in the Radiolarians, and in the syncytium of echinoderms, confronting the nonfunctional, purely form-dependent relationships between formed and forming skeletal components, we see the problem of morphogenesis in all of its present difficulty.

Lecture 33

In the last analysis morphogenesis is governed by the genetic material. This determines the various responses of the respective cells to the external and internal factors to which they are exposed during the course of development. Genetics and cytogenetics demonstrate that certain events in the career of the cell during tissue and organ differentiation depend upon genes located in the chromosomes; and recently important strides have been made toward the understanding of the interplay between the genes and the cytoplasm.

A series of mutations show us that elements of the behavior of the chromosomes in the mitotic cycle (p. 30, 31) and in meiosis (p. 35ff), as well as the formation of the apparatus for cell division (p. 44) are influenced strikingly by individual genes. We may also conjecture that the events prior to postzygotic development also depend on genes: the distribution of materials and processes in the cytoplasm during the growth and maturation period of the oocyte, through which polarity and cleavage sequences are determined. Nevertheless, about the only example we have for this is the famous case of the Mendelian inheritance of dextral or sinistral cleavage in certain snails. Mutations leading to fundamental changes in embryonic cells are probably always lethal as homozygotes. In the major experimental animals of zoological genetics, insects, birds, and mammals, the first developmental events are so obscure and so little expressed in regular sequences of cell division that until now the earliest developmental anomaly that can be associated with genes has been the abnormal operation of the germ layers in a quarter of the progeny of a particular cross. Among the animals with early determination of cell lineage, such as annelids, molluscs, and ascidians, no species has yet proved especially suitable for extensive genetic experiments. It would be somewhat risky to count on the arrival of echinoderm genetics in the near future.

In order to establish the role of an individual gene in the course of development, one must first find the phenocritical phase (HAECKER), the developmental stage in which a departure from normal arises as a result of the substitution of a single allele. This developmental stage, in which a difference between two strains differing only by one allele is determined, can of course be much earlier than the time at which the difference is first seen. Phenogenetics, which studies the genesis of the phenotype, can approach gene-dependent determinative processes closely if they are studied experimentally in the course of development, and especially if organs can be isolated or transplanted into hosts with different genotypes. It is also useful to try to evoke in normal individuals changes in the course of development corresponding to the effect of a mutation; similarly, it is useful to try to restore normal phenotypes in mutant strains. The environmentally induced alteration of the phenotype of one genotype to the phenotype of another genotype has been called phenocopying by RICHARD GOLDSCHMIDT (1935). In many cases, sensitive periods in development have been found during which temperature changes, X rays, or chemicals can alter particular phenotypes; when a quantum alteration corresponds, at least roughly, to the difference caused by a mutant allele, a phenocopy has been produced. To make a finer distinction, one speaks of a true phenocopy when the phenotypic change common to the environmental and genetic effects has exactly the same developmental cause.

Every gene-controlled process in development is integrated with every other. So it is always necessary to ask whether a mutant character arising in one part of the body results from an altered response of the cells in this part caused by the cell's own genes or whether they are changed in response to an influence coming from another part of the body. We can thus distinguish intracellular and intercellular genetic effects or, as HADORN calls them, *autopheny* and *allopheny*.

We shall now consider a few well-analyzed examples which show the role of certain mutations in insect development. "Glassy wing," a loss of scales in the meal moth *Ephestia kühniella* which results from a particular combination of genes is a good example[368], [VI]. It is not expressed in animals raised at 18°, but the longer they are exposed to higher temperatures during development, the greater its frequency (*penetrance*) and intensity (*expressivity*). There is no sensitive period for the expression of this genotype: it doesn't matter whether the high temperature is at the beginning, in the middle, or at the end of development. At high temperatures there is a lasting change in the affected system, apparently in the cytoplasmic metabolism, which the mutant genes enhance, to the extent that the altered phenotype is expressed. The scales formed in the pupae fall off when the moth emerges, so that large, bare, transparent areas are seen on the wings. This glassy wing condition can be mimicked by interfering with metabolism[190]. A lack of linolenic acid in their food produces the same transparent wing areas in wild type moths that are

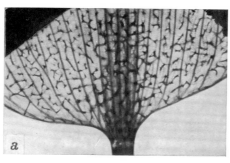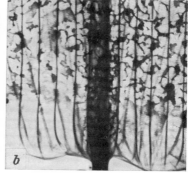

Fig. 575. Electron micrographs of the basal portion of deep scales of *Ephestia kühniella*. a wild type; b mutant *he;* about 5,000×. (Original photos of KÜHN and VON ENGELHARDT)

seen in the mutants raised for varying lengths of time at 25°. The biochemical site – or sites – of action of the genetic and environmental effects is not known, and this is the case for almost all genetic effects which can be phenocopied.

The mutation *he* (lack of color) in the meal moth affects the nature of the cuticle, which is weaker and less elastic in the homozygotes than in the wild type. The color of the scales resulting from reactions of polyphenols with proteins[546] begins to appear later in the mutant than in the wild type; it then develops at normal speed but stops at a particular degree of pigmentation. The scales are frequently folded lengthwise or rolled in at the edges. Electron micrographs reveal disturbances of the trabecular structure during scale development: these supporting elements (p. 364ff) which normally run almost straight from the upper lamellae to the lower, either as branched or unbranched pillars (Fig. 469), lie scattered between the lamellae in many parts of the scales (Fig. 575) in *he/he*. The reduced hardening of the cuticle and the related reduction in pigmentation result, not from a disturbance of the metabolism of the entire body, but only from an autophenic alteration in the epidermal cells; in transplanted vesicles (p. 345ff, Fig. 443), type *he* scales arise in *he/he* implants in wild hosts; conversely, wild implants produce dark, normally formed scales in *he/he* hosts[371].

The mutation *he* affects all the scales, but other mutations in *Ephestia* provide insight into the genetic dependence of the wing pattern and suggest models for the genetic determination of patterns.

The spread of the determination of the field of symmetry in the *Ephestia* wing, which runs its course during hours 48–62 of pupal development (p. 369ff, Fig. 477), is influenced by a variety of mutations. *Syb* enhances it (Fig. 576b), and *Sy* inhibits it (Fig. 576c). The spread of determination moves more slowly in *Sy* wings than in normal wings: a burn (p. 370) at a given age generally stops the spread at an earlier stage in *Sy* animals (Fig. 477k–n) than in wild type pupae. Heat treatments (45°, 45 minutes) cause the field of symmetry to spread in wild type pupae if given from 12 to 36 hours; from 36 to 72 hours they cause it to narrow (Fig. 576d). In *Sy*-animals

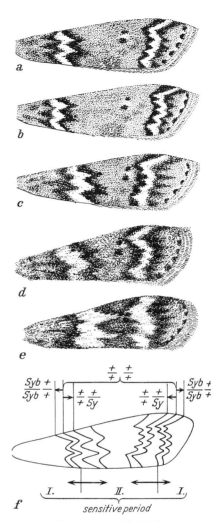

Fig. 576. Spread of the field of symmetry on the upper surface of the forewing in *Ephestia kühniella*. a +/+; b *Syb/Syb*; c *Sy/+* (lethal factor); d, e modification by heat shock; d +/+; e *Sy/+*; f scheme of the effects of various genes and of modifications in the early (*I*) and later (*II*) sensitive periods for the laying out of the field of symmetry

the heat shock can freeze the spread of the field of symmetry at a stage (Fig. 576e) like that seen after a normal wing is burned during the early spread of symmetry (cf. Fig. 477n), but which could not be obtained by the heat treatment of wild type animals. In two sequential sensitive periods, then, the effects of the mutant genes *Syb* and *Sy* can be phenocopied (Fig. 576f). In each case we are dealing with true phenocopies in so far as the heat affects a process resulting from the action of one or another gene. But of course we cannot say that the gene has acted for the first time in this stage of wing development and that the determination of normal or *Syb* or *Sy* symmetry fields results. For *Syb* we can certainly suggest that the tendency for the spread of the field has already been instituted: in reciprocal F_1 hybrids the width of the field depends on whether the mother was $+/+$ or *Syb/Syb:* the dimensions are always displaced in the direction of the maternal strain (Fig. 577). In the F_2's of reciprocal crosses there is no evident similarity to the grandmother; the similarity to the mother in the F_1 thus rests not on a self-perpetuating cytoplasmic difference in the *Ephestia* line used but rather results from differences between *Syb/Syb* oocytes and $+/+$ oocytes. Even before reduction and fertilization, therefore, the cytoplasm is influenced by the homozygous genes present, setting up a causal chain which reaches through embryonic and postembryonic development and expresses itself in the spread of determination across the pupal wing. The individual's own genes also enter this picture, however. Even *Sy* begins to work well before the "final" determination of the wing pattern in the pupae; for *Sy/Sy* is lethal. This mutant is viable only as a heterozygote, which exhibits the normal vigor. The quarter of the progeny with the *Sy/Sy* genotype, in a cross between heterozygotes, die during embryonic development. The wild allele thus participates in a cellular mechanism for which one gene of the pair suffices during embryonic development, while it

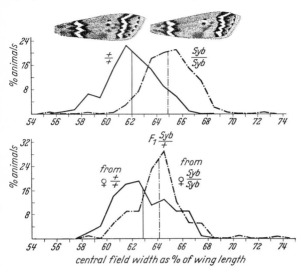

Fig. 577. Results of crosses between wild type and the mutant *Syb* in *Ephestia kühniella*. (After KÜHN and HENKE, 1936)

does not suffice to form a normal field of symmetry. In the present case, it cannot be excluded that the *Sy* mutation is not a single point-mutation, but consists instead of two closely linked genes, one a recessive lethal and the other a visible which is neither dominant nor recessive. For the analysis of developmental mechanics this makes no difference; it is important to see, however, that a single mutation *can* have two very disparate effects.

Whether *Syb*, *Sy*, and heat treatments affect the unknown substance which spreads over the wing, or affect the response of the cells to this substance, we do not know.

The effects of *Syb*, *Sy*, and their $+$ alleles make clear a phenomenon seen in many traits whose phenogenesis has been investigated in detail: the determination of the quantitative expression of a trait, such as the extent of the field of symmetry, is influenced by many genes at various stages of development via the interplay of antagonistic processes, so that the rigidity of determination, and its extent, can be varied quantitatively. The effect of extreme temperatures, which are often used in the analysis of genetic effects, consists apparently of the inhibition of processes during a sensitive period; now, other gene-dependent processes are permitted to play a larger role, and so the level of determination rises or falls.

When the spreading field of symmetry has organized itself into zones determined to become white, black, and the background color, we may speak of a prepattern in the wing surface which consists of a quantitative or qualitative chemodifferentiation. Local differences in the norms of reaction of cells result in the production of the various form-structure-color types which are possible in the differentiation of scales (Figs. 476a, b, 578).

446

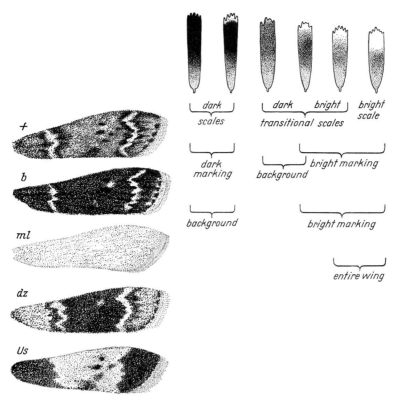

Fig. 578. Scale types and their distribution in the wing pattern of + and the mutants *b*, *ml*, *dz* and *Us*

The alleles *Syb* and *Sy* change only the dimensions of the symmetrical field, but not its organization into the elements of the normal color pattern. The latter, however, are effected by *dz*: the whole central field between the two white marginal stripes is as dark in *dz* animals as the dark interior stripes next to them in the wild type (Fig. 578, *dz*). This mutation also narrows the field; clearly the prepattern is altered, although we do not know whether the basic change is in the nature of the spreading process or the nature of the substrate. The *Us* allele transforms the pattern of the whole wing: The whole field of symmetry is narrower and uniformly pale (Fig. 578, *Us*), except for dark spots of various sizes. Both proximally and distally the wing is black.

The prepattern which results from the spread and organization of the field is itself preceded by an earlier prepattern whose alteration we can recognize by means of still another mutation (p. 363f.): the pattern of the lacunar system which determines the position of the central spots, the marginal spots (p. 369f), the jagged course of the field margins (Fig. 475c), and the arrowheads where the field margins lie above the wing veins (p. 369, Fig. 475). In contrast to the genes affecting a prepattern, others alter the reaction of cells to the prepattern which is already present. In the mutant strain *b/b*, the cells are not competent to respond normally to the qualitative or quantitative gradients in the prepattern; they respond everywhere in an all-or-none fashion either with very dark or very light scales (Fig. 578, *b*). The light elements of the pattern thus stand out sharply from the uniformly black background (Figs. 475b, 578, *b*.) The white of the white bands is determined by the prepattern governed by *Syb*, *Sy*, or their wild alleles. The presence of the prepattern in *b/b* can be seen if the pupae are examined during wing pigmentation:

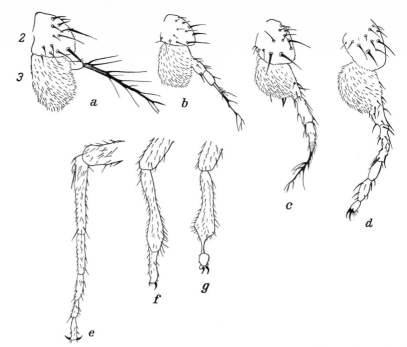

Fig. 579. Effect of *ss*-alleles in *Drosophila melanogaster*. a normal left antenna; b–d transformation into an antennal foot; second and third basal segments of the antenna; e normal tarsus of the left hind leg; f, g tarsal reduction. (After VON FINCK, 1942)

first, the wild pattern appears, and gradually all of the background scales also darken; but with respect to form, these do not quite reach the extreme shapes of the scales in the normally dark areas. Transplantation experiments[371] also show that *b* causes a change in the scales' individual response rather than a remodeling of the prepattern. When a larval skin implant passes through metamorphosis with the host, *b/b* epidermis forms only black scales even in wild type hosts. Evidently only the determining conditions at the presumptive white band sites can cause the *b/b* cells to produce white cells. The mutation *ml* restricts the competence of all cells to the formation of very light scales (Fig. 578, *ml*); the wing has no pattern whatever. In this case also, transplantation experiments have shown that the competence of the cells themselves has been altered. The number of well-known mutations which change the wing pattern shows that many genes must be involved in the laying out of the normal pattern, and the ones we know about may well be just a small sample.

A number of mutations cause various disturbances in abdominal segmentation in *Drosophila*. A few of the mutant genes act in the egg before fertilization with a subsequent impact on the adult structure of the zygote, as can be seen in genetic crosses. Similarly, the same abdominal abnormalities can be produced by treating normal mothers with heat[760]. Thus, phenocopies of the predetermining genetic effects can be achieved during the sensitive period in the developing oocyte. Certain genes influence the organ-specific properties of the imaginal discs. They are not completely determined, as we see from the surprising transformation of the arista of the *Drosophila* antenna into a leg by some chemical agent in an early larval stage (p. 359). This same effect is produced in several degrees by a series of multiple alleles at a locus on the third chromosome. The mutation *ss* (*spineless*) reduces the size of the bristles, although the hairs are unaltered. The dorsal head bristles (Figs. 458a, 580a) are affected most strongly, together with certain bristles on the mesothorax. The whole series, which includes at least seven known mutant alleles, involves extensive pleiotropy, i.e., effects on several diverse characters: in addition to the interference of bristle formation, the arista is transformed into a leg (Fig. 579 b–d) and the tarsal segments of the legs are reduced (Fig. 579f, g). The effects of the various alleles on these phenes are not equally strong; indeed, they cannot be arranged in any single order (Fig. 580). Thus, *ss* has the

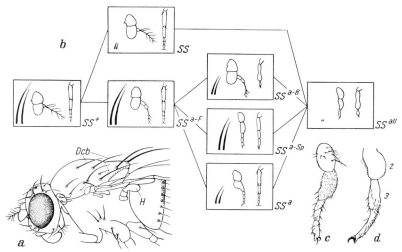

Fig. 580. *Drosophila melanogaster*. a bristle arrangement; b schematic survey of the various effects of the *ss*-alleles; c antennal foot of *ss^{a-Sp}*; d antennal foot of *ss^{all}*. (After von FINCK, 1942)

greatest effect on the bristles, but does not influence the antennae or the legs; *ss^a* ("aristapedia") results in the formation of a complete antennal leg with a complete set of tarsal segments and terminal claws (Figs. 570d, 580), while reducing the bristles, but causes the formation of a well-developed antennal leg and interferes with the organization of the tarsus in the legs (Fig. 580b, c). The allele with the most extreme effects, *ss^{all}*, produces the greatest change in all three phenotypes (Fig. 580): bristles are reduced to a minimum, tarsal segments are highly shrunken (Fig. 579f, g), and the antennae lose all trace of their normal form (Fig. 580d). The antennal disc develops into a leg with terminal claws; the elongated third antennal segment is formed into a tarsus, but it is not sharply distinguished from the tibial segment next to it (the legs also show this lack of normal differentiation), so that the tarsal reduction extends even into the antennal leg. Thus, an organ which is transformed into another now experiences the mutant changes specific to that organ. After the complete conversion of the development of the antennal field of the eye-antenna disc into leg development, it experiences the same influences that affect the thoracic limbs arising from leg discs. Still further changes are associated with certain of the alleles: erection and forking of individual bristles; held-out wings in *ss^{all}*, suggesting a reduction in the thoracic musculature; moreover, *ss^{all}* exhibits the greatly reduced fertility and viability associated with most of the mutant alleles at the *ss* locus.

The expression of the mutant phenotypes of the *ss* series is strongly temperature-dependent in most alleles. The greatest effects are generally achieved only when the animals spend their whole developmental lives, or a particular temperature-sensitive period, at low temperature; in this respect, however, the individual phenes hehave quite diversely with a given allele. Bristle formation is temperature sensitive in *ss^a*, but the transformation of the antenna into a leg takes place at all physiological temperatures. Only the "strongest" allele, ss^{all}, is completely independent of temperature. As for the other alleles, between 16° and 25°, the lower the temperature, the greater the abnormality. The antennae are normal in ss^{a-F} flies raised at 25°. The temperature-sensitive period begins in the last larval instar. The maximum expression is produced by treatment at 16° during the entire last larval instar. If development begins at 25° and third instar larvae are placed at 16°, the later they are transferred, the less effect there is. Conversely, the antennae are normal if the animals begin at 16° and are transferred to 25° at the beginning of the last larval instar. Thus, the course of development of the eye-antenna disc is fully determined in ss^{a-F} during the last larval instar. We have already seen from surgical experiments (p. 358) that regulation can take place within the mosaic of fields within this imaginal disc until this third instar. The use of chemicals and mutant genes shows that the organ-specific determination is still labile until then also. Thus, this developmental pathway is still labile until the onset of the third larval instar, 48 hours after the egg is laid.

The pleiotropic effects of the ss alleles raise a host of questions. First, is the effect intracellular or intercellular? Eye-antenna discs of 48-hour-old ss^{a-F} larvae were transplanted into normal (ss^+) larvae of the same age. Half of the host animals were placed at 25°, the other half at 16°. The implants in both the 25° and the 16° animals formed antennal legs. The genotype of the host, whose organ formation is not temperature dependent, thus had no effect on the differentiation of the implant; the abnormality therefore stems from events caused by the genotype of the cells in the imaginal disc itself. Here, however, unlike the *Ephestia* mutants with altered reactions in the scale-forming cells, it is not the reactive behavior of the individual cells, but rather an alteration in the pattern of determination in the blastema from which the imaginal organs arise. In the cells of the antennal disc, a common position on a spectrum between two functional conditions is assumed, and through this functional condition the further development of the field is determined. Displacement of this position can be caused either by genetic changes or as phenocopies by means of chemical agents.

The nature of the normal and altered developmental processes is unknown. The extreme pleiotropy of the alleles and the temperature-dependence of their effects suggest that the normal gene (ss^+) controls an event which is fundamental to the normal course of morphogenesis in several different theaters. It has been concluded from other cases of multiple alleles, where a variety of effects was observed, that one was dealing with a "complex locus," where mutations could occur in any one of a number of highly similar and closely related genes.

Perhaps a mutation at one of these closely cooperating genes might upset the coordination of developmental processes collaborating at one or more places or times.

A number of other organ-transforming ("homeotic") mutants are known in *Drosophila*. The mutation *pb* (proboscipedia) causes the proboscis to develop a leglike tip; expression is maximal when third instar larvae are kept at 25°. There can be no doubt that quite a variety of developmental processes in mutant flies show differential sensitivity to temperature, which leads to a greater or lesser degree of expression. The *bithorax* mutations (including *tetraptera* − four wings) comprise a number of alleles which participate in organ formation from imaginal discs at a basic level, as can be seen from the structural alterations they produce. The metathorax acquires the characteristic form of the mesothorax to a greater or lesser extent, so that the halteres become winglike. The mutants can be phenocopied if early embryonic stages of wild-type flies are treated at 35° (four hours, Fig. 581f) or with ether (Fig. 581 b–e, g, h). The maximum effect is achieved in both cases three hours after egg laying (at 25°). At this time the embryo is in the blastoderm stage. The frequency and degree of the effect vary considerably among the mutations as well as among phenocopies made under constant culture conditions with a standard temperature treatment. But the degrees of expression seen both in the mutants and in the phenocopies with these

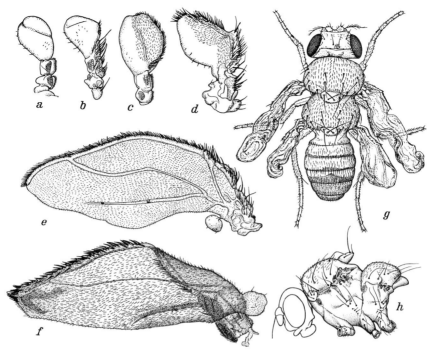

Fig. 581. Phenocopies of the mutant *bithorax* (*tetraptera*) of *Drosophila melanogaster*. a normal haltere; b–f degrees of transformation of the haltere into a winglike structure; g, h extreme modifications of the metathorax; g dorsal view; h thorax seen from the left, metathorax resembles normal mesothorax. Wings and legs have been removed; b–e, g, h after ether treatment; f after heat shock. (a–e, g, h after GLOOR, 1947; f after MAAS, 1948)

two disparate treatments are generally quite similar, and, indeed, the *bithorax* alleles have a temperature-sensitive period early in embryonic development. The infrequent occurrence (low penetrance) of the abnormality in mutants and in phenocopies might be due to a lasting ability to regulate the relevant processes, which take place so early in embryonic development, and which play a decisive role in such a short period of time, as we see from the short duration of the sensitive period.

A number of mutations at various loci which affect the formation of the sex combs in *Drosophila* are of interest with respect to the problem of the genetic control of the individual and the collective determination of certain organs which arise from the imaginal discs. The sex combs are rows of extraordinarily massive bristles on the front legs of males in various *Drosophila* species (Fig. 582d–k). It has been seen in mosaic individuals that male and female first tarsal segments both have a prepattern which permits the formation of sex combs, but only the genetically male epidermis responds to this prepattern with the actual formation of a sex comb. Males with three sets of autosomes and two X chromosomes form sex combs with fewer bristles than diploid males. These cells thus have an intersexual response to the prepattern. Gene mutations at various loci cause the formation of sex combs on the second pair of legs and even on the third pair of legs, although here the sex combs are generally tiny. These genotypes with extra sex combs have hind legs which are abnormally similar ro the front legs in the structure of their segments. So the question arises, do the mutations alter the competence of the cells to respond to local conditions by forming sex combs, or is there a change in the prepattern, the mosaic of determination, which spreads over the epidermis of the leg of the particular thoracic segment at a

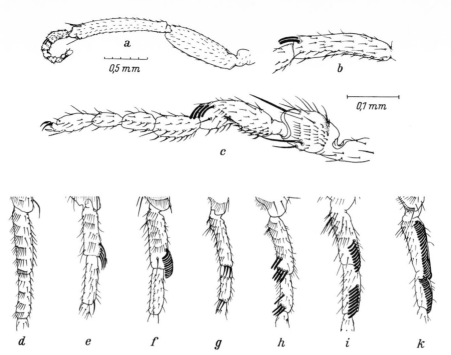

Fig. 582. a–c Homeotic antennal feet of *Drosophila melanogaster* with sex combs, heterozygote for *Antp^Yu* and for a dominant gene for an extra sex comb. a femur, tibia, tarsus whose first segment has a sex comb with two teeth; b greatly enlarged first tarsal segment; c end of the tibia and tarsus, sex comb on first tarsal segment with four teeth; d–k first and second tarsal segments of male forelegs of seven *Drosophila* species: d *ananassae;* e *affinis;* f *melanogaster;* g *helvetica;* h *lutea;* i *subobscura;* k *montium.* (d and h after STERN, 1954, who cites KIKKAWA and PENG, 1938; the others are originals of HANNAH and STERN, 1954)

particular developmental stage? In the first case, one would have to suppose the following: (1) that the sex comb prepattern is qualitatively similar in all the tarsal segments of all the legs; (2) that the stimulus for sex comb formation varies quantitatively and is strongest at the sites where sex combs are normally formed, in the lower part of the first tarsal segments of the first pair of legs; (3) that the tissue with the normal male genotype responds only to the highest intensity of this stimulus; and (4) that the mutations leading to extra sex combs lower the reaction threshold of the male cells so that they now respond also to weaker stimulus intensity at the prepattern. Certain mutant genes which cause extra sex combs on the legs also evoke antennal legs, and these also have sex combs in the males (Fig. 582a–c). This phenomenon makes it more likely that these mutant genes, like the *ss* alleles, alter the prepattern, perhaps by altering some fundamental determinative conditions in the eye-antennal disc and in the meso- and meta-thoracic leg discs, so that they approach the developmental path of the prothoracic leg disc.

The modification of the site and structure of the sex combs leads to a problem in evolutionary genetics: the structure of the sex combs is quite different from species to species in *Drosophila* (Fig. 582d–k) and serves as a useful systematic character. Sex combs may be lacking, they may be small or large on the first tarsal segment or on both the first and second tarsal segments, and the orientation of the "teeth" of the comb also varies. We might well suppose that the steps towards speciation are mutations such as those we have seen in *D. melanogaster* which alter the prepattern, or the responses of cells, or both.

Lecture 34

A remarkable number of mutants have rather insignificant effects on morphology but reduce vitality markedly. This shows that the outwardly visible traits, such as eye color, scale pigmentation, patterns, and number and arrangement of bristles, are only indicators of important alterations in underlying physiological processes. The slightest distortion of the dynamic balance of genetic activities can result in critical periods from which a certain number of individuals cannot escape, perhaps because the mutants are more susceptible to certain environmental conditions than the wild type; and this lability can be so great that in certain mutant strains only a few individuals survive to reproduce. At the end of this spectrum of factors reducing vitality are the mutants which make complete development impossible under any circumstances and lead their bearers to death before the reproductive stage is reached. These *lethal* genes make up the largest class of mutations; and this is true for spontaneous mutations, which arise under normal circumstances, as well as for mutations induced by irradiation and by treatment with chemicals. In *Drosophila*, whose giant salivary chromosomes provide a rigorous cytological control even for very small changes in the chromosomal architecture, it has been established that many lethal factors are in fact alterations of chromosomal structure. In *Drosophila*, the proportion of deficiencies, losses of discrete chromosomal regions among the lethal factors, is very high. Moreover, other chromosomal aberrations, such as duplications (repeats of discrete chromosomal regions), inversions (reversal of a region), and translocations (abnormal placement of a region in its normal chromosome or in a different chromosome), often have lethal consequences. But many other cases appear to be "point" mutations at sites established precisely by crossing over, with no microscopically visible alteration in the chromosome structure. Finally, some cases of back mutation are known, whereby lethal factors are restored to normal. In lower plants and animals, however, one can say nothing about the nature of lethal factors except that they exhibit Mendelian inheritance patterns, i.e., that they are on chromosomes. Since the lethal factors prevent or alter the course of some decisive developmental process, it is very important to spot the first effects of the lethal mutants during development: such observations can lead to conclusions about the action of genes which influence the processes fundamental to normal development. HADORN and his school have led this search with their broad and detailed studies on phenogenesis in lethal genotypes of *Drosophila*.

Many lethal factors are completely recessive. Others lower the vitality of a heterozygote or alter some trait which is not vital, like *Sy* in *Ephestia* (p. 445). This phenotypic alteration in the heterozygote may stem from a developmental disturbance having no clear relationship to the one that kills the homozygote.

There is no sharp line between lethal factors and those which reduce vitality. Often the environment, or the genetic background, makes the difference; and once in a while some individuals manage to complete their development in spite of a "foolproof" lethal genotype.

As a rule, development comes to a standstill or goes astray resulting in death during a definite critical phase characteristic of the particular lethal factor. Accordingly, one can distinguish embryonic larval and pupal lethal factors in the insects; and in addition to such monophasic lethal factors which cause death only during one critical phase, other mutants behave as biphasic or triphasic lethal factors. In Figure 583 the times of death are listed for 7 representative mutants out of a total of 59 nonallelic recessive lethal factors on the second chromosome of *Drosophila melanogaster* which were examined. In the case of monophasic lethal factors all the hymozygous individuals develop until the onset of a critical period in development.

In the case of embryonic lethality, dead embryos were found in at least 25% of the fertilized eggs resulting from a cross of heterozygotes (e.g., 26% in L37, Fig. 583). Only a few individuals in these cultures died in the larval or pupal stage, a number comparable to that found in normal cultures. Often the crisis associated with a lethal factor occurs during the transition between two developmental periods. For example, some of the individuals with a certain lethal genotype die within the egg as fully formed larvae; the rest hatch and then die during the first larval instar, as L12 in Figure 583 illustrates. Of the 59 lethal factors studied, a very high number (24) affected embryonic or very early larval stages. Once a larva had passed the critical onset of the first instar and began to feed and grow, further development often proceeded normally. Ten of the 59 lethal factors studied (e.g., L83, Fig. 583) fall into this category. Another group of

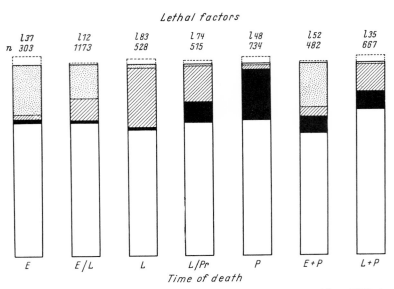

Fig. 583. Representative cases of embryonic lethality (*E*), embryonic-larval transition (*E/L*), larval lethality (*L*), larval prepupal transition (*L/Pr*), pupal lethality (*P*) and diphasic lethality (embryonic and pupal = *E + P*; larval and pupal = *L + P*). Above, numerical designation of the lethal factor. *n* = number of eggs examined from parents heterozygous for the relevant factor. Percentage of embryonic (stippled), larval (diagonally hatched), and pupal (black) mortality. The clear areas enclosed by the dotted lines at the top of the columns represent the proportion of unhatched eggs; these were not included in the total. The clear parts of the columns below correspond to the percentage of the eclosed adults (expected: 75 per cent = 25 per cent +/+ and 50 per cent *l*/+). (After HADORN and CHEN, 1952)

lethal factors is expressed at the end of the larval period shortly before puparium formation or during the transition to adult development. In a few cases, some of the larvae make the transition to the prepupal stage (e.g., L74, Fig. 583); the proportion depends on environmental conditions. If pupation is completed normally, the absence of the normal gene makes itself felt in the succeeding stages of metamorphosis (e.g., L48, Fig. 583), namely during adult development.

The activity of lethal factors is not only phase-specific, but also organ-specific; it always involves changes in particular organ anlagen or tissue systems. The effects are almost always diverse: for each lethal factor studied by HADORN (1945) there is a characteristic pleiotropic pattern of damage. The answer to the question of how *one* gene can influence the formation of a great variety of organs and tissues leads to important conclusions about developmental mechanisms.

Pleiotropy can be either an intracellular phenomenon depending upon a particular function in the cells of many different organs at a particular phase of development; or the termination of development of the whole organism, brought about by a local developmental disturbance in one system, preventing it from providing some requisite for the further development of a variety of other organs. In this latter, intercellular case, the further development of diverse parts is prevented, and we may call this phenomenon secondary pleiotropy. The distinction between these two mechanisms can be made by transplantation experiments.

We can consider a few examples from the great number of exhaustively analyzed lethal factors in *Drosophila*. Naturally, it is well worth knowing how the individual morphogenetic events of embryonic development are controlled by genes. But little is known about this — understandably, for the subtler changes can be seen only in histological sections, while the eggs belonging to the lethal 25 % can be discerned only when they exhibit gross abnormalities or stop developing altogether. Nevertheless, two cases of very early damage have been studied cytogenetically in detail: deficiencies near the left end of the X chromosome (Fig. 584) disturb fundamental processes of embryonic development. These deficiencies range in size from a single band to about 45; and in some cases the same defects have been observed in the absence of a visible deficiency in the chromosome.

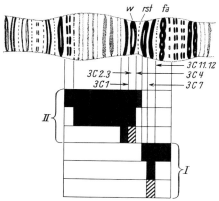

Fig. 584. Region of a deficiency in the X-chromosome of *Drosophila melanogaster* with an early embryonic lethal effect. Black: visible defect in the row of chromomeres. Diagonal hatching: genetic effect without visible alterations in the chromosomal structure. *fa*, *rst*, *w* sites of mutations with visible, but nonlethal, effects on the adults. (After the findings of a number of investigators, from POULSON, 1945)

Deficiencies in the *fa* (*facet*) region (Fig. 584, I) affect the partition of the ectoderm into particular cell types; far too many neuroblasts are formed from the ventral body and head ectoderm, so that very little ectoderm remains for epidermis and other ectodermal structures. The embryo develops a nervous system at least three times as large as normal. It remains exposed on the upper surface; it never organizes into segments, but it does undergo substantial histological differentiation. Epidermis appears only laterally and dorsally; tracheal branches are generally formed and later become normally chitinized. The foregut and hindgut, together with malpighian tubules, are generally present. The midgut anlagen do not unite, and the yolk thus remains uncovered. The mesoderm remains undifferentiated: muscle, heart, and fat body are not formed. Extirpation experiments on insect embryos (p. 333 ff) have demonstrated the dependence of the differentiation of the lower layer on the ectoderm (Fig. 427); so it seems likely that the primary developmental defect is the abnormal organization of the ectoderm in which the inductive effects normally emanating from the ectoderm are lost. This lethal effect begins when a large or even a very small piece of the chromosome is missing, or even when no deficiency is visible (Fig. 584, I); it would seem that a single gene is necessary for this normal morphogenetic step.

In the case of three deficiencies, all of which include the *w* locus (*white eye*, Fig. 584, II), the ectoderm remains normal, but the mesoderm and the endoderm are defective. This becomes evident between 12 and 16 hours after the onset of development. The organization of the mesoderm begins as muscle anlagen appear, only to degenerate. The same is true for the fat body and other mesodermal derivatives. The midgut does not develop into a long, winding tube, but rather remains a large undifferentiated sac, ostensibly because it acquires no mesodermal coat; the gut musculature is also missing. The lethal effect is associated with a gene or genes in the regions designated, according to the terminology of the salivary chromosomes, 3 C 1 to 3 C 2.3 (Fig. 584, II).

Embryos with chromosomal deficiencies embracing both the *w* and *fa* regions look exactly like those with *fa* deficiencies alone, so the gene in region I appears to work before the other (s). This is consistent with the fact that the critical organization of the ectoderm is completed before the consequent differentiation of the lower layer, which depends on still other genes in the course of its further development.

The factor *Kr* in the third chromosome of *D. melanogaster* also causes embryonic death when homozygous. In the heterozygote it causes thoracic malformations (Fig. 585) from which its name, *Kr* (*Kruppel, cripple*), is derived. Homozygous embryos never hatch. They do make the vigorous muscular movements associated with normal hatching, but cannot break out of the chorion, and survive only a few hours after the normal hatching time. The segmentation of these lethal embryos is extremely abnormal. The number of outwardly visible segments, normally nine (Fig. 586a_1–a_3), is reduced to five (Fig. 586b_1–b_3). The epidermis is histologically normal and includes the typical sense organs, hairs, and segmentally arranged sets of rows of chitinous teeth.

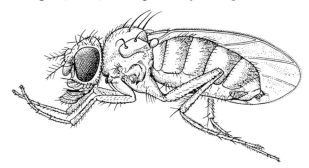

Fig. 585. *Kr*/+ heterozygote adult with thoracic abnormalities. (After GLOOR from HADORN, 1951)

The tracheal system is very defective (Fig. 586b_2, b_3). As to the internal organs, outside of the absence of malpighian tubules, only the nerve cord exhibits great abnormality: it begins to degenerate in the early embryo. The 10–15 hour *Kr*/*Kr* embryos are clearly abnormal in that few or no segments are defined; the tracheae are already irregular and incomplete, even as anlagen; and the neuroblasts that are present along the entire germ band do not assume the regular grouping corresponding to normal segmentation (Fig. 586c_1, c_2). In the central body region they form irregular large and small groups of cells which later degenerate (Fig. 586d_1, d_2). Normal ganglionic masses differentiate only at the front and hind ends. The segmentation of the ectoderm is thus deranged from the start, and the organization of the nervous system collapses in the thorax. When one compares the developmental defects in *Kr*/*Kr* embryos with the effects of the lethal deficiencies previously described, one can get an idea of the time of action of the mutant gene *Kr*: muscles, heart, fat body, and sex combs are normal in size and in histologic differentiation; the mesoderm has therefore differentiated. The organization of the perfectly functional body musculature reflects the abnormal outer organization. Thus, the ectoderm has not failed to induce differentiation in the mesoderm, in contrast to the cases involving deficiencies including the *fa* region. The axial organization of the ectoderm into segments, however, is disturbed before the segmental organization of the mesoderm, which depends on it, can take place. The degeneration of the thoracic groups of neuroblasts has no further influence on the differentiation of the internal organs; but in heterozygotes the disturbance of later organ formation in the thorax can be seen.

The halt in development of larval lethal mutants can have its roots in events much earlier than those which lead to the death of the mutants. A good example of this is the factor *lme* (*lethal-*

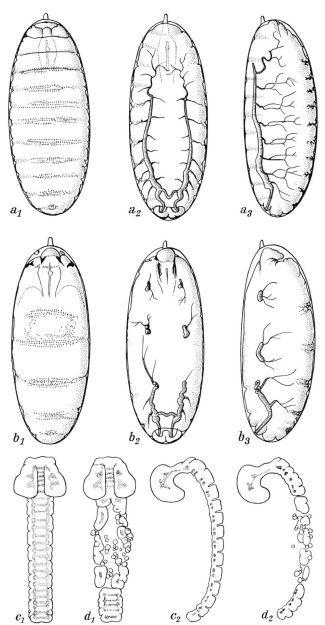

Fig. 586. Homozygous *Kr* mutants and normal controls in *Drosophila melanogaster*. a, b fully differentiated embryos before hatching: ventral, dorsal, and lateral views; a_1–a_3 normal; b_1–b_3 *Kr/kr;* c, d nervous system of 15-hour embryos, dorsal view (reconstructed) and sagittal section; c_1, c_2 normal; d_1, d_2 *Kr/Kr*. (After GLOOR, 1950)

meander) on the second chromosome of *D. melanogaster*, which causes disproportionate organ growth in the larva. Growth is inhibited in the mutant as early as the second larval instar (Fig. 587a). The molt to the third larval instar is generally completed, but the larvae reach

457

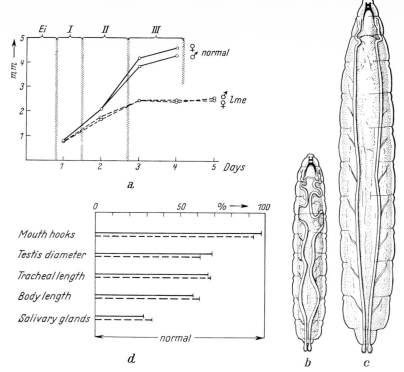

Fig. 587. Growth of homozygous *lme* larvae and normal controls of *Drosophila melanogaster*. a growth curves; b *lme/lme*; c normal controls; both 4 days after egg deposition; d relative organ sizes in the *lme* mutants (solid line) and in the phenocopies (dotted line). (After W. SCHMIDT, 1949, from HADORN, 1951)

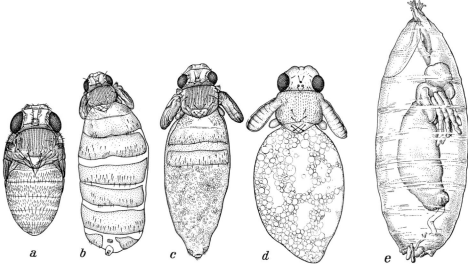

Fig. 588. Final stages in the development of normal pupae and homozygous pupae of the mutant *ltr* (*lethal translucida*) of *Drosophila melanogaster*. a shortly before eclosion; b imaginal differentiation of all the abdominal sclerites and of the genitalia; c metamorphosis of the first three segments and of the external genitalia; d pupa with completely undeveloped abdomen; e shriveled pupa in inflated puparium. (a–c after SOBELS and NIJENHUIS, 1953; d, e after HADORN, 1948, 1949)

only 50–60 % of the normal body length (Fig. 587a–c). The main tracheal trunks grow relatively more than the body as a whole, and so must follow a meandering course. The head is limp, and too large for the little larval body. The mouth hooks are as large as in normal larvae and therefore take up too much space in the little *lme* larva. Puparium formation is never completed; the animals never survive past the third larval instar. Transplantation of imaginal discs and gonads from lethal larvae into normal ones has an important result: the *lme* organs develop normally. Eye-antennal discs produce normal, pigmented eyes and often also antennae, together with head integument, including chitin and bristles, just as implants of normal discs or fragments (cf. p. 357, Fig. 458). The ovaries, which at the time of developmental arrest in four-day old *lme* larvae are reduced to 51 % of their normal size, and the testes, which are reduced to 69 %, reach their normal size in normal hosts and form eggs and sperm. The organs tested are therefore not under the immediate influence of the lethal factor. The *lme* pattern of abnormality can be phenocopied completely: if normal second instar larvae (40–48 hours old) are removed from the medium and placed on moist filter paper, the starved larvae exhibit exactly the same picture seen in *lme* mutants. The relative organ sizes agree completely between phenocopy and mutants (Fig. 587d). We can thus conclude that the *lme* phenomenon is simply a starvation effect. The *lme* larvae do continue to eat and apparently absorb the food from the gut; but from a certain stage on they are no longer capable of assimilating certain nutrients and thus starve in the midst of plenty. If two-day old larvae are starved, they consume the reserves in their fat bodies, which are completely resorbed. Normal larvae can rebuild the fat body when pure protein, pure fat, or pure carbohydrate is offered them. Tissue growth is naturally not possible on a pure fat or pure carbohydrate diet. Starved second instar *lme* larvae can also restore their fat bodies on a diet of carbohydrate or fat. But if only protein is given, the fat body remains in the starvation condition; so it appears that *lme* suffers primarily from a disturbance of protein metabolism which restricts the growth of organs that can develop normally in wild

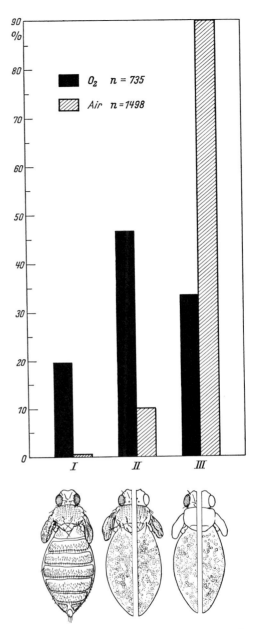

Fig. 589. Metamorphosis of homozygous *ltr* pupae of *Drosophila melanogaster* in air and in pure oxygen. *I* metamorphosis of head, thorax and abdomen; *II* metamorphosis of thorax and head, red eye pigment occasionally lacking; *III* no metamorphosis of the thorax. Generally no significant imaginal differentiation except for rare differentiation of the head. (After SOBELS and NIJENHUIS, 1953, condensed)

hosts. The hemolymph of *lme* larvae is low in amino acids. Gut homogenates from normal larvae digest casein, while those from *lme* cannot break it down. Thus, the starvation in *lme* appears to stem from the absence of a proteolytic enzyme made by the gene *lme*[+] and normally found in the gut[120].

Another mutation, whose effects appear very early, is *ltr* (*lethal-translucida*). The homozygotes can be distinguished from wild type even in the first larval instar. The larvae have reduced fat bodies and are slightly swollen. In the second and third larval instars this abnormality becomes more pronounced. An excess of hemolymph makes the animals bloated and quite transparent.

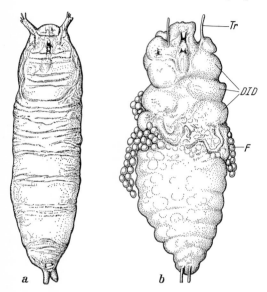

Fig. 590. Homozygous *lgl* prepupae of *Drosophila melanogaster;* a prepupae of an elongated puparium; b after removal; *DID* degenerating imaginal discs; *F* fat body; *Tr* tracheae. (After GLOOR from HADORN, 1948)

Biochemical studies show that the composition of the blood is abnormal in the mutants. It is more dilute than normal. Moreover, the protein content and particularly the globulin content is especially reduced. The puparium is formed about a day late so that the tanned larval integument lifts off and pupation generally takes place inside. At this point development stops in a large proportion of the *ltr* mutants. The animals survive, as respirometric studies show, quite a number of days beyond the time of puparium formation, but only rarely does an animal develop into a fly capable of eclosion. In some of the pupae, metamorphosis takes place in the head and thorax only, but is generally absent in the swollen abdomen (Figs. 588 d, 589); very rarely the epidermis of the abnormal segments metamorphose entirely (Fig. 588 b) or in part. As a rule, tergites are formed only on the most anterior segments, and the genital plates are also formed (Fig. 588 c); these are the parts that develop first in normal animals. The thorax is relatively well developed. Head development varies; often it is not as advanced as the thorax. Sometimes the head capsule is chitinized and the eyes are pigmented, while the thorax shows no adult differentiation (Fig. 589, III). The thorax is significantly smaller than in normal animals, and its musculature is clearly reduced. The excess body fluid appears in the space between the pupa and the puparium. The puparium is therefore turgid and swollen to an abnormal size. The reduced pupal body occupies a relatively small part of the interior (Fig. 588 e).

The proportion of *ltr* pupae that undergo a regional metamorphosis depends heavily on the environment. In pure oxygen, this proportion increases materially. While 90 % of the animals undergo no metamorphosis in air, almost half of the animals form an adult thorax in oxygen, and most of these also develop a head. About 20 % develop completely (Fig. 589). The oxygen is more effective during the first 50 hours of pupal life; then it falls off, and if the pupae are first placed in pure oxygen after 90 hours, their development is not improved over that in air. Accordingly, the failure of adult differentiation seems to rest in part on the insufficient availability of oxygen to the pupae in their abnormal fluid surroundings. But in about 80 % of the pupae, abdominal metamorphosis fails to occur even in oxygen.

Transplantation experiments show that the primary effect of the mutation is not on the imaginal anlagen, which remain undeveloped in the *ltr* pupae: eye discs, which barely grow to half the normal size when they are in regionally metamorphosing animals, do achieve normal structure and size when implanted into normal hosts during the second or third larval instar. Genital discs, which never metamorphose within *ltr* abdomens, carry out entirely normal development when implanted into normal hosts. Accessory glands and associated ducts form normally, and mature motile sperms are produced in the testes. These experiments show that, just as in the case of *lme*, a metabolic disturbance interferes with the program of the determined development of the imaginal discs. The accumulation of hemolymph of abnormal composition is surely an expression

Fig. 591. a Homozygous *frizzle fowl*; b, c heart of a normal hen and of a homozygous *frizzle fowl*. (After LANDAUER, 1942)

of this metabolic anomaly. But the site of the metabolic abnormality, the malfunctioning tissue whose activities are carried out in a normal host, has not been found.

In the mutant *lgl* (*lethal giant larvae*) the larvae exhibit no abnormality but remain in the third larval instar longer than normal, often up to fourteen days. Many die as mature larvae; some of the mutants form an elongated puparium (Fig. 590a); but adult development never begins. While most of the internal organs, the gut, the musculature, the somatic cells of the sex organs, the nervous system, the tracheae, and even the larval epidermis are not affected, the imaginal discs are either retarded or degenerate. If the puparium is opened, parts of the fat body are exposed, since no adult epidermis has been formed from the appropriate imaginal discs (Fig. 590b). The ring gland, the secretory center of the fly, is abnormally small in *lgl/lgl* prepupae. The implantation of a normal ring gland causes a mutant larva to form a puparium. Metamorphosis cannot proceed further, however, since the ectodermal imaginal discs cannot develop, due to an autonomous genetic defect which cannot be corrected by transplantation into a normal host.

Among the vertebrates there are also numerous examples of the activities of mutant genes which upset certain developmental processes. We shall select only a few, particularly from mutants in chickens[322] and in mice[227], on which many excellent studies have been made.

A mutant of the domestic chicken, worldwide and known for centuries, and a characteristic mutant phenotype in other birds as well (pigeons, geese, canaries), is *frizzle fowl*, whose feathers are twisted or curled (Fig. 591a). The factor is expressed more strongly in homozygotes than in heterozygotes; the down is retained longer than usual, and after its loss during the first year, the homozygotes remain nearly bald. The little hooks on the mutant feathers are either missing or

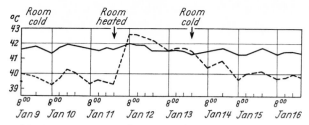

Fig. 592. Influence of the ambient temperature on the average body temperature of two normal hens (—) and two homozygous frizzle fowl (----). (After LANDAUER, 1942)

malformed. The frizzles are more sensitive to the weather than normal chickens; the homozygotes are hard to raise, and if they survive they are frequently sexually immature. If eggs are laid, they often do not develop or else produce chicks that cannot hatch. This defect clearly does not stem from the genome of the zygote but rather from an insufficiency of yolk. The sperms are perfectly good. Extreme frizzle fowls have an abnormally high basal metabolism, ostensibly due to their excessive heat loss, which results from their abnormal plumage. This is illustrated clearly by the fluctuation in their body temperature as a function of ambient temperature (Fig. 592). Food consumption is substantially higher than normal. A secondary trait in frizzle fowl is increased

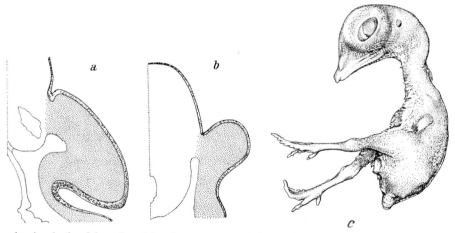

Fig. 593. a, b wing buds of four-day-old embryos. a normal; b *wingless* homozygote; c eleven-day old *wingless* homozygous embryo. (After ZWILLING, 1949)

heart size (Fig. 591 b, c). An increase in the circulation is a compensation for the heat loss, and the demands on the heart lead to a functional hypertrophy. The mutant gene acts autonomously right in the feather rudiments, and its action is not influenced by substances carried to it in the blood from the rest of the body; skin from the back of a frizzle fowl transplanted to a normal host still makes frizzled feathers at every molt, and, conversely, normal skin implanted into a frizzle fowl never forms abnormal feathers.

A recessive lethal mutation with a very diverse and noteworthy organ-specific pattern of defects is called *wingless*. The absence of wings is the most striking outward trait of the homozygotes, which undergo a great deal of development but cannot hatch (Fig. 593c). In addition, the down feathers are abnormal, and internally the lungs, air sacs, metanephros are missing. The

462

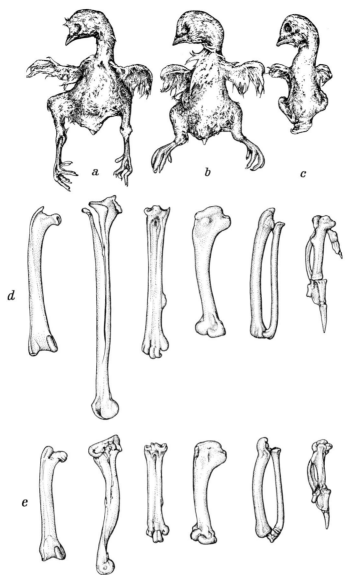

Fig. 594. Mutant *Creeper* (*Cp*) of the domestic fowl. a +/+ ; b +/*Cp*; c *Cp*/*Cp;* d, e long bones; d normal fowl; e +/*Cp* fowl brother of the normal one. Bones from left to right; Femur, Tibia-fibula, Metatarsus, Humerus; Ulna-radius, Metacarpus. (After LANDAUER, a–c 1939; d, e 1931)

legs may form normally or abnormally. In the third day of embryonic development, when the wing buds normally appear, the *wingless* homozygotes can be distinguished. Their wing buds are smaller and lack the characteristic epithelial thickening of normal wing buds (Fig. 593a, b). Experiments have shown that the epidermis has an essential effect on morphogenesis in the mesenchyme (p. 283, Fig. 361f, g). So it is very likely that the failure of wing development can be ascribed to the defective epidermis which fails to induce normal mesenchymal differentiation. As a result, only irregular and insignificant mesenchymal condensations form in the little wing appendages; an apical epidermal thickening never forms. The scapula and coracoid form normally, as they do in wing rudiments whose wing buds have been snipped off (Fig. 361 e). The wing rudiments remain in the four-day stage, and the endodermal epithelial buds do not grow out, while the mesodermal parts appear normal. In addition, the weaker buds do not form at the base of the Wolffian ducts, and the metanephros-forming tissue remains undifferentiated. The failure of metanephris kidney formation is clearly caused by the absence of ureters; for experiments have shown that the differentiation of the metanephric blastema is triggered by the arrival of the

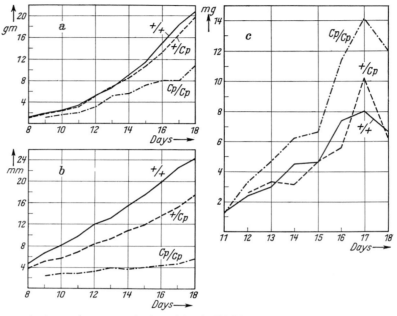

Fig. 595. *Creeper* mutant growth curves. a body weight; b tibial length; c spleen weight. (After LANDAUER, 1939)

ureter. The epithelial layers are involved in all of the diverse defects in the wingless phenotype: bud regions do not grow out normally, and the absence of normal epithelial influences results in the lack of wings and metanephric kidney. It is interesting to note that no single germ layer is affected here; instead, epithelial differentiation is inhibited in the ectodermal, mesodermal, and endodermal layers.

The mutation *Cp* (*Creeper fowl*) produces disproportionate body growth. In the heterozygotes (Fig. 594b) the long bones of the extremeties are too short and sometimes also bent (Fig. 594e); the homozygotes die as embryos, mostly at the end of the third day or the beginning of the fourth. A few breakaways survive this critical period and develop to the end of incubation. They are cripples with shrunken extremities (Fig. 594c) and other abnormalities, especially in the head, particularly the eyes. They never hatch. Even in the heterozygotes hatchability is somewhat reduced.

In the heterozygotes the increase in body weight up to the time of hatching is about normal, but the wing bones grow distinctly slower (Fig. 595b) and show histological abnormalities. From the seventh day on, they display defects in their cartilage differentiation. The periosteal bone formation is delayed at first, but soon catches up with, and even goes beyond, the normal schedule. Endochondral ossification at the ends of the diaphyses begins at the normal time but falls behind rapidly, becoming more marked as the chick grows. The endochondral bones of normal animals consist of a framework of strong trabeculae whose spaces are packed with delicate smaller trabeculae (Fig. 596a). In the *Creepers*, the spongiosa consists only of a few thin trabeculae of the small type in a large mesh net of connective tissue (Fig. 596b). The greater the normal length of a bone, the greater its relative shortening in *Creepers* (Fig. 594d, e). The gross and histological abnormalities of the long bones of *Creeper fowl* agree in all essential points with the symptoms of human chondrodystrophy.

In homozygous *Creeper* embryos, growth and differentiation slows markedly after the 72-hour stage, and the various parts of the embryo are retarded in differing degrees. The head region of the embryo, which normally is developing vigorously at this time, is slowed down most of all. The eye rudiments, which are forming at this time, reach a relatively small size. In this early critical period, then, development comes to a general standstill without any marked disturbance

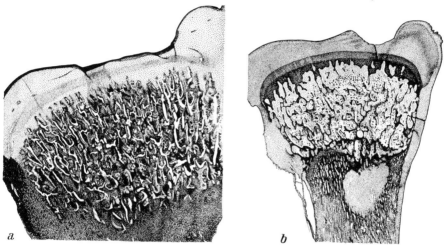

Fig. 596. Distal ends of the humerus of a three-week-old chick. a normal; b $+/Cp$. (After Landauer, 1931)

of organ differentiation. In the few Cp/Cp embryos that develop further, the skeletal components of the extremities slow down in their development at the time of cartilage formation (Fig. 595b). Radius and ulna are more or less fused, as are the tibia and fibula. Cartilage differentiation always remains incomplete. In the epiphyses of the long bones the cartilage cells do not take up their normal arrangement in columns. What little ossification there is now takes place. The scleral cartilage of the eyes is almost completely missing. But not all organs are permanently retarded: the heart becomes smaller than in normal and heterozygous individuals, but then it grows faster, and in spite of the much smaller body of the Cp/Cp embryos, the heart actually becomes larger than normal hearts. The spleen (Fig. 595c) undergoes extraordinary hypertrophy, outgrowing those seen in normal and heterozygous individuals, both in relative and in absolute terms. Due to the defects in the long bones, no bone marrow is formed, and thus the main site of blood cell formation is absent. The spleen compensates for this in part, but in spite of its enlargement it cannot fully replace the missing erythropoietic centers, and anemia sets in, which places increasing demands on the circulation. In the array of abnormalities associated with a lethal mutation, then, we can recognize functional adjustments, although these fail to save the individual's life.

Phenocopies of the *Creeper* mutant can be induced in $+/+$ embryos by various chemical means. Selenium in the food of the laying hens and insulin injected into the yolk sac are two. Deformities in the range seen for heterozygous and for homozygous creeper embryos can be produced according to dose. Apparently, the same fundamental growth process is affected whether the mutant gene or the phenocopying chemicals are involved. The chondrodystrophy of the heterozygotes and the extreme crippling of the homozygous *Creepers* are clearly caused by a general inhibition of growth expressed to different degrees in different tissues; and the death of the *Cp/Cp* embryos results from a disturbance of the normal morphogenetic and functional relationships.

To what extent this inhibition of growth is rooted in the cells of the individual tissues is not yet clear: in tissue cultures of heart fibroblasts of homozygous embryos there was much less growth an in +/+ and +/*Cp* tissues (Fig. 597). Limb buds from +/*Cp* embryos grew into typical chondrodystrophic creeper legs when transplanted to the side or onto the coelom of normal embryos. Wing and leg buds from *Cp/Cp* embryos, whose genotype could be recognized at the beginning of the critical embryonic period, survived in normal hosts beyond the death of the donor. They are extremely abnormal, however, and the best of them resembled the miserable structures of *Cp/Cp* breakaways (Fig. 594 c). The *Cp* allele is therefore directly involved with cartilage formation in the limb rudiments. On the other hand, *Cp/Cp* eye rudiments transplanted into the head of normal embryos reach their normal size; *their* developmental disturbance is therefore secondary (allophenic), and results from the retardation of skull development.

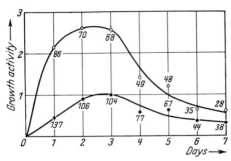

Fig. 597. Growth activity curves of heart tissue culture. −∘− normal embryos; −•− *Cp/Cp* embryos. (After DAVID, 1936)

The mutation *grey lethal* (*gl*) in the house mouse does *not* increase embryonic lethality: the *gl/gl* animals appear in a normal monohybrid F_2 ratio. But during the second week of life, they lose weight (Fig. 598 a), and between the twenty-second and thirtieth day they die. The young *gl/gl* mice are abnormally proportioned. Their skulls are short and flattened, and their jaws are not properly formed (Fig. 598 d). Their limbs are relatively short and crude (Fig. 598 c). Their teeth do not break through their gums. The mutants can eat no solid food, and their weight loss begins with weaning. The little runts can be kept alive up to 42 days by the use of liquid or powdered food. Nevertheless, they begin to lose weight; the lack of teeth, which imposes a certain inefficiency on the chewing process is thus not the only cause of their death. Histologically, the main symptom is a general bone anomaly, first seen in the period of secondary bone formation: ossification stops at an early stage, the primary endochondral spongiosa develops no further, and neither marrow spaces nor a heavily calcified perichondral bony layer appear (Fig. 598 e). Dentine formation is quite incomplete in the teeth, and the roots do not calcify. The formation of the enamel by the epithelial secretory organs is not disturbed, and a well-formed enamel layer arises without its normal substratum of dentine (Fig. 598 k, l). The mutant *gl* provides an excellent example of the assembly of parts that takes place in the development of teeth. The defective cartilage development is not an autophenic malformation of the mesenchyme cells. This is shown by transplation experiments: ribs from *gl/gl* embryos in normal hosts resorb the endochondral spongiosa almost normally. The formation of the bones and teeth depends upon the parathyroid hormone. Indeed, high doses of parathyroid hormone induce spongiosa resorption in *gl/gl* mice. In combined transplants of a *gl/gl* parietal bone plus an extra normal parathyroid gland local bone resorption takes place in normal hosts. But a *gl/gl* parathyroid works just as well in such an experiment. Thus, it is not the function of the parathyroid as such which is affected by the mutations; rather, *gl* causes the production of an inhibitor of the parathyroid hormone[16]. We do not know its source, nor do we know the cause of the complete lack of yellow pigment in the fur of the *gray lethal* mice.

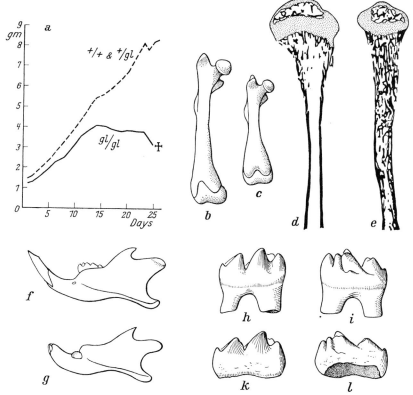

Fig. 598. *Gray lethal* mutant in the house mouse. a growth curves of normal $+/+$ and $+/gl$ and gl/gl mutants; b, c femurs of 28-day-old-mice; b normal; c gl/gl; d, e longitudinal sections through tibias of 23-day-old mice; f normal; g gl/gl; h–l left first molar of the lower jaw; h, i normal; k, l gl/gl; h, k medial view; i, l lateral view. (After Grüneberg, 1935, 1936, 1937)

Now the recessive gene in the mouse *dw* (*dwarf*) is characterized by the intracellular abnormality of one specific tissue: the development of the eosinophil cells of the anterior pituitary is inhibited. The lack of the hormone normally produced by these cells results in the retardation of growth of development of the sex organs and other morphological and physiological disturbances. Daily injection of anterior pituitary extracts from normal mice supports normal growth and leads to sexual maturity.

Lecture 35

As we pose questions about development, we must never lose sight of evolution. Many hereditary factors participate in the development of each organ; for example, more than 40 loci are known, at which mutation interferes with the normal development of the *Drosophila* eye. Many mutations are pleiotropic, another indication that their normal alleles are involved in the formation of numerous organs. The many genetically controlled malformations, including those leading to death, show the danger of mutation for the delicate spatial and temporal interplay of the genetic effects in the course of cell replication, induction, cell movements, and differentiation in structure and function at the cell level and above. And yet, mutations are the building

467

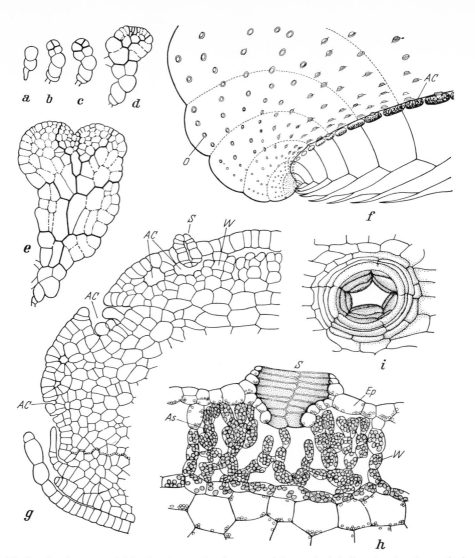

Fig. 599. Thallus development of *Marchantia*. a–e development of the vertical thallus showing the vertical cells the marginal growth in *Marchantia polymorpha;* f diagram of the thallus growth with initials (permanent vertical cells) in the edge of the cap, vertical longitudinal section and view of the right lobe of the thallus; g vertical longitudinal section; border between ventral and dorsal segments dotted, development of the air chambers; h section through an air chamber of *Marchantia planiloba;* i view of the opening of the air chamber of *M. planiloba*. *As* assimilators; *Ep* epidermis; *AC* air chambers; *O* their openings; *S* sleeves; *W* walls separating the air chambers. (a–e after MENGE; f–i after BURGEFF, 1941, 1943)

blocks of speciation; we know no others. In every evolutionary change, selection must establish a new harmonic balance; unbalanced ontogeny is eliminated.

Population genetics is concerned with the relative adaptive values of particular mutants and with the frequency of their occurrence and combinations. But between the alteration of genes and the final phenotype acted upon by selection lies a complex mechanism of genetic effects. A mutation in most cases causes a disturbance of the existing balance. A displacement of the balance of metabolism relative to the surrounding conditions can in some cases be advantageous immediately; it can be an adaptation to particular climatic conditions in which an organism can now take hold, or which it can enter for the first time. As to mutations which lead to morphological changes, including those which distinguish taxa at various levels, the question always arises: how is the organism capable of establishing a new developmental balance leading to the substitution of new traits for old? How are the altered component processes woven into a new developmental fabric?

Paleontology shows us only phenotypic remains; but living plants and animals provide us with many informative examples of speciation right before our eyes.

As for direct experimental studies of evolutionary steps from the developmental point of view, we can see no possibility of creating new genera. The analysis of crosses only rarely goes beyond species borders. We are thus led to consider models.

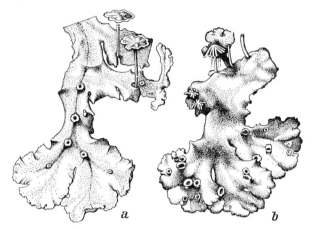

Fig. 600. *Marchantia polymorpha.* a ♂, b ♀ thallus in transition from the sexual to the asexual vegetative condition with gemma cups. (After BURGEFF, 1943)

In plants, mutations are known which change the structural and functional plan to such an extent that certain organs resemble those found in other genera. We will choose a set of examples from *Marchantia*, from the extensive studies of BURGEFF[100].

The typical course of development of the thallus of this liverwort is illustrated in Fig. 599. The first division of the spore separates a green cell from a basal colorless rhizoid, which divides further, lengthens, and anchors the developing plant (Fig. 599a). The green cell divides three more times, forming a four-celled plant (Fig. 599b). One of the upper cells divides obliquely, forming a daughter cap cell (Fig. 599c), which gives off further cells both to the left and right. These cells grow vigorously, and they divide periclinically and anticlinically (Fig. 599d, e). The anterior cells now grow rapidly at their margins, so that two lobes emerge laterally from the cap cell, which gradually sinks into a deep depression (Fig. 599e). Later, the thallus becomes dorsoventrally symmetrical, while the cap cell now gives off dorsal and ventral daughter cells alternately (Figs. 599f, g, 605a). These divisions in turn alternate with longitudinal divisions, resulting in walls lying in the median plane and parallel to it, and forming a cap of identical initials. The ventral cells form scales and rhizoids on the under surface of the thallus, and the dorsal cells form parenchyma with typical air chambers and the other usual structures (Fig. 599f, g). As this two-dimensional growth continues, new initial caps are formed laterally and the thallus branches dichotomously (Fig. 600).

469

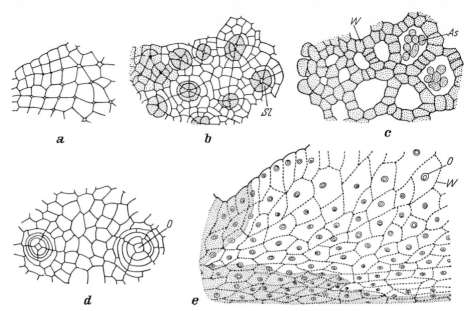

Fig. 601. *Marchantia planiloba;* a, b, d development of the air chambers from above; c optical section through the chambers in the same place as b; e top view of a piece of thallus of a white variegated mutant, *chlorina*, of *M. hybrida*, distribution of chambers. *As* assimilators; *O* chamber openings; *Sl* sleeve initials; *Zw* dividing wall.
(After BURGEFF, 1943)

In the formation of the air chambers, the cell walls of four or five superficial cells separate (Figs. 599 g, 601 a), so that a cavity forms. This extends deeper and deeper between the cells and spreads under the upper surface, the epidermis. Around the narrow openings of the chamber anlagen, a ring of crescent-shaped cells is laid out (Fig. 601 b, d). Then the ring cells divide, forming walls parallel to the surface, so that a barrel-shaped tube is formed (Fig. 599 g–i). Between the air chambers only the walls of one layer of cells remain (Figs. 599 h, 601 c). The cell ring which forms the tube does not arise in the same way as stomata (p. 403), from the divisions of a single stem cell; rather, the tube is formed as part of a pattern of differentiation which extends over the superficial cell layer. In variegated mutants, the borders between white and green regions can run right through the middle of these tubular openings (Fig. 601 e). From the floors of the air chambers rise assimilators, branched cell threads with numerous chloroplasts (Fig. 599 h). In the epidermis, in the dividing walls, in the chamber floors, and in the deeper tissue layer, the chloroplasts are sparser and generally also much smaller (Figs. 599 h, 602 a).

This typical form of *Marchantia* is altered by a number of mutations. We shall consider only a few of these, which produce forms similar to other natural species. The inlet tubes can be reduced to various degrees. The mutant *bavarica* of *Marchantia polymorpha* has none at all (Fig. 602 b). In the mutant *subtilis*, the opening is unusually wide. The epidermis follows the growth of the thallus incompletely and is torn above the separate chambers. In the mutant *dumortieroides*, the epidermis and the intervening walls do not last long, and the floor of the chambers becomes the upper surface of the thallus. The mutant *picturata* of *M. hybrida* (Fig. 603 a) forms chambers filled with green parenchyma throughout large portions of the thallus. These chambers have no opening to the outside, and in transmitted light they appear bright next to the dark air-filled chambers. In the mutant *hyala* of *M. stenolepida*, the thallus has very few chambers (Fig. 603 b). In the wild type, the epidermis which covers the chambers contains poorly developed,

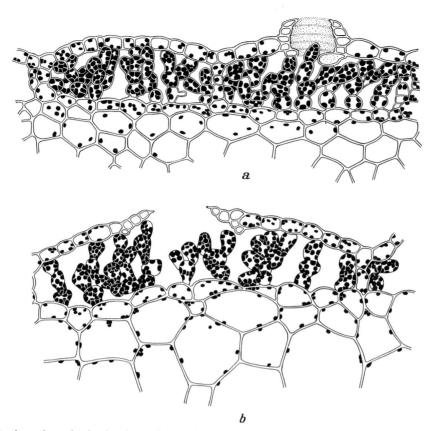

Fig. 602. Sections through air chambers of *M. polymorpha;* a *typica* (= wild type); b mutant *bavarica*. (After BURGEFF, 1943)

almost colorless chromatophores in small numbers, which let the light pass through to the assimilators (Figs. 599 h, 603 c). With its few chambers, *hyala* has more and larger chromatophores in its epidermis (Fig. 603 d); and the almost chamberless combination *M. hybrida picturata-hyala* bears assimilators on the original epidermis, although they are normally present on the chamber floor (Fig. 603 e). These mutant traits correspond to genus characters for the Marchantiales which, according to GOEBEL, have become progressively simpler in their descent from a highly differentiated type corresponding more or less to the present genus *Marchantia*. In the genus *Fegatella*, the inlet tubes are missing; in *F. conica* they are simple openings, as in many *Marchantia* mutants, for example, *bavarica*. The genus *Cyathodium* has chambers with large, simple respiratory pores. The successive steps of chamber reduction illustrated by the mutants *bavarica*, *subtilis*, and *dumortieroides* lead to the genus *Dumortiera*, in which a network of chambers is laid down, only to degenerate, leaving the assimilators exposed. Another route to the loss of chambers leads through the mutant phenotypes such as *picturata*, *hyala*, and *picturata-hyala*, to the *Monosolenia* types. Both in mutant and phylogenetic chamber reduction the epidermis replaces the typical assimilation tissue in the air chambers; its chlorophyll content rises, and it is covered with the papilliform assimilators that are generally found elsewhere. One tissue disappears, and another takes over its function and undergoes similar differentiation. In terms of the total organization these changes are reductions, but for the epidermis, they represent a progressive transformation —

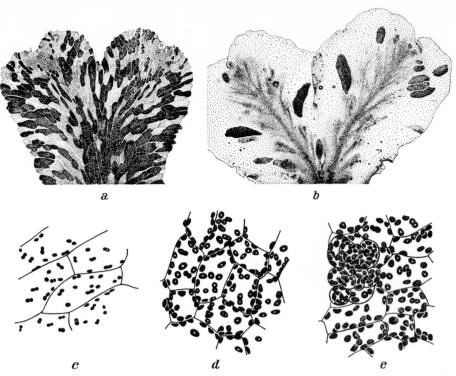

Fig. 603. a, b *Marchantia hybrida*, part of thallus; a mutant *picturata;* b *ultrapicturata;* c–e epidermis seen from above; c, d *stenolepida;* c *typica;* d mutant *hyala;* e *hybrida picturata-hyala* (*vitrea*) with multicellular assimilators on the epidermis. (After Burgeff, 1943)

an excellent example of a regulatory compensation within the norm of reaction, when a particular developmental function is lost. One *Marchantia* mutation demonstrates the appearance of a new trait, which can be called "progressive"; for it makes possible a mode of growth in the liverwort analogous to that seen in the lycopodiums and ferns. In the mutation *blastophora*, unusual cylindrical "blastomes" grow up out of the thallus lobes. They reach a height of 6–8 mm. and a diameter of 1–1.5 mm. These blastomes emerge from a sort of branching of the thallus in the vertical direction (Fig. 604b). In the normal cap region, ventral and dorsal cells are given off alternately (Figs. 599f, 605a). In the mutants the dorsal part of the thallus is thick and forms a second vegetative cap (Fig. 604b), which has a different structure from the primary one which continues to grow below. Since it comes from the upper cells, it forms only assimilation tissue. At the tip, there is a groove rather than a row of initials (Fig. 604c, e). In the groove, tetrahedronal initials are responsible for further growth (Fig. 605b). They do not give off daughter cells laterally in alternation, as do the initials of the primary growing points (Figs. 599f, 605a); rather, mutually perpendicular, vertical walls are formed in alternation with horizontal walls, the latter resulting in cells budding off basally (Fig. 605b). This results in radial growth. The blastome is surrounded by parenchyma containing air chambers, and inside it is pervaded by columns of conducting cells (Fig. 604c). *Marchantia* is thus enabled by a mutation to construct a distinct new organ from its dorsal tissue, which follows, in its growth, the rules for shoots. At times, the blastomes fork: the growing point extends lengthwise in the thallus (Fig. 604f) and pinches in two or rarely in three, and free blastomere branches continue to grow (Fig. 604d). The blastomes can enter a

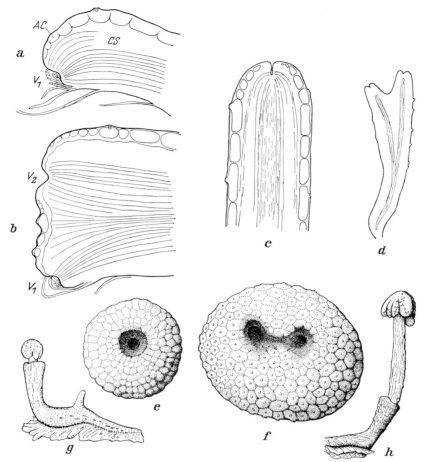

Fig. 604. Mutant *blastophora*. a, b longitudinal section through the thallus; b apical dichotomy; c, d longitudinal sections; c through a simple blastome; d through a bifurcated blastome; e, f top view of a blastome cap, vegetative groove, pattern of air chambers; f onset of bifurcation; g, h blastome with archegonial branches. *AC* air chambers; *CS* conducting strands; V_1 primary, V_2 secondary vegetative grooves, branched upward. (After BURGEFF, 1941, 1943)

vegetative or sexually reproductive phase. On top of the blastome cup, ring-shaped gemma cups sink inward, or else male or female sex organs arise whose structure is radial in contrast to the dorsoventral form of the normal sex organs on the thallus surface (Fig. 600). At the tip of the blastome a knoblike swelling forms, grows, becomes lobed, and rises up on the stalk (Fig. 604 g, h). Through a single mutational step, $+ bl$, a new organ is formed and a substantial alteration of symmetry takes place during differentiation. "For a phylogenetically organized genetics ... the exciting possibility exists of using combinations of mutants in order to develop models of new organisms and to test their fitness in the struggle for existence, as well as their similarity and their relationship to natural forms" (BURGEFF[100, 2]). In the case of animals such mutation models cannot lead to phylogenetic classification on the basis of genetic experiments. But experiments in development can answer this question: how does an organism come to terms with disparities in the gene-controlled differentiation in its various embryonic parts? These disparities have been produced by exnoplastic transplantation.

473

In the Amphibia, embryonic regions can be brought together from phylogenetically and morphologically highly diverse species with divergent development. As we have already seen (e.g., p. 246f, 257), one can transplant anuran material into urodele embryos and raise these chimaeras of two different orders into larvae. Such combinations relate phylogeny to development; they show to what degree general and specialized components have contributed to an organ during its descent. The anuran and urodele larvae are very similar with respect to some organs (brain, spinal cord, eyes, labyrinth) but very different in others. To express the question phylogenetically, how well can an altered part of a system of anlagen serve the unaltered remainder?

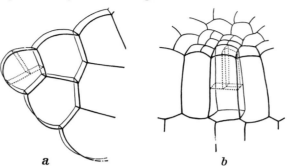

Fig. 605. Diagrams of cap growth. a cap stem cell of the cap surface of the thallus in longitudinal sections (cf. Fig. 599f, g); b initials in the blastome cap. (After Burgeff, 1941)

We never know in this type of experiment what differences are present in the individual characteristics of the united body parts, whether they differ in many or few genes, and to what extent cytoplasmic differences interact with the norm of reaction.

The modern urodeles and anurans have developed along separate pathways since the Paleozoic. This can be seen in fossil adults and is evident from the present larval forms. The phylogenetic distinction between the two amphibian orders is not expressed equally in all organs, however. Thus, the ear labyrinth anlage region can be removed from a *Triturus* larva and replaced unilaterally or bilaterally by a *Bombinator* labyrinth without the ultimate loss of normal movements or orientation reflexes; thus, the labyrinth can be replaced by one from an entirely different order.

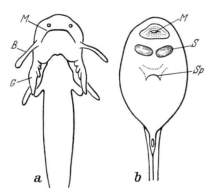

Fig. 606. Position of the adhesive organs in *Triturus* (a) and *Bombinator* larvae (b). *B* balancers; *S* sucker discs; *K* jaw; *M* mouth; *Sp* spiracle. (After Chen and Baltzer, 1954)

So, during the course of evolution, components of development are transmitted unaltered, while species and higher taxa diverge. On the other hand, the larval adhesive organs differ greatly between the anurans of the urodeles (Fig. 606). Anuran larvae have flat, epithelial secretory discs with an inner layer of ordinary epithelial cells and an outer layer of columnar secretory cells which liberate a sticky substance. The larvae of *Triturus* and other urodeles have long, stiff "balancers." These are epithelial tubes with terminal gland cells and loose vascularized mesenchyme inside, derived from the neural crest (p. 243). The mesenchyme produces a supporting layer under the epidermis which stiffens the tube (Fig. 607c). The sites of formation of the two organs are clearly different: the secretory discs lie medially, somewhat behind the mouth (Fig. 606b). The balancers of the urodele larvae arise laterally between the eyes and

474

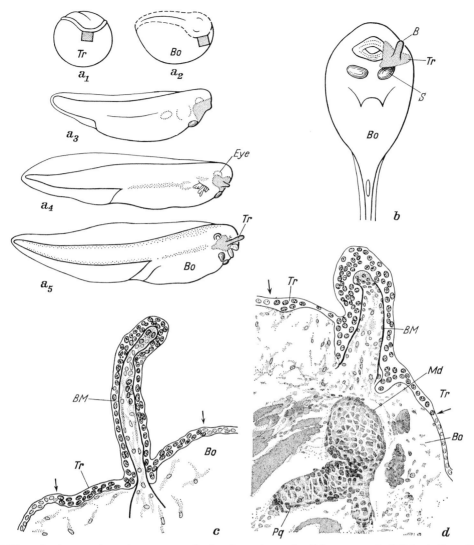

Fig. 607. Xenoplastic transplantation: presumptive head ectoderm of a *Triturus* neurula (*Tr*) is vitally stained and transplanted into a corresponding region of a *Bombinator* neurula (*Bo*). a, b sketches from life; c, d longitudinal sections through chimaeric balancers; *BM* basal membrane; *Md* mandibular bone; *Pq* Palatoquadrate, arrows indicate margins of implant. (After CHEN and BALTZER, 1954)

the upper jaw; in the mature larvae they are located just behind the corners of the mouth (Fig. 606a). Xenoplastic transplants can show to what extent the two types of adhesive organs can be induced in the organization of the head of a larva of a different order, and to what extent they can be formed from cell material of another order.

Presumptive balancer ectoderm from a young *Triturus* neurula was transplanted to a corresponding site of a *Bombinator* neurula (Fig. 607a$_1$, a$_2$). The implant healed over and spread, during the growth of the embryo, toward the mouth between the eyes and jaw region (Fig. 607a$_3$). Here, a chimaeric balancer arose (Fig. 607a$_4$, a$_5$, b): the epidermis was pure *Triturus* material;

the axial mesenchyme came from the *Bombinator* host (Fig. 607c, d). It formed a typical *Triturus* organ beneath the *Triturus* epidermis. When the balancers arose near the edge of the implant, host epidermis was often incorporated into the growing tube. The more *Bombinator* epidermis incorporated, the more restricted was the growth of the chimeric balancer. Thus, the *Triturus* epidermis is the tissue which is critical for normal balancer development; it utilizes *Bombinator* mesenchyme and causes the formation of a typical basal membrane. That the latter comes from the mesenchyme can also be seen from the extension of a "basal funnel" inward from the base

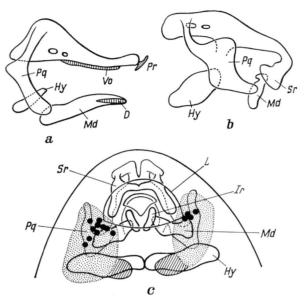

Fig. 608. Head skeleton of larvae. a *Triturus;* b *Bombinator;* c *Bombinator* ventral view. Implant region stippled and sites of chimaeric balancers indicated by black circles. *D* dental; *Hy* hyal; *Ir* Infraostral; *Md* Mandibular; *L* lips; *Pq* palatoquadrate; *Pr* premaxillary; *Sr* suprarostral; *Vo* Vomer. (a, b after WAGNER, 1949; c after CHEN and BALTZER, 1954)

of the balancer (Fig. 607c, d). The xenoplastically assembled balancers in these chimeras do not occupy the normal *Triturus* position laterally on the head behind the eyes (Fig. 606a), but, rather, a place on the ventral surface of the *Bombinator* head near the little mouth (Fig. 607b) and, in large implants, extending into the region between eyes and jaws. This suggests that the site of the balancer is derermined by the mesenchyme which forms the mandibular arch (Fig. 607a, d) and, to be sure, the inducing region is situated where the mandibular and palatoquadrate elements meet. The site of this joint in *Bombinator* is displaced forward relative to *Triturus* (Fig. 608a, b), corresponding to the smaller mouth and shorter mandibular in *Bombinator*. The inductive mechanism is the same, but the topographical change in the inducer causes a difference in the localization of its effect on the development of the implant.

Chimaeras can also be produced in the opposite direction. If undetermined belly skin from *Rana esculenta* is transplanted into the presumptive mouth region of a young *Triturus* neurula (Fig. 609a), it forms sucker discs (Fig. 609b, c) in response to the inductive influence of the host's head, and depending upon the size of the implant, a true frog mouth (Fig. 609d, e) forms in the *Triturus* embryo.

To express these results in an evolutionary context, we can say the following: with the acquisition of one or the other adhesive apparatus, what is new in the norm of reaction can be confined

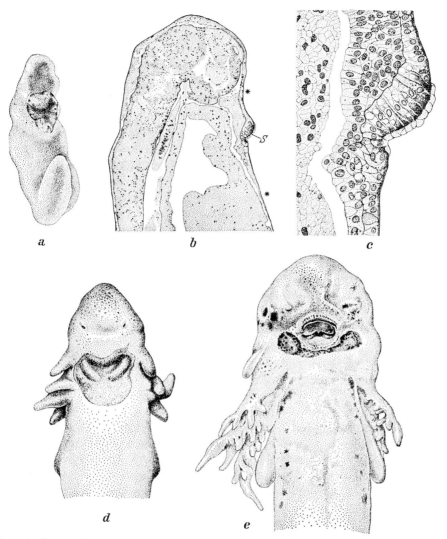

Fig. 609. Xenoplastic transplants: ventral skin of *Rana esculenta* implanted into *Triturus* gastrula. a tail bud stage; b later stage, longitudinal section; c adhesive disc region of the embryo of b greatly magnified; d, e urodele larva with an implant of anuran epidermis in the mouth region; e with sucker discs, tadpole mouth with horny jaws and horny spines on the lips. * margins of the implanted epidermis. S sucker discs. (After SPEMANN and SCHOTTÉ, 1932 and SPEMANN, 1936)

to an alteration of potency in the epidermis; the induction system and the reactive competence of the mesenchyme are maintained.

Deeper insight into organization is provided by xenotransplantation of pieces of neural crest from which the ectomesenchyme for parts of the head skeleton and teeth is derived (cf. p. 243). The visceral skeleton of the head differs between young urodele and anuran larvae conforming to the different layouts of the mouth (Figs. 608a, b, 610a, b). When a large piece of neural fold from a *Bombinator* neurula is put in place of a corresponding piece of a *Triturus* embryo (Fig. 610c), chimeric skeletons arise which have an almost complete *Bombinator* skeleton on one side of the *Triturus* host at the site of the operation (Fig. 610d). The final structure is thus essentially a function of the ectomesenchyme. A species-specific process of the *Bombinator* palatoquadrate serves as a site of attachment for a muscle derived from *Triturus* material. Most impressive of all is the similarity between host and implant in the case of tooth formation. The anuran larva has no teeth; it chews with horny plates and spikes. The tooth-bearing bones are not yet laid down in anuran larvae; they appear first during metamorphosis. *Triturus* larvae, on the other hand, have these bones together with teeth (Fig. 610a). These develop from a conical

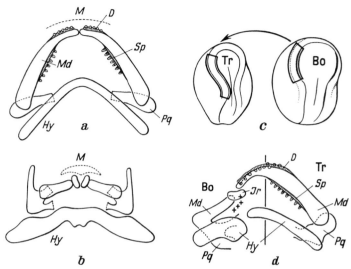

Fig. 610. Xenotransplantation. a larval skeleton of *Triturus;* b of *Bombinator;* c operative scheme, neurulae. *Bo* from *Bombinator; Tr* from *Triturus;* d chimaeric skeleton in a *Triturus* larva, left part made from *Bombinator* ectomesenchyme; small circles normal *Triturus* teeth. Crosses chimaeric teeth. *D* dental; *Hy* hyal; *Ir* infrarostral; *M* mouth; *Md* mandibula; *Pq* palotoquadrate; *Sp* splenial. (After WAGNER, 1949, simplified from BALTZER, 1952)

epidermal enamel organ and an ectomesenchymal dentine papilla. In the chimaera there are chimaeric teeth (Fig. 610d): the *Triturus* epidermis produces the enamel organ, while the *Bombinator* ectomesenchyme provides the dentine papilla. The mesenchymal papillae arise in the epidermal region near the infraostral and supraostral bones in *Triturus*. The mesenchyme papilla is generally a somewhat broader cone than in *Triturus*, and under these conditions it looks like the tooth papilla of a metomorphosed *Bombinator*. It is covered from base to tip with a layer of colorless, hard substance, evidently pre-dentine.

Phylogenetically, the urodele larval type would appear to be more primitive than that of the anurans, whose toothless, horny jaws appear to be secondarily acquired specializations, like many other characteristics of theirs. The potency for tooth formation is blocked in the anuran larvae, but the barrier is removed later in their ontogeny. The potency is expressed when the larval chewing apparatus disappears. The ectomesenchyme of certain skeletal elements near the mouth still shows the power to induce tooth rudiments when competent ectoderm is placed where it can accomplish what the anuran ectoderm no longer can do; the ectomesenchyme also produces papillae for the enamel organs, which in turn induce the formation of pre-dentine.

These transordinal chimaeras tell us that during divergent development a foundation of common properties remains, and that newly arisen species- and order-specific processes can make use of certain potencies of this foundation in a "regulative" way. This means that the story of evolution is dominated by a "great plasticity and combinability of component processes" (BALTZER[13, 297]).

The organism makes use of these developmental potencies in order to transform its body plan and functional organization under the regulative influence of selection, when the occasion is offered by environmental changes and by the occurrence of mutations. "Evolution is a creative response of the living substance," as DOBZHANSKY has said so well[170, 96]. The word "creative" has no metaphysical implications and suggests nothing further than the fact that the organism, as it changes, employs its developmental mechanism to bring forth from a disharmonious situation something new and balanced that was not present before. As an organism comes to terms with the disturbances of a previous balance and with the utilization of the genetic possibilities by means of gradual transformation, it pursues a phylogenetic path to a higher order of structure and function; and in this evolution, with the elaboration of sophisticated instrumentation in the nervous system, there emerges the apparatus of thought, enduring and engaged awareness. The emergent mind now has properties inaccessible to the essentially physicochemical methods used in the study of the developmental process leading to its emergence.

Lecture 36

It is the task of developmental genetics to elucidate the role of the genetic information in the complex of developmental processes leading from egg to adult. We have already seen an association of visible traits, as well as developmental tendencies, with hereditary factors, as shown in genetic crosses. But even though we have been able to follow genetically controlled changes in the expression of a character back to earlier stages in development, we are in most cases still fairly far from the primary actions of genes. Models providing deeper insight into the mode of action of individual genes and the interplay of various genes have been useful up till now only on the chemical level. One area close both to chemical and histological levels is the genetic control of eye pigments, which are laid down in granules. Precursors of the pigments in the Lepidoptera and Diptera arise not only in the cells which form the pigment, but also earlier in various tissues from which they are released into the blood. So it is possible to supply a missing precursor to the eye cells of a mutant which cannot form a particular pigment, and this is done by a normal implant. Certain intracellular defects of the mutant cells can also be corrected by contributions of other cells.

In *Ephestia*, the mutant allele reduces pigment formation: the eyes of a/a individuals are red instead of the normal dark brownish black; the testes are colorless instead of the normal brownish violet; and the skin of a/a has no pigment, while that of the wild type is reddish. If the male genital disc of a wild-type larva is implanted into the next-to-last larval instar of an a/a individual, the integument will be reddish in the next larval instar; the host testes will be pigmented, and the eyes in the adult male moth will be dark. In *Drosophila*, too, there are eye color mutations which can be corrected by transplantation. When one finds the active substance and determines its chemical nature, one can establish the synthetic step dependent upon a particular gene and lacking or defective in the mutant. This has been done with a series of mutants. The main eye pigments of insects belong to a certain group of natural substances called ommochromes, whose major precursor is tryptophan. The a^+ allele in *Ephestia* (equals v^+ in *Drosophila*) causes the formation of an enzyme which converts tryptophan into kynurenine (Fig. 611). When kynurenine is added to a mutants, it substitutes for the a^+ gene. But for the formation of the pigments additional genes are necessary; they participate further along the pathway and are known from certain mutants. One such gene, cn^+, governs the conversion of kynurenine into 3-hydroxy-kynurenine (Fig. 611). If 3-hydroxy-kynurenine is injected into cn mutants, normal pigment formation takes place, whether a^+ is present or not, for 3-hydroxy-kynurenine is a later step in the synthetic pathway. The eye pigment granules themselves contain RNA and protein as well as pigments; thus, further steps in eye pigmentation exist, and these, too, can be blocked by mutations[366]. The formation of the structural protein is interfered with by the mutation *wa* (*white eyes*), and this blocks pigmentation entirely: in *wa/wa* eye cells the granules are missing. In *a/a* individuals the granules are present just as in the wild type. Their dimensions and location appear identical to those of granules in histological preparations of wild eyes after the pigment has been dissolved away. The addition of normal implants to *wa/wa* animals does not help, for the wa^+ gene has only an intracellular effect, and we call its action "autonomous." The pigment precursors, kynurenine and 3-hydroxy-kynurenine, are released into the blood by various tissues of *wa* mutants just as in normal individuals. This makes it possible to repair genetic blocks in ommochrome formation by implanting *wa/wa* tissues into *a/a* hosts. The following experiment is the most impressive: When a piece of the eye anlage of a *wa/wa* pupae is placed in the eye anlage of a young *a/a* pupae, the host eyes become pigmented. The *wa/wa* eye anlage has an even greater effect than that of a similarly sized piece of wild genotype; for the wild type binds the pigment precursors on the granules and forms its own pigment, while essentially all of the precursors diffuse out of the *wa/wa* cells. This example provides a clear model of sequences of genetically controlled processes, illustrating both the successive steps in the formation of a phenotype and the junctions of separate pathways. Figure 611 illustrates the steps in the conversion of tryptophan to 3-hydroxy-kynurenine and an admittedly oversimplified scheme of granule formation, leading eventually to the final pigment granule.

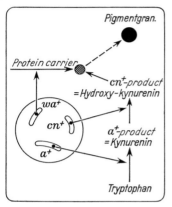

Fig. 611. Diagram of the effects of several genes known to be active in eye pigment formation in *Ephestia*. Effects of the genes $a^+ (= v^+)$, cn^+ and wa^+ in a pigment-forming ommatidial cell.
(After KÜHN, 1964)

The great number of genetically controlled steps involved in biosynthetic and developmental pathways is well illustrated by the results of experiments on nutritional mutations on bacteria and on the fungus *Neurospora crassa*. Even the synthesis of a single amino acid is not a simple affair: at least seven genes are involved in a series of steps leading to arginine synthesis. How many pathways of this type must form the sort of network necessary to produce species- and organ-specific protein structures; and how many other fundamental metabolic processes depend similarly on genes and chains of gene products! No wonder that most mutations produced by irradiation are lethal. The production of morphogenetic substances and of structure requires genetically controlled pathways to assure the availability of specific materials; and now we must add the problem of structure itself. We have seen it already in the pigment formation model: The carrier protein in the eye cells has the form of granules of certain dimensions; particular enzyme substances must be bound to these granules in order to carry out the sequence of reactions from precursor to pigment.

Experiments on the chemistry of gene action all suggest that each gene has only one chemical effect; most genetic effects on the chemical level have to do with enzyme-controlled reactions. These must be numerous and diverse in the various differentiated cells. This raises the question of how just the right set of genes happens to be operating in each cell. Are the vast number of genes on the chromosomes continually active, producing primary products which only in certain cases and under certain circumstances couple into the cellular machinery and thus participate in metabolic events and morphogenesis? This does not seem very likely. The real question seems to be this: how are genes turned on and off?

Numerous observations show that the chromosomes have visible responses, actual morphological changes, to particular conditions in their cellular environment. The onset of endomitosis (p. 71f.) and the degree of polyploidy or polyteny (p. 20ff., 72ff.) in certain animal and plant tissues depends upon the position of the cells in the developing organism. Influences from other parts of the organism or from its environment can reach the nucleus through the cytoplasm. Visible changes in the cytoplasm often precede nuclear changes.

Chromatin diminution in *Ascaris megalocephala* is a classic case of a response of the chromosomes to the constitution of the cytoplasm in which they happen to lie. In the polar-differentiated

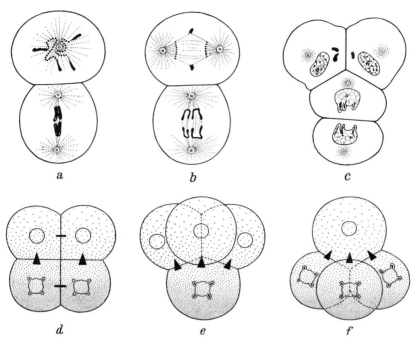

Fig. 612. *Ascaris megalocephala*. a–c behavior of the chromosomes during the second cleavage division; d–f diagram explaining the determination of diminution or nondiminution in the four blastomeres which arise simultaneously after dispermic fertilization. The black arrow heads indicate the axes of the gradient. (After BOVERI, 1899, 1910)

egg of this nematode (Fig. 193), the first cleavage spindle lies in the main axis of the egg. The yolk-poor, somewhat larger animal cell, or dorsal cell, is the "first somatic cell," which forms only ectoderm; the vegetal or ventral cell is the first "stem cell," or germ line cell. In the second division the spindle in the ventral cell lies once more in the animal-vegetal axis, but in the dorsal cell it is perpendicular to this axis (Fig. 612a, b). In the dorsal cell, the ends of the long chromosomes are pushed into the equatorial plate, and the central regions break up into a certain number of small chromosomes. These participate in further development, but the ends gradually disappear in the cytoplasm. During mitosis in the stem cells the original chromosomes remain intact, and in both daughter nuclei the chromosome ends protrude in little nuclear appendages. The "diminished" nuclei containing the numerous small chromosomes have a smooth surface. In three further divisions, the derivatives of the first stem cell become somatic cells with a particular determination; in each of these, the chromosomes always undergo diminution at their next division. Only in the germ line, through the first germ cell and all the way to mature gametes, do

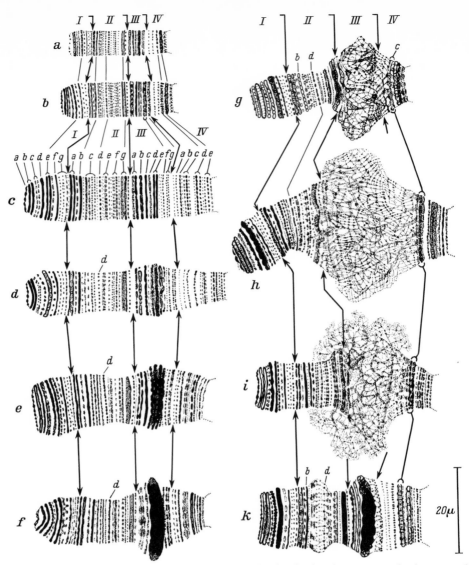

Fig. 613. One end of the C-chromosome from the salivary glands of *Rhynchosciara angelae* larvae. a–d growth during the larval stages; d–i development of a puff in region III; k regression of the puff, appearance of a functional structure in region II. (After BREUER and PAVAN, 1954)

the chromosomes retain their original integrity. A variety of observations shows now that the behavior of the chromosomes — diminished or undiminished — depends upon the cytoplasmic region in which they are found during division. Cases of dispermic fertilization are striking: when two sperms enter an egg, both sperm asters divide, and a tetrapolar spindle arises. The chromosomes are distributed to four daughter nuclei, and the egg divides into four blastomeres. Of these one or two or three may behave as stem cells; the abnormal embryo contains one, two, or three ventral cell sequences, while only one is found in the normal embryo. Accordingly, of the four blastomeres which arise simultaneously, from three down to one may become exclusively ecto-

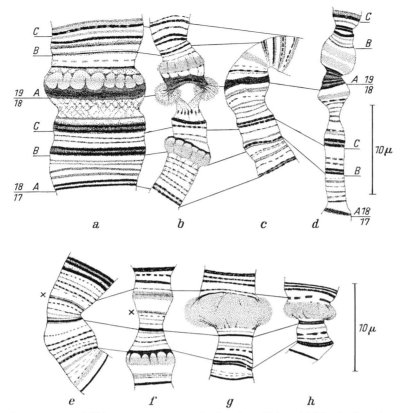

Fig. 614. Giant chromosomes of *Chironomus tentans*. a–d subregions 18 A to 19 C in the first chromosome; a from the salivary glands; b from the Malpighian tubules; c from the hindgut; d from the midgut; e–h region 9 of chromosome III; e, g from the Malpighian tubules; f, h from the hindgut; e, f from the larvae; g, h from the pupa. (After BEERMANN, 1952)

derm-forming somatic cells according to whether one, two, or three stem cells have been produced. In the somatic cells, the nuclei diminish at the next division, and in the stem cells they do not.

This phenomenon can be explained in terms of the possible positions of the four-poled spindle in the egg and the resulting locations of the four nuclei in the qualitatively and quantitatively different zones of polar differentiated egg cytoplasm (Fig. 612 d–f); their resulting behavior would then determine the further careers of the cells. Boveri said in 1910: "The case of *Ascaris* strikes me as a simple paradigm of nucleo-cytoplasmic interactions during development, where even the tiniest inequality within the cytoplasm reverberates back and forth, resulting in the most extreme disparities among the resulting cells[73, 191]."

We shall now take a closer look at the giant chromosomes of the Diptera: Here, tissue- and function-specific structural modifications in the banding pattern are abundant. Sites of nucleolus formation (p. 12ff.) and the varying appearance of heterochromatic regions (p. 17ff.) have long been known. Now, however, structural changes of euchromatic discs, corresponding more or less to gene loci, have been discovered and investigated from several points of view. At certain places in the chromosomes the bundles of chromonemata loosen, and there arise more or less extended swellings, the so-called puffs (or, when very large, Balbiani rings) (Fig. 613). In the midge *Rhynchosciara angelae*, for example, structural changes in the salivary chromosomes can be followed

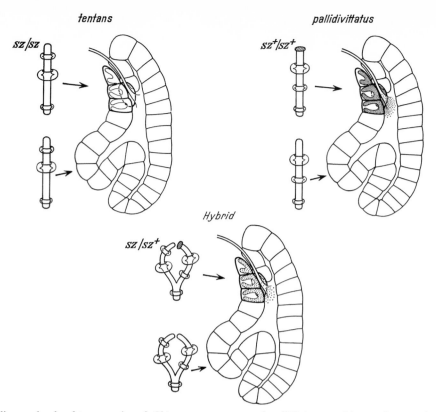

Fig. 615. Salivary glands of two species of *Chironomus tentans* and *pallidivittatus* with certain granules which are lacking in *tentans* and which are fewer in number in the hybrids. Similarly a Balbiani ring is lacking in *tentans*. Its formation depends upon sz^+. In the hybrid there is incomplete pairing due to an inversion. (After BEERMANN, 1963)

through the developmental stages in a naturally synchronized army of larvae. Figure 613 shows the increase in polyteny at the end of a certain chromosome as the larva matures (a–d). A large puff now forms in region III (d–i); later, it shrinks (k). But in region II, whose appearance has previously remained constant, there now arises a smaller puff between the subregions *b* and *d* (Fig. 613k). In the Chironomids individual organs and developmental stages can be characterized by the puffing patterns of their giant chromosomes. In Figure 614a–d, the region from 18A to 19C of the first chromosome is illustrated from various tissues of the same *Chironomus tentans*: salivary glands, malpighian tubules, hindgut, and midgut. The magnification is the same in all cases. While only one large puff is present in the salivary gland chromosome, there are three swellings of various sizes in the malpighian tubules. In the hindgut cells, only one band has a somewhat diffuse appearance in region 19A; in the midgut cells, two bands in this region are relatively wide. In region 9 of the third chromosome the chromosomes of the larval malpighian tubules and hindgut are clearly different (Fig. 614e, f): two bands which are small and compact in the malpighian tubules are broad and diffuse in the hindgut. In the pupae, one of these bands forms a substantial puff in both organs (Fig. 614g, h).

In any event, the structural pattern of the chromosomes corresponds to the activity pattern of genes. In many cases, the origin of a puff has been localized to a single band, but generally this functional structure extends over several bands. The maximum activity of a Balbiani ring can in the course of metamorphosis spread over some 20 bands[435]. Clearly, a region including a number of genes has become active.

When the presence of certain structures or products clearly depends upon certain genes, we have an opportunity to learn a lot more about gene action. Here is one example: two closely related *Chironomus* species, *pallidivittatus* and *tentans*, can be crossed successfully in spite of numerous inversions. They differ in the structure of the secretion of certain specialized cells of the salivary glands, and also by the presence or absence of a Balbiani ring in the fourth chromosome which is only found in *pallidivittatus* (Fig. 615). The inheritance of the secretory granules is simple and Mendelian; the site of the gene on the genetic map corresponds to the physical site of the Balbiani ring on the chromosome. In the hybrid, the puff forms only on one chromosome of the pair and the number of secretory granules is clearly reduced (Fig. 615).

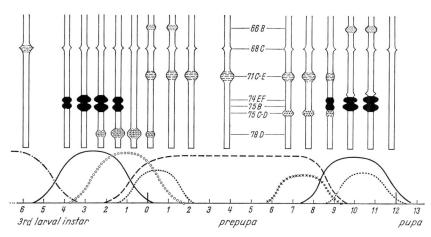

Fig. 616. Diagram of the sequence of puffs in a region of chromosome III in the salivary gland of *Drosophila melanogaster* during the latter part of the third larval instar and the whole prepupal stage. (After H. J. BECKER, 1959)

The course of events in metamorphosis provides an example of the turning on of certain genes. Toward the end of the last larval instar, there are many large and small changes of the type normally associated with puffing in the salivary gland chromosomes, and the alteration of the puffing pattern continues through the prepupal stage until the pupa is formed. During this period, puffing activity persists for a long time in some places; in others, it may be reduced and reappear later; in still others it appears only once (Fig. 616). The puff formation is synchronized exactly with the developmental stages. If a particular puff arises in a certain stage, its reappearance can be caused experimentally. This is shown by transplantation experiments: from young prepupae in whose salivary chromosomes the puff at region 78 D has contracted completely, the salivary glands are implanted into a larva; as the host develops, the puff appears anew in the implant (Fig. 617).

The puffing pattern associated with metamorphosis can be induced in young larvae by the injection of the molting hormone ecdysone (p. 338ff.). Long before molting is evident morphologically and biochemically, the formation of certain puffs has been triggered by ecdysone. In *Chironomus tentans*, among others, two puffs stand out, one on the first chromosome and another on the fourth. The first appears only 15–30 minutes after the injection of ecdysone, and the other follows 15–30 minutes later. In the normal pupal molt an interval of several days separates the activation of the first puffing site from the second. This difference can be under-

485

stood in terms of a gradual rise in the ecdysone concentration in the blood during normal development until it finally reaches the level necessary to induce the second puff. The size of the puff and the duration of the functional condition are at first dependent upon the concentration of ecdysone (Fig. 618). The further behavior of the two puffs is different: one remains until the end of the molt, but the other shrinks within 72 hours, even though the ecdysone concentration remains high. The experimental induction of molting also causes other changes in the puff pattern of the salivary gland chromosomes and these appear later (six hours to two days after injection) and show no direct dependence on ecdysone concentration. They are temporary correlates of changes in cellular metabolism during molting. The events associated with metamorphosis differ

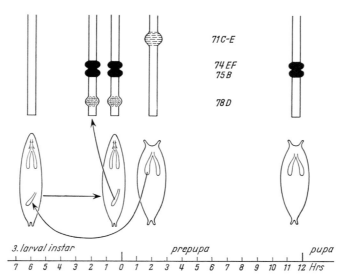

Fig. 617. Salivary gland transplantation in *Drosophila melanogaster*. Left to right: host larva at the time of transplantation, host larva later in development, prepupa, late prepupa; above, proximal end of left arm of chromosome III from each of the stages. To the left of one host chromosome the corresponding implant chromosome is shown. (After BECKER, 1962)

among the various tissues and thus activate different genes. But the first puffing reaction at a particular place (I 18C) is hormone-specific and not tissue-specific. It appears in the nuclei in the cells of the hindgut and in those of the malpighian tubules, which suggests that the primary hormone reaction is the same in all tissues. This puff induction is the earliest demonstrated effect of ecdysone; but one cannot yet decide whether the hormone acts directly on a gene or whether there is an intermediate which acts in turn on the chromosomes.

RNA-specific staining has already made it clear (p. 19) that RNA is formed in the puffs: methyl green-pyronin stains the ordinary bands green and the puffs bright red; orcein-fast green stains "resting" bands reddish brown and "functional" regions, even including tiny swellings, green; metachromatic toluidine blue stains the RNA-rich functional structures red, distinguishing them from the remaining blue.

Radioisotope experiments have provided conclusive proof of the formation of RNA in the puffs[503]: Tritiated uridine, an RNA precursor, is injected, and activity appears after only a few minutes in the chromosomes and in the nucleolus when an autoradiograph of the salivary glands is made. This activity, which increases over the next two hours, is restricted to the puffs, and its intensity is proportional to the size of the puff and therefore greatest in a Balbiani ring (Fig. 619).

The nucleolus acquires RNA from its chromosomal organizing region until it is filled. Later, the marked RNA can be detected in the nuclear sap and in the cytoplasm. We have already seen the striking movements of RNA in the nurse cells of insect ovaries (p. 314, Fig. 401a, b). Within certain Balbiani rings the formation of RNA-containing granules has been observed, and these have been seen also in the pores of the nuclear membrane. The mechanism of their transport is unknown.

Entirely different results are obtained by the injection of radioactive amino acids, e.g., tritiated leucine, for incorporation into protein. In this case, great activity — protein synthesis — is seen in the cytoplasm after only a few minutes, but the nucleus remains unmarked for a long time (Fig. 620).

These results relating to the functional structure of chromosomes establish a relationship between developmental cytology and molecular genetics which begins to clarify the primary molecular process in development at the genetic level. It is well established that the genetic substance of chromosomes is DNA (p. 26f). The genetic information is contained in a sequence of bases in the DNA chains (p. 19, Fig. 20). Each gene determines the amino acid sequence of a particular polypeptide. The relationship between base sequence and amino acid sequence is called the genetic code[145, 742], and its mechanism of expression, or deciphering, is the following: (1) the transcription of the nucleotide sequence of the DNA into a corresponding sequence in RNA; and (2) the *translation* of the base sequence of messenger RNA into an amino acid sequence in a polypeptide. This second step takes place on the ribosomes to which the messenger RNA molecules attach. The nucleotides on the messenger RNA are read in threes, each triplet determining a particular amino acid via a specific adapter, the transfer RNA. It is a spectacular example of one of the unifying

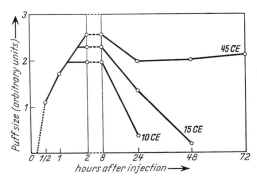

Fig. 618. Size changes in puff 18C of chromosome I of the salivary gland of *Chironomus tentans* after the injection of 10, 15, and 45 *Calliphora*-units (CE) of ecdysone extract. The dotted line represents the change which has not been measured but is believed to take place in the first thirty minutes. Before the injection no puff was present. (After CLEVER, 1962)

principles of biology, its common chemistry, that throughout nature, the genetic code employs the same four bases in DNA and the same twenty amino acids in proteins.

This whole process is central to the continuous procession of cellular events: the replication of the information and the transcription of RNA are enzymatically catalyzed and require a supply of nucleotide building blocks. Similarly, amino acids are required for the formation of polypeptide chains, which in turn form the wide variety of enzymes in the cell. Thus, the expression of a piece of information does not represent an absolute beginning; rather, it is induced to play its part by specific influences in the developmental mechanism of the organism, working with a supply of materials which its activity may eventually renew.

The problem of determination in development embraces further questions, too: what is the origin of the local conditions in an embryo which activate certain genes in certain cells? How is the determination of cells achieved and maintained? Most differentiated animal cells hold fast to a lasting condition of determination which restricts their competence and the competence of their derivatives. Epidermal cells from developing Lepidopterans can often multiply in transplant vesicles, retaining their competence to respond to hormones with further developmental steps but remaining determined in a regionally specific way (p. 351f.). In cultures of bird and mammalian tissues, the cells retain their determination; they remain, for example, only epidermal cells or myoblasts. With the achievement of a particular functional condition, does the nucleus lose its competence for certain reactions, that is, can only certain genes still be turned on? Does rigid determination, therefore, depend upon a permanent restriction of the information? Here we have a clue, since it is not wise to generalize from the rare cases of chromosome elimination. Nuclei have been removed from somatic cells of amphibians in various stages of development and transplanted into enucleated eggs to see whether they could support complete normal development[188,189,228]: in this refined experimental technique a group of cells are removed from a donor

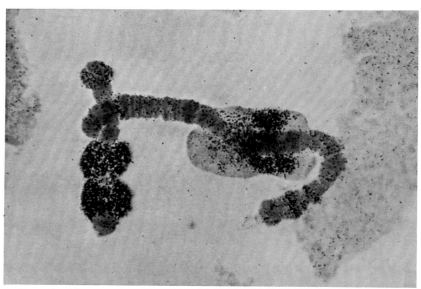

Fig. 619. Autoradiograph of ³H-uridine incorporation indicating RNA synthesis in *Chironomus tentans* chromosomes. Chromosome III with nucleus. Chromosome IV with three Balbiani rings. After thirty minutes incubation. (After original microphotograph of PELLING)

and a few are drawn up individually into a micropipette; the cytoplasm is broken up, and the nucleus is injected into the egg. Nuclei from gastrulae, neurulae, and even from the endoderm of swimming larvae could lead to the formation of essentially normal larvae and even sexually mature adults. Thus, it is certain that the nuclei of some tissues need not forfeit some of their talents in order to reach a certain stage; rather, they can replace the egg nucleus, and their derivatives can satisfy all the demands of the developmental steps which the various cells must pass through. The very delicate method of isolation and implantation of nuclei prevents one from concluding, from the many embryos that do not develop normally, that some nuclei have lost their full competence in embryonic or differentiated tissue. The nuclear membrane or some of the nuclear contents can be damaged; a nucleus may already be in an active condition from which it cannot be recalled by the particular commands of the egg cytoplasm. These experiments are continuing in the hands of several investigators, and the results are being awaited with excitement. The regeneration of entire plants (e.g., p. 395ff., Fig. 508) from differentiating tissue cells shows that their nuclei, too, have lost none of their information content and none of their ability to use it.

We are thus led to a determined condition of the cytoplasm. Predetermination in the growing oocyte (Fig. 507) can govern the reactions of cells over long periods of the development of the individual. In dauermodifications the consequences of transient environmental influences can last for many cell generations in single-celled organisms and for several individual generations in multicellular organisms. In cases where a dauermodification is crossed with an ordinary individual, the relevant trait is inherited matrilineally. The norm of reaction of the cells is in all of these cases controlled by alterations of the cytoplasm, which, as long as they exist, permit the expression of a certain set of genes only. The continuity of determination cannot depend upon a single finite supply of a substance which is distributed when the cells divide. Such a supply

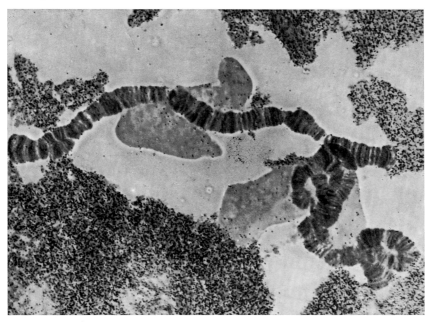

Fig. 620. Autoradiograph of ³H-leucine uptake indicating protein synthesis in the cytoplasm. After three hours incubation. (After original microphotograph of Pelling)

would be diluted continuously and its effect would disappear quickly. In cases of fixed determination, therefore, as in dauermodifications, cytoplasmic components with altered properties must replicate. Thus, in order to understand the nature of determination, one is led to the possibility that certain cytoplasmic structures are capable of self-replication and that their relative numbers and properties can be altered by appropriate conditions.

That part of the hereditary mechanism lies in the cytoplasm cannot be doubted[105, 107, 237a, 446, 447, 483, 484]. Differences between reciprocal hybrids combined with the non-mendelian inheritance of certain hereditary traits, and the retention of cytoplasmic specificity in spite of the presence of foreign nuclei over many generations, show that hereditary differences can be lodged in places other than the chromosomes. Numerous genetic experiments, particularly in plants and in chlorophyll-free protists, and some even in insects, show that extranuclear hereditary factors exist in addition to the plastids. A cytoplasmic property which shows extranuclear inheritance has been called the plasmotype, or plasmon, by von Wettstein. The name "plasmagene" has been given to the bearer of properties inherited in an extranuclear fashion. The random mixing of green and colorless plastids in the cells of variegated leaves provides us with a model for the distribution of cytoplasmic hereditary factors, where the number of each type in an ancestral cell can determine the distribution of the trait in the progeny. The behavior of some respiratory mutants in yeast (the extrachromosomal "petites") is consistent with this concept[181]. From the number of cell generations which lead to a complete segregation of the opposing cytoplasmic hereditary factors one can conclude something about the total number of units involved in each alternative form.

The question now arises as to whether self-replicating elements can be seen as distinct structures in the cytoplasm visible in the light microscope or in the electron microscope. Structures of various orders of size can be considered, from the mitochondria, whose longest dimension generally ranges from 0.5 to 3 microns, to the various microsomal components of 50–150 millimicrons. They all have a protein framework and contain enzymes; their content of DNA or RNA, which bears the genetic information in plant viruses, suggests a comparison with viral derivatives and raises the possibility that these cytoplasmic particles can replicate themselves just as viruses do. Many observations have been made on the multiplication of the mitochondria. The apparent cytoplasmic transmission of metabolic mutations in *Neurospora* and in yeast[181] have for some time aroused interest and perhaps can now be understood in terms of these recent findings. The mitochondria of normal yeast contain a series of cytochromes; mutant mitochondria show various changes in one or more of these key electron transport proteins.

What sort of role do the DNA- or RNA-containing cytoplasmic structures play in morphogenesis[77, 393]? The concentration of mitochondria and of ribosomes in embryonic cells rises several fold during determination and morphogenetic movement. One thus suspects them of possessing the two fundamental properties of genes, continuity and an influence on the cell. Until recently, ribosomal RNA itself was considered as a possible template for the synthesis of the various proteins. Now we know that the ribosomes are cell organelles on whose surface the messenger RNA attaches, there serving as a template for the synthesis of a specific protein. The aggregation of ribosomes into polysomes is common and appears to be due to the attachment of several ribosomes to one messenger during the normal course of synthesis. Ribosomal RNA comes from the nucleolus, but we are not sure of the immediate source of the ribosomal protein. The ribosomes appear to be the same throughout the tissues of an organism with respect to their RNA and protein content; indeed, ribosomes from the most diverse organisms are relatively interchangeable for *in vitro* experiments in protein synthesis. The specificity clearly lies in the messenger RNA. For example, ribosomes from *E. coli* enable Tobacco Mosaic RNA to form Tobacco Mosaic Virus coat protein in cell-free systems[682]. But the total absence of phylogenetic specificity has not been established, nor has the total phylogenetic range.

Developmental problems remain which are independent of the answers to these questions: Although the cytoplasmic particles are distributed only approximately evenly during ordinary cell division, the segregation of particles does take place during the development of plants and animals. During the maturation period and fertilization, a pattern of cytoplasmic regions is established in many animal eggs, so that the blastomeres can be distinguished in the light microscope and in the electron microscope by differences in cytoplasmic inclusions. Later, too, these particles are often segregated in a polar fashion during differential divisions, and cells of various prospective fates exhibit cytoplasmic structure with corresponding populations of particles[393, 394, 711, 712]. At any rate, the enrichment of certain particles has a developmental significance, although we cannot yet go so far as to identify this significance with a morphogenetic determination. One of the recent investigations employing the electron microscope has come to the conclusions: "It is reasonable to consider such patterns as *symptomatic* rather than *causative* events in embryonic differentiation."[712, 407] Again and again, a determining role appears probable for the egg cortex, whose nature remains unknown.

The mechanism of determination is still obscure. How gene products affect morphogenesis is unknown. This problem reaches even into virus genetics, where nucleic acid molecules are packed by bacteriophages in bacterial cells into a coat of chemically diverse, specifically structured proteins. It would be hoped that a simple morphogenetic model will come from work in this area and in the confluence of protein chemistry and submicroscopic cell structure and find application throughout development.

Everywhere, developmental problems are finding meaningful expression in terms of structure at lower levels of organization: cells, organelles, macromolecules, and even small molecules and functional groups.

Developmental physiology, in conjunction with genetics and biochemistry, has led to a wealth of results to date. But if we look back to the fundamental questions which we saw posed in the first lecture of this series, we must admit that we have not moved a great distance. Each step forward makes us see farther down the road, where the going will be rougher, if anything. But this we should not regret: what is known already belongs to everyone, but at the border of the unknown, lies the exciting territory of the explorer.

References

Comprehensive articles and those with extensive bibliographies are marked with a star.

1 ABBATE, C., and G. ORTOLANI: The development of *Ciona* eggs after partial removal of cortex or ooplasma. *Acta Embryol. Morphol. Exp.* **4**(1961).

2 ADELMANN, H. B.: Experimental studies on the development of the eye. IV. The effect of the partial and complete excision of the prechordal substrate on the development of the eyes of *Amblystoma punctatum.* *J. Exp. Zool.* **75** (1937).

2a★ Advances in Morphogenesis. (ed.) ABERCROMBIE, M., and BRACHET, J.: **1** (1961).

3 AHRENS, W.: Über das Auftreten von Nukleolenchromosomen mit endständigem Nukleolus in der Oogenese von *Mytilicola intestinalis. Z. wiss. Zool.* **152** (1939).

4 ANCEL, P., et P. P. VINTEMBERGER: Recherches sur le déterminisme de la symétrie bilatérale dans l'oeuf des Amphibiens. *Bull. Biol. Fr. Belg.*, Suppl. **31** (1948).

5★ ANDRES, G. M.: Eine experimentelle Analyse der Entwicklung der larvalen Pigmentmuster von fünf Anurenarten. *Zoologica (Stuttgart)* **40** (1963).

6 ANIKIN, A. W.: Das morphogene Feld der Knorpelbildung. *Wilhelm Roux' Arch. Entwicklungsmech. Organismen* **114** (1929).

7 ASCHOFF, J.: Exogene und endogene Komponenten der 24-Std-Periodik bei Tier und Mensch. *Naturwissenschaften* **42** (1955).

8 BAIRATTI, A., u. E. LEHMANN: Über die submikroskopische Struktur der Kernmembran bei *Amoeba proteus. Experientia (Basel)* **8** (1952).

9 BALINSKY, B. I.: Growth and cellular proliferation in the early rudiments of the eye and the lens. *Quart. J. Micr. Sci.* **93** (1952).

10 —— An Introduction to Embryology. Philadelphia and London: Saunders, 1960.

11 BALTZER, F.: Chimären und Merogone bei Amphibien. *Rev. Suisse Zool.* **57** (1950).

12 —— Entwicklungsphysiologische Betrachtungen über Probleme der Homologie und Evolution. *Rev. Suisse Zool.* **57** (1950).

13 —— Experimentelle Beiträge zur Frage der Homologie. *Experientia (Basel)* **8** (1952).

14 BANKI, Ö.: Die Entstehung der äußeren Zeichen der bilateralen Symmetrie am Axolotlei nach Versuchen mit örtlicher Vitalfärbung. 10. Internat. Congr. Zool. 1929.

15 BARBER, H. N.: The rate of movement of chromosomes on the spindle. *Chromosoma (Ber.)* **1** (1939/40).

16 BARNICOT, N. A.: The local action of the parathyreoid and other tissues on bone in intercerebral grafts. *J. Anat.* **82** (1948).

17 BARRINGTON, E. J.: An Introduction to General and Comparative Endocrinology. Oxford: Clarendon Press, 1963.

18 BAUER, H.: Der Aufbau der Chromosome und seine Abänderungen. *Jena Z. Med. Naturw.* **75** (1942).

18a ——, R. DIETZ u. C. RÖBBELEN: Die Spermatocytenteilungen der Tipuliden. III. Mitteilung. *Chromosoma* **12** (1961).

19 BEADLE, G. W.: Further studies of asynaptic maize. *Cytologia (Tokyo)* **4** (1932/33).

20 —— A gene for sticky chromosomes in *Zea mays. Z. indukt. Abstamm.- u. Verb.-Lehre* **73** (1933).

21 BEAMS, H. W., and R. L. KING: The effect of ultracentrifuging chick embryonic cells, with special reference to the "resting" nucleus and the mitotic spindle. *Biol. Bull.* **71** (1936).

22 —— —— An experimental study on mitosis in the somatic cells of wheat. *Biol. Bull.* **75** (1938).

23 BECHER, E.: Über den feineren Bau der Skelettsubstanz bei Echinoiden, insbesondere über statische Strukturen in derselben. *Zool. Jahrb., Abt. allg. Zool. Physiol. Tiere* **41** (1924).

24 BECHER, S.: Untersuchungen über nichtfunktionelle Korrelation in der Bildung selbständiger Skelettelemente und das Problem der Gestaltbildung in einheitlichen Protoplasmamasse. *Zool. Jahrb., Abt. allg. Zool. Physiol. Tiere* **31** (1911).

25 —— Über doppelte Sicherung, heterogene Induktion und assoziativen Induktionswechsel. *Zool. Jahrb., Suppl.* **15**, 3 (1912).

26 BECKER, H. J.: Die Puffs der Speicheldrüsenchromosomen von *Drosophila melanogaster. Chromosoma* I. **10** (1959); II. **13** (1962).

27 BECKWITH, C. J.: The genesis of the plasma-structure in the egg of *Hydractinia echinata. J. Morph.* **25** (1914).

28 BEERMANN, W.: Chromomerenkonstanz und spezifische Modifikationen der Chromosomenstruktur in der Entwicklung und Organdifferenzierung von *Chironomus tentans. Chromosoma* **5** (1952).

29 —— Der Nukleolus als lebenswichtiger Bestandteil des Zellkerns. *Chromosoma* **11** (1960).

30 —— Cytologische Aspekte der Informationsübertragung von den Chromosomen in das Cytoplasma. In: Induktion und Morphogenese, 13. Colloqu. Ges. Physiol. Chemie 1962.

31★ —— Riesenchromosomen. In: Protoplasmatologia, Bd. 6D. Wien: Springer, 1962.

32 BEHRENDS, J.: Über die Entwicklung des Lakunen-, Ader- und Tracheensystems während der Puppenruhe im Flügel der Mehlmotte *Ephestia kühniella. Z. Morph. u. Ökol. Tiere* **30** (1935).

33 BEISSENHIRTZ, H.: Experimentelle Erzeugung von Mehrfachbildungen bei Planarien. *Z. wiss. Zool.* **132** (1928).

34 BELAR, K.: Untersuchungen an *Actinophrys sol*. I. Die Morphologie des Formwechsels. *Arch. Protistenk.* **46** (1922).

35 —— Untersuchungen an *Actinophrys sol*. II. Beiträge zur Physiologie des Formwechsels. *Arch. Protistenk.* **48** (1924).

36 —— Beiträge zur Kenntnis des Mechanismus der indirekten Kernteilung. *Naturwissenschaften* **15** (1927).

37 —— Beiträge zur Kausalanalyse der Mitose. II. Untersuchungen an den Spermatocyten von *Chorthippus (Stenobothrus) lineatus* PANZ. *Wilhelm Roux' Arch. Entwicklungsmech. Organismen* **118** (1929); III.-V. Untersuchungen an den Staubfadenhaarzellen von *Tradescantia virginica. Z. Zellforsch.* **10** (1929).

38 —— Zur Teilungsautonomie der Chromosomen. Aus dem wissenschaftlichen Nachlaß von KARL BELAR, herausgeg. von W. HUTH. *Z. Zellforsch.* **17** (1933).

39 BENEDEN, E. VAN, et C. JULIN: Le segmentation chez les Ascidiens dans ses rapports avec l'organisation de la larve. *Arch. Biol.* **5** (1884).

40 BERGNER, A. D., J. L. CARTLEDGE, and A. F. BLAKESLEE: Chromosome behavior due to a gene which prevents metaphase pairing in *Datura. Cytologia (Tokyo)* **6** (1935).

41 BERGQUIST, H.: Studies on the cerebral tube in vertebrates. The neuromeres. *Acta Zool. (Stockholm)* **33** (1952).

42 BERNARD, F.: Recherches sur la morphogénèse des yeux composés d'arthropodes. *Bull. Biol. Fr. Belg., Suppl.* **23** (1937).

43* BERTALANFFY, L.: Theoretische Biologie, Bd. 1 Wachstum. Berlin: Borntraeger, 1932; Bd. 2 Stoffwechsel. Berlin: Borntraeger, 1942.

44 BESSERER, S.: Das Wachstum der Speicheldrüsen- und Epidermiskerne in der Larvenentwicklung von *Chironomus. Biol. Zentralbl.* **75** (1956).

45 BETH, K.: Ein- und zweikernige Transplantate zwischen *Acetabularia mediterranea* und *Acicularia schenckii. Z. indukt. Abstamm.- u. Vererb.-Lehre* **81** (1943).

46 BIER, K.: Endomitose und Polytänie in Nährzellkernen von *Calliphora erythrocephala. Chromosoma* **8** (1957).

47 —— Unterschiedliche Reproduktionsraten im Eu- und Heterochromatin: Ein Weg zur Kerndifferenzierung. *Verh. Deut. Zool. Ges.* 1961.

48 —— Synthese, interzellulärer Transport und Abbau von Ribonukleinsäure im Ovar der Stubenfliege *Musca domestica. J. Cell. Biol.* **16** (1963).

49 —— Autoradiographische Untersuchungen zur Dotterbildung. *Naturwissenschaften* **49** (1963).

50 —— Autoradiographische Untersuchungen über die Leistungen des Follikelepithels und der Nährzellen bei der Dotterbildung und Eiweißsynthese im Fliegenovar. *Wilhelm Roux' Arch. Entwicklungsmech. Organ.* **154** (1963).

51 BLAUSTEIN, W.: Histologische Untersuchungen über die Metamorphose der Mehlmotte *Ephestia kühniella. Z. Morph. u. Ökol. Tiere* **30** (1935).

52 BOCK, F.: Experimentelle Untersuchungen an kolonielbildenden Volvocaceen. *Arch. Protistenk.* **56** (1926).

53* BODENSTEIN, D.: Embryonic development. Regeneration. The role of hormones in moulting and metamorphosis. In: Insect Physiology, ed. by K. D. ROEDER. New York and London: Wiley, 1953.

54 ——, and A. ABDEL-MALEK: The induction of aristopedia by nitrogen mustard in *Drosophila virilis. J. Exp. Zool.* **111** (1949).

55 BOFILL-DEULOFEN, J.: Die argyrophilen Faserstrukturen in mesenchymalen Gewebekulturen von verschiedener Herkunft und von verschiedener Wachstumsgeschwindigkeit. *Z. Zellforsch.* **14** (1932).

56 BONDI, C.: Osservazioni sulla rigenerazione in *Dugesia lugubris. Arch. Zool. Ital.* **40** (1955).

57 BONNER, J. T.: A descriptive study of development of the slime mold *Dictyostelium discoideum. Amer. J. Bot.* **31** (1944).

58 —— Evidence for the formation of cell aggregates by chemotaxis in the development of the slime mold *Dictyostelium discoideum. J. Exp. Zool.* **106** (1947).

59 —— The demonstration of acrasin in the later stages of the development of the slime mold *Dictyostelium discoideum. J. Exp. Zool.* **110** (1949).

60 —— Observations on polarity in the slime mold *Dictyostelium discoideum. Biol. Bull.* **99** (1956).

61* —— Morphogenesis. Princeton: Univ. Press, 1952.

62 —— Epigenetic development in the cellular slime moulds. *Symp. Soc. Exp. Biol.* **17** (1963).

63 ——, and D. ELDREDGE: A note on the rate of morphogenetic movement in the slime mold *Dictyostelium discoideum. Growth* **9** (1945).

64 ——, and E. B. FRASCELLA: Mitotic activity in relation to differentiation in the slime mold *Dictyostelium discoideum. J. Exp. Zool.* **121** (1952).

65 BONNER, JAMES, and A. W. GALSTON: Principles of Plant Physiology. San Francisco: Freeman, 1952.

66 BONNER, J. T., and M. K. SLIPKIN: A study of the control of differentiation: The proportions of stalk and spore in the slime mold *Dictyostelium discoideum*.

67* BOUNHIOL, J. J.: Recherches expérimentales sur le déterminisme de la métamorphose chez les lépidoptères. *Bull. Biol. Fr. Belg., Suppl.* **24** (1938).

68 ——— Intervention de plusieurs glandes endocrines dans la métamorphose des insectes lépidoptères. Assoc. Franç. Avance. Sci. (Biarritz) (1947).

69* ——— Endocrinologie de la métamorphose des lépidoptères. *Bull. Biol. Fr. Belg. Suppl.* **33** (1948).

70 BOVERI, T.: Die Entwicklung von *Ascaris megalocephala* mit besonderer Rücksicht auf die Kernverhältnisse. Festschr. KUPFFER. Jena: G. Fischer, 1899.

71 ——— Zellen-Studien IV. Über die Natur der Centrosomen. *Jena Z. Naturwiss.* **35** (1901).

72 ——— Die Polarität von Ovocyte, Ei und Larve des *Strongylocentrotus lividus. Zool. Jahrb., Abt. Anat. u. Ontog.* **14** (1901).

73 ——— Die Potenzen der *Ascaris*-Blastomeren bei abgeänderter Furchung. Festschr. R. HERTWIG, Bd. 3. Jena: G. Fischer 1910.

74 BOYSEN-JENSEN, P.: Untersuchungen über Determination und Differenzierung. *I. Kgl. Danske Vidnsk. Selsk. Biol. Medd.* **18** (1950).

75 BRACHET, J.: La localisation des protéines sulfhydrilées pendant le développement des Amphibiens. *Bull. Acad. Sci. Belg.* **24** (1938).

76* ——— Embryologie chimique, 2nd Ed. Paris: Masson et Cie., 1947.

77* ——— Le rôle des acides nucleiques dans la vie de la cellule et de l'embryon. *Actual. Biochim.* **16** (1952).

78 ———, and H. CHANTRENNE: Protein synthesis in nucleated and nonnucleated halves of *Acetabularia mediterranea*, studies with carbon-14-dioxide. *Nature (London)* **168** (1951).

79 BRAGG, A. N.: The organisation of the early embryo of *Bufo cognatus* as revealed especially by the mitotic index. *Z. Zellforsch.* **28** (1938).

80 BRAUNS, A.: Untersuchungen zur Ermittlung der Entstehung der roten Blutzellen in der Embryonalentwicklung der Urodelen. *Wilhelm Roux' Arch. Entwicklungsmechan. Organ.* **140** (1940).

81 BREFELD, O.: *Dictyostelium mucoroides*, ein neuer Organismus aus der Verwandtschaft der Mycomyceten. *Abh. Senckenberg. naturforsch. Ges.* **7** (1869).

82 BREUER, M. E., and C. PAVAN: Salivary chromosomes and differentiation. IX. Intern. Congr. Genet. 1954.

83 BRIEN, P.: La pérennité somatique. *Biol. Rev. Cambridge. Philos. Soc.* **28** (1953).

84 BRØNDSTED, H. V.: The time-graded regeneration field in planarians and some of its cyto-physiological implications. Proc. VIIth Sympos. Colston Res. Soc. 1954.

85 ——— Experiments on the time-graded regeneration field in planarians. *Biol. Medd. Kgl. Dan. Vidensk. Selsk.* **23** (1956).

86* BÜCKMANN, D.: Entwicklungphysiologie der Arthropoden. Postembryonale Entwicklung. *Fortschr. Zool.* **14** (1962).

87 BULLOUGH, W. S., and M. JOHNSON: The energy relations of mitotic activity in adult mouse epidermis. *Proc. Roy. Soc. Ser. B. Biol. Sci.* **138** (1951).

88 BÜNNING, E.: Die entwicklungsphysiologische Bedeutung der endogenen Tagesrhythmik bei den Pflanzen. *Naturwissenschaften* **33** (1946).

89 ——— Der tagesperiodische Verlauf der Substanzproduktion und speziell der Zellulosesynthese in Beziehung zur photoperiodischen Reaktionsweise. *Planta (Berl.)* **39** (1951).

90 ——— Morphogenesis in plants. *Surv. Biol. Progr.* **2** (1952).

91* ——— Über den Tagesrhythmus der Mitosehäufigkeit in Pflanzen. *Z. Bot.* **40** (1952).

92* ——— Entwicklungs- und Bewegungsphysiologie der Pflanzen, 2nd, 3rd Eds. Berlin, Göttingen, Heidelberg: Springer 1952.

92a* ——— Regeneration bei Pflanzen. In: Handbuch der allgemeinen Pathologie, Bd. 6, Teil I. Berlin, Göttingen, Heidelberg: Springer, 1955.

93 ——— Polarität und inäquale Teilung des pflanzlichen Protoplasten. *Protoplasmatologia* **8**, 9a (1957).

94* ——— Die physiologische Uhr, 2nd Ed. Berlin-Göttingen-Heidelberg: Springer, 1963.

95 ———, u. F. BIEGERT: Die Bildung der Spaltöffnunginitialen bei *Allium cepa. Z. Bot.* **41** (1953).

96 ———, u. H. SAGROMSKY: Die Bildung des Spaltöffnungsmusters in der Blattepidermis. *Z. Naturforsch.* **3b** (1948).

97 ———, u. D. v. WETTSTEIN: Polarität und Differenzierung an Mooskeimen. *Naturwissenschaften* **40** (1953).

98 BURGEFF, H.: Progressive Mutationen bei der Lebermoos-Gattung *Marchantia, Biol. Zentralbl.* **61** (1941).

99 ——— Konstruktive Mutationen bei *Marchantia. Naturwissenschaften* **29** (1941).

100 ——— Genetische Studien an *Marchantia*. Jena: G. Fischer, 1943.

101 BUTENANDT, A.: Biochemie der Gene und Genwirkungen. *Naturwissenschaften* **40** (1953).

102 —— u. P. KARLSON: Über die Isolierung eines Metamorphose-Hormons der Insekten in kristallisierter Form. *Z. Naturforsch.* **9** b (1954).

103 CAMUS, G.: Recherches sur le rôle des bourgeons dans les phénomènes de morphogénèse. *Rev. Cytol. Biol. vég.* **11** (1949).

104 CARNIEL, K.: Endständige Nukleolen und Zahl der Nukleolenchromosomen bei *Rhoea discolor. Öst. bot. Ges.* **107**, 403 (1960).

105* CASPARI, E.: Cytoplasmatic inheritance. *Advanc. Genet.* **2** (1948).

106* —— Pleiotropic gene action. *Evolution* **6** (1952).

107 —— The role of genes and cytoplasmatic particles in differentiation. *Ann. N. Y. Acad. Sci.* **60** (1955).

108 CASPERSSON, T.: Über den chemischen Aufbau der Strukturen des Zellkerns. *Skand. Arch. Physiol., Suppl.* **8** (1936).

109 —— Die Eiweißverteilung in den Strukturen des Zellkerns. *Chromosoma* **1** (1939/40).

110* CAZAL, P.: Les glandes endocrines rétro-cérébrales des insectes. *Bull. Biol. Fr. Belg., Suppl.* **32** (1948).

111 CHAMBERS, R., and E. L. CHAMBERS: Nuclear and cytoplasmatic interrelations in the fertilization of the *Asterias* egg. *Biol. Bull.* **96** (1949).

112 CHASE, H. Y.: The origin and nature of the fertilization membrane in various marine ova. *Biol. Bull.* **69** (1935).

113 CHATTON, E.: *Pleodorina california* à Banyuls sur mer. *Bull. Sci. Fr. Belg., Ser.* **7**, 44 (1911).

114* Chemie der Genetik. 9. Colloqu. Ges. Physiol. Chem. 17.19.4. 1958 Mosbach. Beiträge von RIS, SIEBERT, ALFERT, WACKER, KAUDEWITZ, WEIDEL, WALDENSTRÖM. Berlin-Göttingen-Heidelberg: Springer, 1959.

115 CHEN, P. S.: Xenoplastische Transplantationen des Chorda- und Myotommaterials zwischen *Triton alpestris* und *Bombinator pachypus* im Gastrula- und Neurulastadium. *Wilhelm Roux' Arch. Entwicklungsmech. Organ.* **147** (1955).

116 —— Über den Nucleinsäure- und Proteinstoffwechsel der Frühentwicklung bei Seeigeln. *Vierteljahresschr. naturforsch. Ges. Zürich* **104** (1959).

117* —— On the problem of regional differences in the biochemical pattern during embryonal development. Sympos. Germ. Cells Development 1960.

118 —— Changes in DNA and RNA during embryonic urodele development. *Exp. Cell Res.* **21** (1960).

119 ——, u. F. BALTZER: Chimärische Haftfäden nach xenoplastischem Ektodermaustausch zwischen *Triton* und *Bombinator. Wilhelm Roux' Archiv Entwicklungsmech. Organ.* **147** (1954).

120 ——, u. E. HADORN: Zur Stoffwechselphysiologie der Mutante *letalmeander* (*lme*) bei *Drosophila melanogaster. Rev. Suisse Zool.* **62** (1955).

121* CHILD, C. M.: The physiological gradients. *Protoplasma* **5** (1928).

122 —— Differential reduction of vital dyes in the early development of echinoderms. *Wilhelm Roux' Arch. Entwicklungsmech. Organ.* **135** (1937).

123 CHUANG, H. H.: Spezifische Induktionsleistungen von Leber und Niere im Explantatversuch. *Biol. Zentralbl.* **58** (1938).

124 CLARK, F. J.: Cytogenetic studies of divergent meiotic spindle formation in *Zea mays. Amer. J. Bot.* **27** (1940).

125 CLAUSSEN, P.: Über Eientwicklung und Befruchtung bei *Saprolegnia monoica. Ber. deut. bot. Ges.* **26** (1907).

126 CLEFFMANN, G.: Über die Beteiligung von Sulfhydrylen an biologischen Prozessen. *Ergeb. Biol.* **21** (1959).

127 CLEMENT, A. C., and F. E. LEHMANN: Über das Verteilungsmuster von Mitochondrien und Lipoidtropfen während der Furchung des Eies von *Ilyanassa obsoleta. Naturwissenschaften* **43** (1956).

128 CLEVELAND, L. R.: The centrioles of *Pseudotrichonympha* and their role in mitosis. *Biol. Bull.* **69** (1935).

129 —— Origin and development of the achromatic figure. *Biol. Bull.* **74** (1938).

130 —— The whole life cycle of chromosomes and their coiling systems. *Trans. Amer. Philos. Soc., N. S.* **39** (1949).

131 —— Studies on chromosomes and nuclear division. *Trans. Amer. Phil. Soc., N. S.* **43** (1953).

132 CLEVER, U.: Über experimentelle Modifikationen des Geäders und die Beziehungen zwischen den Versorgungssystemen im Schmetterlingsflügel. Untersuchungen an *Galleria mellonella. Wilhelm Roux' Arch. Entwicklungsmech. Organ.* **151** (1959).

133 —— Der Einfluß der Sinneszellen auf die Borstenentwicklung bei *Galleria mellonella. Wilhelm Roux' Arch. Entwicklungsmech. Organ.* **152** (1960/61).

134 —— Genaktivitäten in den Riesenchromosomen von *Chironomus tentans* und ihre Beziehungen zur Entwicklung. I. *Chromosoma* **12** (1961); II. **13** (1962).

135 —— Über das Reaktionssystem einer hormonalen Induktion. Untersuchungen an *Chironomus tentans. Verh. Deut. Zool. Ges.* 1961.

136 —— Untersuchungen an Riesenchromosomen über die Wirkungsweise der Gene. *Mater. Med. Nordmark* **14** (1962).

496

137 CLEVER, U.: Von der Ecdysonkonzentration abhängige Genaktivitätsmuster in den Speicheldrüsenchromosomen von *Chironomus tentans*. *Develop. Biol.* **6** (1963).

138 COLWIN, A. L., and L. H. COLWIN: Morphology of fertilisation: Acrosome filament formation and sperm entry. In: The Beginnings of Embryonic Development. Washington: AAAS, 1957.

139 CONKLIN, E. G.: The embryology of *Crepidula*. *J. Morphol.* **13** (1897).

140 —— Organization and cell lineage of the ascidian egg. I. *Acad. Nat. Sci. Philadelphia* **13** (1905).

141 —— The development of centrifuged eggs of Ascidians. *J. Exp. Zool.* **60** (1939).

142 COOPER, K. W.: Cytogenetic analysis of the major heterochromatic elements (especially X and Y) in *Drosophila melanogaster* and the theory of "heterochromatin". *Chromosoma* **10** (1959).

143 COSTELLO, D. P.: Ooplasmic segregation in relation to differentiation. *Ann. N.Y. Acad. Sci.* **49** (1948).

144* COUNCE, S. J.: The analysis of insect embryogenesis. *Annu. Rev. Entomol.* **6** (1961).

145 CRICK, F. H.: Über den genetischen Code. *Angew. Chem.* **75** (1963).

145a CURTIS, A. S. G.: Die Zellrinde. *Endeavour* **32** (1963).

146 CZEIKA, G.: Größenunterschiede der Chromosomen von *Kleinia spinulosa*. *Chromosoma* **11** (1960).

147 CZIHAK, G.: Entwicklungsphysiologische Untersuchungen an Echiniden. *Wilhelm Roux' Arch. Entwicklungsmech. Organ.* **154** (1962/63).

148 DA CUNHA, A. X.: O desenvolvimento das glândulas sexuais na *Ephestia kühniella*. Mem. e Estodos Mus. Zool. Univers. Coimbra 1942.

149 DALCQ, A.: Étude des localisations germinales dans l'oeuf vierge d'Ascidie par des expériences de mérogonie. *Arch. Anat. micr.* **28** (1932).

150 —— Étude micrographique et quantitative de la mérogonie double chez *Ascidella scabra*. *Arch. Biol.* **49** (1938).

151 —— Contribution à l'étude du potentiel morphogénétique chez les Anoures. I. Expériences sur la zone marginale dorsale et le plancher du blastocoele. *Arch. Biol.* **51** (1940).

152 —— Le potentiel morphogénétique du pronéphros présomptif dans la gastrula des Amphibiens. *Bull. Acad. Roy. Méd. Belg. VI.* **5** (1940).

153* —— L'oeuf et son dynamisme organisateur. Paris: A. Michel, 1941.

154 —— Initiation à l'embryologie générale. Liège et Paris: Masson, 1952.

155 ——, et J. PASTEELS: Potentiel morphogénétique, régulation et «axial gradients» de Child. *Bull. Acad. Roy. Méd. Belg.* **6**, 3 (1938).

156 DALYELL, I. G.: Observations on some interesting phenomena in animal physiology, exhibited by several species of Planariae. Edinburgh 1814.

157 D'AMELIO, V., and M. P. CEAS: Distribution of protease activity in the blastula and early gastrula of *Discoglossus pictus*. *Experientia (Basel)* **13** (1957).

158 DAN, K., and T. ONO: A method of computation on the surface area of the cell. *Embryologia* **2** (1954).

159 DANIEL, J. F., and E. A. YARWOOD: The early embryology of *Triturus torosus*. *Univ. Calif. Publ. Zool.* **43** (1939).

160 DARLINGTON, C. D.: Meiosis in *Agapanthus* and *Kniphofia*. *Cytologia (Tokyo)* **4** (1932/33).

161* —— Recent Advances in Cytology, 2nd Ed. Philadelphia: Blakiston, 1937.

162 —— Misdivision and the genetics of the centromere. *J. Genet.* **37** (1939).

163 ——, and L. LA COUR: Nucleic acid starvation of chromosomes in Trillium. *J. Genet.* **40** (1940).

164* ——, and K. MATHER: The Elements of Genetics. New York: Macmillan, 1950.

165 DAVID, P. R.: Studies on the creeper fowl X. *Wilhelm Roux' Arch. Entwicklungsmech. Organ.* **135** (1936).

166 DEARING, W. H.: The material continuity and individuality of the somatic chromosomes of *Amblystoma tigrinum*, with special reference to the nucleolus as a chromosomal component. *J. Morphol.* **56** (1934).

167* DETWILER, S. R.: Neuroembryology. New York: Hafner, 1936.

168 Developing cell systems and their control, ed. DOROTHEA RUDNICK with contributions by J. T. BONNER, P. E. TARDENT, A. A. MOSCONA, B. O. PHINNEY, C. A. WEST, C. A. CILLEY, M. SINGER, H. STERN, A. B. NOVIKOFF, T. S. WORK. New York: Ronald 1960.

169 DIETZ, R.: Polarisationsmikroskopische Befunde zur chromosomeninduzierten Spindelbildung bei der Tipulide *Pales crocata* (Nematocera). *Verh. Deutsch. Zool. Ges.* 1962.

170 DOBZHANSKY, T.: Evolution und Umwelt. In: Hundert Jahre Evolutionsforschung, Herausg. G. HEBERER u. F. SCHWANITZ. Stuttgart: G. Fischer, 1960.

171 DRIESCH, H.: Der Restitutionsreiz. *Vortr. u. Aufs. Entwicklungsmech.* H. **7**. 1909.

172 DUBOIS, F.: Contribution a l'étude de la migration des cellules de régénération chez les planaires dulcioles. *Bull. Biol. Fr. Belg.* **83** (1949).

173 DU BUY, H. G., and R. A. OLSON: The presence of growth regulators during the early development of *Fucus*. *Amer. J. Bot.* **24** (1937).

174* DUSPIVA, F.: Zur Biochemie der normalen Wirbeltierentwicklung. *Naturwissenschaften* **42** (1955).

175 Egelhaaf, A.: Cytologisch-entwicklungsphysiologische Untersuchungen zur Konjugation von *Paramecium bursaria*. *Arch. Protistenk.* **100** (1905).

176* —— Genphysiologie. Biochemische Genwirkungen. *Fortschr. Zoll.* **15** (1962).

177 Eidmann, H.: Lehrbuch der Entomologie. Berlin: Parey, 1941.

178* Eigenschaften und Wirkungen der Gibberelline. Sympos. Oberhess. Ges. Natur- u. Heilkunde. Naturw. Abt. (Herausg. R. Knapp). Berlin, Göttingen, Heidelberg: Springer 1962.

179 Ekmann, G.: Beobachtungen über den Bau durch halbseitige obere Urmundlippe induzierter Embryonen bei *Triton*. *Ann. Acad. Sci. Fenn. A* **45** (1936).

180 Engelmann, F., u. M. Lüscher: Zur Frage der Auslösung der Metamorphose bei Insekten. *Naturwissenschaften* **43** (1956).

181* Ephrussi, B.: Remarks on Cell Heredity. Genetics in the 20th Century. New York: Macmillan, 1951.

182 Esser, H.: Untersuchungen zur Entwicklung des Puppenflügels von *Ephestia kühniella*. *Wilhelm Roux' Arch. Entwicklungsmech. Organ.* **153** (1961).

183 Fankhauser, G.: Über die Beteiligung kernloser Strahlungen (Cytaster) an der Furchung geschnürter *Triton*-Eier. *Rev. Suisse Zool.* **36** (1929).

184 —— Cytological studies on egg fragments of the salamander *Triton*. V. Chromosome number and chromosome individuality in the cleavage mitoses of merogonic fragments. *J. Exp. Zool.* **68** (1934).

185 Farinella-Ferruzza, N.: Lo sviluppo embrionale della ascidie doppa trattamento con LiCl. *Pubbl. Sta. Zool. Napoli* **26** (1955).

186 —— The transformation of a tail into limb after xenoplastic transplantation. *Experientia (Basel)* **12** (1956).

187 Finck, E. v.: Die Allelenserie des Gens *ss („spineless")* bei *Drosophila melanogaster*. *Biol. Zentralbl.* **62** (1942).

188 Fischberg, M., J. B. Gurdon, and T. R. Elsdale: Nuclear transplantation in *Xenopus laevis*. *Nature (London)* **181** (1958).

189 ——, and A. Blackler: Nuclear changes during the differentiation of animal cells. *Sympos. Soc. Exp. Biol.* **17** (1963).

190 Fraenkel, G., and M. Blewett: Linoleic acid, vitamin E and other fat-soluble substances in the nutrition of certain insects, *Ephestia kühniella, E. elutella, E. cautella* and *Plodia interpunctella. J. Exp. Biol.* **22** (1946).

191 Frey-Wyssling, A.: Das Streckungswachstum der pflanzlichen Zellen. *Arch. Julius Klaus-Stift. Vererbungsforsch., Sozialanthropol. Rassenhyg.* Erg.-Bd. zu **20** (1945).

191a* —— Submicroscopic Morphology of Protoplasm and Its Derivatives. London: Elsevier, 1948.

192 Friedrich-Freksa, F. H.: Bei der Chromosomenkonjugation wirksame Kräfte und ihre Bedeutung für die identische Verdoppelung von Nukleoproteinen. *Naturwissenschaften* **28** (1940).

193 Fukuda, S.: Role of the prothoracic gland in differentiation of the imaginal characters in the silkworm pupa. *Annot. Zool. Japan* **20** (1941).

193a* Gallien, L.: Problèmes et concepts de l'embryologie expérimentale. Paris: Gallimard, 1958.

194 Gautheret, R.-J.: La culture des tissus végétaux. Paris: Hermann, 1945.

195 —— La culture des tissus végétaux et les phénomènes d'histogénèse. *Anneé Biol.* **26** (1950).

196 Geigy, R.: Erzeugung rein imaginaler Defekte durch ultraviolette Eibestrahlung bei *Drosophila melanogaster*. *Wilhelm Roux' Arch. Entwicklungsmech. Organ.* **125** (1931).

197 —— Action de l'ultraviolet sur le pole germinal dans l'oeuf de *Drosophila melanogaster*. *Rev. Suisse Zool.* **38** (1931).

198* Geitler, L.: Chromosomenbau. Berlin: Borntraeger, 1938.

199 —— Das Heterochromatin der Geschlechtschromosomen bei Heteropteren. *Chromosoma* **1** (1939).

200 —— Die Entstehung der polyploiden Somakerne der Heteropteren durch Chromosomenteilung ohne Kernteilung. *Chromosoma* **1** (1939/40).

201* —— Das Wachstum des Zellkernes in tierischen und pflanzlichen Geweben. *Ergeb. Biol.* **18** (1941).

202* —— Endomitose und endomitotische Polyploidisierung. In: Protoplasmatica, Bd. VI. Wien: Springer 1953.

203 Gerisch, G.: Die Zelldifferenzierung bei *Pleodorina californica* und die Organisation der Phytomonadinenkolonien. *Arch. Protistenk.* **104** (1959/60).

204 —— Zellfunktionen und Zellfunktionswechsel in der Entwicklung von *Dictyostellum discoideum*. I. Zell. agglutination und Induktion der Fruchtkörperpolarität. *Wilhelm Roux' Arch. Entwicklungsmech. Organ.* **152** (1960).

205 —— III. Mitteilung. Getrennte Beeinflussung von Zelldifferenzierung und Morphogenese. *Wilhelm Roux-Arch. Entwicklungsmech. Organ.* **153** (1961a).

206 —— Zellkontaktbildung vegetativer und aggregationsreifer Zellen von *Dictyostelium discoideum*. *Naturwissenschaften* **48** (1961b).

207 GLOOR, H.: Zur Entwicklungsphysiologie und Genetik des Letalfaktors *crc* bei *Drosophila melanogaster*. *Arch. Klaus-Stift. Vererbungsforsch. Sozialanthropol. Rassenhyg.* **20** (1945).

208 —— Phänokopie-Versuche mit Äther an *Drosophila*. *Rev. Suisse Zool.* **54** (1947).

209 —— Biochemische Untersuchungen am Letalfaktor "*lethal-translucida*" (*ltr*) von *Drosophila melanogaster*. *Rev. Suisse Zool.* **56** (1949).

210 —— Schädigungsmuster eines Letalfaktors (*Kr*) von *Drosophila melanogaster*. *Arch. Julius Klaus-Stift Vererbungsforsch. Sozialanthropol. Rassenhyg.* **25** (1950).

211* GLUECKSOHN-WAELSCH, S.: Lethal factors in development. *Quart. Rev. Biol.* **28** (1953).

212 GOEBEL, K.: Organographie der Pflanzen, 3. Aufl., Teil 3, 2. Hälfte. Jena: G. Fischer 1933,.

213 GOERTTLER, K.: Die Formbildung der Medullaranlage bei den Urodelen. *Wilhelm Roux' Arch. Entwicklungsmech. Organ.* **106** (1925).

214 GORGONE, I.: Larve giganti uniche ottenute da due uova die Ascidie fuse allo stadio di 8-blastomeri. *Accad. Naz. Lincei, Cl. Fisiche, Ser. VIII,* **30** (1961).

215 GRAFL, I.: Cytologische Untersuchungen an *Sauromatum guttatum*. *Österr. Bot. Z.* **89** (1940).

216 GRASSÉ, P.-P., N. CARASSO et P. FAVARD: Les ultrastructures au cours de la spermiogenèse d'escargot *(Helix pomatia)*. *Ann. Sci. Natur. Zool. Biol. Anim.* **18** (Ser. II) (1956).

217 GREGG, J. R.: Morphogenesis and metabolism of gastrula-arrested embryos in the hybrid. In: The Beginnings of Embryonic Development, ed. by A. TYLER, R. C. VON BORSTEL and C. B. METZ. Washington: AAAS 1957.

218* GRELL, K. G.: Protozoologie. Berlin, Göttingen, Heidelberg: Springer, 1956.

219 —— Studien zum Differenzierungsproblem an Foraminiferen. *Naturwissenschaften* **45** (1958).

220 —— Untersuchungen über die Fortpflanzung und Sexualität der Foraminiferen. IV. *Petellina corrugata*. *Arch. Protistenk.* **104** (1959).

221 —— Zur Determination der Zellkerne bei der Foraminifere *Rotaliella heterocaryotica*. *Naturwissenschaften* **47** (1960).

222 —— Nachweis der sexuellen Differenzierung bei *Patellina corrugata* durch Teilbildanalyse eines Films. *Z. Naturforsch.* **15** b (1960).

223 —— Morphologie und Fortpflanzung der Protozoen (einschließlich Entwicklungsphysiologie und Genetik). *Fortschr. Zool.* **14** (1962).

224 GRÜNEBERG, H.: A new sub-lethal colour mutation in the house mouse. *Proc. Roy. Soc.* B **118** (1935).

225 —— *Grey-lethal*, a new mutation in the house mouse. *J. Hered.* **27** (1936).

226 —— The relations of endogenous and exogenous factors in bone and tooth development. The teeth in the *grey-lethal* mouse. *J. Anat.* **71** (1939).

227* —— The Genetics of the Mouse. s'Gravenhage: Mart. Nijhoff, 1952.

228 GURDON, J. B.: Adult frogs derived from the nuclei of single somatic cells. *Develop. Biol.* **4** (1962).

229 GUSTAFSON, I., and M. -B.HJELTE: The amino acid metabolism of the developing sea urchin egg. *Exp. Cell Res.* **2** (1951).

230 HADORN, E.: Gene action in growth and differentiation of lethal mutants of *Drosophila*. *Symp. Soc. Exp. Biol.* **2** (1948).

231 —— Zur Entwicklungsphysiologie der Mutante „*lethal translucida*" *(ltr)* von *Drosophila melanogaster*. *Rev. Suisse Zool.* **56** (1949).

232 —— Physiogenetische Ergebnisse der Untersuchungen an Drosophilablastemen aus letalen Genotypen. *Rev. Suisse Zool.* **57** (1950).

233 —— Developmental action of lethal factors in *Drosophila*. *Advan. Genet.* **4** (1951).

234 —— Regulation and differentiation within field, districts in imaginal discs of *Drosophila*. *J. Embryol. Exp. Morphol.* **1** (1953).

235* —— Letalfaktoren in ihrer Bedeutung für Erbpathologie und Genphysiologie der Entwicklung. Stuttgart: Thieme, 1955.

235a —— Differenzierungsleistungen wiederholt fragmentierter Teilstücke männlicher Genitalscheiben von *Drosophila melanogaster* nach Kultur in vivo. *Develop. Biol.* **7** (1963).

235b —— Bedeutungseigene und bedeutungsfremde Entwicklungsleistungen proliferierender Primordien von *Drosophila* nach Dauerkultur in vivo. *Rev. Suisse Zool.* **71** (1964).

236 ——, u. G. BERTANI u. J. GALLERA: Regulationsfähigkeit und Feldorganisation der männlichen Genitalimaginalscheibe von *Drosophila melanogaster*. *Wilhelm Roux' Arch. Entwicklungsmech. Organ.* **144** (1949).

237 ——, u. P. S. CHEN: Untersuchungen zur Phasenspezifität der Wirkung von Letalfaktoren bei *Drosophila melanogaster*. *Arch. Julius-Klaus-Stift. Vererbungsforsch. Sozialanthropol. Rassenhyg.* **27** (1952).

237a* HAGEMANN, R.: Plasmatische Vererbung. Jena: G. Fischer, 1964.

238 HÅKANSSON, A.: Notes on the giant chromosomes of *Allium nutans*. *Bot. Notis* **110** (1957).

239* HALDANE, J. B. S.: The Biochemistry of Genetics. London: Allen & Unwin, 1954.

240 HÄMMERLING, J.: Über formbildende Substanzen bei *Acetabularia mediterranea*, ihre räumliche und zeitliche Verteilung und ihre Herkunft. *Wilhelm Roux' Arch. Entwicklungsmech. Organ.* **131** (1934).

241 ―――― Über Genomwirkungen und Formbildungsfähigkeit bei *Acetabularia*. *Wilhelm Roux' Arch. Entwicklungsmech. Organ.* **132** (1934).

242 ―――― Über die Bedingungen der Kernteilung und der Cystenbildung bei *Acetabularia mediterranea*. *Biol. Zentralbl.* **59** (1939).

243 ―――― Ein- und zweikernige Transplantate zwischen *Acetabularia mediterranea* und *A. crenulata*. *Z. indukt. Abstamm- u. Vererb.-Lehre* **81** (1943).

244 ―――― Neuere Versuche über Polarität und Differenzierung bei *Acetabularia*. *Biol. Zentralbl.* **74** (1955).

245 ―――― Über mehrkernige Acetabularien und ihre Entstehung. *Biol. Zentralbl.* **74** (1955).

246 ―――― *Spirogyra* und *Acetabularia*. (Ein Vergleich ihrer Fähigkeiten nach Entfernung des Kernes.) *Biol. Zentralbl.* **78**, 703–709 (1959).

247 HARDER, R.: Zur Frage nach der Rolle von Kern und Protoplasma im Zellgeschehen und bei der Übertragung von Eigenschaften. *Z. Bot.* **19** (1927).

248 ―――― Fortpflanzung der Gewächse. In: Handwörterbuch der Naturwissenschaft, Bd. 4. Jena: Fischer, 1934.

249 ―――― Über die photoperiodisch bedingte Organ- und Gestaltbildung bei den Pflanzen. *Naturwissenschaften* 33 (1946).

250 ―――― Die stoffliche Grundlage einiger Gestaltungsvorgänge bei der Pflanze. *Naturw. Rundschau* **6** (1953).

251 ――――, u. E. GALL: Über die Trennung der Blühhormon- und Metaplasinwirkung bei *Kalanchoë Blossfeldiana* durch Narkose. Nachr. Akad. Wiss. Göttingen, math.-physik. Kl. (1945).

252 ――――, u. H. v. WITSCH: Über den Einfluß der Tageslänge auf den Habitus, besonders die Blattsukkulenz und den Wasserhaushalt von *Kalanchoë Blossfeldiana*. *Jahrb. wiss. Bot.* **89** (1940).

253 ―――― ――――, u. O. BODE: Über die Erzeugung einseitig und allseitig verlaubter Infloreszenzen durch photoperiodische Behandlung von Laubblättern (Untersuchungen an *Kalanchoë Blossfeldiana*.) *Jahrb. wiss. Bot.* **90** (1942).

254 HARRISON, R. G.: Experiments in transplanting limbs and their bearing upon the problems of the development of nerves. *J. Exp. Zool.* **4** (1907).

255 ―――― Experiments on the development of the fore-limb of *Amblystoma*, a self-differentiating equipotential system. *J. Exp. Zool.* **25** (1918).

256 ―――― On relations of symmetry in transplanted limbs. *J. Exp. Zool.* **32** (1921).

257 ―――― The effect of reversing the mediolateral or transverse axis of the fore-limb bud in the salamander embryo (*Amblystoma punctatum* LINN). *Wilhelm Roux' Arch. Entwicklungsmech. Organ* **106** (1925).

258 HARTMANN, M.: Untersuchungen über die Morphologie und Physiologie des Formwechsels der Phytomonadinen (Volvocales). I. *Arch. Protistenk.* **39** (1918); III. **43** (1921); IV. **49** (1924).

259 ―――― Über experimentelle Unsterblichkeit von Protozoen-Individuen. Ersatz der Fortpflanzung von *Amoeba proteus* durch fortgesetzte Regenerationen *Zool. Jahrb., Abt. allg. Zool. Physiol. Tiere* **45** (1928).

260* ―――― Die Sexualität, 2. Aufl. Stuttgart: Fischer, 1956.

261 HARVEY, E. B.: The development of half and quarter eggs of *Arbacia punctata* and of strongly centrifuged whole eggs. *Biol. Bull.* **62** (1932).

262 ―――― Parthenogenetic merogony or cleavage without nuclei in *Arbacia punctata*. *Biol. Bull.* **71** (1936).

263 HASITSCHKA-JENSCHKE, G.: Bemerkenswerte Kernstrukturen im Endosperm und im Suspensor zweier Helobiae. *Österr. Bot. Z.* **106** (1959a).

264 ―――― Vergleichende karyologische Untersuchungen an Antipoden. *Chromosoma* **10** (1959b).

265 HAUPT, W.: Untersuchungen über den Determinationsvorgang der Blütenbildung bei *Pisum sativum*. *Z. Bot.* **40** (1952).

266 ―――― Die Entstehung der Polarität in pflanzlichen Keimzellen, insbesondere die Induktion durch Licht. *Ergeb. Biol.* **25** (1962).

267 HEATLEY, N. G., and P. E. LINDAHL: The distribution and nature of glycogen in the amphibian embryo. *Proc. Roy. Soc. London* **122** (1937).

267a HECKMANN, K.: Paarungssystem und genabhängige Paarungsdifferenzierung bei dem hypotrichen Ciliaten *Euplotes vannus*. *Arch. Protistenk.* **106** (1963).

268* ―――― Der Zellkern der Protozoen, Tatsachen und Probleme. Verh. Anat. Ges. 1963. *Anat. Anz.* Ergänzungsheft **113** (1964).

269 HEITZ, E.: Die Ursache der gesetzmäßigen Zahl, Lage, Form und Größe pflanzlicher Nucleolen. *Planta* (Berl.) **12** (1931).

270 ―――― Die somatische Heteropyknose bei *Drosophila melanogaster*. *Z. Zellforsch.* **20** (1934).

271 ―――― Die Polarität keimender Moossporen. *Verh. schweiz. naturforsch. Ges.* **168–170** (1940).

272 ―――― Die keimende *Funaria*-Spore als physiologisches Versuchsobjekt. *Ber. deut. bot. Ges.* **60** (1942).

273 Henke, K.: Untersuchungen an *Philosamia cynthia* Drury zur Entwicklungsphysiologie des Zeichnungsmusters auf dem Schmetterlingsflügel. *Wilhelm Roux' Arch. Entwicklungsmech. Organ.* **128** (1933).

274 —— Einfache Grundvorgänge in der tierischen Entwicklung. *Naturwissenschaften* **35** (1948).

275 —— Die Musterbildung der Versorgungssysteme im Insektenflügel. *Biol. Zentralbl.* **72** (1953).

276 ——, u. H.-J. Pohley: Differentielle Zellteilungen und Polyploidie bei der Schuppenbildung der Mehlmotte *Ephestia kühniella. Z. Naturforsch.* **7**b (1952).

277 Herbst, C.: Weiteres über die morphologische Wirkung der Lithiumsalze und ihre theoretische Bedeutung. *Mitt. zool. Stat. Neapel* **11** (1893).

278★ Hess, O.: Entwicklungsphysiologie der Mollusken. *Fortschr. Zool.* **14** (1962).

279 Hirschler, I.: Organisation und Genese des Ei-Nährzellen-Verbandes im Ovarium von *Macrothylacia rubi* L. — Lepidoptera. *Biol. Zentralbl.* **62** (1942).

280 —— Osmiumschwärzung perichromosomaler Membranen in den Spermatocyten der Rhynchoten-Art *Palomena viridissima. Naturwissenschaften* **30** (1942).

281 Hoffmann-Berling, H.: Die glycerin-wasserextrahierte Telophasenzelle als Modell der Zytokinese. *Biochim. Biophys. Acta* **15** (1954).

282★ —— Physiologie der Bewegungen und Teilungsbewegungen tierischer Zellen. *Fortschr. Zool.* **11** (1958).

283 Hohl, K.: Experimentelle Untersuchungen über Röntgeneffekte und chemische Effekte auf die pflanzliche Mitose. Stuttgart: O. Thieme, 1949.

284 Holman, R. M., and W. W. Robbins: A Textbook of General Botany. Madison U. S. Armed Forces Inst. 1944.

285 Holtfreter, J.: Der Einfluß von Wirtsalter und verschiedenen Organbezirken auf die Differenzierung von angelagertem Gastrulaektoderm. *Wilhelm Roux' Arch. Entwicklungsmech. Organ.* **127** (1933a).

286 —— Die totale Exogastrulation, eine Selbstablösung des Ektoderms von Entomesoderm. *Wilhelm Roux' Arch. Entwicklungsmech. Organ.* **129** (1933b).

287 —— Organisationsstufen nach regionaler Kombination von Entomesoderm mit Ektoderm. *Biol. Zentralbl.* **53** (1933c).

288 —— Regionale Induktionen in xenoplastisch zusammengesetzten Explantaten. *Wilhelm Roux' Arch. Entwicklungsmech. Organ.* **134** (1936).

289 —— Differenzierungspotenzen isolierter Teile der Urodelengastrula. *Wilhelm Roux' Arch. Entwicklungsmech. Organ.* **138** (1938a).

290 —— Differenzierungspotenzen isolierter Teile der Anurengastrula. *Wilhelm Roux' Arch. Entwicklungsmech. Organ.* **138** (1938b).

291 —— Studien zur Ermittlung der Gestaltungsfaktoren in der Organentwicklung der Amphibie. *Wilhelm Roux' Arch. Entwicklungsmech. Organ.* **139** (1939).

292 —— Gewebeaffinität, ein Mittel der embryonalen Formbildung. *Arch. exp. Zellforsch.* **23** (1939).

293 —— Properties and functions of the surface coat in amphibian embryos. *J. Exp. Zool.* **93** (1943).

294 —— Neural differentiation of ectoderm through exposure to saline solution. *J. Exp. Zool.* **95** (1944).

295 —— Neuralization and epidermization of gastrula ectoderm. *J. Exp. Zool.* **98** (1945).

296 —— Some aspects of embryonic induction. *Growth* **15**, Suppl. (1951).

297 Hovanitz, W., A. R. T. Denues, and R. M. Sturrock: The internal structure of isolated chromosomes. *Wasmann Collector* **7** (1949).

298 Hörstadius, S.: Über die Determination des Keimes bei Echinodermen. *Acta Zool. (Stockholm)* **9** (1928).

299 —— Über die Determination im Verlaufe der Eiachse bei Seeigeln. *Publ. Stat. Zool. Napoli* **14** (1935).

300 —— Determination in the early development of the seaurchin. *Collect. Net. (Woods Hole)* **11** (1936).

301 —— Über die zeitliche Determination im Keim von *Paracentrotus lividus. Wilhelm Roux' Entwicklungsmech. Organ.* **135** (1937).

302 —— Weitere Studien über die Determination im Verlaufe der Eiachse bei Seeigeln. *Wilhelm Roux' Arch. Entwicklungsmech. Organ.* **135** (1937).

303 —— Investigations as to the localisation of the micromere-, the skeleton- and the entoderm-forming material in the unfertilized egg of *Arbacia pustulata. Biol. Bull.* **73** (1937).

304 —— Experiments on determination in the early development of *Cerebratulus lacteus. Biol. Bull.* **73** (1937).

305 —— Schnürungsversuche an Seeigelkeimen. *Wilhelm Roux' Arch. Entwicklungsmech. Organ.* **183** (1938).

306 —— The mechanics of sea urchin development, studied by operative methods. *Biol. Rev.* **14** (1939).

307 —— Über die Entwicklung von *Astropecten aranciacus. Pubbl. Sta. Zool. Napoli* **17** (1939). *Ann. Sci. Sér.* 11, **18** (1956).

308 —— Experimental researches on the developmental physiology of the sea urchin. *Pubbl. Sta. Zool. Napoli*, Suppl. **21** (1949).

309★ —— The Neural Crest. London, New York, Toronto: Oxford Univ. Press 1950.

310 Hörstadius, S.: Induction and inhibition of reduction gradients by the micromeres in the sea urchin egg. *J. Exp. Zool.* **120** (1952).

311 —— Reduction gradients in animalized and vegetalized sea urchin eggs. *J. Exp. Zool.* **129** (1955).

312 —— The effect of sugars on differentiation of larvae of *Psammechinus miliaris*. *J. Exp. Zool.* **142** (1959).

313 —— J. J. Lorch, and J. F. Danielli: Differentiation of the sea urchin egg following reduction of the interior cytoplasm in relation to the cortex. *Exp. Cell Res.* **1** (1950).

314 ——, u. A. Wolsky: Studien über die Determination der Bilateralsymmetrie des jungen Seeigelkeimes. *Wilhelm Roux' Arch. Entwicklungsmech. Organ* **135** (1936).

315* Hughes, H.: The Mitotic Cycle. New York and London: Butterworth 1952.

316 Hughes-Schrader, S.: A study of the chromosome cycle and the meiotic division-figure in *Llaveia bouvari*, a primitive coccid. *Z. Zellforsch.* **13** (1931).

317* —— Cytology of coccids (Coccoidea-Homoptera). *Advan. Genet.* **2** (1948).

318 —— The "pre-metaphase stretch" and kinetochore orientation in phasmids. *Chromosoma* **3** (1950).

319 ——, and H. Ris: The diffuse spindle attachment of coccids, verified by the mitotic behavior of induced chromosome fragments. *J. Exp. Zool.* **87** (1941).

320 Huskins, C. L.: The subdivision of the chromosomes and their multiplication in non dividing tissues: Possible interpretation in terms of gene structure and gene action. *Amer. Natur.* **81** (1947).

321 —— and S. G. Smith: Meiotic chromosome structure in *Trillium erectum* L. *Ann. Bot.* **49** (1941).

322* Hutt, F. B.: Genetics of the Fowl. New York: McGraw Hill 1949.

323* Huxley, J. S., and G. R. De Beer: The Elements of Experimental Embryology. Cambridge: Univ. Press, 1934.

324 Induktion und Morphogenese. 13. Colloqu. Ges. Physiol. Chem. Berlin, Heidelberg, Göttingen: Springer, 1962.

325 Ishida, I., and I. Yasumasu: Changes in protein specifity determined by protective enzymes during embryonic development of the sea urchin and the fresh-water fish. I. *Fac. Sci. Univ. Tokyo* **8** (1957).

326 Jacobj, W.: Über das rhythmische Wachstum der Zellen durch Verdoppelung ihres Volumens. *Wilhelm Roux' Arch. Entwicklungsmech. Organ.* **106** (1925).

327 Jantsch, B.: Entwicklungsphysiologische Untersuchungen am Blatt von *Iris japonica*. *Z. Bot.* **47** (1959).

328 Janssens, F. A.: La chiasmatypie dans les insectes. *Cellule* **34** (1924).

329 Karlson, P.: Chemische Untersuchungen über die Metamorphosehormone der Insekten. *Ann. Sci. Nat., Sér. XI*, **18** (1956).

330* —— Chemie und Biochemie der Insektenhormone. *Angew. Chem.* **75** (1963).

331 ——, u. H. Hoffmeister: Zur Chemie des Ecdysons. *Justus Liebigs Ann. Chem.* **662** (1963).

332 Kaudewitz, F.: Zur Entwicklungsphysiologie von *Daphnia pulex*. *Wilhelm Roux' Arch. Entwicklungsmech. Organ.* **144** (1950).

333 Kaufmann, B. P.: Somatic mitoses of *Drosophila melanogaster*. *J. Morph.* **56** (1934).

334 —— Morphology of the chromosomes of *Drosophila ananassae*. Cytologia Fujii. Jub. Vol. 1937.

335 Keil, E.: Studien über Regenerationserscheinungen an *Polycelis nigra*. *Wilhelm Roux' Arch. Entwicklungsmech. Organ.* **102** (1924).

336 Keyl, H. G.: Erhöhung der chromosomalen Replikationsrate durch Mikrosporidieninfektion in Speicheldrüsen von *Chironomus*. *Naturwissenschaften* **47** (1960).

337 —— Die cytologische Diagnostik der Chironomiden. III. *Arch. Hydrobiol.* **58** (1961).

338 Klebs, G.: Die Bedingungen der Fortpflanzung bei einigen Algen und Pilzen. Jena: G. Fischer, 1896.

339 —— Willkürliche Entwicklungsveränderung bei Pflanzen. Jena: G. Fischer, 1903.

340 Knaben, N.: Oogenese bei *Tischeria angusticolella* Dup. *Z. Zellforsch.* **21** (1934).

341 Knapp, E.: Entwicklungsphysiologische Untersuchungen an Fucaceen-Eiern. *Planta (Berl.)* **14** (1931).

342 Kniep, H.: Beiträge zur Keimungs-Physiologie und Biologie von *Fucus*. *Jahrb. wiss. Bot.* **44** (1907).

343* Koecke, H. U.: Entwicklungsphysiologie der Vögel. *Fortschr. Zool.* **16** (1963).

344 Koehler, O.: Über die Abhängigkeit der Kernplasmarelation von der Temperatur und vom Reifezustand der Eier. *Arch. Zellforsch.* **8** (1912).

345 Koelitz, W.: Längsteilung und Doppelbildungen bei *Hydra*. *Zool. Anz.* **35** (1910).

346 Köhler, W.: Die Entwicklung der Flügel bei der Mehlmotte *Ephestia kühniella*, mit besonderer Berücksichtigung des Zeichenmusters. *Z. Morph. u. Ökol. Tiere* **24** (1932).

347* Korschelt, E., u. K. Heider: Lehrbuch der vergleichenden Entwicklungsgeschichte der wirbellosen Tiere, Allgem. Teil. Jena: G. Fischer, 1910.

348* —— Regeneration und Transplantation. I. Regeneration. Berlin: Borntraeger, 1927.

349 Krause, G.: Die Regulationsfähigkeit der Keimanlage von *Tachycines* (Orthoptera) im Extraovatversuch. *Wilhelm Roux' Arch. Entwicklungsmech. Organ.* **139** (1939).

350* —— Die Eitypen der Insekten. *Biol. Zentralbl.* **59** (1939).

351 KRAUSE, G.: Schnittoperationen im Insektenei zum Nachweis der komplementären Induktion bei Zwillings-bildung. *Naturwissenschaften* **39** (1952).

352 —— Die Aktionsfolge zur Gestaltung des Keimstreifs von *Tachycines* (Saltatoria), insbesondere das morphogenetische Konstruktionsbild bei Duplicitas parallela. *Wilhelm Roux' Arch. Entwicklungsmech. Organ.* **146** (1953).

353* —— Induktionssysteme in der Embryonalentwicklung von Insekten. *Ergeb. Biol.* **20** (1958).

354 —— Zum Verhalten explantierter Lavalgewebe und Embryonalanlagen des Seidenspinners *Bombyx mori.* *Zool. Anz.* **26**, Suppl. Verh. Dtsch. Zool. Ges. 1962 (1963).

354a —— Schichtenbau und Segmentierung junger Keimanlagen von *Bombyx mori* (Lepidoptera) in vitro ohne Dottersystem. *Wilhelm Roux' Arch. Entwicklungsmech. Organ.* **155** (1964).

354b —— Preformed ooplasmic reaction systems in insect eggs. Symp. Germ Cells Develop. 1960.

355 KUHL, W., u. G.: Neue Ergebnisse zur Cytodynamik der Befruchtung und Furchung des Eies von *Psammechinus miliaris.* *Zool. Jahrb., Abt. Anat u. Ontog.* **70** (1950).

356 KÜHN, A.: Über die Beziehungen zwischen Plasmateilung und Kernteilung bei Amöben. *Zool. Anz.* **48** (1916).

357 —— Untersuchungen zur kausalen Analyse der Zellteilung. I. Zur Morphologie und Physiologie der Kern-teilung von *Vahlkampfia bistadialis.* *Wilhelm Roux' Arch. Entwicklungsmech. Organ.* **46** (1920).

358* —— Entwicklungsphysiologisch-genetische Ergebnisse an *Ephestia kühniella.* *Z. indukt. Abstamm.- u. Vererb.-Lehre* **73** (1937).

359 —— Zur Entwicklungsphysiologie der Schmetterlingsmetamorphose. Intern. Entom.-Kongr. Bd. 2., 1939.

360 —— Die Ausprägung organischer Formen in verschiedenen Dimensionen und die Grundfragen der Ent-wicklungsphysiologie. *Naturwissenschaften* **31** (1943).

361 —— Konstruktionsprinzipien von Schmetterlingsschuppen nach elektronenmikroskopischen Aufnahmen. *Z. Naturforsch.* **1** (1946).

362 —— Die Wirkung der Mutation *Va (Venis abnormibus)* bei *Ptychopoda seriata.* *Z. indukt. Abstamm.-u. Vererb.-Lehre* **82** (1948).

363 —— Genetisch bedingte Mosaikbildungen bei *Ephestia kühniella.* *Z. Vererbungsl.* **91** (1960).

364 —— Grundriß der Vererbungslehre, 4. Aufl. Heidelberg: Quelle & Meyer 1965.

365 ——, u. M. AN: Elektronenoptische Untersuchungen über den Bau von Schmetterlingsschuppen. *Biol. Zentralbl.* **65** (1946).

366 ——, u. B. BERG: Parabiosen als Mittel zur Aufklärung von Genwirkungen. *Biol. Zentralbl.* **81** (1962).

367 ——, u. M. V. ENGELHARDT: Über die Determination des Symmetriesystems auf dem Vorderflügel von *Ephestia kühniella.* *Wilhelm Roux' Arch. Entwicklungsmech. Organ.* **130** (1933).

368 ——, u. K. HENKE: Genetische und entwicklungsphysiologische Untersuchungen an der Mehlmotte *Ephestia kühniella.* A. Ges. Wiss. Göttingen, math.-physik. Kl., N. F. 15, I–VII (1929); VII–XII (1932); XIII, XIV (1936).

369 ——, u. H. PIEPHO: Über hormonale Wirkungen bei der Verpuppung der Schmetterlinge. Nachr. Ges. Wiss. Göttingen, math.-physik. Kl., N. F. Fachg. VI (1936).

370 —— —— Die Reaktionen der Hypodermis und der Versonschen Drüsen auf das Verpuppungshormon bei *Ephestia kühniella.* *Biol. Zentralbl.* **58** (1938).

371 —— —— Über die Ausbildung der Schuppen in Hauttransplantaten von Schmetterlingen. *Biol. Zentralbl.* **60** (1940).

372 KUNO, M.: On the nature of the egg surface during cleavage of the sea urchin egg. *Embryologia* **2** (1954).

373 KUPKA, E., u. F. SEELICH: Die anaphasische Chromosomenbewegung. Bei Beitrag zur Theorie der Mitose. *Chromosoma* **3** (1950).

374 KUSKE, G.: Untersuchungen zur Metamorphose der Schmetterlingsbeine. *Wilhelm Roux' Arch. Entwicklungs-mech. Organ.* **154** (1963).

375* KÜSTER, E.: Die Pflanzenzelle in: Experimentelle Zellforschung, 2. Aufl. Jena: G. Fischer, 1949.

376 LANDAUER, W.: The "Frizzle" character of fowls. *J. Hered.* **21** (1930).

377 —— Untersuchungen über das Krüperhuhn II. *Z. mikrosk.-anat. Forsch.* **25** (1931).

378 —— Studies on the creeper fowl. XII *Storrs Agr. Exp. Stat. Bull.* **232** (1939).

379 —— Hereditary abnormalities and their chemically induced phenocopies. Growth Symp. **12** (1948).

380 ——, and L. C. DUNN: Studies on the creeper fowl I. *J. Genet.* **23** (1930).

381* LANG, A.: Entwicklungsphysiologie. *Fortschr. Bot.* **11** (1944); **12** (1949).

382* —— Entwicklungsphysiologie. *Fortschr. Bot.* **23** (1961).

383 LANG, P.: Regeneration bei Planarien. *Arch. mikr. Anat.* **79** (1912).

384 —— Heteromorphose und Polarität bei Planarien. *Arch. mikr. Anat.* **82** (1913).

385 LA SPINA, R.: Development of *Ascidia malaca* egg fragments produced by centrifugation. *Acta Embryol. Morph.* **4** (1961).

386 Le Calvez, J.: Recherches sur les foraminifères. I. Développement et reproduction. *Arch. Zool. exp.* **80** (1938).

387 —— In (3 R) *ss*^Ar: Mutation « *Aristapedia* » hétérozygote dominante, homozygote léthale chez *Drosophila melanogaster*. (Inversion dans le bras du chromosoms III.) *Bull. Biol.* **82** (1948).

388 Lehmann, F. E.: Neuere experimentelle Forschungen über die Morphogenese des Nervensystem der Wirbeltiere. *Z. ges. Neurol. Psychiat.* **115** (1928).

389 —— Mesodermisierung des präsumptiven Chordamaterials durch Einwirkung von Lithiumchlorid auf die Gastrula von *Triton alpestris*. *Wilhelm Roux' Arch. Entwicklungsmech. Organ* **136** (1937).

390 —— Regionale Verschiedenheiten des Organisators von *Triton*, insbesondere in der vorderen und hinteren Kopfregion, nachgewiesen durch phasenspezifische Erzeugung von lithiumbedingten und operativ bewirkten Regionaldefekten. *Wilhelm Roux' Arch. Entwicklungsmech. Organ.* **138** (1938).

391 —— Über die Struktur des Amphibieneies. *Rev Suisse Zool.* **49** (1942).

392★ —— Einführung in die physiologische Embryologie. Basel: Birkhäuser, 1945.

393 —— Die Morphogenese in ihrer Abhängigkeit von elementaren biologischen Konstituenten des Plasmas. *Rev. Suisse Zool. Fasc. suppl.* **57** (1950).

394 —— Plasmatische Eiorganisation und Entwicklungsleistung beim Keim von *Tubifex* (Spiralia). *Naturwissenschaften* **43** (1956).

395 —— Der Feinbau der Organoide von *Amoeba proteus* und seine Beeinflussung durch verschiedene Fixierstoffe. *Ergeb. Biol.* **21** (1959).

396★ —— Die Physiologie der Mitose. *Ergeb. Biol.* **27** (1964).

397 Lehn, H.: Teilungsfolgen und Determination von I-Zellen für die Cnidenbildung bei *Hydra*. *Z. Naturforsch.* **6** b (1951).

398 —— Die histologischen Vorgänge bei der Reparation von Hydren aus Aggregaten kleiner Fragmente. *Wilhelm Roux' Arch. Entwicklungsmech. Organ.* **146** (1953).

399 Leikola, A.: The mesodermal and neural competence of isolated gastrula ectoderm studied by heterogenous inductors. *Ann. Zool. Soc. Zool. Botan. Fennicae Vanamo* **25** (1963).

400 Lender, Th.: Le rôle inducteur du cerveau dans la régéneration des yeux d'une planaire d'eau douce. *Bull. Biol. Fr. Belg.* **86** (1952).

401 —— Recherches expérimentales sur la nature de l'inducteur de la régéneration des yeux de la planaire *Polycelis nigra*. *J. Embryol. Exp. Morph.* **4** (1956).

402 Lerche, W.: Untersuchungen über Entwicklung und Fortpflanzung in der Gattung *Dunaliella*. *Arch. Protistenk.* **88** (1937).

403 Lesley, M., and B. Frost: Mendelian inheritance of chromosome shape in *Matthiola*. *Genetics* **12** (1927).

404 Levan, A.: The effect of colchicine on root mitoses in *Allium*. *Hereditas (Lund)* **24** (1938).

405 Lillie, F. R.: Observations and experiments concerning the elementary phenomena of embryonic development in *Chaetopterus*. *J. Exp. Zool.* **3** (1906).

406 —— Polarity and bilaterality of the annelid egg. Experiments with centrifugal force. *Biol. Bull.* **16** (1909).

407 Lindahl, P. E.: Zur Kenntnis des Ovarialeis bei dem Seeigel. *Wilhelm Roux' Arch. Entwicklungsmech. Organ.* **126** (1932).

408 —— Zur Kenntnis der physiologischen Grundlagen der Determination im Seeigelkeim. *Acta Zool. (Stockholm)* **17** (1936).

409 ——, u. L. O. Öhmann: Weitere Studien über Stoffwechsel und Determination im Seeigelkeim. *Biol. Zentralbl.* **58** (1938).

410 —— Zur Kenntnis der Entwicklungsphysiologie des Seeigeleies. *Z. vergl. Physiol.* **27** (1940).

411 Lindh, N. O.: The metabolism of nucleic acids during regeneration in *Euplanaria polychroa*. *Ark. Zool. (Stockh.) Andra Ser.*, **9** (1956).

412 Lison, L., et J. Pasteels: Études histophotométriques sur la teneur en acide désoxyribonucléique du noyau aux cours du développement embryonaire chez l'oursin *Paracentrotus lividus*. *Arch. Biol.* **62** (1951).

413 Lund, E. J.: Electrical control of organic polarity in the egg of *Fucus*. *Bot. Gaz.* **76** (1923).

414 Lus, J.: Studies on regeneration and transplantation in Turbellaria. *Bull. Soc. Nat. Moskau, Sect. Biol. exp.* **1** (1924).

415 Lüscher, M.: Experimentelle Untersuchungen über die larvale und imaginale Determination im Ei der Kleidermotte. *Rev. Suisse Zool.* **51** (1944).

416 Maas, A.-H.: Über die Auslösbarkeit der Temperaturmodifikationen während der Embryonalentwicklung von *Drosophila melanogaster*. *Wilhelm Roux' Arch. Entwicklungsmech. Organ.* **143** (1948).

417 Mahr, E.: Struktur und Entwicklungsfunktionen des Dotterentoplasmasystems im Ei des Heimchens *(Gryllus domesticus)*. *Wilhelm Roux' Arch. Entwicklungsmech. Organ.* **152** (1960).

418 —— Bewegungssysteme in der Embryonalentwicklung von *Gryllus domesticus*. *Wilhelm Roux' Arch. Entwicklungsmech. Organ.* **152** (1960).

419 MAKINO, S.: On the chromocentre observed through the mitotic cycle of somatic cells in *Drosophila virilis*. *Cytologia (Tokyo)* **10** (1940).

420★ MANGOLD, O.: Das Determinationsproblem. III. Das Wirbeltierauge in der Entwicklung und Regeneration. *Ergeb. Biol.* **7** (1931).

421 —— Über die Induktionsfähigkeit der verschiedenen Bezirke der Neurula von Urodelen. *Naturwissenschaften* **21** (1933).

422★ —— Experimente zur Entwicklungsphysiologie des Urodelenkopfes. *Verh. Anat. Ges.* **54**. Verslg 1957.

423★ —— Grundzüge der Entwicklungsphysiologie der Wirbeltiere mit besonderer Berücksichtigung der Mißbildungen auf Grund experimenteller Arbeiten an Urodelen. Aus: LUIGI GEDDA, De genetica medica, Roma 1961.

424 —— Molchlarven ohne Zentralnervensystem und ohne Ektomesoderm. *Wilhelm Roux' Arch. Entwicklungsmech. Organ.* **152** (1961).

425★ —— Freie Linsen in augenlosen Köpfen und Isolaten von *Triturus alpestris*. *Acta Morphol. Acad. Sci. Hung.* **10** (1961).

426 ——, u. F. SEIDEL: Homoplastische und heteroplastische Verschmelzung ganzer Tritonkeime. *Wilhelm Roux' Arch. Entwicklungsmech. Organ.* **111** (1927).

427 ——, u. C. v. WOELLWARTH: Das Gehirn von *Triton*. Ein experimenteller Beitrag zur Analyse seiner Determination. *Naturwissenschaften* **37** (1950).

428 MASCHLANKA, H.: Physiologische Untersuchungen am Ei der Mehlmotte *Ephestia kühniella*. *Wilhelm Roux' Arch. Entwicklungsmech. Organ.* **137** (1938).

429 MATSUURA, H.: Chromosome studies on *Trillium kamtschaticum* PALL. XIII. The structure and behavior of the kinetochore. *Cytologia (Tokyo)* **11** (1941).

430 MAXIMOW, A.: Development of argyrophile and collagenous fibres in tissue cultures. *Proc. Soc. Exp. Biol. Med.* **25** (1928).

431 MAYER, B.: Über das Regulations- und Induktionsvermögen der halbseitigen oberen Urmundlippe von *Triton*. *Wilhelm Roux' Arch. Entwicklungsmech. Organ.* **133** (1935).

432 McCLENDON, J. F.: The segmentation of eggs of *Asterias forbesii* deprived of chromatin. *Wilhelm Roux' Arch. Entwicklungsmech. Organ.* **26** (1908).

433 McCLINTOCK, B.: The relation of a particular chromosomal element to the development of the nucleoli in *Zea mays*. *Z. Zellforsch.* **21** (1934).

434 MECHELKE, F.: Reversible Strukturmodifikationen der Speicheldrüsenchromosomen von *Acricotopus lucidus*. *Chromosoma* **5** (1953).

435 —— Das Wandern des Aktivitätsmaximums im BR_4-Locus als Modell für die Wirkungsweise eines komplexen Locus. *Naturwissenschaften* **48** (1961).

436 MEDEM, F.: Befruchtungs-Stoffe (Gamone) bei Wirbellosen und Wirbeltieren. *Z. Sex.-Forsch.* **1** (1950).

437★ MELCHERS, G., u. A. LANG: Die Physiologie der Blütenbildung. *Biol. Zentralbl.* **67** (1948).

438 MENGE, F.: Die Entwicklung der Keimpflanzen von *Marchantia polymorpha* und *Plagiochasma rupestre*. *Flora (Jena)* **24** (1930).

439 METZ, C. W.: Monocentric mitosis with segregation of chromosomes in *Sciara* and its bearing on the mechanism of mitosis. *Biol. Bull.* **64** (1933).

440 —— Factors influencing chromosome movements in mitosis. *Cytologia (Tokyo)* **7** (1936).

441★ METZ, C. B.: Specific egg and sperm substances and activation of the egg. In: The Beginnings of Embryonic Development. Washington: AAAS, 1957.

442 MEVES, F.: Über die Entwicklung der männlichen Geschlechtszellen von *Salamandra maculosa*. *Arch. mikr. Anat.* **48** (1896).

443 —— Die Spermatocytenteilung bei der Honigbiene (*Apis mellifica* L.) nebst Bemerkungen über Chromatinreduktion. *Arch. mikr. Anat.* **70** (1907).

444 —— u. J. DUESBERG: Die Spermatocytenteilung bei der Hornisse (*Vespa crabro* L.). *Arch. mikr. Anat.* **71** (1908).

445 MEYER, G.: Über die photoperiodische Beeinflußbarkeit des Habitus von *Sedum kamtschaticum*. *Biol. Zentralbl.* **66** (1947).

446★ MICHAELIS, P.: Wege und Möglichkeiten zur Analyse des Plasmatischen Erbgutes. *Biol. Zentralbl.* **73** (1954).

447★ —— Probleme, Methoden und Ergebnisse der Plasmavererbung. *Naturwissenschaften* **50** (1963).

448 MICHEL, K.: Die Kern- und Zellteilung im Zeitrafferfilm. Die meiotischen Teilungen bei der Spermatogenese der Schnarrheuschrecke *Psophus stridulus* L. Zeiss. *Nachr.* **4** (1943).

449 MIKAMI, Y.: Reciprocal transformation of parts in the developing eyeside, with special attention to the inductive influence of lens-ectoderm on the retinal differentiation. *Magazine (Tokyo)* **51** (1939).

450 MINGANTI, A.: Transplantations d'un fragment du territoire somitique présomptif de la jeune gastrula chez l'Axolotl et chez *Triton*. *Arch. Biol.* **60** (1949).

451* MIRSKY, A. E.: Austauschvorgänge zwischen Zellkern und Zellplasma. *Naturwissenschaften* **50** (1963).

452 ——, and H. RIS: The chemical composition of isolated chromosomes. *J. Gen. Physiol.* **31** (1947).

453 MITCHISON, J. M.: Cell membranes and cell division. *Symp. Soc. Exp. Biol.* **6** (1952).

454 MOFFETT, A. A.: Chromosome studies in *Anemone*. I. A new type of chiasma behaviour. *Cytologia (Tokyo)* **4** (1933).

455 MONNÉ, L., and S. HÄRDE: On the cortical granules of the sea urchin egg. *Ark. Zool. Ser. II,* **1**, 127 (1951).

456 MONROY, A.: A preliminary electrophoretic analysis of proteins and protein fractions in sea urchin eggs and their changes on fertilization. *Exp. Cell Res.* **1** (1950).

457 MONTI, R.: La genetica dei coregoni italiani e la loro viaiabilità in relazione coll'ambiente. *Arch. Zool. Ital.* **18** (1933).

458* MOORE, A. R., and A. S. BURT: On the locus and nature of the forces causing gastrulation in the embryos of *Dendraster excentricus. J. Exp. Zool.* **82** (1939).

459 MORGAN, T. H.: Regeneration of heteromorphic tails in posterior pieces of *Planaria simplicissima. J. Exp. Zool.* **1** (1904).

460* —— Regeneration, 2nd Ed. Übers. von M. MOSZKOWSKI. Leipzig: W. Engelmann, 1907.

461 —— Experimental Embryology. New York: Columbia, 1927.

462 —— The formation of the antipolar lobe in *Ilyanassa. J. Exp. Zool.* **64** (1933).

463 —— The rhythmic in form of the isolated antipolar lobe of *Ilyanassa. Biol. Bull.* **98** (1935).

464 ——, u. A. M. BORING: The relation of the first plane of cleavage and the grey crescent to the median plane of the embryo of the frog. *Wilhelm Roux' Arch. Entwicklungsmech. Organ.* **16** (1903).

465 ——, and G. B. SPOONER: The polarity on the centrifuged egg. *Wilhelm Roux' Arch. Entwicklungsmech. Organ.* **28** (1909).

466 MOSER, F.: Studies on a cortical layer response to stimulating agents in the *Arbacia* egg. *J. Exp. Zool.* **80** (1939).

467 MÜLLER, H. J.: Die Saisonformbildung von *Araschnia levana*, ein photoperiodisch gesteuerter Diapause-Effekt. *Naturwissenschaften* **42** (1955).

468 MÜLLER, W.: Experimentelle Untersuchungen über Stockentwicklung, Polypendifferenzierung und Sexual-chimären bei *Hydractinia echinata. Wilhelm Roux' Arch. Entwicklungsmech. Organ.* **155** (1964).

469 MUTH, F. W.: Untersuchungen zur Wirkungsweise der Mutante „*kfl*" bei der Mehlmotte *Ephestia kühniella* Z. *Wilhelm Roux' Arch. Entwicklungsmech. Organ.* **153** (1961).

470 NAVASHIN, M.: Chromosome alterations caused by hybridisation and their bearing upon certain general genetic problems. *Cytologia (Tokyo)* **5** (1934).

471* NEEDHAM, J.: Biochemistry and Morphogenesis. Cambridge: Univ. Press, 1950.

472 NICHOLAS, J. L.: The effect of the rotation of the area surrounding the limb bud. *Anat. Rec.* **23** (1922).

473 NIENBURG, W.: Die Polarisation der *Fucus*-Eier durch das Licht. *Wiss. Meeresunters. Helgoland, N. F.* **15** (1923–1930).

474 —— Die Einwirkung des Lichts auf die Keimung der Equisetumspore. *Ber. deut. bot. Ges.* **42** (1924).

475 —— Die Entwicklung der Keimlinge von *Fucus vesiculosus* usw. *Wiss. Meeresunters. Helgoland, N. F.* **21** (1928–1933).

476* NIEUWKOOP, P. D.: Neural competence and neural fields. *Rev. Suisse Zool.* **57**, Suppl. 1 (1950).

477 NOLL, W.: Embryonalentwicklung von *Biophytum dendroides* D. C. *Planta (Berl.)* **24** (1935).

478 NYHOLM, M., L. SAXÉN, S. TOIVONEN and T. VAINIO: Electron microscopy of transfilter neural induction. *Exp. Cell Res.* **28** (1962).

479 OEHLKERS, F.: Neue Versuche über zytologisch-genetische Probleme (Physiologie der Meiosis). *Biol. Zentralbl.* **57** (1937).

480 —— Meiosis and crossing over. *Biol. Zentralbl.* **60** (1940).

481 —— Der Einfluß der Plastiden auf den Ablauf der Meiosis. *Naturwissenschaften* **28** (1940).

482 —— Zytologische und zytogenetische Untersuchungen an *Streptocarpus. Z. Bot.* **39** (1944).

483* —— Neue Überlegungen zum Problem der außerkaryotischen Vererbung. *Z. indukt. Abstamm.- u. Vererb.-Lehre* **84** (1952).

484* —— Außerkaryotische Vererbung. *Naturwissenschaften* **40** (1953).

485 —— Blattstecklinge als Indikatoren für blütenbildende Substanzen. *Z. Naturforsch.* **10** (1955).

486 —— Veränderungen in der Blühbereitschaft vernalisierter Cotyledonen von *Streptocarpus*, kenntlich gemacht durch Blattstecklinge. *Z. Naturforsch.* **11** (1956).

487 OLIVE, Z. W.: A monograph of the Acrasieae. *Proc. Boston Soc. Nat. Hist.* **30** (1902).

488* OLTMANNS, F.: Morphologie und Biologie der Algen. Jena: G. Fischer, 1922.

489 ORTOLANI, G.: Risultati definitivi sulla distribuzione dei territori presuntivi degli organi nel germe die ascidie allo stadio VIII, determinata con le marche al carbone. *Pubbl. Sta. zool. Napoli* **25** (1953).

490 —— The presumptive territory of the mesoderm in the Ascidian germ. *Experientia (Basel)* **11** (1955).

491 Ortolani, G.: Il territorio precoce della corda nelle ascidie. *Acta Embryol. Morph. Exp. (Palermo)* **1** (1957).

492 —— Cleavage and development of egg fragments in ascidians. *Acta Embryol. Morphol. Exp.* **1** (1958).

493 Pallas, P. S.: Miscellanea zoologica. *Lagd. Batav.* **1779**.

493a Pappas, G. D., and P. W. Brandt: Helical structures in the nuclei of free-living amebas. 4. Intern. Kongr. Elektronenmikroskopie Berlin 1958. Verh. II, 1960.

494 Pascher, A.: Über die Beziehungen zwischen Lagerform und Standortsverhältnissen bei einer Gallertalge (Chrysocapsales). *Arch. Protistenk.* **68** (1929).

495 Pasteels, J.: Recherches sur les facteurs initiaux de la morphogénèse chez les amphibiens anoures. I. Résultars de l'expérience de Schultze et leur interprétation. *Arch. Biol.* **49** (1938).

496 —— III. Effects de la rotation de 135° sur l'oeuf insegmenté, muni de son croissant gris. *Arch. Biol.* **51** (1940).

497 —— Un aperçu comparatif de la gastrulation chez les chordées. *Biol. Rev.* **15** (1940).

498 —— Observations sur la localisation de la plaque préchordale et de d'entoblaste présomptifs au cours de la gastrulation chez *Xenopus laevis*. *Arch. Biol.* **60** (1949).

499★ —— Centre organisateur et potentiel morphogénétique chez les batraciens. *Bull. Soc. Zool. Fr.* **76** (1951).

500 Pätau, K.: Chromosomenmorphologie bei *Drosophila melanogaster* und *Drosophila simulans* und ihre genetische Bedeutung. *Naturwissenschaften* **23** (1935).

501 ——, and D. Srinivasachar: The DNA-content of nuclei in the meristem of onion roots. *Chromosoma* **10** (1959).

502 Pauli, M. E.: Die Entwicklung geschnürter und centrifugierter Eier von *Calliphora erythrocephala* und *Musca domestica*. *Z. wiss. Zool.* **129** (1927).

503 Pelling, C.: Ribonucleinsäure-Synthese der Riesenchromosomen. Autoradiographische Untersuchungen an *Chironomus tentans*. *Chromosoma* **15** (1964).

504 Peltrera, A.: La capacità dell' uovo di *Aplysia limacina* L. studiata con centrifugazione e con reazioni vitali. *Pubbl. Sta. zool. Napoli* **18** (1940).

505 Penners, A.: Schultzescher Umdrehversuch an ungefurchten Froscheiern. *Wilhelm Roux' Arch. Entwicklungsmech. Organ.* **116** (1929).

506 —— Neue Experimente zur Frage nach der Potenz der ventralen Keimeshälfte von *Rana fusca*. *Z. wiss. Zool.* **148** (1936).

507 ——, u. W. Schleip: Die Entwicklung der Schultzeschen Doppelbildung aus dem Ei von *Rana fusca*. *Z. Zool.* **130** (1927); **131** (1928).

508 Pentz, S.: Über die Augendifferenzierung bei *Polycelis nigra*. *Wilhelm Roux' Arch. Entwicklungsmech. Organ.* **154** (1963).

509★ Pflugfelder, O.: Entwicklungsphysiologie der Insekten, 2. Aufl. Leipzig: Akadem. Verl.-Gesellschaft Geest & Portig, 1958.

510 Pfützner-Eckert, R.: Entwicklungsphysiologische Untersuchungen an *Dictyostelium mucoroides* Brefeld. *Wilhelm Roux' Arch. Entwicklungsmech. Organ.* **144** (1950).

511 Piech, K.: Über die Entstehung der generativen Zelle bei *Scirpus uniglumis* durch „freie Zellbildung". *Planta (Berl.)* **6** (1928).

512 Piepho, H.: Wachstum und totale Metamorphose an Hautimplantaten bei der Wachsmotte *Galleria mellonella*. *Biol. Zentralbl.* **58** (1938).

513 —— Über die experimentelle Auslösbarkeit überzähliger Raupenhäutungen und vorzeitiger Verpuppung an Hautstücken bei Kleinschmetterlingen. *Naturwissenschaften* **26** (1938).

514 —— Über die Oxydation-Reduktionsvorgänge im Amphibienkeim. *Biol. Zentralbl.* **58** (1938).

515 —— Untersuchungen zur Entwicklungsphysiologie der Insektenmetamorphose. Über die Puppenhäutung der Wachsmotte. *Wilhelm Roux' Arch. Entwicklungsmech. Organ.* **141** (1942).

516 —— Über das Ausmaß der Artunspezifität von Metamorphosehormonen bei Insekten. *Biol. Zentralbl.* **69** (1950).

517 —— Über die Hemmung der Falterhäutung durch Corpora allata. *Biol. Zentralbl.* **69** (1950).

518 ——, u. H. Meyer: Reaktionen der Schmetterlingshaut auf Häutungshormone. *Biol. Zentralbl.* **70** (1951).

519 Pohley, H.-J.: Über das Wachstum des Mehlmottenflügels unter normalen und experimentellen Bedingungen. *Biol. Zentralbl.* **78** (1959).

520 —— Experimentelle Beiträge zur Lenkung der Organentwicklung des Häutungsrhythmus und der Metamorphose bei der Schabe *Periplaneta americana* L. *Wilhelm Roux' Arch. Entwicklungsmech. Organ.* **151** (1959).

521 —— Experimentelle Untersuchungen über die Steuerung des Häutungsrhythmus bei der Mehlmotte *Ephestia kühniella*. *Wilhelm Roux' Arch. Entwicklungsmech. Organ.* **152** (1960).

522 —— Untersuchungen über die Veränderung der Metamorphoserate durch Antennenamputation bei *Periplaneta americana*. *Wilhelm Roux' Arch. Entwicklungsmech. Organ.* **153** (1962).

523 POHLEY, H.-J.: u. H. ESSER: Über das Verhalten mutanter Schuppen auf den Hinterflügeln der Mehlmotte *Ephestia kühniella* nach Puppenbestrahlungen. *Z. Vererbungsl.* **89** (1958).

523a POLLISTER, A. W., and J. A. MOORE: Tables for the normal development of *Rana sylvatica. Anat. Record.* **68** (1937).

524 POPOFF, M.: Experimentelle Zellstudien. *Arch. Zellforsch.* **1** (1908); **3** (1909).

524a POULSON, D. F.: Chromosomal control of embryogenesis in Drosophila. *Amer. Naturalist.* **79** (1945).

525 QUERNER, H.: Untersuchungen über die Flügelform der Mehlmotte *Ephestia kühniella,* insbesondere den Faktor *kfl (kurzflügelig). Biol. Zentralbl.* **67** (1948).

526 RANZI, S.: Die Proteine in der Embryonalentwicklung. *Verh. Deut. Zool. Ges.* 1949.

527 ———, u. E. TAMINI: Die Wirkung von NaSCN auf die Entwicklung von Froschembryonen. *Naturwissenschaften* **27** (1939).

528 RAPER, R. B.: The communal nature of the fruiting process in the *Acrasieae. Amer. J. Bot.* **27** (1940).

529 RASHEVSKY, N.: Mathematical Biophysics. Chicago: The University of Chicago Press, 1938.

530 RAVEN, C. P.: Über die Potenz von Gastrulaektoderm nach 28stündigem Verweilen im äußeren Blatt der dorsalen Urmundlippe. *Wilhelm Roux' Arch. Entwicklungsmech. Organ.* **137** (1938).

531 ——— Lithium as a tool in the analysis of morphogenesis in *Limnaea stagnalis. Experientia (Basel)* **8** (1952).

532 ——— Morphogenesis: The analysis of molluscan development. London-New York-Los Angeles: Pergamon, 1958.

533 REHM, M.: Die zeitliche Folge der Tätigkeitsrhythmen der inkretorischen Organe von *Ephestia kühniella* während der Metamorphose und des Imaginallebens. *Wilhelm Roux' Arch. Entwicklungsmech. Organ* **145** (1952).

533a* REINERT, J.: Wachstum. *Fortschr. Bot.* **26** (1964).

534 REITH, F.: Die Entwicklung des *Musca*-Eies nach Ausschaltung verschiedener Eibereiche. *Z. wiss. Zool.* **126** (1925)

535 RESENDE, F., A. DE LEMOS-PEREIRA et A. CABRAL: Sur la structure des chromosomes dans les mitoses des méristèmes radiculaires. III. L'action de la température sur la structure chromosomique. *Port. Acta Biol. A* **1** (1944–1946).

536* ——— Karyokinesis. *Port. Acta Biol.* **2** (1947).

537* ——— Über passive und aktive funktionelle Strukturänderungen bei pflanzlichen und tierischen Chromosomen. *Port. Acta Biol. A* **7** (1963).

538 REVERBERI, G.: The mitochondrial pattern in the development of the ascidian egg. *Experientia (Basel)* **12** (1956).

539 ——— Selective distribution of mitochondria during the development of the egg of *Dentalium. Acta Embryol. Morph. exp.* **2** (1958).

540* ——— The embryology of ascidians. *Advanc. Morphogenesis* **1** (1961).

541* ——— L'uova delle ascidie e le sue potenze organoformativo. *R. C. Ist. Sci. Camerino* **2** (1961).

542 ———, e I. GORGONE: Gigantic tadpoles from ascidian eggs fused. *Acta Embryol. Morphol. exp.* **5** (1962).

543 —, e A. MINGANTI: Su alcune recenti ricerche di embryologia spermentale delle ascidie. *Riv. Biologia (Perugia), N.S.* **45** (1953).

544 ———, G. ORTOLANI, and N. FARINELLA-FERRUZZA: The causal formation of the brain in the ascidian larva. *Acta Embryol. Morph. exp.* **3** (1960).

545 RHUMBLER, L.: Trajektorien-Modell zur Demonstration einer automatischen Entstehung der Innentrajektorien eines fötalen Femurknochens durch dessen Oberflächenwachstum. *Verh. Deut. Zool. Ges.* **24** (1914).

546* RICHARDS, A. G.: Structure and function of the integument. Chemical and physical properties of cuticle. In: Insect Physiology, ed. by K. D. ROEDER. New York and London: Wiley, 1953.

547 RICHTER, G.: Die Auslösung kerninduzierter Regeneration bei gealterten kernlosen Zellteilen von *Acetabularia* und ihre Auswirkung auf die Synthese von Ribonucleinsäure und Cytoplasmaproteinen. *Planta (Bierl.)* **52** (1959).

548 ——— Die Auswirkung der Zellkern-Entfernung auf die Synthese von Ribonucleinsäure und Cytoplasmaproteinen bei *Acetabularia mediterranea. Biochim. Biophys. Acta* **34** (1959).

549 RIES, E.: Die Verteilung von Vitamin C, Glutathion, Benzidin-Peroxydase, Phenolase (Indophenolblauoxydase) und Leukomethylenblau-Oxydoreduktase während der frühen Embryonalenentwicklung verschiedener wirbelloser Tiere. *Pubbl. Sta. zool. Napoli* **16** (1937).

550 ——— Histochemische Sonderungsprozesse während der frühen Embryonalentwicklung verschiedener wirbelloser Tiere. *Arch. exp. Zellforsch.* **22** (1939).

551 ——— Versuche über die Bedeutung des Substanzmosaiks für die embryonale Gewebedifferenzierung bei Ascidien. *Arch. exp. Zellforsch.* **23** (1939).

552 Ries, E.: u. M. Gersch: Die Zelldifferenzierung und Zellspezialisierung während der Embryonalentwicklung von *Aplysia limacina* L., zugleich ein Beitrag zu Problemen der vitalen Färbung. *Pubbl. Sta. Zool. Napoli* **15** (1936).

553 Ris, H.: The anaphase movement of chromosomes in the spermatocytes of the grasshopper. *Biol. Bull.* **96** (1949).

554 ――― Die Feinstruktur des Kerns während der Spermatogenese. In: Chemie der Genetik 9. Coll. Ges. für Physiol. Chemie 1958. Berlin, Göttingen, Heidelberg: Springer, 1959.

555 Risler, H.: Kernvolumenänderungen in der Larvenentwicklung von *Ptychopoda seriata*. *Biol. Zentralbl.* **69** (1950).

556 Robert, A.: Recherches sur le développement des Troques. *Arch. Zool. Exp. gen. (3)* **10** (1902).

557 Rosin, S.: Zur Frage der Pigmentmusterbildung bei Urodelen. *Rev. Suisse Zool.* **47** (1940).

558 Rothert, W., u. L. Jost: Gewebe der Pflanzen. In: Handwörterbuch Naturwissenschaft, 2. Aufl., Bd. 5. Jena: G. Fischer, 1934.

559 Rottmann, E.: Die Rolle des Ektoderms und Mesoderms bei der Formbildung der Kiemen und Extremitäten von *Triton*. *Wilhelm Roux' Arch. Entwicklungsmech. Organ.* **124** (1931).

560 ――― Der Anteil von Induktor und reagierendem Gewebe an der Entwicklung der Amphilinse. *Wilhelm Roux' Arch. Entwicklungsmech. Organ.* **139** (1939).

561 ――― Die Bedeutung der Zellgröße für die Entwicklung der Amphibienlinse. *Wilhelm Roux' Arch. Entwicklungsmech. Organ.* **140** (1940).

562 ――― Neuere Untersuchungen über das Induktionsgeschehen in der tierischen Entwicklung. *Decheniana Verh. naturhist. Ver. Rheinlande Westfalens* **100** (1941).

563 ――― Über den Auslösungscharakter des Induktionsreizes bei der Linsenentwicklung. *Biol. Zentralbl.* **62** (1942).

564 Ruud, G.: Die Entwicklung isolierter Keimfragmente frühester Stadien von *Triton taeniatus*. *Wilhelm Roux' Arch. Entwicklungsmech. Organ.* **105** (1925).

565 ―――, u. H. Spemann: Die Entwicklung isolierter dorsaler und lateraler Gastrulahälften, ihre Regulation und Postgeneration. *Wilhelm Roux' Arch. Entwicklungsmech. Organ.* **52** (1922).

566 Runnström, J.: Plasmabau und Determination bei dem Ei von *Paracentrotus lividus*. *Wilhelm Roux' Arch. Entwicklungsmech. Organ.* **113** (1928).

567 ――― Über Selbstdifferenzierung und Induktion bei dem Seeigelkeim. *Wilhelm Roux' Arch. Entwicklungsmech. Organ.* **117** (1929).

568 ――― Zur Entwicklungsmechanik des Skelettmusters bei dem Seeigelkeim. *Wilhelm Roux' Arch. Entwicklungsmech. Organ.* **124** (1931).

569 ――― The mechanism of fertilization in metazoa. *Advanc. Enzymol.* **9** (1949).

570 ――― Some results and views concerning the mechanism of initiation of development in the sea urchin egg. *Pubbl. Sta. Zool. Napoli* **21**, Suppl. (1949).

571 ――― The cell surface in relation to fertilization. *Symp. Soc. Exp. Biol.* **6** (1952).

572* Runnström, I.: Die Analyse der primären Differenzierungsvorgänge im Seeigelkeim. *Verh. Deut. Zool. Ges.* 1954.

573 Sachs, R. M., C. Bretz and A. Lang: Cell division and gibberellic acid. *Exp. Cell. Res.* **18** (1959).

574 ――― ――― ――― Shoot histogenesis. The early effects of gibberellin upon stem elongation in two rosette plants. *Amer. J. Bot.* **46** (1959).

575 Sager, R., and M. R. Ishida: Chloroplast DNA in *Chlamydomonas*. *Proc. Nat. Acad. Sci. U.S.A.* **50** (1963).

576 Sagromsky, H.: Weitere Beobachtungen zur Bildung des Spaltöffnungsmusters in der Blattepidermis. *Z. Naturforsch.* **4** b (1949).

577 Salord, J.: The action of temperature on the chromosomic structure of *Trillium*. *Port. Acta Biol. A* **2** (1949).

578 Sander, K.: Analyse des ooplasmatischen Reaktionssystems von *Euscelis plebejus* Fall. (Cicadina) durch Isolieren und Kombinieren von Keimteilen I. u. II. *Wilhelm Roux' Arch. Entwicklungsmech. Organ.* **151** (1959/60).

579 ――― Umkehr der Keimstreifpolarität in Eifragmenten von Euscelis (Cicadina). *Experientia (Basel)* **17** (1961).

580 ――― Über den Einfluß von verlagertem Hinterpolmaterial auf das metamere Organisationsmuster im Zikadenei. *Verh. Deut. Zool. Ges.* 1961.

581 Sangel, P.: Sur les conditions de la régénération normale du pharynx chez la planaire *Dugesia (Euplanaria) lugubris*. *Bull. Biol.* **85** (1951).

582 Sato, T.: Beiträge zur Analyse der Wolffschen Linsenregeneration. I. *Wilhelm Roux' Arch. Entwicklungsmech. Organ.* **122** (1930); II. **130** (1933); III. **133** (1935).

583 ――― Über die linsenbildende Fähigkeit des Pigmentepithels bei *Diemyctylus pyrrhogaster*. *Embryologia* **1** (1951).

584 SATO, T.: Versuche zur Formbildung bei *Acetabularia. Wilhelm Roux' Arch. Entwicklungsmech. Organ.* **151** (1960).

585 SAUER, G.: Beziehungen zwischen der Bewegungsmöglichkeit und dem Regenerationsvermögen von Blastodermbereichen im Grillenei. *Verh. Deut. Zool. Ges.* 1961.

586 ────── Die Regenerationsfähigkeit des Keimanlagenblastoderms von *Gryllus domesticus* beim Beginn der Wirkung des Differenzierungszentrums. *Zool. Jahrb. Abt. Anat. u. Ontog.* **79** (1961).

587 SAUNDERS, J. W.: The proximo-distal sequence of origin of the wing and the role of the ectoderm. *J. Exp. Zool.* **108** (1948).

588 ────── Analyses of the role of the apical ridge of ectoderm in the limb bud in the chick. *Anat. Rec.* **105** (1949).

589 SAXÉN, L., and S. TOIVONEN: The two-gradient hypothesis in primary induction. The combined effect of two types of inductors mixed in different ratios. *J. Embryol. Exp. Morphol.* **9** (1961).

589a★ ────── Primary Embryonic Induction. London: Logos Press, 1962.

590 SCHAFFSTEIN, G.: Untersuchungen über den Feinbau der Prophasechromosomen in der Reduktionsteilung von *Lilium martagon.* **22** (1935).

591 SCHARRER, B.: Neurosecretion XI. The effect of nerve section on the intercerebralis-cardiacum-allatum system of the insect *Leucophaea maderae. Biol. Bull.* **102** (1952).

592★ SCHARRER, E., and B. SCHARRER: Neuroendocrinology. New York and London: Columbia Univ. Press, 1963.

593 SCHECHTMANN, A. M.: The mechanism of amphibian gastrulation *(Hyla regilla). Univ. Calif. Publ. Zool.* **51** (1942).

594 SCHLEIP, W.: Die Herkunft der Polarität des Eies von *Ascaris megalocephala. Wilhelm Roux' Entwicklungsmech. Organ.* **100** (1924).

595★ ────── Die Determination der Primitiventwicklung. Leipzig: Akademische Verlagsges., 1929.

596 SCHLENK, W., u. H. KAHMANN: Ein Verfahren zur Messung der Spermatozoenbewegung. *Pflügers Arch. ges. Physiol.* **236** (1935).

597 SCHLÖSSER, L. A.: Geschlechtsverteilung und fakultative Parthogenese bei Saprolegniaceen. *Planta (Berlin)* **8** (1929).

598 SCHLOTE, F. W.: Über den Feinbau von Chromosomen. I. Zur Cytologie von *Escherichia coli. Arch. Mikrobiol.* **40** (1961).

599 SCHMID, W.: Analyse der letalen Wirkung des Faktors *lme (letalmeander)* von *Drosophila melanogaster. Z. indukt. Abstamm. u. Vererb.-Lehre* **83** (1949).

600 SCHMIDT, W. J.: Die Skelettstäbe der Stachelhäuter als Biokristalle. *Zool. Jahrb., Abt. allg. Zool. Physiol. Tiere* **47** (1930).

601 ────── Die Doppelbrechung der Kernspindel und ihre Feinstruktur, ein Argument für die Zugfasertheorie und die Fadenmolekeltheorie. *Biodynamica* **22** (1936).

602★ ────── Die Doppelbrechung von Karyoplasma, Zytoplasma und Metaplasma. *Protoplasma (Wien)* (Monogr.) **11** (1937).

603 SCHNEIDER, L.: Elektronenmikroskopische Untersuchungen der Konjugation von *Paramecium. Protoplasma* **56** (1963).

604★ SCHNEIDERMAN, H. A., and L. J. GILBERT: The chemistry and physiology of insect growth hormones In: Cell, Organism and Milieu (17th Growth Symposium), ed. by D. RUDNICK. New York: Ronald Press, 1959.

605 SCHNETTER, M.: Physiologische Untersuchungen über das Differenzierungszentrum der Honigbiene. *Wilhelm Roux' Arch. Entwicklungsmech. Organ.* **131** (1934).

606 ────── Morphologische Untersuchungen über das Differenzierungszentrum in der Embryonalentwicklung der Honigbiene. *Z. Morph. u. Ökol. Tiere* **29** (1934).

607 ────── Die Entwicklung von Zwerglarven in geschnürten Bieneneiern. *Verh. Deut. Zool. Ges.* **1936.**

608 SCHÖNMANN, W.: Der diploide Bastard *Triton palmatus* ♀ × *Salamandra* ♂. *Wilhelm Roux' Arch. Entwicklungsmech. Organ.* **138** (1938).

609 SCHRADER, F.: Recent hypotheses on the structure of spindles in the light of certain observations in Hemiptera. *Z. wiss. Zool.* **142** (1932).

610 ────── On the reality of spindle fibres. *Biol. Bull.* **67** (1934).

611 ────── The structure of the kinetochore at meiosis. *Chromosoma* **1** (1939).

612 ────── Chromatin bridges and irregularity of mitotic coordination in the pentatomid *Peromatus notatus. Biol. Bull.* **81** (1941).

613★ ────── Mitosis. The Movements of Chromosomes in Cell Division. 3. print. New York: Columbia University Press, 1949.

614★ ────── A critique of recent hypotheses of mitosis. Sympos. Cytol. Michigan State College Press 1951.

615★ ────── Mitose. Die Bewegungen der Chromosomen bei der Zellteilung. Wien: Denticke, 1954.

510

616 SCHRADER, F.: and C. LEUCHTENBERGER: A cytochemical analysis of the functional interrelations of various cell structures in *Arvelius albopunctatus* (DE GEER). *Exp. Cell Res.* **1** (1950).

617* SCHRAMM, G.: Die Biochemie der Viren. Berlin, Göttingen, Heidelberg: Springer, 1954.

618 SCHULZE, K. L.: Cytologische Untersuchungen an *Acetabularia mediterranea* und *Acetabularia wettsteinii*. *Arch. Protistenk.* **92** (1939).

619 SCHWINCK, ILSE: Veränderung der Epidermis, der Pericardialzellen und der Corpora allata in der Larven-Entwicklung von *Panorpa communis* unter normalen und experimentellen Bedingungen. *Wilhelm Roux' Arch. Entwicklungsmech. Organ.* **145** (1951).

620 SCOTT, A. C.: Haploidy and aberrant spermatogenesis in a coleopteran, *Micromalthus debilis* LE CONTE. *J. Morph.* **59** (1936).

621 SEIDEL, F.: Untersuchungen über das Bildungsprinzip der Keimanlage bei der Libelle *Platycnemis penipes* I–V. *Wilhelm Roux' Arch. Entwicklungsmech. Organ.* **119** (1929).

622 —— Die Potenzen der Furchungskerne im Libellenei und ihre Rolle bei der Aktivierung des Bildungszentrums. *Wilhelm Roux' Arch. Entwicklungsmech. Organ.* **126** (1932).

623* —— Entwicklungsphysiologie des Insekten-Keims. *Verh. Deut. zool. Ges.* 1936.

624* —— Entwicklungsphysiologie der Tiere. Berlin: Borntraeger, 1953.

625* —— Entwicklungsphysiologische Zentren im Eisystem der Insekten. *Verh. Deut. Zool. Ges.* **1960** (1961).

626* ——, E. BOCK u. G. KRAUSE: Die Organisation des Insekteneies. *Naturwissenschaften* **28** (1940).

627* SERRA, J. A.: Relations entre la chimie et la morphologie nucléaire. *Bol. Soc. Broteriana, Sér. II,* **16** (1942).

628 —— Compound-helix and monomer structure of chromosomes. *Revista Portug. Zoologia e Biol. gener.* **2** (1959).

629* SHARP, L. W.: Fundamentals of Cytology. New York and London: McGraw Hill, 1943.

629a SIEGEL, R. W.: New results on the genetics of mating types in *Paramecium bursaria*. *Genet. Res.* **4** (1963).

630 SILBER, H., and V. HAMBURGER: The production of *duplicitas cruciata* and multiple heads by regeneration in *Euplanaria tigrina*. *Physiol. Zool.* **12** (1939).

631 SIRLIN, J. L., S. K. BRAHMA, and C. H. WADDINGTON: Studies on embryonic induction using radioactive tracers. *J. Embryo. exp. Morphol.* **4** (1956).

632 SLIZYNSKI, B. M.: Production of structural changes in somatic chromosomes of *Drosophila melanogaster*. *Nature (London)* **159** (1947).

633 SIVICKIS, P. B.: A quantitative study of regeneration in *Dendrocoelum lacteum*. *Arb. Ungar. Biol. Forschunginst., II. Abt.,* **4** (1931).

634 —— A quantitative study of regeneration along the main axis of the triclad body. *Arch. Zool. Ital.* **16** (1931).

635 SMITH, F. H.: Anomalous spindles in *Impatiens pallida*. *Cytologia (Tokyo)* **6** (1935).

636 SMITH, G. E.: On the orientation of stomata. *Ann. Bot. (London)* **49** (1935).

637 SNOW, R.: A hormone for correlative inhibition. *New Phytol.* **39** (1940).

638 SOBELS, F. H., u. L. E. NIJENHUIS: An investigation into the metamorphosis of the mutant *lethal-translucida* of *Drosophila melanogaster*. *Z. indukt. Abstamm.- u. Vererb.-Lehre* **85** (1953).

639 SOKOLOW, N. O., G. G. TINIAKOW, and J. E. TROFIMOW: On the morphology of the chromosomes in Gallinaceae. *Cytologia (Tokyo)* **7** (1936).

640 SONNEBORN, T. M.: Sex, sex inheritance and sex determination in *Paramecium aurelia*. *Proc. Nat. Acad. Sci. U.S.A.* **23** (1937).

641* —— Recent advances in the genetics of *Paramecium* and *Euplotes*. *Advan. Genet.* **1** (1947).

642 —— The role of the genes in cytoplasmic inheritance. In Genetics in the 20th Century. New York: Macmillan, 1951.

643* —— Breeding systems, reproductive methods, and species problems in protozoa. In: The Species Problem. Washington: AAAS, 1957.

644 —— Bearing of protozoan studies on current theory of genic and cytoplasmic actions. Proc. XVI. Internat. Zool. Congr. 1963.

645 SPEK, J.: Die amöboiden Bewegungen und Strömungen in den Eizellen einiger Nematoden während der Vereinigung der Vorkerne. *Wilhelm Roux' Arch. Entwicklungsmech. Organ.* **44** (1918).

646 SPEMANN, H.: Entwicklungsphysiologische Studien am Tritonei. I. *Wilhelm Roux' Arch. Entwicklungsmech. Organ.* **12** (1901); II. **15** (1902); III. **16** (1903).

647 —— Zur Entwicklung des Wirbeltierauges. *Zool. Jahrb. Abt. allg. Zool. Physiol. Tiere* **32** (1912).

648 —— Über die Entwicklung umgedrehter Hirnteile bei den Amphibienembryonen. *Zool. Jahrb. Suppl.* **15** (1912).

649* —— Experimentelle Beiträge zur Theorie der Entwicklung. Berlin: Springer, 1936.

650 ——, u. H. MANGOLD: Über Induktion von Embryonalanlagen durch Implantation artfremder Organisatoren. *Wilhelm Roux' Arch. Entwicklungsmech. Organ.* **100** (1924).

651 SPEMANN, H., u. O. SCHOTTÉ: Über xenoplastische Transplantation als Mittel zur Analyse der embryonalen Induktion. *Naturwissenschaften* **20**, 463 (1932).

651a STEINICKE, F.: Zur Polarität von *Bryopsis*. *Bot. Arch.* **12** (1925).

652 STEINMANN, P.: Autotomie, ungeschlechtliche Fortpflanzung durch Teilung, Transplantation und Regeneration der Tricladida. In: Bronn's Klassen und Ordnungen. Leipzig 1916.

653 ——— Das Verhalten der Zellen und Gewebe im regenerierenden Tricladenkörper. *Verh. naturforsch. Ges. Basel* **36** (1925).

654 ——— Prospective Analyse von Restitutionsvorgängen. I. *Wilhelm Roux' Arch. Entwicklungsmech. Organ.* **108** (1920); II. **112** (1927).

655 ——— Transplantationsversuche mit vital gefärbten Tricladen. *Rev. Suisse Zool.* **40** (1933).

656 ——— Vitale Färbungsstudien an Planarien. 2. Teil. *Rev. Suisse Zool.* **40** (1933).

657* STERN, C.: Gene and character in: Genetics, Paleontology and Evolution. Princeton: Princeton Univ. Press, 1949.

658 ——— Genes and origin of pattern. IX. Internat. Congr. Genet. 1954.

659 STONE, L. S.: Further experiments on the extirpation and transplantation of mesectoderm in *Amblystoma punctatum. J. exp. Zool.* **44** (1926).

660 ——— Lens regeneration in adult newt eyes related to retina pigment cells and the neural retina factor. *J. Exp. Zool.* **139** (1958).

661 STOSSBERG, M.: Die Zellvorgänge bei der Entwicklung der Flügelschuppen von *Ephestia kühniella. Z. Morph. u. Ökol. Tiere* **34** (1938).

662 STRAUB, J.: Untersuchungen zur Physiologie der Meiosis VII. Die Abhängigkeit der Chiasmenbildung bei *Vicia faba* und *Campanula persicifolia* von äußeren Bedingungen. *Z. Bot.* **32** (1937).

663 ——— Neuere karyologische Probleme und Ergebnisse. IV. Die Spiralstruktur der Chromosomen. *Z. Bot.* **33** (1938).

664* STUBBE, H.: Genmutation I. Allgemeiner Teil. In: Handbuch der Vererbungswissenschaft, II. F. Berlin: Borntraeger, 1938.

665* SWANN, M. M.: The nucleus in fertilization, mitosis and cell division. *Symp. Soc. Exp. Biol.* **6** (1952).

666 TARDENT, P.: Axiale Verteilungs-Gradienten der interstitiellen Zellen bei *Hydra* und *Tubularia* und ihre Bedeutung für die Regeneration. *Wilhelm Roux' Arch. Entwicklungsmech. Organ.* **146** (1954).

667* ——— Principles governing the process of regeneration in hydroids. 18th Growth Sympos. Soc. Study of Developm. and Growth. 1960.

668* ——— Regeneration in the hydrozoa. *Biol. Rev. (Cambridge)* **38** (1963).

669 ———, u. H. EYMANN: Experimentelle Untersuchungen über den regenerationshemmenden Faktor von *Tubularia. Wilhelm Roux' Arch. Entwicklungsmech. Organ.* **151** (1959).

670 ———, u. R. TARDENT: Wiederholte Regeneration bei *Tubularia. Pubbl. Sta. zool. Napoli* **28** (1956).

671 TEISSIER, G.: La croissance embryonnaire de *Chrysaora hysocella. Arch. Zool. exp. gen.* **69** (1929).

672 TEISSIER, M. G.: Étude expérimentale du développement de quelques hydraires. *Ann. Sci. nat., Sér. X*, **14** (1931).

673 The beginning of embryonic development, Sympos. (Ed. A. TYLER, R. C. V. BORSTEL and C. B. METZ). Washington: AAS, 1957.

674 TIEDEMANN, H., U. BECKER u. H. TIEDEMANN: Über die primären Schritte bei der embryonalen Induktion. *Embryologia* **6** (1961).

675 ———, K. KESSELRING, U. BECKER u. H. TIEDEMANN: Über die Induktionsfähigkeit von Microsomen und Zellkernfraktionen aus Embryonen und Leber von Hühnern. *Develop. Biol.* **4** (1962).

676 ——— Induktion und Morphogenese. 13. Coll. Ges. Physiol. Chemie. Berlin, Göttingen, Heidelberg: Springer, 1963.

677 ———, u. H. TIEDEMANN: Das Induktionsvermögen gereinigter Induktionsfaktoren im Kombinationsversuch. *Rev. Suisse Zool.* **71** (1964).

678 TOIVONEN, S.: Über die Leistungsspezifität der abnormen Induktoren im Implantationsversuch bei *Triton. Ann. Acad. Sci. fenn.* A **55** (1940).

679 ——— Stoffliche Induktoren. *Rev. Suisse Zool.* **57** (1950).

680 ———, L. SAXÉN u. I. VAINO: Über die Natur der deuterenkephalen Leistung in der embryonalen Induktion. *Wilhelm Roux' Arch. Entwicklungsmech. Organ.* **154** (1963).

681 TÖNDURY, G.: Beiträge zum Problem der Regulation und Induktion: Umkehrtransplantation des mittleren Materialstreifens der beginnenden Gastrula von *Triton alpestris. Wilhelm Roux' Arch. Entwicklungsmech. Organ.* **134** (1936).

681a TREMBLEY, A.: Mémoires pour servir à l'histoire d'un genre de Polypes d'eau douce. Leiden 1774.

682 TSUGITA, A. H., H. FRAENKEL-CONRAT, M. W. NIRENBERG, and J. H. MATTHAEI: Demonstration of the messenger rote of viral RNA. *Proc. Nat. Acad. Sci. U.S.A.* **48** (1962).

683* TWITTY, V. C.: Growth correlations in amphibia, studied by the method of transplantation. *Cold Spring Harbor Symp. Quant. Biol.* **2** (1934).

684 —— The developmental analysis of specific pigment patterns. *J. Exp. Zool.* **99** (1945).

685 —— Eye. In: Analysis of Development (Ed. WILLIER, WEISS, HAMBURGER). Philadelphia: Saunders, 1955.

686* TYLER, A.: Fertilization and immunity. *Physiol. Rev.* **28** (1948).

687 UBISCH, L. v.: Untersuchungen über Formbildung. IV. Über Plutei ohne Darm, Fortsätze ohne Skelett und die Grenzen der virtuellen Keimbezirke. *Wilhelm Roux' Arch. Entwicklungsmech. Organ.* **129** (1933).

688 —— Über die Organisation des Seeigelkeims. *Wilhelm Roux' Arch. Entwicklungsmech. Organ.* **134** (1936).

689 —— Über Keimesverschmelzungen an *Ascidiella aspersa. Wilhelm Roux' Arch. Entwicklungsmech. Organ.* **138** (1938).

690 —— Über die Entwicklung von Ascidienlarven nach frühzeitiger Entfernung organbildender Keimbezirke. *Wilhelm Roux' Arch. Entwicklungsmech. Organ.* **139** (1939).

691 —— Weitere Untersuchungen über Regulation und Determination im Ascidienkeim. *Wilhelm Roux' Arch. Entwicklungsmech. Organ.* **140** (1940).

692 —— Die Entwicklung der Monascidien. *Verh. Kon. Ned. Akad. Wetensch., Afd. Natuurk.* 2. R. **49** (1952).

693 UMBACH, W.: Entwicklung und Bau des Komplexauges der Mehlmotte *Ephestia kühniella. Z. Morph. u. Ökol. Tiere* **28** (1934).

694 URSPRUNG, H.: Fragmentierungs- und Bestrahlungsversuche zur Bestimmung vom Determinationszustand und Anlageplan der Genitalscheiben von *Drosophila melanogaster. Wilhelm Roux' Arch. Entwicklungsmech. Organ.* **151** (1959).

695 —— Einfluß des Wirtsalters auf die Entwicklungsleistung von Sagittalhälften männlicher Genitalscheiben von *Drosophila melanogaster. Develop. Biol.* **4** (1962).

696 VAINIO, T., L. SAXÉN, S. TOIVONEN, and J. RAPOLA: The transmission problem in primary embryonic induction. *Exp. Cell Res.* **27** (1962).

696a VANDERKLOOT, W. G.: The control of neurosecretion and diapause by physiological changes in the brain of the *cecropia* silkworm. *Biol. Bull.* **109** (1955).

697 VASSEUR, E.: Demonstration of a jellysplitting enzyme at the surface of the sea urchin spermatozoon. *Exp. Cell Res.* **2** (1951).

698 VEJDOVSKY, F., u. A. MRAZEK: Umbildung des Cytoplasmas während der Befruchtung und Zellteilung. *Arch. mikr. Anat.* **62** (1903).

699 VOGT, M.: Zur labilen Determination der Imaginalscheiben von *Drosophila.* I. Verhalten verschiedener Imaginalanlagen bei operativer Defektsetzung. *Biol. Zentralbl.* **65** (1946).

700 VOGT, W.: Morphologische und physiologische Fragen der Primitiventwicklung, Versuche zu ihrer Lösung mittels vitaler Farbmarkierung. *S.-B. Ges. Morph. u. Physiol. Münch.* **35** (1933/34).

701 —— Gestaltungsanalyse am Amphibienkeim mit örtlicher Vitalfärbung. I. *Wilhelm Roux' Arch. Entwicklungsmech. Organ.* **106** (1925); II. **120** (1929).

702 VOIGT, W.: Künstlich hervorgerufene Neubildung von Körperteilen bei Strudelwürmern. *S.-B. Niederrhein. Ges. Bonn* 1899.

703 WACHS, H.: Neue Versuche zur Wolffschen Linsenregeneration. *Wilhelm Roux' Arch. Entwicklungsmech. Organ.* **39** (1914).

704 —— Über Augenoperationen an Amphibienlarven. *S.-B. Ges. naturforsch. Freunde Berl.* 1920.

705* WADDINGTON, C. H.: The Epigenetics of Birds. Cambridge University Press 1952.

706 —— Modes of gastrulation in vertebrates. *Quart. J. Micr. Sci.* **93** (1952).

707* —— Principles of Embryology. London: Allen & Unwin 1956.

708 WAGNER, G.: Das Wachstum der Epidermiskerne während der Larvenentwicklung von *Calliphora erythrocephala* MEIGEN. *Z. Naturforsch.* **6**b (1951).

709 WAKAYAMA, K.: Contributions to the cytology of fungi. II. Cytological studies in *Morchella deliciosa* FR. *Cytologia (Tokyo)* **2** (1931).

709a WARDLAW, C. W.: Morphogenesis in Plants. London: Methuen, 1952.

710 WEBER, H.: Grundriß der Insektenkunde, 3. Aufl. Stuttgart: G. Fischer, 1954.

711* WEBER, R.: Submicroscopical and biochemical characteristics of morphodynamic units in spiral cleaving eggs. Sympos. Germ Cells Devel. 1960.

712 —— Electron microscopy in the study of embryonic differentiation. *Sympos. Int. Soc. Cell Biol.* **1** (1962).

713 WEBSTER, G., S. L. WHITMAN, and R. L. HEINTZ: Cytoplasmic formation of ribosomes. *Plant. Physiol.* **37**, XX Suppl. (1962).

714 WEILER-STOLT, B.: Über die Bedeutung der interstitiellen Zellen für die Entwicklung und Fortpflanzung mariner Hydroiden. *Wilhelm Roux' Arch. Entwicklungsmech. Organ.* **152** (1960).

715 WEISS, P.: Experimentelle Organisierung des Gewebswachstums in vitro. *Biol. Zentralbl.* **48** (1928).

716 ——— Erzwingung elementarer Strukturverschiedenheiten am in vitro wachsenden Gewebe. *Wilhelm Roux' Arch. Entwicklungsmech. Organ.* **116** (1929).

717* ——— Principles of Development. A Text in Experimental Embryology. New York: Holt, 1939.

718 ——— Perspectives in the field of morphogenesis. *Quart. Rev. Biol.* **25** (1950).

719 ———, and B. GARBER: Shape and movement of mesenchyme cells as functions of the physical structure of the medium. Contributions to a quantitative morphology. *Proc. Nat. Acad. sci. U.S.A.* **38** (1952).

720 WERZ, G.: Über Strukturierung der Wuchszonen von *Acetabularia mediterranea*. *Planta (Berl.)* **55** (1960).

721 WETTSTEIN, D. V.: Beeinflussung der Polarität und undifferenzierte Gewebebildung aus Moossporen. *Z. Bot.* **41** (1953).

722 WETTSTEIN, F. V.: Morphologie und Physiologie des Formwechsels der Moose auf genetischer Grundlage. II. *Bibl. Genet.* **10** (1928).

723 WHITAKER, D. M.: Some observations on the eggs of *Fucus* and upon their mutual influence on the determination of the developmental axis. *Biol. Bull.* **61** (1931).

724 ——— The effects of ultra-centrifuging and of pH on the development of *Fucus* eggs. *J. Cell. Comp. Physiol.* **14** (1939).

725 WIEDBRAUCK, H.: Wiederholung der Metamorphose von Schmetterlingshaut. *Biol. Zentralbl.* **72** (1953).

726* WIGGLESWORTH, V. B.: The Physiology of Insect Metamorphosis. Cambridge: University Press, 1954.

727 ——— Formation and involution of striated muscle fibres during the growth and moulting cycles of *Rhodnius prolixus* (Hemiptera). *Quart. J. Micr. Sci.* **97** (1956).

728* ——— The action of growth hormones in insects. *Symp. Soc. Exp. Biol.* **11** (1957).

728* ——— Insect Hormones. *Endeavour.* **24** (1965).

729 WILHELMI, H.: Ein Beitrag zur Theorie der organischen Symmetrie. *Wilhelm Roux' Arch. Entwicklungsmech. Organ.* **46** (1920).

730 WILLIAMS, C. M.: Physiology of insect diapause. I. *Biol. Bull.* **90** (1946); II. **93** (1947).

731 ——— Physiology of insect diapause. III. The prothoracic glands in the *Cecropia* silkworm. *Biol. Bull.* **94** (1948).

732 ——— Physiology of insect diapause. IV. The brain and prothoracic glands as an endocrine system in the *Cecropia* silkworm. *Biol. Bull.* **103** (1952).

733* ——— Morphogenesis and metamorphosis of insects. *Harvey Lect. Ser.* **47** (1952).

734 ——— The juvenile hormone. I. Endocrine activity of the corpora allata of the adult *Cecropia* silkworm. *Biol. Bull.* **116** (1959).

735 WILSON, E. B., and A. P. MATHEWS: Maturation, fertilization and polarity of the echinoderm egg. *J. Morph.* **10** (1895).

736 ——— Experimental studies in germinal localization. I. The germ regions in the egg of *Dentalium*. II. Experiments on the cleavage mosaic in *Patella* and *Dentalium*. *J. Exp. Zool.* **1** (1904).

737* WILSON, E. B.: Cell in Development and Heredity, 3rd Ed. New York: Macmillan, 1925.

738 WILSON, I. Y.: Temperature effect on chiasma frequency in the bluebell *Endymion nonscriptus*. *Chromosoma* **10** (1959).

739 WITSCH, H. V.: Über den Zusammenhang zwischen Membranbau, Wuchsstoffwirkung und der Sukkulenzzunahme von *Kalanchoë blossfeldiana* im Kurztag. *Planta (Berl.)* **31** (1941).

740 WITTEK, M.: La vitellogénèse chez les amphibiens. Arch. Biol. **63** (1952).

741 WITTMANN, C.: Untersuchungen an Schultzschen Doppelbildungen von *Rana fusca* und *Triton taeniatus*. *Z. wiss. Zool.* **134** (1929).

742* WITTMANN, H. G.: Übertragung der genetischen Information. *Naturwissenschaften* **50** (1963).

743 WOELLWARTH, C. V.: Die Induktionsstufen des Gehirns. *Wilhelm Roux' Arch. Entwicklungsmech. Organ.* **145** (1952).

744 ——— Zur Frage der Induktionsfelder in jungen Embryonalstadien der Amphibien. *Verh. Deut. Zool. Ges.* 1957.

745 ——— Die Rolle des Neuralleistenmaterials und der Temperatur bei der Determination und der Augenlinse. *Embryologia* **6** (1961).

746 WOERDEMAN, W. M., and C. P. RAVEN: Experimental embryology in the Netherlands. (Monogr. Progr. Res. Holland. 10.) New York and Amsterdam: North Holland 1946.

747 WOLFF, E.: Les phénomènes d'induction dans la régéneration des planaires d'eau douce. Rev. Suisse Zool. **60** (1953).

748 ———, et T. LENDER: Sur le rôle organisateur du cerveau dans la régéneration des yeux chez und planaire d'eau douce. *C. R. Acad. Sci. (Paris)* **230** (1950).

749 WOODLAND, W.: The skeletoblastic development of the plate and anchor specules of *Synapta*. *Quart. J. Micr. Sci., N.S.* **51** (1907).

750 YAJIMA, H.: Studies on embryonic determination of the Harlequin fly, *Chironomus dorsalis*. I. *J. Embryol. Exp. Morph.* **8** (1960).

751 YAMADA, T.: Der Determinationszustand des Rumpfmesoderms im Molchkeim nach der Gastrulation. *Wilhelm Roux' Arch. Entwicklungsmech. Organ.* **136** (1938).

752 ―――― Beeinflussung der Differenzierungsleistung des isolierten Mesoderms von Molchkeimen durch zugefügtes Chorda- und Neuralmaterial. *Okajimas Folia Anat. Jap.* **19** (1940).

753 YATSU, N.: Experiments on germinal localization in the egg of *Cerebratulus*. *Coll. Sci. Tokyo* **27** (1910).

754 YOSII, R.: Über den Determinationszustand der Raupenhaut bei *Ephestia kühniella*. *Biol. Zentralbl.* **64** (1944).

755* ZALOKAR, M.: Sites of ribonucleic acid and protein synthesis in *Drosophila*. *Exp. Cell Res.* **19** (1960).

756 ZECH, L.: Cytochemische Messungen an den Zellkernen der Foraminiferen *Patellina corrugata* und *Rotaliella heterocaryotica*. *Arch. Protistenk.* **107** (1964).

757 ZEIGER, K.: Zur Strukturanalyse der Chromosomen von *Salamandra*. *Z. Zellforsch.* **20** (1934).

758 ZEPF, E.: Über die Differenzierung des Sphagnumblattes. *Z. Bot.* **40** (1952).

759 ZEUTHEN, S.: Oxygen consumption during mitosis; experiments on fertilized eggs of marine animals. *Amer. Naturalist* **83** (1949).

760 ZIMMERMANN, W.: Über genetisch und modifikatorisch bedingte Störungen der Segmentierung bei *Drosophila melanogaster*. *Z. indukt. Abstamm.- u. Vererb.-Lehre* **86** (1954).

761 ZÜRN, K.: Die Bedeutung der Plastiden für den Ablauf der Meiosis. *Jahrb. wiss. Bot.* **85** (1937).

762 ZWILLING, E.: The role of epithelial components in the developmental origin of the "wingless" syndrome of chick embryos. *J. Exp. Zool.* **111** (1949).

Author Index

Subject Index

Systematic Index